1 Week

The person charging this material is responsible for its return to the library from which it was withdrawn on or before the **Latest Date** stamped below.

Theft, mutilation, and underlining of books are reasons for disciplinary action and may result in dismissal from the University.

To renew call Telephone Center, 333-8400

UNIVERSITY OF ILLINOIS LIBRARY AT URBANA-CHAMPAIGN

OCT 31 1985

OCT 31 REC'D

APR 20 1988

APR 18 REC'D

APR 18 1991

MAY 03 REC'D

NOT TO BE TAKEN FROM LIBRARY
FOR ROOM USE ONLY

REFERENCE
NOT FOR LOAN

NOT TO BE TAKEN FROM LIBRARY
FOR ROOM USE ONLY

REFERENCE
NOT FOR LOAN

L161—O-1096

FREEZE DRYING PROCESSES
FOR THE FOOD INDUSTRY

FREEZE DRYING PROCESSES FOR THE FOOD INDUSTRY

Marcia H. Gutcho

NOYES DATA CORPORATION
Park Ridge, New Jersey, U.S.A.
1977

Copyright © 1977 by Noyes Data Corporation
   No part of this book may be reproduced in any form
   without permission in writing from the Publisher.
Library of Congress Catalog Card Number: 77-71930
ISBN: 0-8155-0662-7
Printed in the United States

Published in the United States of America by
Noyes Data Corporation
Noyes Building, Park Ridge, New Jersey 07656

FOREWORD

The detailed, descriptive information in this book is based on U.S. patents issued since the early sixties that deal with freeze drying processes for the food industry.

This book serves a double purpose in that it supplies detailed technical information and can be used as a guide to the U.S. patent literature in this field. By indicating all the information that is significant, and eliminating legal jargon and juristic phraseology, this book presents an advanced, technically oriented review of freeze drying techniques adaptable to food processing.

The U.S. patent literature is the largest and most comprehensive collection of technical information in the world. There is more practical, commercial, timely process information assembled here than is available from any other source. The technical information obtained from a patent is extremely reliable and comprehensive; sufficient information must be included to avoid rejection for "insufficient disclosure." These patents include practically all of those issued on the subject in the United States during the period under review; there has been no bias in the selection of patents for inclusion.

The patent literature covers a substantial amount of information not available in the journal literature. The patent literature is a prime source of basic commercially useful information. This information is overlooked by those who rely primarily on the periodical journal literature. It is realized that there is a lag between a patent application on a new process development and the granting of a patent, but it is felt that this may roughly parallel or even anticipate the lag in putting that development into commercial practice.

Many of these patents are being utilized commercially. Whether used or not, they offer opportunities for technological transfer. Also, a major purpose of this book is to describe the number of technical possibilities available, which may open up profitable areas of research and development. The information contained in this book will allow you to establish a sound background before launching into research in this field.

Advanced composition and production methods developed by Noyes Data are employed to bring our new durably bound books to you in a minimum of time. Special techniques are used to close the gap between "manuscript" and "completed book." Industrial technology is progressing so rapidly that time-honored, conventional typesetting, binding and shipping methods are no longer suitable. We have bypassed the delays in the conventional book publishing cycle and provide the user with an effective and convenient means of reviewing up-to-date information in depth.

The Table of Contents is organized in such a way as to serve as a subject index. Other indexes by company, inventor and patent number help in providing easy access to the information contained in this book.

15 Reasons Why the U.S. Patent Office Literature Is Important to You —

1. The U.S. patent literature is the largest and most comprehensive collection of technical information in the world. There is more practical commercial process information assembled here than is available from any other source.

2. The technical information obtained from the patent literature is extremely comprehensive; sufficient information must be included to avoid rejection for "insufficient disclosure."

3. The patent literature is a prime source of basic commercially utilizable information. This information is overlooked by those who rely primarily on the periodical journal literature.

4. An important feature of the patent literature is that it can serve to avoid duplication of research and development.

5. Patents, unlike periodical literature, are bound by definition to contain new information, data and ideas.

6. It can serve as a source of new ideas in a different but related field, and may be outside the patent protection offered the original invention.

7. Since claims are narrowly defined, much valuable information is included that may be outside the legal protection afforded by the claims.

8. Patents discuss the difficulties associated with previous research, development or production techniques, and offer a specific method of overcoming problems. This gives clues to current process information that has not been published in periodicals or books.

9. Can aid in process design by providing a selection of alternate techniques. A powerful research and engineering tool.

10. Obtain licenses — many U.S. chemical patents have not been developed commercially.

11. Patents provide an excellent starting point for the next investigator.

12. Frequently, innovations derived from research are first disclosed in the patent literature, prior to coverage in the periodical literature.

13. Patents offer a most valuable method of keeping abreast of latest technologies, serving an individual's own "current awareness" program.

14. Copies of U.S. patents are easily obtained from the U.S. Patent Office at 50¢ a copy.

15. It is a creative source of ideas for those with imagination.

CONTENTS AND SUBJECT INDEX

INTRODUCTION 1

PART I. FREEZE DRYING PRACTICES

THE FREEZING OPERATION 3
   Use of Liquefied Gas and Other Fluid Refrigerants. 3
      Use of Liquid Carbon Dioxide. 3
      Dual Use of Liquid Refrigerant 5
      Addition of Liquid Nitrogen with Reclamation and Use of Evolved Gas. . .10
      Injecting Solution to Be Frozen into Bottom of a Body of Denser
        Refrigerant 13
   Frozen Mass 17
      Perforated Frozen Slab of Product 17
      Frozen Material Comminuted to Desired Size Within Vacuum Chamber. . .19
      Instantaneous Freezing Under High Vacuum into Porous Chunks. 20
      Freezing Material onto Heat Transferring Means 24
      Frozen Extrusions. 26

HEAT TRANSFER MECHANISMS 29
   Radiant Heating 29
      Simultaneously Applying Radiant Heat and Blowing Cold Dry Air. 29
      Improving Efficiency of Radiant Heat by Coating Conveyor Belt. 31
      Regelation of Frozen Particles on Conveyor Belt 34
      Radiant and Conduction Heat Sources in Series 36
   Contact Heating 39
      Inflatable Platen Contact Heating 39
      Heat Input into Product at -10°F by Heat Pumping 42
      Material Swept Downward on Stacked Heating Plates. 49
   Arrangement. 50
      Rectangular Container Contacted by Parallel Planar Heating Elements. . . .50
      Receptacle for Materials and Heating Apparatus Arranged in Vertical
        Direction. 52
      Chamber Containing Material to Be Dried in Stacked Arrangement over
        Cold Trap 54

Other...57
   Sensing Glow Discharge in Microwave Heating 57
   Flow of Refrigerant Through Internal Passages in Both Freezing and
     Sublimation Steps...59

MAINTAINING OPTIMUM CONDITIONS FOR SUBLIMATION 64
  Use of Inert Gas 64
   Introduction of Inert Dry Gas to Heating Zone......................64
   Periodic Pulsing of Dry Inert Gas Through Drying Chamber............66
   Alternate Pulsing of the Two Drying Chambers with Inert Gas.........69
   Addition of Helium to Drying Atmosphere..........................71
   Use of Inert Gas During Terminal Portion of Drying..................73
   Use of Flash Evaporator for Readily Condensable Heat Carrier Heptane...75
   Gas Circulated Through Bed of Frozen Particles Mixed with Molecular
     Sieve..78
  Pressure Modifications..80
   Increasing Pressure After Predetermined Portion of Cycle Is Completed...80
   Pressure Modification via Introduction of Incondensable Gases to Achieve
     Temperature Control...83
   Pressure Controlled by Temperature of Material Being Dried 86
   Adjusting Feed Rate to Maintain Constant Pressure Within Drying
     Chamber..88
  Other Methods 89
   Determining Temperature of Ice Within Material Being Dried..........89
   Throttling Flow of Water Vapor from Container....................91
   Removal of Thin Surface Layer as It Dries with Scraping Means.........94
   Continuous Removal of Fully Dried Layer 97

DEICING 100
  Use of Tubular Internal Freeze Drying Condenser.....................100
   Baffle Arrangement for Separation of Condensable and Noncondensable
     Vapors...100
   Discharge End of Tube Elevated from Horizontal...................103
  Other Techniques 105
   Deicing in Tetra-2-Ethylhexyl Silicate Condensing Medium 105
   Condensing Water Vapor in Low Temperature Condensing Spray.......107
   Vacuum Dryer with Vapor Absorbent Means in Unitary Construction...109
   Freeze Dryer Utilizing an Expendable Refrigerant 112

MONITORING THE FREEZE DRYING CYCLE........................115
  Monitoring Electrical Properties 115
   Monitoring Drying by Measuring Variation in Resistivity 115
   Monitoring Dielectric Properties Without Disturbing Drying
     Conditions..118
  Measuring Weight Loss 120
   Use of Food Product Weight as Parameter in Controlling Dehydration...120
   Automatic Control in Response to Weight Loss 125
   Recording Apparatus for Measuring Weight Loss....................127
  Other Monitors..129
   Controlling Temperature of Heating Means.......................129
   Determination of Practical End Point of Drying....................132

PART II. EQUIPMENT FOR FREEZE DRYING

CONTAINERS 135
   Extended Surface Trays 135
      Ribbed Tray for Freeze Drying 135
      High Thermal Conductivity Tray 136
   Vented Elements to Facilitate Bulk Drying in Containers 137
      Wedge Shaped Venting Element 137
      Venting Channel Member in Form of a Partition 139
      Container with Vapor Venting Passages for Drying Granulated
         Material 140
   Other Equipment 143
      Assembly of Stacked Containers Rotatable over 90° Arc 143
      Housing Means for Drying in Sealable Containers 146
      Drying in Small Bottles with Freedom from Contamination 148

APPARATUS MODIFICATIONS 150
   Combination Freezer-Freeze Dryer 150
      Double Walled Container 150
      Feed Frozen on Continuous Belt 152
      Isolating Valve Mechanism in Shield Between Heater and Condenser 154
      Vertical Tube Dryer for Freeze Drying Liquids 157
   Vibratory Installations 159
      Vibration by Means Outside Vessel 159
      Vibrating Housing Within Vacuum Chamber 160
      Supplying Continuous Apparatus with Material at Proper Rate 162
      Multistage Vibrating Conveyor 164
      Vibrated Support plus Filter to Retain Fine Particles 167
   Avoiding Loss of Fines During Drying 170
      Deflecting Plates in Drying Chamber for Removal of Entrained
         Particles 170
      Filter Tray Assembly for Confining Material During Freeze Drying 172
      Use of Filter Which Functions as Heatings and Agitating Means 173
   Other Innovations 175
      Providing for Thermal Expansion of Shelves 175
      Helical Drying Bed 177
      Cooling Chamber Separated from Heating Chamber by Partition 178
      Preventing Uptake of Atmospheric Oxygen in Dried Product 181
   Economies in Equipment Cost 183
      Use of Multicylindrical Vacuum Chambers with Removable Product
         Cars 183
      Use in Baffles in Internal Condenser to Prevent Stagnation of Air 187
      Compact Plant Using Horizontally Elongated Conveyor in Parallel,
         Spaced, Superimposed Relationship 190

APPARATUS FOR CONTINUOUS FREEZE DRYING PROCESSES 195
   Drying Tubes 195
      Rotating Drying Tube of Polygonal Cross Section 195
      Rotating Drying Tube with Fins and Baffles 197
   Drying Chambers 199
      Upright Drying Chamber with Vacuum Locks at Entrance and Exit 199
      Elongated Drying Chamber Straddled by Inlet and Outlet Lock
         Chambers 201

Heating Arrangements..204
   Parallel Heating Plates Defining Tortuous Path.....................204
   Efficient Access to Heat via Use of Screw Conveyor.................210
   Eliminating Need to Supply Heat from External Heat Source..........214
   Use of High Frequency Oscillators or Infrared Radiators............222
Other Equipment 228
   Shield Structure Between Freezing and Drying Regions...............228
   Elongated Vapor Passage Between Sublimation and Precipitation
     Chambers..234
   Detachable Conveyor and Heater Means for Easy Cleaning.............237
   Pair of Traps Alternately in Communication with Vacuum Manifold....240

PART III. FREEZE DRYING OF SPECIFIC FOODSTUFFS

FREEZE DRYING OF COFFEE 245
Coffee Extraction 245
Removal of Undesirable Solids from Coffee Extract..................247
   Clarification in a Desludger Type Centrifuge.......................247
   Filtration Through Medium Duty Type Filter.........................248
   Removal of Tars and Waxes..248
Freeze Concentration of Coffee Extract.............................250
   Growing Large and Uniform Ice Crystals.............................250
   Multistage Freeze Concentration of Extract.........................251
   Minimizing Loss During Washing of Ice Crystals in Continuous
     Process...253
   Recovery of Coffee Adhering to Ice.................................253
   Processing Ice Stream to Recover Trapped Soluble Coffee............254
   Use of Coated Aluminum or Tin Heat Transfer Surface................256
   Freeze Concentration of Extract, Followed by Freeze Drying.........256
Freezing Step Modifications 257
   Formation of Extrudable Slush......................................257
   Extruding Slush into Cold Gaseous Atmosphere.......................258
   Slow Partial Initial Freezing, Quick Completion....................259
   Separation of Ground Frozen Extract into Coarse and Fine Fractions.261
   Composite Porous Material Consisting of Continuous Phase and
     Dispersed Phase...262
   Shock Freezing of Sprayed Drops with Cryogenic Gas.................263
   Spray Freezing Using Two-Fluid Nozzle..............................265
   Particles of Concentrate Frozen into Prills by Refrigerated Gas....267
Facilitating Separation of Frozen Extract from Retaining Surface...269
   Freezing Precoat of Aqueous Film...................................269
   Cooling Below -50°F and Flexing Retaining Surface..................271
Drying Modifications...272
   Increasing Drying Rate by Disrupting Surface of Frozen Slab........272
   Increasing Drying Rate by Compressing Frozen, Granulated Material
     into a Porous Slab..273
   Rapid, High Temperature Freeze Drying..............................274
   Two-Phase Freeze Drying at Different Temperatures and Pressures....277
   Use of Auxiliary Heaters to Avoid Formation of Liquid..............277
   After-Dryer for Drying Dust Recovered in Process...................279
   Spray Freeze Drying System...281
Color of Coffee Product 283
   Darker Coffee by Controlled Melt-Back..............................283

Darker Color by Slow Freezing............................285
Slow Freezing Followed by Subdivision to Coarse Particle Size........287
Temperature and Freezing Rate Dependent on Concentration of
    Extract..289
Freezing in a Plurality of Layers on a Continuous Belt..............290
Coating Frozen Particles of Extracts with Soluble Coffee Powder.......291
Extract Dispersed into Immiscible Liquid Refrigerant...............292
Frozen Ground Particles Screened and Fines Dried as Bottom Layer....292
Product of Controlled Density 294
    Freezing Slushed and Foamed Extract 294
    Slurry Containing Ice of Controlled Cyrstal Size and Mother Liquor of
        40 to 50% Soluble Solids...................................295
Aromatized Coffee 296
    Extract Aromatized Before Freezing.............................296
    Part of Extract Stream Aromatized and Freeze Dried................299
    High Aroma Extract via Ultrasonic Cold Extraction 300
    Roll Milling of Instant Coffee-Aromatizing Oil Blend into Flakes.......302
    Freeze Drying Mixture of Spray Dried Coffee and Steam Distillate of
        Flavor and Aroma Constituents............................304
    Extraction of Aromatic Fatty Constituents with Gaseous or
        Liquid CO_2 306
Aroma and Flavor Stability...................................307
    Coating Freeze Dried Beads with Fat 307
    Pressure and Temperature Lowered for Four Hours During Vacuum
        Drying..308
    Addition of Ungelatinized Modified Corn Starch to Extract Before
        Freezing..310
    Retaining Volatile Aromatics During Foam Freezing Operation by
        Avoiding Evaporative Cooling.............................311

FREEZE DRYING OF OTHER FOODSTUFFS.......................313
Meats..313
    Heat-Conductive Pin System for Drying Sliced Material 313
    Heater Rack 314
    Infrared Sublimation......................................317
    Use of Cryogenic Gas System 320
    Frozen Particles on Porous Carrier Contacted with Gaseous Drying
        Agents..322
    Drying Gas Maintained at Subatmospheric Pressure and Controlled
        Temperature and Velocity................................323
    Repeated Cycles of Gas Pressure Variation During Drying 328
    Freeze Drying at Atmospheric Pressure..........................330
    Controlled Humidity Freeze Drying Process 331
    Fluidized Bed Freeze Drying.................................334
    Fluidized Bed of Solid Discrete Sodium Chloride Particles 336
    Bacon Which Does Not Spatter on Cooking 339
    Mechanical and Chemical Tenderization of Steak Before Cooking 340
Milk Products 342
    Whole Milk Cakes 342
    Nonhygroscopic Honey-Milk Product 342
    Separate Drying of Serum Portion and Coagulated Residue 344
Juices 345
    Alternately Raising and Lowering Temperatures of Hard Frozen Mass...345

Use of Electromagnetic Induction Heating 346
Increasing Surface Area via Use of Conductive Teflon-Coated Spheres ...347
Fluidized Bed at Atmospheric Pressure 348
Fluidized Bed of Foodstuff in Particulate Form 351
Use of Inert Gas to Eliminate Oxidation of Cells of Citrus Fruits 353
Use of a High Carbohydrate Additive in Processing Apple and Grape
    Juice 354
Potatoes 356
    Diced Potatoes—Freeze Drying Followed by Air Drying.............256
    Diced Potatoes—Two Blanchings Prior to Freezing Step 358
    French Fried Potatoes—Improving Permeability of Crust After Frying...359
    French Fried Potatoes—Density Fractionation Before Freezing 361
    Dehydrated Fried Potato Cake 362
Vegetables..364
    Mushrooms—Reducing Enzymatic Degradation and Discoloration 364
    Carrots—Color Stabilization with Ascorbic and Erythorbic Acids 365
    Avocado—Heating Single Particle Layer of Frozen Material 366
    Peas—Alternate Vacuum Freezing Dehydration and Air Dehydration
        Steps..368
    Green Beans—Simultaneous Compaction and Freeze Drying..........369
    Salads—Use of Critical Proportion of Emulsion Type Dressing 371
    Low Calorie Dietary Fruit or Vegetable Foodstuff 373
Fruit..374
    Blueberries—Frozen Berry Punctured Before Freeze Drying 374
    Fruit for Dry Cereal Mix—Use of Slow Freezing Technique 376
    Preliminary Intense Heat Treatment of Frozen Surface................377
Artificially Sweetened Fruit 379
    Frozen Fruit—Sweetened, Refrozen and Freeze Dried.................379
    Freeze Dried Fruit Impregnated to High Level of Sweetness...........380
Desserts and Candy Centers..382
    Sour Cream Fruit Product..382
    Puddings, Jellies, Pie Fillings.....................................383
    Ice Milk Confections..384
    Gelatin Based Desserts..385
    Gelatin of Improved Solubility in Cold Water......................386
Tea 387
    Extract Chilled and Filtered Before Freeze Concentration to Remove
        Tars...387
    Isolation of Cream of Tea for Later Use 388
    Sublimation at Reduced Pressure in Gaseous Medium................389
Other Food Products..390
    Rice—Use of Two Cycles of Consecutive Thawing and Freezing 390
    Peanut Butter 391
    Oxyster and Shrimp Soup Material 392
    Foodstuff Extracts—Addition of High Molecular Weight Organic
        Substances...393
    Liquid Foodstuffs—Thin Film of Liquid in Revolving Apparatus 393

COMPANY INDEX..394
INVENTOR INDEX 396
U.S. PATENT NUMBER INDEX 399

INTRODUCTION

Freeze drying is recognized as the best method of producing dried material of high quality. The freeze drying process has several advantages which make it desirable for food processing. By maintaining the material in the frozen state until it is dry, shrinkage and migration of dissolved constituents are eliminated. Physical and chemical changes are inhibited thereby minimizing loss of volatile components. The freeze dried products have a porous texture, and are readily reconstituted (rehydrated) to their original size and shape. There is obtained a product of good flavor and appearance, and a high preservation of nutrients. In addition, freeze dried foods are adaptable to simple packaging, storage and shipment, and can be kept for long periods.

Basically, freeze drying involves the following steps. The material to be treated is frozen, and while in the frozen state, subjected to a vacuum. The ice in the material sublimes (is transformed directly to vapor without going through the liquid state). The evolved water vapor is condensed on a refrigerated coil, pumped directly to the outer atmosphere, or is absorbed in a suitable medium. Drying is promoted by supplying the latent heat of sublimation from an appropriate source. Drying of different types of foods requires different processing conditions.

The cost and practicality for commercial freeze drying operation is directly related to the time required for processing the food. The freeze drying process is limited by the high cost of the operation and the long drying time. 25% of the drying takes place in the first 5% of the drying cycle, 50% in the first 15%. Thereafter the rate slows down to a relatively low level compared to the initial rate of drying. The growing demand for freeze dried food products has led to the need for more efficient means of performing all the operations necessary to obtain a freeze dried product.

This book describes different types of freezing techniques, a variety of heat transfer means including contact and radiant heating as well as arrangements of heating elements and frozen products, processes for reducing the drying cycle time through pressure modification and the use of inert gas, and even methods for carrying out freeze drying at atmospheric pressure. Processes for monitoring the freeze drying cycle to enhance both efficiency and economy are included. There is strong emphasis on equipment and equipment arrangements designed to reduce costs. Since batch processes are slow and expensive, the continuous freeze drying processes, which are more practical for commercial use, are of particular interest.

Aside from processes which are applicable to foods in general, there are methods devoted to specific products. One section pertains to the preparation of high quality freeze dried coffee, a commercially desirable product. Methods for the preparation of freeze dried meats, fruits and vegetables, milk products, juices, potatoes, and tea are among those included. These are of great interest to the armed forces.

Part I. Freeze Drying Practices

THE FREEZING OPERATION

USE OF LIQUEFIED GAS AND OTHER FLUID REFRIGERANTS

Use of Liquid Carbon Dioxide

A.S. Guerard; U.S. Patent 3,673,698; July 4, 1972 describes a process for the freeze drying of liquids. This can be accomplished by mixing the liquid carbon dioxide at a pressure of about 600 psi absolute and at a temperature of about 40°F with the solution to be processed so that freezing of the water in the solution does not occur. Following creation of the comingled feed and carbon dioxide at a temperature above the freezing point of water at the existing pressure, the pressure on the mixture is then rapidly reduced as by release through an expansion valve.

To achieve this, the mixture can be flashed into a vessel wherein the gaseous carbon dioxide is also separated from the solid crystals which form. The finely divided crystals which are formed consist of carbon dioxide and frozen feed stock. The solid crystals are then fed to a second vessel wherein the water ice and carbon dioxide are removed as by a suitable procedure. The final crystalline material is of such small particle size that the solid material behaves as a liquid, thereby making possible the utilization of the fluid bed technique to fluidize the crystals. Dry gaseous carbon dioxide is passed through the fluidized bed, the water present as ice subliming to water vapor which is carried away by the carbon dioxide gas which is fed to maintain the fluidized bed and the carbon dioxide which sublimes from the crystals.

Referring to Figure 1.1, the solution to be converted to a solid state is contained in feed tank **6** where it is maintained in a homogenous state by the mixer **7**. Additional material is provided through the fill connection **8** as desired. Liquid material is withdrawn from the feed tank through line **9** and is passed through a heat exchanger **11** wherein heat exchange occurs with a stream of water vapor and carbon dioxide vapor in line **12**. The feed material in line **9** is cooled by heat exchange **10** to about 40°F. The material in line **9** is forced by pump **14** through line **16** to a mixer **17**.

FIGURE 1.1: PROCESS FOR FREEZE DRYING WITH CARBON DIOXIDE

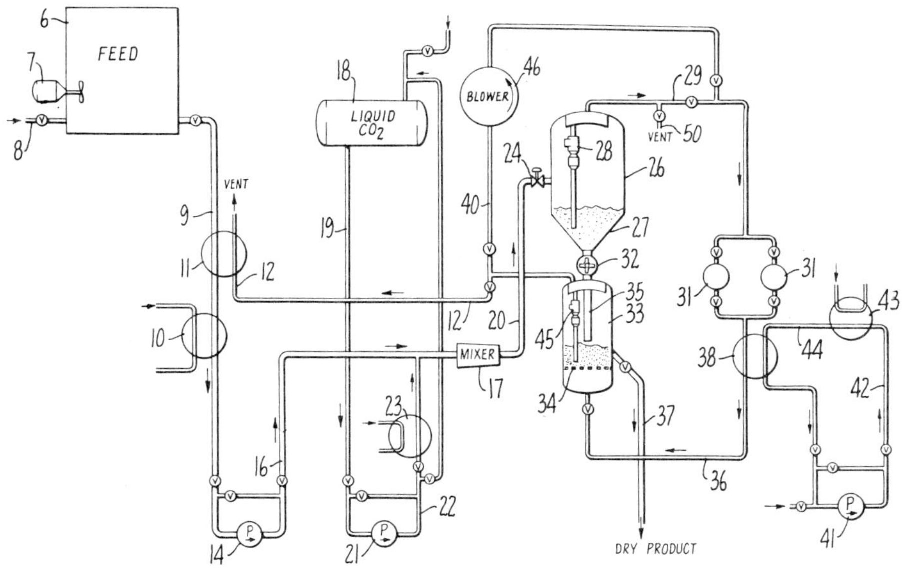

Source: U.S. Patent 3,673,698

Liquid carbon dioxide is derived from a tank **18** and is fed through line **19** to a pump **21** which in turn forces the liquid carbon dioxide on through line **22** through heat exchanger **23** into line **16** ahead of the mixer **17**. The mixer serves to mix the two liquid streams quite thoroughly before they are passed through line **20** to expansion valve **24** provided immediately adjacent to a first pressure vessel **26**. In the pressure vessel crystals are formed of solid carbon dioxide, water ice and the frozen solid contained in the water fed to the process. The crystals settle in the cone-shaped bottom **27** of vessel **26**.

The carbon dioxide which does not solidify is removed through cyclone separators **28** and is passed through line **29** to dryers **31**, a portion being vented as desired as at **50**. Preferably the dryers are provided in parallel and contain a conventional drying agent such as silica gel, phosphorus pentoxide, activated alumina, calcium sulfate, or magnesium perchlorate, as well as those materials conventionally known as molecular sieves which are aluminosilicates or zeolites, the crystal of which contains minute pores and which have the ability to absorb relatively large volumes of water vapor. The dryers are operated sequentially, one being reactivated while the other is in use.

The crystals which collect in the bottom of vessel **26** are fed through a power-operated star valve **32** and pipe **35** into a second vessel **33** which includes a grid or screen **34** adjacent to the lower end thereof. Dry carbon dioxide gas from the dryers **31** is fed through line **36** into the bottom of vessel **33** to pass upwardly and fluidize and maintain the solid particles on the screen **34** in a fluidized state. Conventional cyclones or centrifugal separators **45** are provided in vessel **33** to prevent removal of solids from this vessel. A portion of the sublimed water vapor and carbon dioxide vapor is vented through line **12** from the upper portion of the vessel 33 while the solid dried material is removed as the product through line 37. That portion of the mixed stream of sublimed water vapor and carbon dioxide which is not vented through line 12 is removed through line 40 and is passed to line 29 to ensure the presence of an adequate quantity of carbon dioxide for sublimation of the ice in vessel 33. Line 40 derives its stream of water vapor and carbon dioxide from cyclone separators 45 in vessel 33. Blower 46 is provided to increase the pressure in line 40 to a value equal to that in line 29.

The temperature of the stream of carbon dioxide gas issuing through from the dryers 31 through line 36 is controlled by utilizing heat exchanger 38. Methanol, glycol or other low freezing point liquid is forced through a circulatory system which includes pump 41, line 42 passing to a heat exchanger 43 and thence through line 44 to the heat exchanger 38 to regulate the temperature of the dry carbon dioxide stream passing on to vessel 33.

Example: In a typical operation utilizing a relatively heavy aqueous coffee infusion or syrup, the syrup was fed continuously through line 9 by pump 14 and was delivered to line 16 at a pressure of 600 psi absolute and 40°F. Carbon dioxide derived from vessel 18 at a temperature of −20°F and 200 psi was fed to pump 21 and through heat exchanger 23 for delivery to line 16 at the same temperature and pressure, that is, 40°F and 600 psi absolute. The liquid mixture so formed was in the ratio of 1.8 pounds of carbon dioxide per pound of water present in the aqueous solution. The resultant mixture was then fed to the expansion valve and was permitted to flash in vessel 26 which was at a pressure of approximately 20 psi. The pressure in vessel 26 exceeded that in vessel 33 by the amount sufficient to provide the pressure drop through the dryers and piping. The dried coffee product was removed through line 37.

Dual Use of Liquid Refrigerant

The process developed by *J.L. Achucarro; U.S. Patent 3,323,225; June 6, 1967* relates to a freeze drying apparatus in which the substance under process is conveniently prefrozen in situ on a drying assembly in the vacuum chamber by using a fluid refrigerant which subsequently is used to condense the water vapor produced by the sublimation of the ice in the frozen substance under subatmospheric pressure.

When some part of any apparatus is kept at a temperature below that of the ambient, however good the thermal insulation of that part may be, there is an unavoidable amount of heat gained from the surroundings. This heat gained from the surroundings together with the thermal energy initially stored in the substance to be freeze dried, containers, assemblies and other components of the apparatus, through the medium of the gas evolved in the sublimation of solid carbon dioxide immersed in a liquid refrigerant or in the evaporation of some liquid gas used as refrigerant, is made to do work in driving some of the liquid refrigerant through a double cooling coil automatically and with a periodic reversion of flow direction for the purpose of prefreezing the substance under process. When the substance is frozen, the flow through the double cooling coil is stopped, leaving the assembly and the substance ready for drying.

Drying of the frozen substance is carried out under subatmospheric pressure, and the liquid refrigerant which formerly was used for prefreezing, is commonly now used in the condenser vessels for condensing the water vapor produced by the drying substance.

The apparatus shown in Figure 1.2 comprises a vertical, cylindrical chamber **1** closed at both ends by two flat circular discs **2** and **3**. Figure 1.2a shows a sectional elevation view of the cylindrical chamber and Figure 1.2b shows an automatic valve and part of the condenser vessel attached to it.

The whole arrangement is such that a relatively large central portion of the interior of the chamber is reserved for the assembly **14** for prefreezing and subsequent drying with heating means, thermostats, thermocouples or any other temperature sensing devices and other accessories.

The condenser vessels **15** and **16** have hinged or removable circular lids **27** and **28** provided with seal gaskets **29** and relief valves **30** and **31** similar to those commonly used for motor car radiators. By means of knurled screws **32** in their peripheries these lids **27** and **28** can seal the condenser vessels and maintain a certain internal pressure above atmospheric, determined by the setting of the corresponding relief valves. The relief valve **30** of the condenser vessel **15** directly connected to the automatic valve container **23** is set to open at a higher differential pressure than the valve **31** of the other condenser vessel **16**. Flexible gas outlet pipes **33** and **34** are fitted to the bodies of the relief valves **30** and **31**, the relief valve also fulfills the function of overflow pipe.

The method by which the apparatus operates is as follows. The substance under process is loaded on to the appropriate assembly **14** for prefreezing and drying inside the cylindrical chamber, the temperature sensing devices placed in position and the chamber door, not shown, closed.

Both relief valve caps **89** and **90** are removed from their corresponding relief valves **30** and **31** and the lids **27** and **28** of both condenser vessels **15** and **16** opened. The vent valve **47** inside the condenser vessel **16** is closed and the vent valve **46** inside the other condenser vessel **15** is opened.

Any of the usual liquefied gases can be used as liquid refrigerant **91** or alternatively a suitable liquid refrigerant cooled by direct contact with subliming solid carbon dioxide **53** may be used. Where solid carbon dioxide **53** is used, the amount introduced into each condenser vessel **15** and **16** should be approximately the same and the total quantity used at least sufficient to complete prefreezing without further reloading.

Now, whether solid carbon dioxide is used or not, liquid refrigerant **91** is poured into both condenser vessels **15** and **16** up to similarly positioned level marks (not shown) provided in them. To accomplish this, it may be necessary to pour in the liquid in stages; letting the initial fierce bubbling and violent gas evolution subside before adding more liquid refrigerant **91** to both condenser vessels.

Both the lid **28** of the condenser vessel **16** and its corresponding relief valve cap **90** are secured in the closed position. The pressure exerted by the gases collected above the liquid refrigerant level in the condenser vessel **16** will cause the transfer of liquid refrigerant **91** from this condenser vessel **16** to the other condenser vessel **15** through the double cooling coil **38**.

FIGURE 1.2: FREEZE DRYING APPARATUS

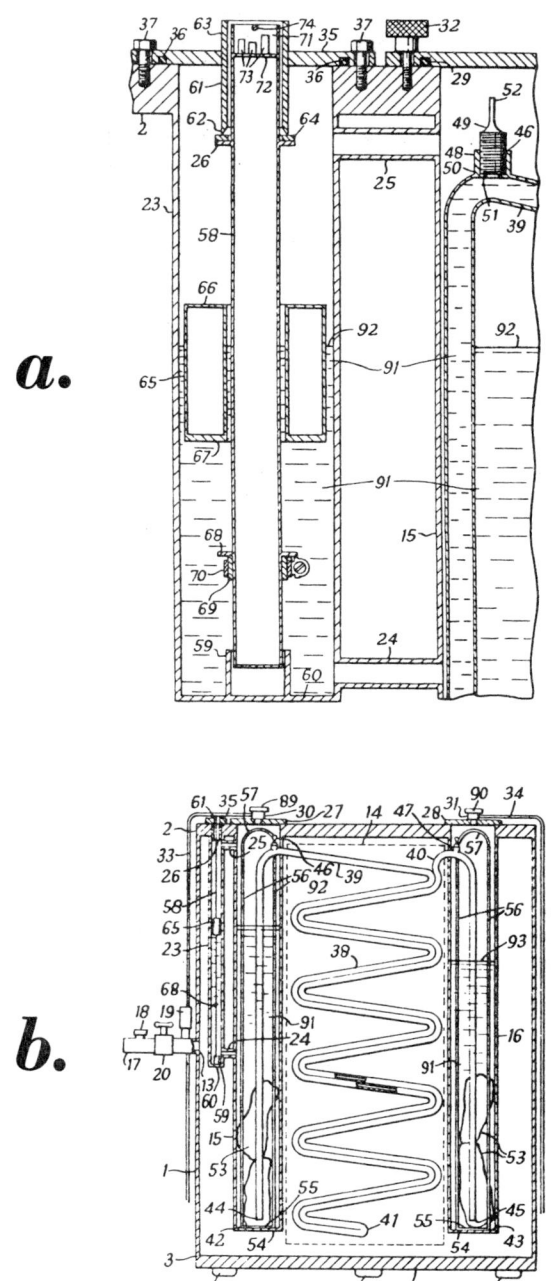

Source: U.S. Patent 3,323,225

When the rising level of the liquid refrigerant **91** in the condenser vessel **15** reaches the body **48** of the vent valve **46**, this valve is subsequently closed and the relief valve **31** on the lid **28** of the other condenser vessel **16** uncapped promptly. This double operation performed in the prescribed sequence ensures both that no air is entrapped in this part **39** of the double cooling coil **38** and that the liquid refrigerant **91** in the condenser vessel **15** does not overflow.

Now the lid **28** of the condenser vessel **16** is opened to gain access to the vent valve **47** which is subsequently opened. The lids of both condenser vessels and their corresponding relief valve caps are secured in the closed position. At this stage the automatic valve **26** is also in the closed position. Eventually liquid refrigerant **91** will be syphoned from the condenser vessel **15** to the condenser vessel **16** through the double cooling coil **38**. Directly the automatic valve **26** opens, the relief valve **31** and the lid **28** are opened. If the level **93** of the liquid refrigerant **91** in the condenser vessel **16** has reached the body of the vent valve **47**, this valve is closed.

On the other hand if the level **93** of the liquid refrigerant **91** has not reached the body of the vent valve, more liquid refrigerant is added to bring that level to the vent valve, which is subsequently closed. If there is an excess of liquid refrigerant in the cooling system, the excess is automatically expelled through the overflow pipe **34** prior to the opening of the automatic valve **26**.

After closing the vent valve **47**, the lid **28** of the condenser vessel **16** and both relief valve caps **89** and **90** are secured in the closed position and the cooling system commences its automatic operation.

When using one of the common liquid gases as liquid refrigerant, it is scarcely necessary to take any special precautions to displace any entrapped air in the double cooling coil **38** since at any moment during the prefreezing period this coil will contain substantial amounts of boiled off gas. In this case the preparation of the cooling system for prefreezing of the substance under process can be simplified considerably in the following way.

Once the substance under process is loaded in the chamber **1** and the chamber door closed, both relief valves **30** and **31** are uncapped and the lids **27** and **28** of both condenser vessels **15** and **16** opened; the vent valves **46** and **47** are closed. Then solid carbon dioxide **53**, where this substance is used, is lowered into both condenser vessels by means of the crates **54**, and liquid refrigerant **91** is poured into both condenser vessels up to the level marks provided. The lids of both condenser vessels and both relief valve caps are secured in the closed position.

In brief, the cyclic operation of the automatic valve can be described as follows. When the liquid **91** in the valve container **23** reaches its high level point, the valve **26** closes. When the level **92** is dropping, the valve **26** remains closed. When the liquid **91** reaches its low level point, the valve **26** opens. When the level **92** is rising, the valve **26** remains open.

The volume of liquid refrigerant **91** involved in this alternating displacement depends on the capacity between the high and low levels of the liquid refrigerant in the condenser vessel **15** and in the valve container **23**. This capacity can be adjusted by varying the position in height of the supporting platform **68** for the free floater **65**, and should not be less than the internal volume of the double cooling coil **38** so that the liquid refrigerant inside the double cooling coil has a chance to be renewed at each displacement.

Should it be necessary to gain access to the inside of either condenser vessel **15** and **16** or valve container **23** during the prefreezing period for topping up with liquid refrigerant and/or solid carbon dioxide, or for any other reason, the reciprocating automatic transfer of liquid refrigerant **91** between both condenser vessels can be stopped at any moment by uncapping the relief valve **31**. During the prefreezing period no attempt should be made to open any condenser vessel lid **27** and **28** without first removing its corresponding relief valve cap **89** and **90** and the relief valve cap **89** should not be open unless the condenser vessel **16** is open or its corresponding relief valve cap **90** has been removed.

Where solid carbon dioxide is used to cool the liquid refrigerant, the liquid refrigerant flowing through the double cooling coil **38** extracts heat from the tubes **39** and **40** and hence warms up gradually as it flows. The further the liquid refrigerant is from its source or inlet measured along its path, the warmer it becomes. Also, other conditions remaining unchanged, the heat extraction capacity of a given flow of liquid refrigerant depends on its temperature, i.e., on its distance from the inlet measured along its path. At any point along the length of double cooling coil and at any moment when the liquid refrigerant is flowing through one of the tubes **39, 40** in one direction, it is flowing also through the other tube **40, 39** in the opposite direction.

The total distance from the prevailing liquid refrigerant inlet **44, 45** of any point on the double cooling coil **38** measured along the direction of flow separately for the two tubes **39** and **40** forming the double cooling coil **38** and then added up together is a constant value for all points on the double cooling coil **38**. This condition tends to provide a uniform heat extraction per unit length of double coil **38** all along its length. This effect is enhanced considerably by the additional, periodic reversion of flow and counterflow.

Where the liquid refrigerant **91** is one of the usual liquid gases, there will be a mixture of the liquid and gas phases flowing through the tubes **39** and **40** forming the double cooling coil **38**. This mixture will be the richer in its liquid component the nearer it is to the prevailing liquid refrigerant inlet **44** and **45** and therefore, in this case also, both the disposition of the double cooling coil **38** and the periodic reversion of flow and counterflow of the liquid refrigerant will promote the desirable uniformity of heat extraction all along the length of the double cooling coil.

In most cases, once the cooling system has been initially supplied with the right amounts of liquid refrigerant and solid carbon dioxide, where this substance is used, there will be no need for further intervention until prefreezing of the substance under process is completed.

The rise in the temperature of sublimation of the solid carbon dioxide or in the boiling point of the liquid refrigerant arising from the superatmospheric pressures prevalent in the cooling system during the prefreezing period is relatively small and its effects negligible.

At the end of the prefreezing period the chamber is evacuated to a suitable subatmospheric pressure by a vacuum pump, not shown, and the flow of liquid refrigerant through the double cooling coil is stopped by uncapping both relief valves as described above. The lids of both condenser vessels are opened and both vent valves are opened as well.

At this stage it is not normally necessary to take any action to empty the double cooling coil of liquid refrigerant, but if so desired it can easily be done by temporarily resealing the condenser vessel **16** with its lid **28** and relief valve cap **90**. Now, or at any convenient time during the subsequent drying period, more liquid refrigerant and/or more solid carbon dioxide is added to one or both condenser vessels if so required. The lids of the condenser vessels are closed but the relief valves should be left uncapped and the vent valves opened.

Heat is applied to the frozen substance by the drying assembly **14** and the water vapor coming from the ice in the frozen substance will condense in the form of ice on the convex cylindrical wall and circular base of the condenser vessels and valve container **23**.

At the end of the drying period the heating means are turned off and the vacuum pump, not shown, is stopped or isolated from the chamber by the isolation valve **20** while air or an inert gas at atmospheric pressure is admitted into the chamber **1** through the gas or air inlet valve **18** which is interchangeable with the pressure gauge head **19**. The chamber door, not shown, is opened and the dried substance unloaded.

Addition of Liquid Nitrogen with Reclamation and Use of Evolved Gas

R.G. Gidlow; U.S. Patent 3,458,941; August 5, 1969; assigned to *The Pillsbury Company* describes an apparatus for drying products in a frozen condition consisting of a chamber for freezing the products under cryogenic conditions by exposing the food products to liquid nitrogen, ducts for collecting the gas that is evolved in the process of freezing the material and a chamber for drying the product. The collected gas is circulated through the drying chamber in heat and mass transfer relationship with the frozen product. In this manner an inexpensive supply of an inert gas of low humidity is made available for the drying operation. The low oxygen content of the gas reduces oxidative damage to the product as it is dried.

The process is described with reference to Figure 1.3. A freezing chamber **10** consists of an initial preliminary cooling section **12**, a secondary cooling station **14** and a final cooling section **16** through which products **18** that are to be frozen and subsequently dried are transported by endless belt conveyors **20**, **22** and **24** driven in the proper direction to transfer the products from right to left as seen in the drawing. The housing of the cryogenic freezing apparatus **10** preferably consists of a vacuum-insulated double wall cylinder.

As the products enter the precooling section **12** they pass under a rubber door **26** and travel beneath a plurality of openings **28** through which a cool substantially oxygen-free gas is supplied as shown by dotted lines **30**. Refrigerated nitrogen gas is introduced through a duct **32** from a blower **34**. This gas consists of the nitrogen gas evolved during subsequent exposure of the product to liquid nitrogen in the final freezing section **16**.

FIGURE 1.3: ADDITION OF LIQUID NITROGEN WITH RECLAMATION AND USE OF EVOLVED GAS

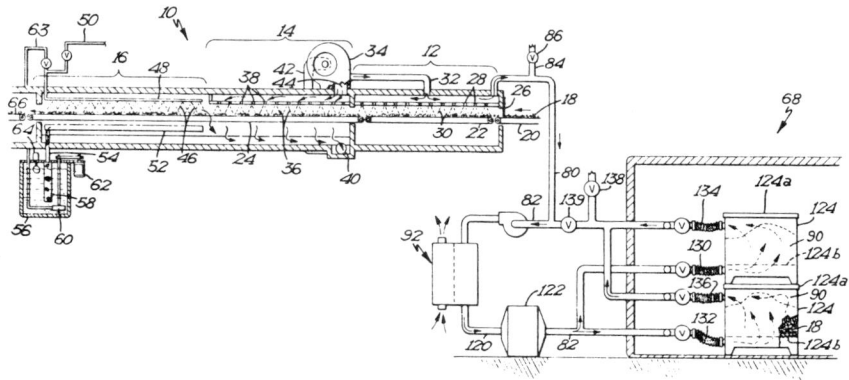

Source: U.S. Patent 3,458,941

As the product travels through section **12**, the downward flow of refrigerated nitrogen rapidly chills the surface of the product until it reaches a temperature somewhat above freezing. The product then passes out of the section **12** and enters section **14**. The precooled product on the conveyor is chilled further as it travels through the secondary cooling section **14** again by downward flow of refrigerated gaseous nitrogen as shown at **36** that was introduced from the blower **34** through openings **38**. The cool gas after flowing downwardly over the conveyor **24** is exhausted through a duct **40** that is connected to the inlet of the blower **34** by a duct **42**. A portion of the refrigerated nitrogen expelled by blower **34** is passed to section **12** through line **32** and the remaining portion is reintroduced to the section **14** through a duct **44**.

In the final cooling section **16**, the product passes beneath sprays of nitrogen **46** in a liquid state. Liquid nitrogen is introduced through a line **48** from a supply duct **50**. The liquid nitrogen then flows over the conveyor **24** to a collection pan **52** and through a pipe **54** to an external reservoir **56**. A filter **58** is provided at the lower end of pipe **54** for removing sediment. A pump **60** driven by motor **62** returns liquid nitrogen through a duct **63** to the feed line **48** under the control of a liquid level controller **64**. The liquid nitrogen supplied to the cryogenic freezer is transferred through the pipes **48** and **50** at a pressure on the order of 5 to 7 psig by the centrifugal pump **60**.

A portion of the excess liquid is collected in the pan **52** and is recirculated through duct **62** to the spray line **48**. The remaining portion flashes to a gas when contact is made with the product and it is this evolved gas that is recirculated through the preliminary cooling sections **12** and **14** and is then exhausted from the precooling section **12** to the drying compartment **68**. The product is then carried to the drying compartment where the drying operation is performed. During the transfer to the drying compartment care must, of course, be taken to maintain the product in a frozen condition.

Relatively cold nitrogen gas substantially free of oxygen is exhausted from the precooling chamber **12** through a duct **80**, which is insulated to prevent heating of the inert gas. Excess gas flowing out of the precooling section **12** can, if desired, be vented to the atmosphere through an exhaust duct **84** by a metering valve **86**. The gas flowing through duct **80** passes into a circulation duct **82**. The nitrogen introduced to the circulation duct is maintained at a moisture level below saturation by a desiccation unit **92** so that the ice core is not in equilibrium with its surrounding environment. During the process, heat is transferred to the ice core from the surrounding nitrogen, thus causing sublimation of moisture up to the saturation point of the surrounding gas.

The dried gaseous nitrogen is exhausted from the desiccating unit **92** through an exhaust section **120** of duct **82**. It then passes through a heat exchange **122** of known construction where it is either cooled or warmed as required. The gaseous nitrogen fed to the bins is ordinarily held at the same temperature as the gaseous nitrogen in the chamber **68** which is usually about +10°F to -20°F. A plurality of bins designated **124** are provided within chamber **68**.

In operation, the desiccated gas is introduced through flexible hoses **130** and **132**. It then circulates freely through the interstices between the pieces in bins **124** and is exhausted to duct **82** through hoses **134** and **136**. Gas is vented to the atmosphere by valve **138** at the same rate it is introduced through duct **80**. The flow rate through duct **82** is controlled by valve **139**. As this takes place, a dried porous layer will begin to form on the surfaces of each of the particles as the moisture contained in the frozen center portion of each piece is removed by sublimation.

Each bin consists of an airtight enclosure formed from an imperforate material defined by side walls, a bottom and a removable cover **124a** through which the product is introduced and removed and a false bottom formed from perforated material **124b**. The walls, cover and bottom of the bins can be formed from any suitable material such as sheet metal.

During operation, the desiccated gaseous nitrogen passes into the bins through the flexible hoses **130** and **132** respectively, thence upwardly through the perforate false bottom **124b** through the passages between the pieces of products that are being dried. The moisture-laden gaseous nitrogen is then exhausted through ducts **134** and **136** and is returned to the inlet of the desiccating unit **92** through line **82**. The gaseous nitrogen is recycled continuously throughout operation.

As the gas is transferred from the freezing section to the drying section, care is taken to maintain its temperature with as little heat gain as possible. By insulating the duct work the cold temperature of the evolved gas is utilized for maintaining the dried products under refrigeration and in this way the overall thermal efficiency is improved.

The temperature of the material being dried after leaving the freezer is preferably on the order of about -20°F to +10°F. During the drying operation within the chamber **68**, the temperature of the material being dried is preferably maintained at about 0°F to +10°F.

Example: Frozen cut green beans of commercial grade are placed in a wire mesh container 4 inches in diameter and 10 inches high. The beans are frozen by introducing them into a freezing apparatus shown in Figure 1.3 to expose them first to nitrogen gas at a temperature of about −100°F to precool them and finally in the last stage of the freezing operation to liquid nitrogen gas at a temperature approaching −340°F for a total time of 5 minutes. The beans are then collected in bins of the type shown without being allowed to thaw and transferred to a drying chamber where the hoses provided on the bins are connected to a gas outlet leading from the freezing unit. The evolved nitrogen gas is then continuously recirculated through the bin, through a desiccating unit and a heat exchanger at the rate of about 200 cubic feet per minute. The gas transferred through the product in the bins is maintained at a temperature of −5°F and at a dew point of −50°F.

Initial Wt, g	Final Wt, g	Air Temp, °F	Dew Point, °F	Drying Time, hrs
781.1	384.0	0 to +12	−95 to −30	493
783.1	369.9	−4 to 0	−90 to −30	1,269

Upon rehydration in water the color and shape is completely restored and the beans had excellent flavor and texture.

Injecting Solution to Be Frozen into Bottom of a Body of Denser Refrigerant

In a process described by *H.A. Sauer; U.S. Patent 3,484,946; December 23, 1969; assigned to Bell Telephone Laboratories, Inc.* moisture-containing substances, particularly aqueous salt solutions, preparatory to freeze drying are frozen by introduction into the lower region of a body of nonmiscible refrigerant having a molecular density greater than the substance. A suitable freezing and freeze drying chamber for practicing this process includes a set of wire mesh baskets which collect the frozen substance and thereafter in the drying step slowly rotate to promote uniform drying. Primary drying is effected at atmospheric pressure. Purging of residual moisture, if slight enough, is achieved by repeated pressure excursions back and forth across the vapor boundary at increasing temperatures in the presence of a drying gas.

The process is described with reference to the apparatus of Figure 1.4. The interior of chamber 1 is lined to contain a refrigerant such as Freon E1. A bottom extension of chamber 1 defines a drain channel 4 in which is mounted an elongated solution injector 5 connected to the solution in bath 2 through suitable piping 6. In its upper end is mounted a tube 7 consisting of a stainless steel tube whose diameter may be as little as 1 mil.

The bottom of channel 4 terminates at an inlet duct 8 which serves both as a refrigerant drain and an entry point for temperature controlled drying gas. The injector mounting member 9 is partially enveloped in a heater 10. The region of refrigerant between the outlet of tube 7 and approximately 1 inch above same must be carefully temperature controlled to avoid coalescing of the droplets when ejected. Similarly, in order to prevent the solution stream in injector 5 from freezing, the injector body is temperature controlled. Accordingly, at the orifice of tube 7 the temperature is maintained just above the freezing point of the solution, but about 1 inch above, the refrigerant temperature is well below the freezing point of the solution. These controls are achieved through a regulator **30** shown in Figure 1.4c, which receives inputs monitoring thermocouples **11** and **12** suitably placed in the critical region, as well as thermocouple **13** which is the heater control. The output of regulator **30** is coupled to heater **10**. Thus, a sharp temperature gradient just above the capillary outlet results in prevention of coalescing.

Surrounding the refrigerant chamber is a thin gas envelope **14** which extends to a point just above the level of tube **7**. At this point an entrance port **15** is connected to the envelope for the supply of cooling or heating medium from a suitable heat exchanger and circulator **31** shown in Figure 1.4d. Envelope **14** substantially surrounds the body of refrigerant and is connected to an exit port **16** that directs the heat exchange medium to conventional recovery facilities.

FIGURE 1.4: APPARATUS FOR FREEZE-FREEZE DRYING

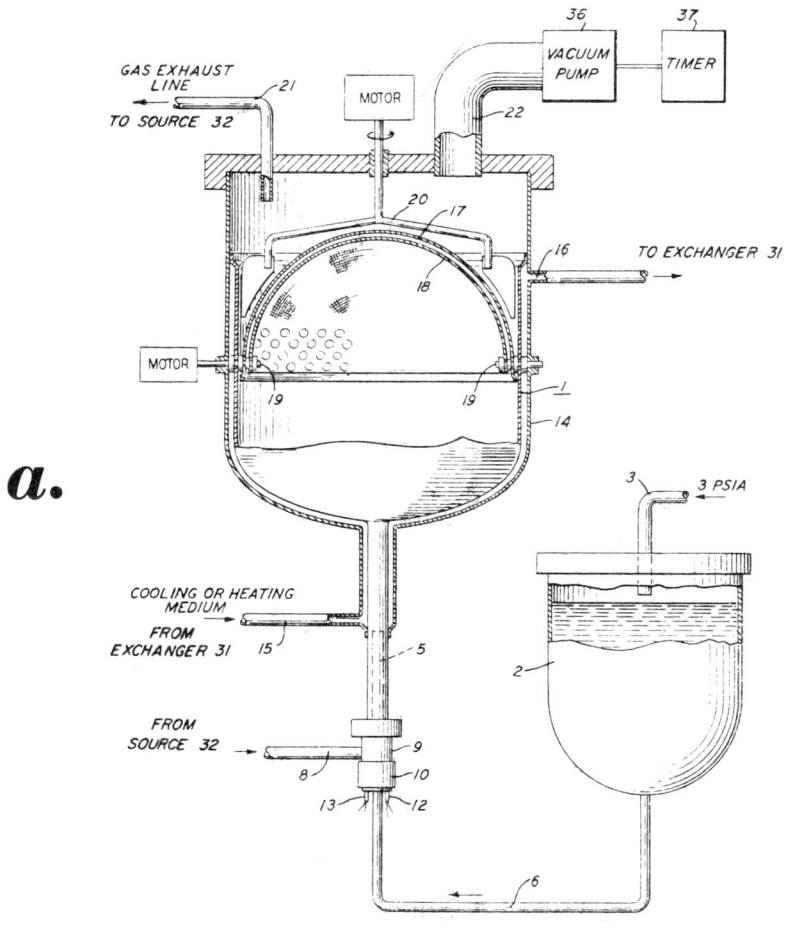

Unitary Freeze-Freeze Dryer

(continued)

FIGURE 1.4: (continued)

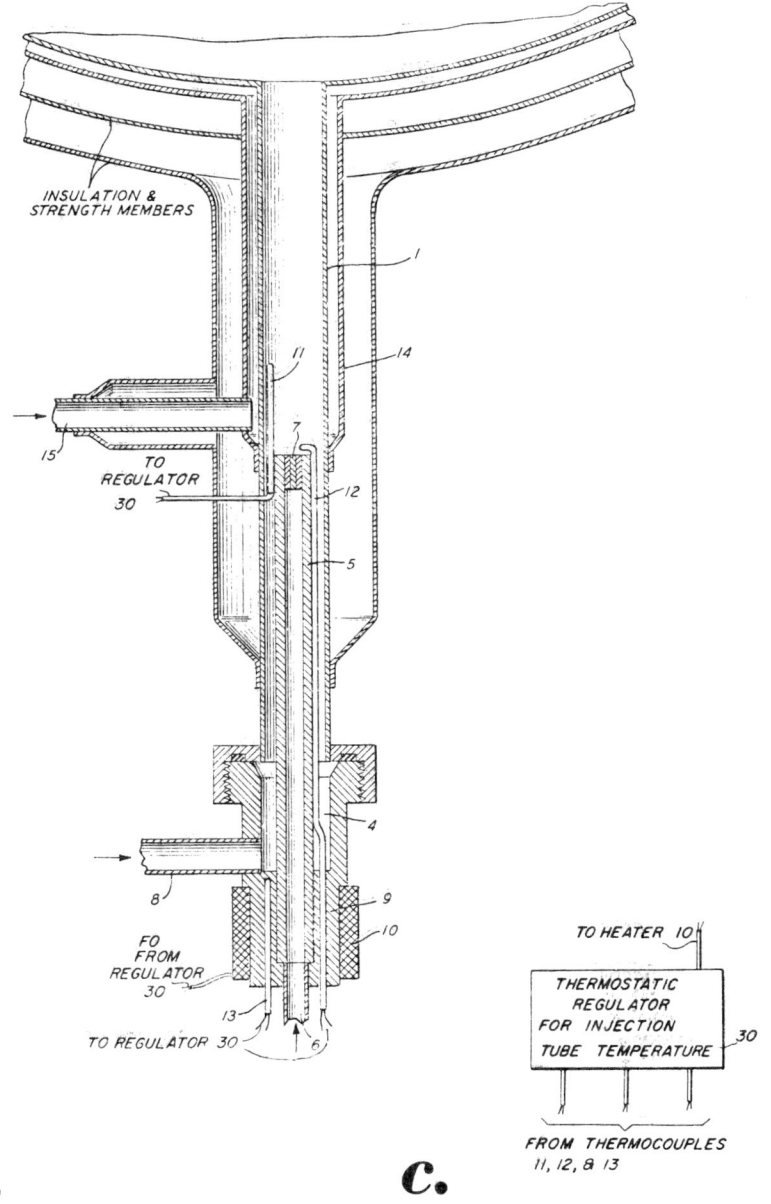

b.

Bottom Extension of Chamber 1 of
Figure 1.4a in Detail

c.

Block Diagram of Equipment
Ancillary to Apparatus

(continued)

FIGURE 1.4: (continued)

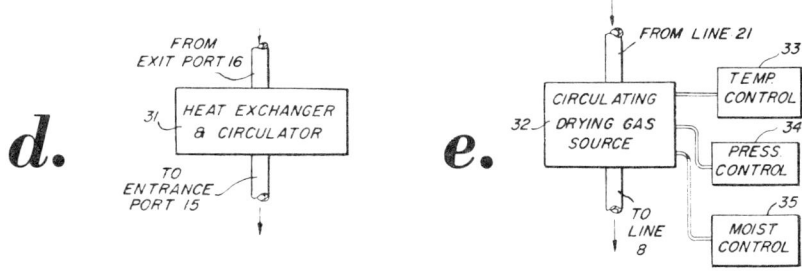

(d)(e) Block Diagram of Ancillary Equipment

Source: U.S. Patent 3,484,946

In practicing the freezing stage of the process, the solution to be frozen is forced under controlled pressure of, for example, 3 psia from bath **2** and into injector **5** where at the tube **7** droplets of solution are formed and injected into the neck region of the refrigerant. Their diameter is principally dependent on the diameter of the tube and the solution pressure.

Since the droplets of solution are less dense than the refrigerant, the droplets ascend through the refrigerant and upwardly where they collect at the upper surface of the refrigerant on the concave side of partially immersed screen **18**. The drying process may be practiced within chamber **1** by rotating outer screen **17** 180°, draining the refrigerant out duct **8** and applying through duct **8** a drying gas such as air or nitrogen at the temperature of the refrigerant which typically is −40°C.

The screens **17** and **18** are slowly rotated around axes **19** while a moderate flow of dry gas, for example dry air or dry nitrogen, is passed through the drying vessel between inlet **8** and exhaust **21**. The gas temperature is initially at the temperature of the refrigerant, and the pressure inside the chamber is maintained at near atmospheric. The gas temperature is then slowly increased to approximately −5°C, which is a little below the slightly depressed freezing point of water, and there maintained until the frozen droplets are virtually dry.

This residual water, if determined to be sufficiently slight, may be removed in a surprisingly short time relative to other known methods by the atmospheric pumping procedure. Its practice calls for the controlled application of heat in envelope **14** as well as to the drying gases entering through port **15** by conventional apparatus including a gas source **32** with temperature control **33**, pressure control **34** and moisture content control **35**, all of which are depicted in block diagram form in Figure 1.4e. Additionally, the required negative pressure excursions are effected by vacuum pump **36** operating with timer **37** and connected to the interior of chamber **1** by vacuum line **22**.

It is important to realize that, in the practice of this phase of the process, the residual water remaining in the almost dry material actually transits back and forth between the gas and liquid states. It therefore is important that a liquid stage will not be deleterious to the end product. The final highest temperature reached in practicing the above procedure is often the maximum safe temperature for the material in question.

FROZEN MASS

Perforated Frozen Slab of Product

M.L. Brewster; U.S. Patent 3,482,326; December 9, 1969 describes the preparation of a perforated slab of frozen product for dehydration in a vacuum drying process. Perforations provide greatly increased vapor path area and decreased vapor path travel, the latter being as much or more in the lateral direction as normal to the upper and lower slab surfaces. A freezing tray having a plurality of upright pins for forming such perforations, as the product is frozen into the slab form, is described with reference to Figure 1.5. Figure 1.5a shows a top plan view of the freezing tray; Figure 1.5b is a cross-sectional view along line 2-2 of Figure 1.5a, showing the drying tray disposed above the frozen slab, preparatory to transfer of the frozen product to the drying tray; Figure 1.5c is a fragmentary perspective view of the frozen product slab.

FIGURE 1.5: PERFORATED SLAB OF FROZEN PRODUCT

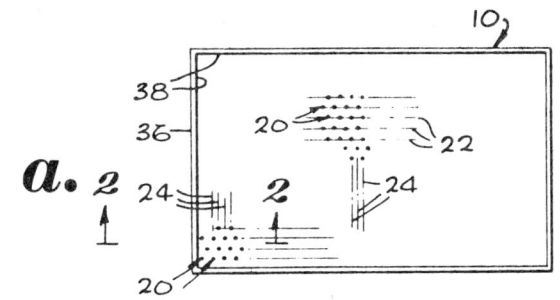

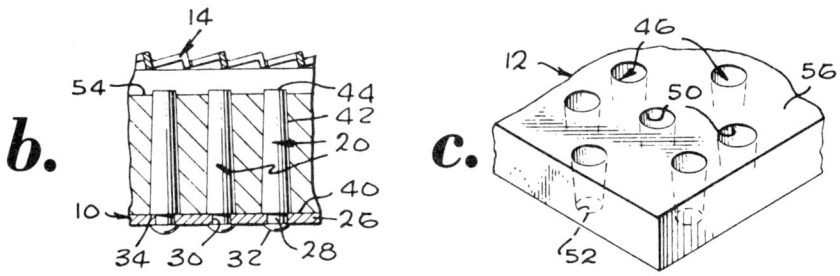

Source: U.S. Patent 3,482,326

The freezing tray 10 is provided with a plurality of vertical pins 20 of identical configuration and arranged in a uniform pattern preferably comprising staggered rows and columns, indicated generally by broken lines at 22 and 24, respectively, wherein the individual pins are equally spaced from each other, i.e., each pin is circular in horizontal cross section and its axial center is equidistant from the axial centers of each adjacent pin. Each pin is secured in an upright position to the bottom plate 26 of the tray as by the illustrated reduced-diameter shank portion 28 extending through a hole 30 in the plate and having an upset or swaged head 32.

Preferably, the tray sidewalls 36 are upwardly outwardly sloped, and their interior surfaces 38, as well as the interior surface 40 of the bottom plate 26 and lateral surface 42 of pins 20, preferably are coated with Teflon or a similar inert material to minimize binding or sticking of the frozen product slab 12 thereto whereby the slab may be transferred therefrom by virtue of its weight alone when the freezing tray 10 is turned over upon the drying tray 14. Partially for the same reason of preventing binding or sticking during slab transfer, each pin 20 is tapered upwardly inwardly from its body bottom extremity 34 to its upper extremity 44.

Release of the frozen slab from the freezing tray is accomplished by merely turning the tray over, preferably directly upon a drying tray for transfer thereto, whereby the previously upper surface of the frozen slab becomes the bottom surface thereof, and vice versa. Thus, the frozen slab as it rests upon the drying tray has the same plurality of holes as the number of pins in the freezing tray, such holes having the same configuration as that of the pins albeit vertically reversed or inverted.

Thus, with the preferred tapered configuration of the pins, the holes are similarly tapered and constitute basically vertical passages having top and bottom apertures, the top apertures being larger than the bottom apertures. The drying tray is preferably of an expanded metal construction to provide maximum open area for the absorption of heat from radiant heating platens disposed both above and below the drying tray, as well as to maximize the VPA (vapor path area) and minimize the VPT (vapor path travel).

The preferred pattern arrangement of the holes or perforations 46 (and therefore of the pins 20) is such that the hole centers are located at the apices of equilateral triangles, such triangles preferably having ½-inch sides, i.e., the center of each hole or pin is spaced ½ inch from the center of its six nearest adjacent holes or pins, such uniform pattern being most easily obtained by the staggered rows and columns 22 and 24 previously indicated. With such preferred pattern and spacing, pins 20 preferably are at least 1 inch high and have diameters of about 5/16 inch and 1/8 inch at ends 34 and 44, respectively.

Since the walls of the frozen slab defining the holes therethrough constitute surfaces of the frozen slab, the aggregate surface area of the slab is greatly increased over the normal surface presented by a solid slab. Such increased surface area simultaneously attains correspondingly improved characteristics of both radiant heat absorptivity and also vapor emissivity.

Frozen Material Comminuted to Desired Size Within Vacuum Chamber

The work of *G.-W. Oetjen, F.-J. Schmitz and H. Eilenberg; U.S. Patent 3,612,411; October 12, 1971; assigned to Leybold-Heraeus-Verwaltung GmbH, Germany* relates to a method and apparatus in which the product, after being frozen outside the vacuum drying chamber into relatively large chunks is brought into the chamber and comminuted into particles of appropriate size for drying. Fine particles, of less than the optimum size for drying, which are produced during the grinding or comminuting process, are screened out before the particles are inserted into the vacuum chamber.

This method takes advantage of the fact that through use of the vacuum in the vacuum drying chamber the latent heat of evaporation is not added to the material before it comes to the heated drying beds. This is because the absolute pressure of the vacuum space is lower than the water-vapor pressure of the product which is equivalent to the temperature where the product is treated. Accordingly, thawing cannot occur. The method is described with reference to Figure 1.6.

FIGURE 1.6: COMMINUTION WITHIN VACUUM CHAMBER

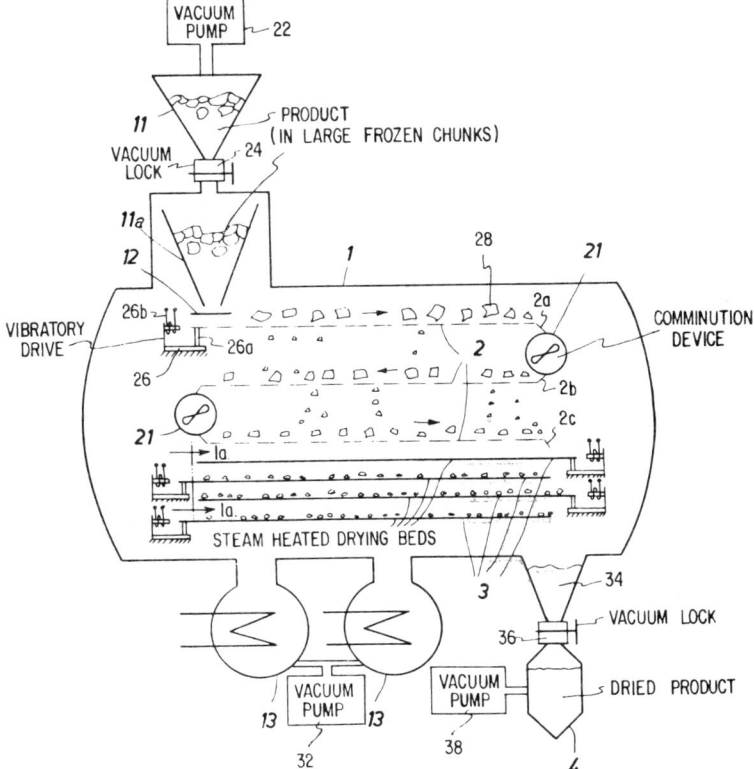

Source: U.S. Patent 3,612,411

In operation the product to be freeze dried is added in the form of large frozen chunks into the feed hopper **11** while the vacuum lock **24** is closed. Vacuum pump **22** is activated to evacuate the feed hopper to the subatmospheric pressure which prevails in the vacuum chamber **1**. The lock is then opened and the large chunks continuously pass into the internal feed hopper **11a**. Particles continually flow out of the discharge opening **12** in the bottom of the internal feed hopper, appropriate mechanism (such as a stirrer) being supplied to assist this flow if desired.

The large chunks move along the first screening and conveyor section **2a** in response to the forward action of its vibratory drive mechanism **26**. Particles which are already at the desired optimum size for freeze drying pass through the openings **28** and fall onto the second screening conveyor **2b**, from which they thereafter fall through to the third screening conveyor section **2c** and eventually to the drying beds **3**. The larger chunks on the vibratory screening conveyor are turned to the first comminution device **21** which breaks most of them to the desired particle size for freeze drying.

The comminuted particles are moved along conveyor screening section **2b**. During this movement most pass to the lowermost screening conveyor section **2c**. Any particles which do not pass through the openings are further comminuted in the second comminution device **21** and are carried along the third screening conveyor section **2c** to the vibratory drying beds **3**. The particles move successively along these beds, and while moving are subjected to the heating action of the double-walled drying beds.

The heat supplied here by the double-walled drying beds sublimates the ice present in the particles since the vacuum pump **32** and the condensers **13** are so arranged as to maintain the pressure within vacuum chamber **1** below that at which thawing can take place.

The dried product discharged from the vibratory drying beds is continuously accumulated in hopper **34** and periodically passed through the vacuum lock **30** to chamber **4** after it has been evacuated by vacuum pump **38**. Thereafter the dried product is removed from chamber **4**.

Instantaneous Freezing Under High Vacuum into Porous Chunks

The process of *A. van Gelder; U.S. Patent 3,477,137; November 11, 1969; assigned to Sun-Freeze, Inc.* is directed to preparation of liquids, with or without entrained solids, for subsequent dehydration, by feeding the product into a closed chamber under high vacuum with rapid vapor removal so that there is instantaneous freezing of the material into porous ice chunks, which are simultaneously broken into smaller sizes from pea size to powder, the frozen form being porous in nature throughout.

The frozen pieces are then moved through the chamber, subjected to the same vacuum vapor removal, care being exercised to insure that no more than 80% of the moisture content is removed by the time it reaches refrigerated storage or passes on for further and final dehydration. The entry of the material must be from the bottom of the treating chamber through an elongated opening, the feed being drawn by the suction of the low pressure in the chamber, interrupted only, if desired, by a proportioning device. The process is described with reference to Figure 1.7.

FIGURE 1.7: INSTANTANEOUS FREEZING UNDER HIGH VACUUM INTO POROUS CHUNKS

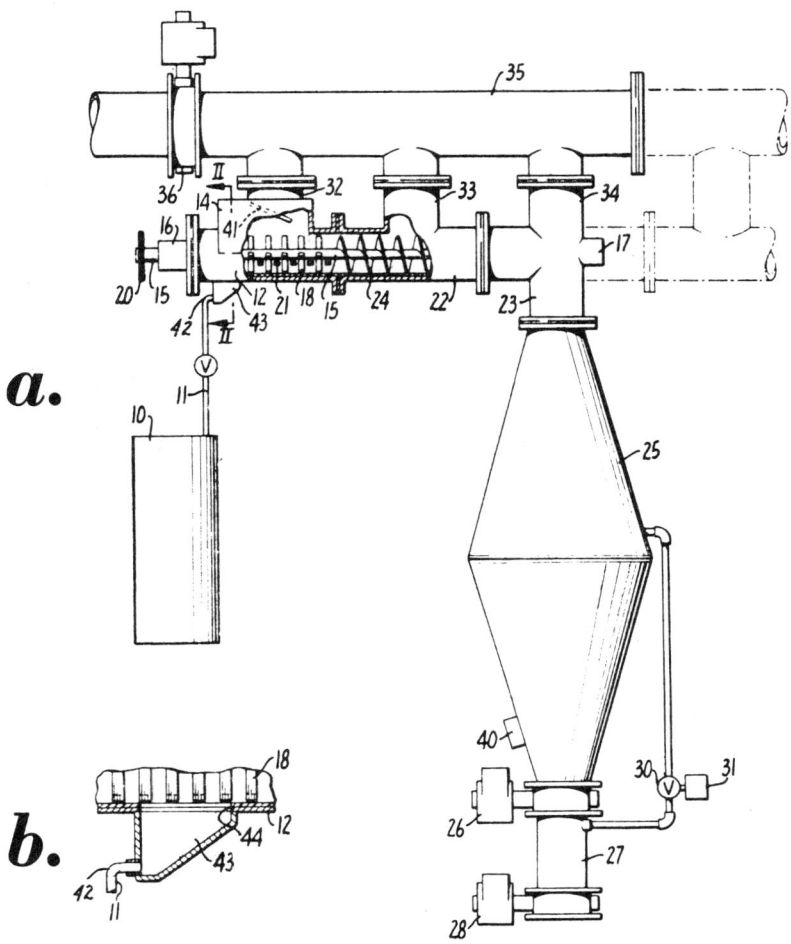

(a) Schematic Vertical Layout
(b) Elevational View of Inlet

Source: U.S. Patent 3,477,137

The flow of liquid to chamber **12** is adequate due to the very low pressure or high vacuum within the chamber. To accomplish this feed flow, which will not be clogged with ice and stop the continuous operation, a slot feed entry **44** has been devised which produces a unique and unexpected result. Axially of the chamber **12** is a shaft **15**, which shaft is journaled in vacuum-tight bearings **16** and **17**. Within the chamber **12** spikes **18** are radially secured to the shaft **15** at the product entry and rotate with the shaft. The shaft **15** is continuously rotated by the sprocket **20** connected to any suitable prime mover for positive continuous and substantially uniform rotation at the speed setting. Also at the entry end of chamber **12** and cooperating with the rotating spikes **18** are radially upstanding stationary spikes **21** which are secured at spaced intervals from each other on the interior so as to permit the rotating spikes **18** on the shaft **15** to pass therebetween in close proximity thereto. Within the chamber **14**, and shielding the vapor outlet **32** is an angular baffle **41**.

Operation of the device is as follows. The liquid, whether it is clear, colloidal or solids in suspension, is fed from the supply tank **10** upwardly through the line **11**, turns right angle at **42** and enters into the feed chamber **43**, through the elongated slot **44** and into the chamber **12** and swept by the pins **18**. At this point the liquid passes from room temperature into the vacuum maintained in the system of from 2,000 to 100 microns of mercury.

At the moment of entry through the slot **44** there is a flash freezing or instantaneous freezing, the product almost literally explodes into ice clusters or chunks of a porous nature, to the extent that at least 15% of the vapor is eliminated from the product at the moment of entry into chamber **12**. The rapid freezing is caused by a massive and rapid evaporation which process removes the heat from the liquid resulting in a differential vapor pressure between the liquid and the vacuum. This evaporation continues until the dew point of the operating pressure is reached. For example, at a pressure of 100 microns of mercury the product will continue evaporating until it has reached a temperature of about $-40°F$, which temperature is reached almost instantaneously upon delivery to the tube.

The removal of water vapor continues in chambers **12, 22** and **23** until the frozen product has 80% or less of the moisture removed. In liquids with entrained solids, if more than 80% of the moisture is removed, then the product becomes taffy and cannot be moved. On the other hand, when a liquid is undergoing treatment, if more than 80% of the moisture is removed, it becomes so light and feathery that it is removed by the vapor flow through **33** and **34** and is lost. The vapor from chamber **12** is taken off through the connection **32** to the manifold **35** and thence to the condenser. There is a momentary buildup of ice chunks, the expanded volume of which is temporarily accommodated by the box **14**. The water vapor, in leaving the ice chunks so rapidly, creates vapor channels throughout the entire chunk which becomes sponge-like with a noticeable porosity.

The rotation of the pins **18** in combination with the stationary pins **21** continuously breaks up any buildup of the chunks, which are readily friable, into granules sized from powder to pea size. Each granule, like the chunk, is laced with channels left by the removal of the vapor so that the particles themselves are also porous. Because of the explosive and ebullient nature of the instantaneous freezing of the product, a baffle **41** is provided at the top of chamber **14**, shielding the opening to vapor take-off **32**, to prevent inadvertent removal of the frozen product and separate the frozen product from the vapor take-off.

The Freezing Operation

The manner of delivery of the product to the chamber **12** is of unexpected and critical importance. In the first place the slot **44** provides a self-clearing or releasing means which does not freeze up and stop operation. Certain areas within the slot may freeze and block part of the passage for a short time, but the warmer entering liquid through the remainder of the slot soon frees the material and thus self-cleaning is accomplished. This phenomenon is aided by the observed nature of the instantaneous freezing of the entering material. The instantaneous freezing under high vacuum is so violent and explosive that only the top portion of each increment of feed is thus frozen, leaving a liquid base forming the next increment. This means that there is always liquid feed in the chamber **43** and very nearly always in the feed slot **44**.

Under the operating conditions defined herein it would be quite impossible for the feed to be otherwise than at the bottom or otherwise than a slot. The material enters the chamber **12** at substantially zero velocity and is instantaneously subjected to the operating conditions. There is no appreciable or deleterious buildup of a frozen skin inside the chamber **12** or frozen chunks within the chamber and blades or spikes **18** and **21** are not primarily scrapers but perform their real function of breaking up the immediately formed ice chunks into smaller sized pieces and particles.

The broken ice chunks in the form of frozen particles or pieces are then delivered by the screw conveyor **24** through the chamber **22** where they are again subjected to removal of water vapor under vacuum. It is to be observed that there is constant movement of the particles and pieces through the chambers **12** and **22** due to the vapor velocity plus the continued decrease in specific gravity. The vapor from chamber **22** leaves through the connector **33** to the common manifold **35** and thence to the condenser (not shown). This same movement by screw conveyor **24** through chamber **23** continues and at the same time is being subjected to vapor removal under vacuum with high velocity. No more than 80% of water vapor is removed, and so this is not to be construed as a dehydrating method or apparatus.

Upon delivery at the end of chamber **23** the frozen porous particles drop into a holding tank or bin **25** and pack down in the bottom thereof. Any heat coming from the outside of the tank may cause some melt, but this melt is refrozen and the water vapor of the refreeze is removed through the tank as this tank is under the same vacuum and connected by connector **34** to the common exhaust manifold **35**. The delivery of the frozen porous product from the tank is accomplished through the air lock valve **26** into the passage **27**. Since the tank is already under the same vacuum conditions as the entire system, in order to equalize the vacuum in the tank with that of the conveyor line **27**, it is necessary to have a valve **30** operated by a solenoid to keep the negative pressures equalized in the system. It is important to note that the product delivered at **23** can be connected directly to a continuous dehydrating system.

The pressure of the pack in the bottom of the tank is sufficient to block the free flow of the frozen porous particles into the line **27**. A vibrator **40** may be attached to the outside of the tank adjacent to the air lock valve **26** to loosen any packing of the material. The frozen porous products in granular form are released through air lock valve **28** for subsequent treatment and dehydration.

Freezing Material onto Heat Transferring Means

P.D. Porta; U.S. Patent 3,289,314; December 6, 1966; assigned to Edwards High Vacuum International Limited, England describes a method of freeze drying material which includes the steps of placing the material in its unfrozen state in conducting and preferably surrounding relation to a heat transferring means and freezing the material onto the heat transferring means.

Further steps may include placing the frozen material together with the heat transferring means in an evacuable chamber, reducing the pressure in the chamber and supplying heat to the frozen material by conduction from the heat transferring means.

The heat transferring means may be a substantially plane sheet having formed therein a fluid passage or passages extending between an inlet and an outlet, the sheet also having formed therein a plurality of holes, the step of freezing the material may be carried out by passing a cooling fluid through the fluid passage or passages, the frozen material transfixing the sheet through the holes formed therein, and the step of supplying heat to the frozen material in the chamber may be carried out by passing a heating fluid through the fluid passage or passages.

Alternatively the heat transferring means may be an interstitial heater grid of electrically conducting material operable to conduct heat to the frozen material, or an interstitial support grid having thereon or wound into it an electrical heater element or elements operable to conduct heat to the frozen material and the step of freezing the material may be carried out by placing the material together with the heat transferring means in thermal contact with an independent cooling means, the frozen material transfixing the heat transferring means through its interstices.

In any of the methods the further step of supplying heat to the frozen material when in the chamber by a second heat transferring means may be included and the second heat transferring means may be operable to supply heat to the frozen material by radiation. The process is described with reference to Figure 1.8.

FIGURE 1.8: FREEZING MATERIAL ONTO HEAT TRANSFERRING MEANS

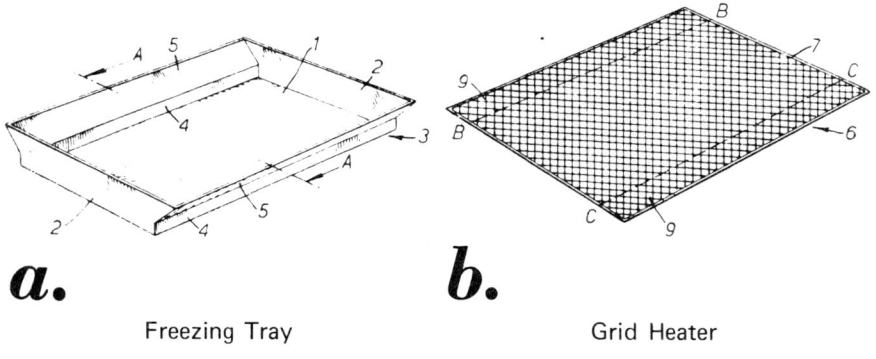

a. Freezing Tray   *b.* Grid Heater

(continued)

The Freezing Operation

FIGURE 1.8: (continued)

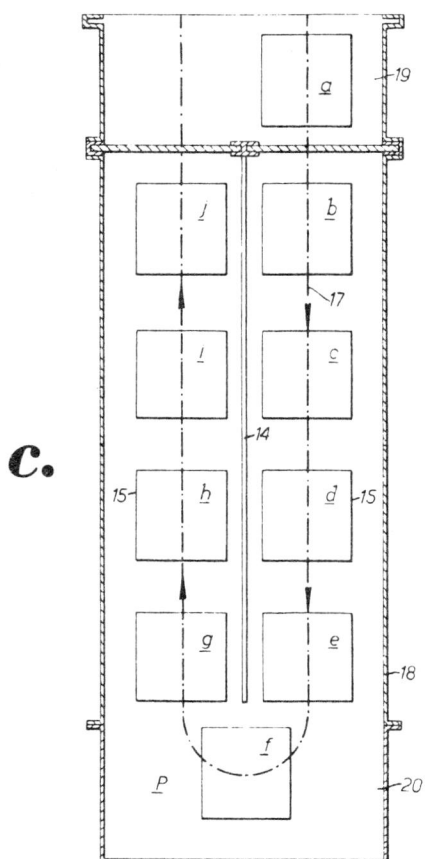

c.

Schematic Horizontal Section Through Vacuum Chamber

Source: U.S. Patent 3,289,314

The foodstuff is placed in the tray **1** up to the level of the vertical portions of the side walls **4**. The heater grid **8** is folded along lines **BB, CC** so that flange portions **9** extend upwardly and outwardly. The grid is placed in the tray over the foodstuff and more foodstuff placed over the heater up to two-thirds of the depth of the tray. The foodstuff is now able to transfix the heater grid. The tray with its contents is now placed in a refrigerator and the foodstuff frozen.

The material could be in the form of a liquid or a slurry. Alternatively, slices or slabs of a wet solid could be placed on either side of the first heating means before freezing. Water or a suitable aqueous solution or an oil or wax could be used so that when frozen it acted as a cement adhering to or transfixing the first heating means.

The frozen slab is removed from the tray and the flange portions 9 folded back to their original positions coplanar with the main grid 8. The frozen slab is loaded into a carriage in the loading chamber 19 isolated from the vacuum chamber 20 with flanges 9 supported on two electrodes. The chamber 19 is isolated from the atmosphere and evacuated. The chamber 19 is connected to the chamber 20 and the loaded carriage moved into the evacuated chamber 20. The radiant heating plates are operated as the carriage passes through the first compartment up to the point P. During this period the surface portions of each slab become dried. Once past the point P the radiant heating plates are not operated but the embedded grid heaters are. The central regions of each slab are now dried as the carriage passes through the second compartment of the chamber 20. The carriages are unloaded when they reach chamber 19 once more.

Frozen Extrusions

The freeze drying method developed by *T.R. Folsom; U.S. Patent 3,583,075; June 8, 1971; assigned to FMC Corporation* comprises the suitable preparation of frozen bodies having extended area and thin section, exposing these bodies to heat coming almost entirely through the surrounding vapor while suspending the bodies so that little heat is conducted to them through the supporting members and so that most of the external area can receive heat and give off vapor rapidly.

The method and apparatus is mainly concerned with the processing of masses of a thickness between about 1/8 of an inch to about 1/2 of an inch. Much of the technique is capable of extension to thinner or thicker masses, but this region of thickness is chosen because a reasonable drying time, say from 1 hour to 8 hours, can be readily attained by convenient practice and with a wide variety of substances.

The open suspension of the material to be dried on supporting members such as thin wires rather than upon trays or shelves or container walls is recommended because the former provides for proper heating and exposure whereas the latter causes serious limitation to heat input and vapor escape. Although ice is a better thermal conductor than is dilated vapor, introduction of the major portion of the heat by means of extensive contact with an ice body with a heat-supplying surface is erratic and the possible heat input is limited directly and indirectly by this primitive technique.

This open exposure can be accomplished by supporting the frozen material on members which are prevented from conducting appreciable heat to the body, and which are designed so as to avoid interfering with the arrival of heat from the distant heat source and so as to avoid interfering with escape of vapor, especially from those surfaces of the body receiving heat.

Freezing must be done in such a way that proper exposure and heating can be carried out economically. The extrusion method provides a means for introducing the material immediately to vacuum in a suitable geometric form, without thawing at the surface at any time, and continuously, and with ideal crystalline conditions at the frozen surface.

The Freezing Operation

Since essentially all of the surfaces which are brought to exposure are the same surfaces which, by this method, have been used to extract heat during freezing, and since freezing results in the growth of needle-shaped ice crystals normal to the freezing surfaces, ice crystals are arranged normal to the exposure surface. Therefore, as drying proceeds, tubular passages normal to the outer surface are formed by the vaporization of the needles and permit most easy escape of vapor from the interior. The method is uniquely characterized by the beneficial aspect that almost all of the exposed surface has been previously a freezing surface.

Figure 1.9 illustrates a device for the production of frozen extrusions in the form of a continuous solid rod having circular cross section. A cylinder **90** with a smooth bore and closed end **102** is surrounded for part of its length with a jacket **91** containing a refrigerant **R**, to define a heat exchanger. Circular fins **92** are provided to enhance the heat transfer from the cylinder **90** into the refrigerant. The refrigerant is illustrated here as being circulated as in the conventional flood system method where some such refrigerant as ammonia is used. (A cold brine or other secondary cooling fluid could be used in the jacket for the same purpose.) The refrigerant is shown entering through pipe **P1** and leaving through pipe **P2**. There is no contact between refrigerant **R** and the substance being frozen.

FIGURE 1.9: FROZEN EXTRUSIONS

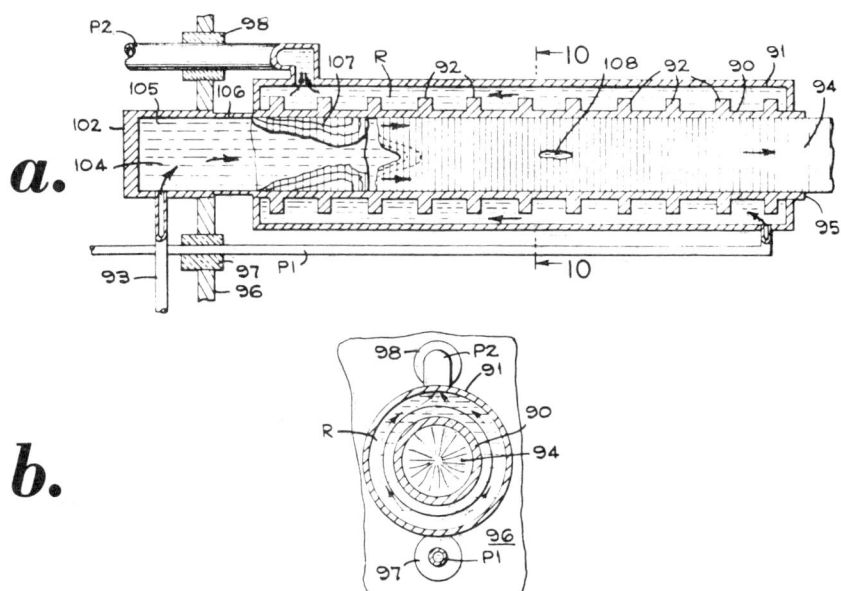

(a) Device for Production of Frozen Extrusions in Form of a Continuous Solid Rod Having a Circular Cross Section
(b) View Along Line **10-10** of Figure 1.9a

Source: U.S. Patent 3,583,075

The liquid to be frozen is introduced into the freeze chamber defined by the bore of cylinder **90** through pipe **93** under such a pressure as to force the solid frozen material **94** out the distal end **95**.

The freezing cylinder jacket and pipes are insulated as follows. The wall of the vacuum chamber is represented by **96**, shown in section and bounding the atmosphere on its left side and the vacuum on its right side. Pipes **P1** and **P2** penetrate the wall **96** through seating and insulating plugs **97** and **98** respectively. Therefore, all members to the right of wall **96** and plugs **97** and **98** are surrounded by vacuum and are thus effectively insulated from thermal losses. Polishing the outer surfaces of the jacket, cylinder and pipes improves this insulation.

Since the freezing cylinder penetrates the vacuum chamber **96**, and since the frozen plug seals off the fluid material, a means is thus provided for introducing the material to be treated into the vacuum chamber without interrupting or spoiling the vacuum. Freezing cylinders can be constructed to produce frozen extrusions having rectangular shape. In addition, hollow extrusions can be produced by the addition of a suitable core.

For any given substance there is a more or less critical temperature below which it is unprofitable to reduce the temperature of the refrigerant. This temperature varies greatly with the constitution of the substance and is best ascertained by test.

It is often found possible to add a small amount of certain materials to a substance before processing which effectively prevent matrix collapse. Electrolytes should be avoided because of depression of freezing points caused; colloidal material, such as albumins, in amounts of 10% on a dry solids basis are very effective when permitted. The most effective additive for fruit juices and similar substances is a smooth suspension containing particles of size ranging from the true colloidal up to the palpable or even coarser.

As an example of this mechanical matrix support method, to 10 to 20 parts pure orange juice is added 1 part by wet weight of orange rind, or preferably the outer yellow portion only of the rind for the sake of flavor and other benefits. The rind solids are reduced to a smooth cream in a comminutor together with the juice, or alone. This mixture dries much faster than pure juice and stands much abuse because of reinforcement of the matrix structure.

HEAT TRANSFER MECHANISMS

RADIANT HEATING

Simultaneously Applying Radiant Heat and Blowing Cold Dry Air

The method developed by *G.C.W. van Olphen; U.S. Patent 3,270,428; September 6, 1966* entails the positioning of food to be dried in a confined dehydration zone and simultaneously applying radiant heat to the food and blowing cold dry air thereover in an amount and at a temperature to maintain the radiantly heated food below 32°F whereby ice within the food will be sublimated and entrained in the passing air.

Preferably, the air is exposed to a spray of coolant at a temperature substantially below 32°F which not only cools the air but removes the moisture therefrom in the form of small ice particles whereupon the cooled, demoisturized air can be recirculated through the dehydration zone. The method is described in greater detail with reference to Figure 2.1.

The apparatus includes a chamber which is substantially enclosed by rectangularly related end and side walls **13**, a top and a bottom. A central partition **22** extends through the chamber, terminating short of the end walls to thus form a continuous conduit through which an air stream can pass. On one side of the central partition, a food dehydration zone **5** is formed and on the other side of the partition, an air-conditioning or more particularly an air-reconditioning zone **2** is provided.

Adjacent the food dehydration zone, spaced openings **9** and **12** are provided in the side wall **13**, one to permit ingress of the food to be dried and the other to provide egress thereof. The food is preferably placed on trays **10** which are in turn stacked to leave an air space above the food.

A driven rotary fan **15** is positioned at the entrance end of the air conditioning zone **2** of the continuous conduit formed around the central partition **22** to direct air through the zone **2** and then through the dehydration zone **5** over the food supported on the trays **10**. The flow of air is in countercurrent relationship to the movement of the food supporting trays so that the food having the lowest moisture content is subjected to the cold fully demoisturized air first entering the dehydration zone **5** to thus provide a maximum differential between the moisture content of the food and the moisture content of the air.

FIGURE 2.1: DIAGRAMMATIC TOP PLAN VIEW OF APPARATUS WITH ITS TOP REMOVED

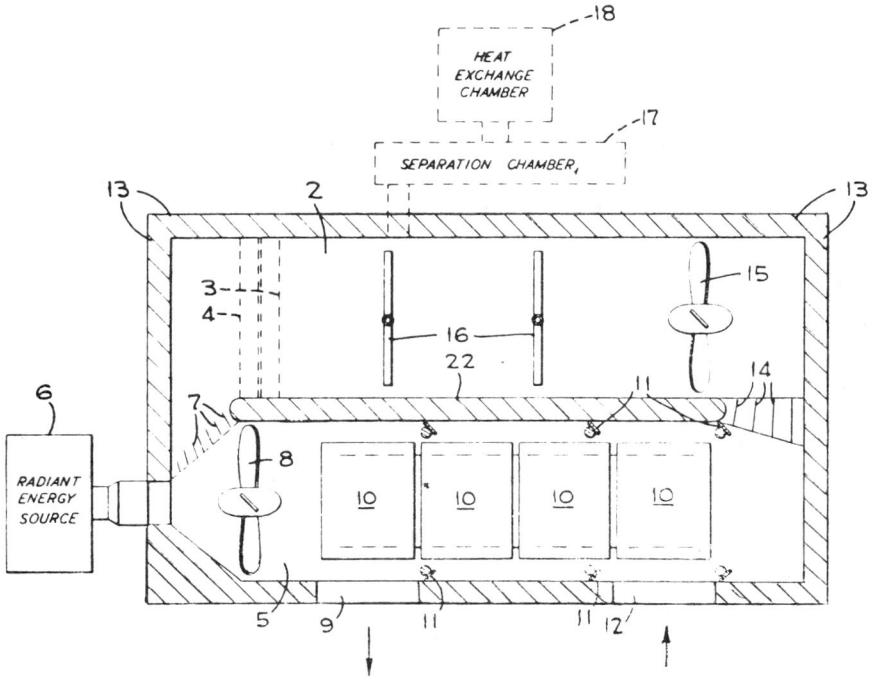

Source: U.S. Patent 3,270,428

To maximize the vapor pressure differential and thus expedite the rate of dehydration, radiant energy is directed against the food in the dehydration zone. Preferably highly-directive radiant energy at microwave frequencies is generated by a magnetron or similar source indicated at **6** to pass longitudinally through the dehyration zone **5** in a direction also countercurrent to the motion of the food trays therethrough.

An air-impelled metal-bladed fan **8** adjacent the radiant-energy source disperses the energy so as to fill the entire zone. The food resting on the cart adjacent the exit opening **9** in the side wall **13** is subjected to the maximum radiant energy as well as the lowest temperature and driest air to thus facilitate the removal of the final moisture therefrom. Adjacent the sides of the other carts, metal deflectors **11** are pivotally supported to provide a controlled deflection of the radiant high frequency energy inwardly. Pivotal adjustment of these deflectors can be utilized to control the amount of radiant energy supplied to the food on any of the carts. The amount of radiant energy so supplied is of course related to the temperature and rate of flow of the cold-dehumidified air, the relationship being such that the food remains below 32°F so that the moisture removal is entirely a sublimation process. The latent heat of sublimation of ice being well established, a preliminary calculation of the amount of moisture to be removed from the food will enable the establishment of the correct air temperature, air flow rate and radiant energy to be applied to assure that the emergent food is below the requisite moisture content.

Preferably, the air entering the dehydration zone 5 is at or below 0°F and the amount of radiant energy applied to the food must then be sufficient not only to supply the latent heat of sublimation but to keep the food at a relatively higher temperature, such as 30°F. Such temperature differential between food and the air obviously increases the vapor pressure differential between and thus allows the sublimation process to take place at a relatively rapid rate.

The air, after its passage over the food on the carts, moves around the end of the central partition 22 and is given further impetus by the rotary fan 15 for continued motion through the air-conditioning zone 2. Since the air is not only demoisturized but is recooled in this air-conditioning zone, preferably appertured metal baffles 7, 14 are disposed between the ends of the central partition and the end walls 13 of the chamber so as to preclude entry of radiant energy into such air-conditioning zone.

In the air-conditioning zone, the blown subfreezing air with the moisture entrained is directed through two or more sprays of coolant which emanate from a plurality of nozzles 16 adjacent the top of the air-conditioning zone and form transverse fluid curtains against which the air is projected.

The coolant has complete fluidity over a wide temperature range and, dependent upon the rate of air flow, is projected into the air-conditioning zone at a temperature between −30° and −90°F. Preferably, low viscosity organic silicates such as tetra(2-ethylbutyl) silicate or tetra(2-ethylhexyl) silicate are utilized as the coolant. Beyond the curtains of coolant, at the remote end of the air-conditioning zone 2, air-permeable curtains 3, 4 are supported to allow egress of the cold demoisturized air for return to the dehydration zone 5 but to preclude removal of ice or liquids from the air-conditioning zone.

The coolant and entrained moisture in the form of small ice particles flow from the bottom of the air-conditioning zone through a conduit onto one end of a vibrating screen (not shown) mounted for longitudinal reciprocation in a small coolant ice separation chamber 17.

Improving Efficiency of Radiant Heat by Coating Conveyor Belt

The object of this process of *H.L. Smith, Jr.; U.S. Patent 3,266,169; August 16, 1966; assigned to Hupp Corp.* is the improvement of the capacity and efficiency of radiation heated type conveyor belt dehydrators and heat treating equipment without addition of radiant heating capacity. This is accomplished by materially reducing the radiation reflectivity and improving the coefficient of radiant energy absorption of the inner side of the belt opposite the product carrying side so it will absorb radiation at a much higher rate than heretofore, with resultant material increase of the temperature of the belt and the heat supplied to the material layer on the opposite side of the belt without increase of heat input, and by coordination of exposure times and temperatures with utilization of improved radiators designed for radiation wavelength emissivities providing peak heat transfer coordinated with the radiation and heat absorptive characteristics and the ultimate desired characteristics and uses of the end products.

The efficiency of radiant heat applying apparatus such as the dehydrator is improved by enhancing the absorptivity of the conveyor belt upon which the material being dehydrated or treated is supported, and by the selection of radiant heat radiating at an optimum peak wavelength.

This is accomplished by coating the inside of the conveyor belt with a suitable heat resistant material which increases the radiation absorption and emissivity coefficients of the belt surface, will effectively transfer the absorbed heat by conduction of heat through the belt to the outer material supporting surface, and which will adhere for a reasonable time to the belt through bending use and wide variations in temperatures.

Such coating finishes may be applied in any suitable manner, as by chemical means, such as anodizing, by brushing, spraying or rolling, and subsequently may be baked or heat treated or electrically deposited on the belt surface. Examples of suitable coatings may be the colored silicone varnishes, lamp black applied in a vehicle, black enamel, lacquer or shellac; and a particularly desirable coating for many uses may be applied to the belt in accordance with the ebonizing process disclosed in U.S. Patent 2,394,899 which provides a smooth, black oxide film or skin about one-hundred-thousandth of an inch thick to the surface of either chromium-nickel or chromium stainless steel belt materials.

The blackened skin which retains the corrosion resistance of stainless steel is applied, after conventional cleaning steps of the parts to be treated, by immersing the parts in a molten bath of dichromates at a temperature of 730° to 750°F for approximately 15 to 30 minutes, followed by cooling and rinsing.

In the freeze drying process the product to be dried is reduced to a fine frozen powder by freezing and comminuting it or by spraying it into a flowing stream of cold, inert gas. The frozen powder is then delivered to a storage vessel from which it is fed onto a vibrating feeder which spreads it in a very thin layer of uniform thickness onto the belt of a continuous conveyor sealed in a vacuum chamber and provided with the process radiant heat transfer mechanism. As the belt passes through the vacuum chamber, radiant heat rapidly effects sublimation of the water from the frozen product.

About 500 minutes is required to dry a one inch thick layer of food product. Assuming that a continuous conveyor in accord with the process having a 20" wide belt and a product feed rate of 100 cubic inches per minute is employed to freeze dry a desired food product, a belt having a capacity of 50,000 cubic inches must be 2,500 inches (208' 4") long and must be operated at a speed of 5 inches per minute to insure that the product is dried for the full 500 minutes.

The following table shows the results that can be obtained by taking advantage of the heat transfer phenomenon by reducing the thickness of the product layer.

a practice eminently practical with the apparatus of the process. In the following table, product feed rate is 100 cubic inches per minute; belt width, 20 inches.

Layer Thickness, in	Drying Time, min	Belt Length, in (ft-in)	Belt Speed, in/min
1.0 (1)	500	2,500 (208-4)	5
0.5 (½)	125	1,250 (104-2)	10
0.25 (¼)	32	625 (52-1)	20
0.125 (⅛)	8	320 (26-8)	40
0.0625 (1/16)	2	160 (13-4)	80
0.03125 (1/32)	½	80 (6-8)	160

The process is described with reference to Figure 2.2. Figure 2.2 illustrates diagrammatically a process and apparatus for applying radiant heat to articles. A thin endless stainless steel or similar metal belt **10** is trained about a pair of spaced rotatably mounted rollers **14** and **18** either one or both of which may be driven in the direction indicated by the arrows by an appropriate source of power (not shown).

FIGURE 2.2: VACUUM FREEZE DRYING APPARATUS

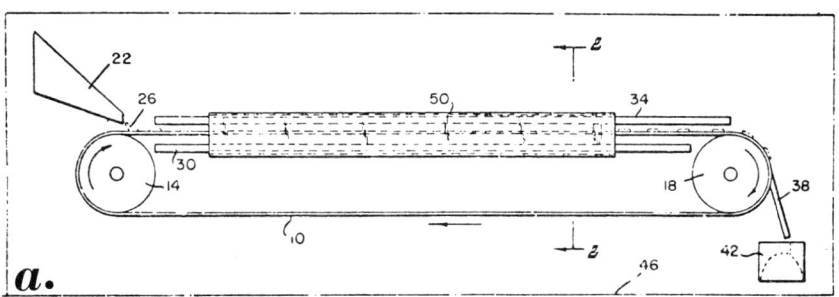

Front Elevational View of Heat Applying Apparatus

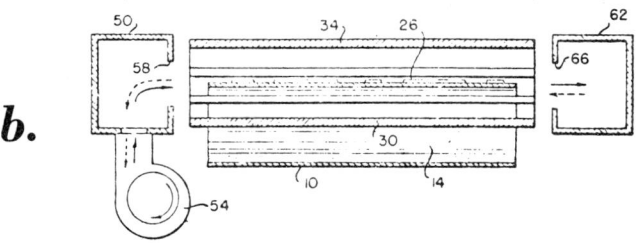

Side Elevational View Along Line 2–2 of Figure 2.2a

Source: U.S. Patent 3,266,169

A dispenser **22** (a hopper with control outlet) is located above to distribute material **26** to be heated onto the outer surface of belt **10**. Radiators **30** and **34** are mounted in the apparatus substantially parallel to a leg of the belt between rollers **14** and **18** and spaced therefrom.

As the material on belt **10** emerges from between radiators **30** and **34** it is passed over the surface of cooling drum **18**, scraped from the belt by doctor blade **38** or other suitable means, and delivered to hopper **42**. Radiators **30** and **34** may be of any well known type.

The dehydrating apparatus may be enclosed in a housing **46** (shown in phantom lines) if it is desired to carry out the dehydration in a subatmospheric environment.

Instead of providing a desiccant however, and to improve the speed and longevity of operation without shutdown, cross ventilating means are provided for removing vaporized or sublimated substances from the heating zone between radiators **30** and **34**. As shown diagramatically in Figure 2.2b this may be accomplished by providing an air or gas plenum **50** extending substantially throughout the length of a leg of belt **10**.

One side of plenum **50** receives the gas or air from any suitable source such as blower **54** and guides it as indicated by the solid line arrows out opening **58** across the heating zone. Plenum **62** on the opposite side of belt **10** collects the vapor laden gases as they emerge from between the radiators through opening **66** and conducts them to exhaust. Or where desirable blower **54** may be reversed to apply suction to plenum **50** thereby reversing the gas and vapor flow as shown in the dotted line arrows so that plenum **62** becomes the inlet and **50** becomes the outlet plenum.

Regelation of Frozen Particles on Conveyor Belt

This process of *H.L. Smith, Jr.; U.S. Patent 3,324,565; June 13, 1967; assigned to Hupp Corp.* is an extension of the work described above in U.S. Patent 3,266,169. In a continuous type freeze drying apparatus in which the product to be dried is reduced to a frozen powder and spread in a layer on a conveyor disposed in a vacuum chamber, considerable difficulty may be experienced with the loose particles bouncing or flying off the conveyor.

This can result in a build-up of powdered product within the vacuum vessel great enough to materially reduce the efficiency and/or prevent proper operation of the apparatus.

This problem can be solved by bonding the particles to each other and to the conveyor after they are spread on it to form a more or less continuous thin layer or sheet of frozen product frozen to the conveyor. This is accomplished by employing the phenomenon of regelation by which pieces of ice will freeze to one another when pressed together, even at subfreezing temperatures and subatmospheric pressures.

This is accomplished by passing the conveyor of the apparatus on which a layer of the frozen powder has been deposited, through the nip between a pressure roll and a backup roll. This provides the necessary pressure to bond the particles of the product to be dried to each other and to the conveyor. The net result is a generally continuous sheet of product frozen to the conveyor, eliminating the problem of product flying off the conveyor and building up in the vacuum chamber.

As shown in Figure 2.3, rolls **138** and **139** are disposed in parallel, spaced apart relationship with roll **138** being spring-loaded or otherwise biased in a downward direction so that the distance L between the center lines CL_1 and CL_2 of the two rolls is less than the sum of the radii R_1 and R_2 of the two rolls. The two rolls are rotatably supported in suitable bearings (not shown) and may be driven by the conveyor belt **128** running therebetween or by any other type of drive system which may be desired.

FIGURE 2.3: APPARATUS AND METHOD FOR FREEZE DRYING

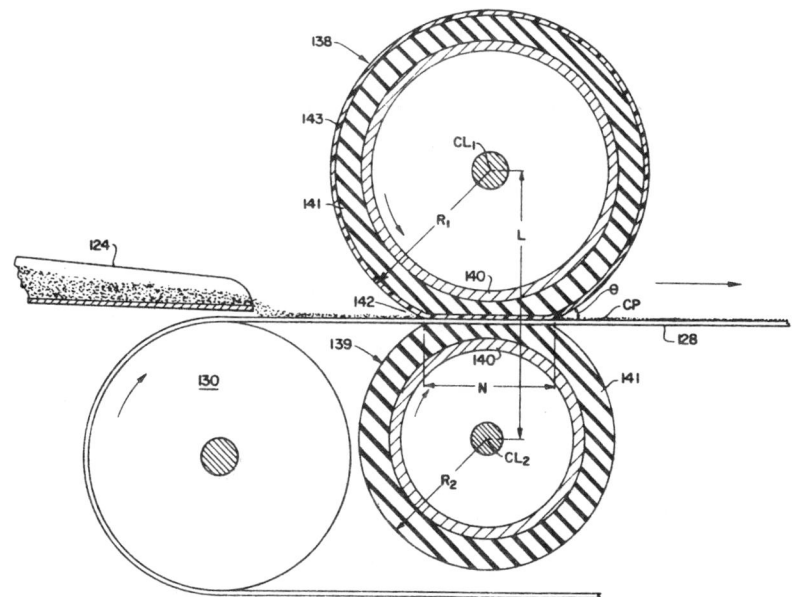

Source: U.S. Patent 3,324,565

Both of the rolls include a cylindrical inner member **140** surrounded by a sleeve **141** of resilient, elastically deformable material. Because dimension L is less than the sum of radii R_1 and R_2, there is area contact between the two rolls and conveyor belt **128** through a span indicated by reference character N.

As the loose frozen powder passes through the nip **142** between the two rolls it is increasingly compressed, bonding the particles together and to conveyor **128** as shown by reference character CP. The readily deformable eleastomeric sleeves **141** on the rolls make it practicable to exert pressure on the product layer over a relatively large area. This insures that sufficient pressure is exerted on all particles in the product layer for a sufficiently long time to produce the desired bonding. In addition, the large pressure area results in the exertion of substantially uniform pressure on all areas of the layer, compacting the entire layer to substantially the same thickness, even if the frozen powder is not uniformly distributed across conveyor **128** by feeder **124**.

To reduce the tendency of the compressed layer to adhere to pressure roll **138**, the elastomeric sleeve **141** of the latter is preferably surrounded by a thin sleeve **143** of Teflon or other material with nonsticking properties.

Proper adhesion of the frozen product to conveyor **128** may, in some instances, require that the temperature of the conveyor be at a given level as it approaches the nip **142** between rolls **138** and **139**. Regulation of the temperature of the conveyor may be readily accomplished by circulating a liquid of the proper temperature through the interior of conveyor belt supporting roller **130**. Similarly, if deemed necessary, temperature changes affecting the formation of the layer bonded to conveyor **128** may be regulated by circulating liquid of appropriate temperature through roll **138** or roll **139** or both.

As the frozen product bonded to the conveyor is carried from rolls through the vacuum vessel toward the end of the conveyor trained over a roller, it is heated by radiators disposed on opposite sides of the upper run of the conveyor. The interior of the vacuum vessel is maintained at a pressure below 0.180 inch of mercury absolute, a typical pressure being on the order of 0.0126 inch of mercury absolute.

At this pressure the water in the product will, upon the application of heat, pass directly from the frozen state to the vapor state, i.e., sublimate, at temperatures of $-20°F$ or higher. The radiators therefore cause the moisture in the frozen product to sublime.

Radiant and Conduction Heat Sources in Series

W. Nerge, U. Hackenberg and H. Ehlers; U.S. Patent 3,270,433; September 6, 1966 are concerned with more efficient transfer of heat to the drying product in a freeze drying process without danger of heat damage to the drying product. One feature of this process is the provision in a freeze drying apparatus of sources of both radiant and conduction heating for the product to be dried and wherein the radiant and conduction heat sources are in series.

The radiant heat source is adapted to supply heat energy to the conduction heat source. The source of conduction heat is adapted to serve as a container for the product to be dried. The combination food container and conduction heat source is adapted to completely shield the drying product from direct radiation from the radiant heat source.

The radiant heat from the radiant heat sources is applied to at least two sides of the combination product container and conduction heat source. The portions of the product containers which are subject to direct radiation from the radiant heat source are of unitary construction so as to provide good heat conductivity throughout the container body. The process is described with reference to Figure 2.4.

FIGURE 2.4: RADIANT AND CONDUCTION HEAT SOURCES IN SERIES

a.

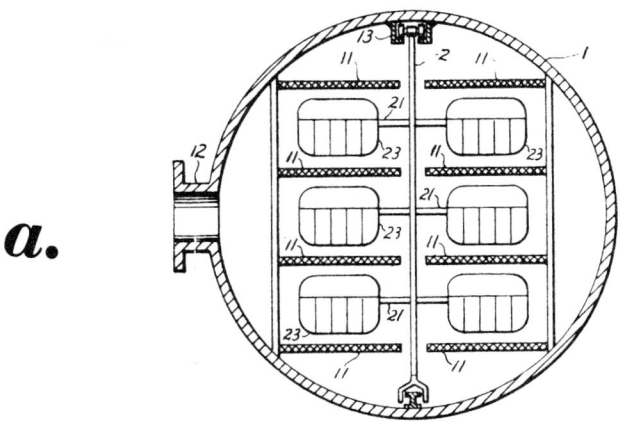

Cross Section of Freeze Drying Apparatus

b.

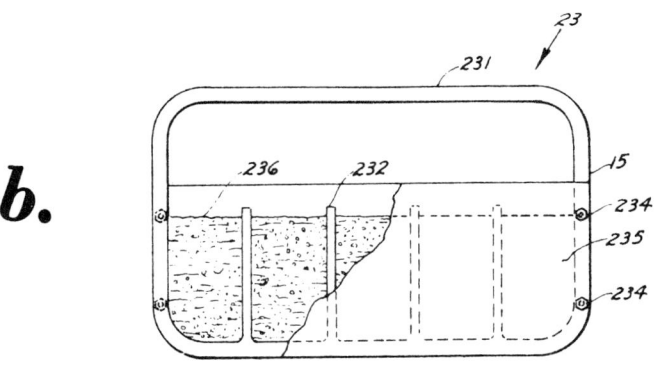

Enlarged Cross Section of Product Container of Figure 2.4a

Source: U.S. Patent 3,270,433

In Figure 2.4a there is shown a cylindrical vacuum freeze drying chamber **1** enclosing a radiant heat assembly **11**, which includes a plurality of parallel arranged heating plates **11** supported on opposite sides of the chamber. Mounted from the top of the chamber, there is a fixed rail **13** which supports a hanging rack **2**, containing a plurality of parallel moveable product container supports **21**, which interleave with the parallel radiation heater elements **11**.

A flanged tubulation **12** communicates with the chamber **1** and is adapted for connection to conventional vacuum pumping equipment (not shown). In Figure 2.4b there is shown a single container **23** which is constructed of a good heat conducting material. A unitary annular housing **231** having a plurality of inwardly projecting thin walls **232** contains a material **236** to be dried. The open ends of the housing **231** are partially closed by bolt **234**, attached end plates **235**, which extend from the bottom of the housing to about two-thirds its height.

In operation, the vacuum chamber **1** is evacuated via tubulation **12** and the radiant heaters **11** suitably energized by, for example, electrical current to relatively high temperatures (for example, 150°C). The resulting heat energy is then radiated through the vacuum space to the top and bottoms of container housings **23**. The heat is easily passed by conductor throughout the unitary housings **23** including the projecting walls **232**.

The housings then transfer heat by both conduction and convection (via the water vapor produced) to the surrounded frozen product **236** so as to cause sublimation therein. The sublimated water vapor coming from the product is removed through the openings in the ends of housings **231** above the end plates **235**. Thus the heat is serially conducted from the radiant heat source **11** through the conduction heat sources **23** and into the frozen product.

Inasmuch as the walls of container **23** are not brought to a temperature higher than that which is permissible for the material to be dried, the radiation from tray wall to the material is harmless to the latter. The top of the housing **23** not being in direct contact with the product easily conducts the received heat thereto via the unitary side wall portions **15** of the housing **23**.

The heating elements **11** can advantageously be coated on their upper surfaces with a layer which emits infrared rays to a large extent. The bottom sides of the heating elements can be left either untreated or coated with a layer of low emissivity. In this way a major portion of the radiant heat will be directed to the bottom portions of the housings **23**, which are in direct contact with the product **236**. Similarly the tops of the housings can be coated for low emissivity and the bottoms for high emissivity.

The temperature of the containers **23** can be measured (by thermal elements, resistance thermometers, contact thermometers, etc., in the tray bottom, for example); and the temperature thus determined can be used to control the temperature of the radiant heaters **11**. In this way the temperature of the containers can be maintained to evenly transmit any desired amount of heat to the material **236**.

In this way the temperature of the containers can be maintained at a suitable maximum temperature (for example, between 40° and 80°C) which is determined according to the characteristics of the material being dried. However, the material to be dried is shielded on all sides by the containers against direct radiation from the radiators **11** operating at above maximum temperature. Also, a sufficiently open cross section is available above the material for the removal of water-vapor from the drying material.

CONTACT HEATING

Inflatable Platen Contact Heating

The process by *H.A. Oldenkamp and R.F. Small; U.S. Patent 3,199,217; August 10, 1965; assigned to FMC Corporation* is carried out by heating the products in a vacuum chamber by direct conduction or contact heating applied to large surfaces of the products by means of heated platens, one of which is inflatable to establish top and bottom heat transfer contact with the product.

The products are arranged in layers and are supported on the inflatable, heated platen which in turn, rests on a fixed heated platen. The pressure applied to the products by inflation of each inflatable heated platen can be precisely controlled so that the mechanical contact pressure with the products will not cause the pressure on the ice in the products to exceed the pressure of the triple point of water. The process is described with reference to Figure 2.5.

FIGURE 2.5: INFLATABLE PLATEN CONTACT HEATING

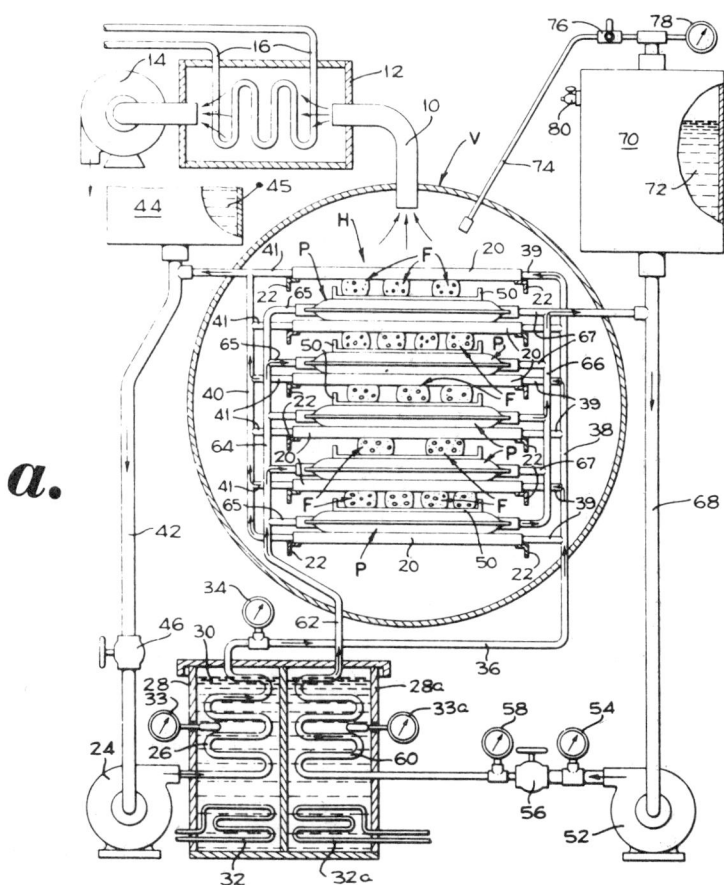

(continued)

FIGURE 2.5: (continued)

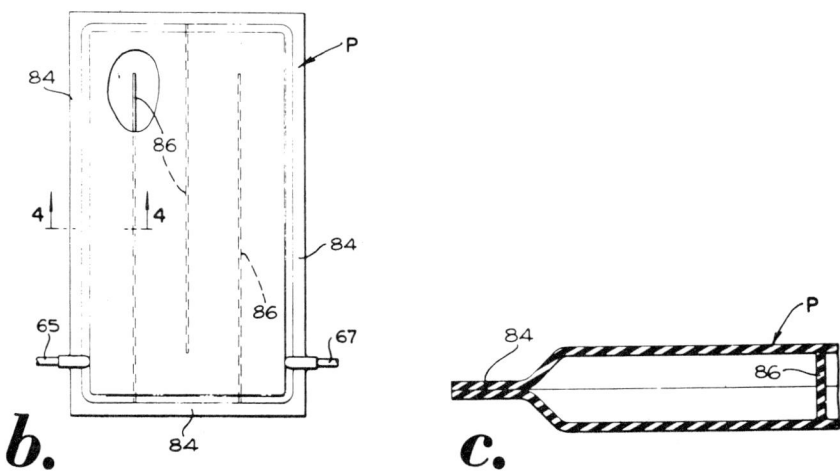

(a) Schematic Diagram of Freeze Drying Systems
(b) Plan, Partially Broken Away, of Inflatable Platen
(c) Enlarged Section Along Line 4—4 of Figure 2.5b

Source: U.S. Patent 3,199,217

In order to withdraw the air trapped in the vacuum chamber at the beginning of a cycle, and to withdraw vapor sublimed from the ice in the product during drying, an exhaust line 10 leads from the vacuum chamber to a vapor condensing chamber 12.

Connected to the chamber 12 is a vacuum pump 14 which maintains the vacuum in the chamber and also removes any noncondensed water vapor. In order to condense water vapor sublimed from the ice in the product, refrigerated coils 16 extend into the condensing chamber 12, each coil being connected to the usual refrigeration unit (not shown).

After the air has been exhausted from the vacuum chamber V, sublimation begins and the water vapor sublimed from the ice in the product will condense on the refrigeration coils in the form of ice crystals. The usual defrosting and drain means for the condensing unit will be provided.

In the freeze drying system two liquid heating and circulating systems are provided, one for the fixed shelves or platens and the other for the inflatable platens. These systems cooperate to provide a heating unit indicated generally at H, which unit supplies heat to the product and makes it possible to complete the drying process in a much shorter period of time than would be required if no heat were supplied to the product during the process.

The heating unit H includes a plurality of hollow aluminum shelves or platens 20 which are mounted in fixed spatial relation within the vacuum chamber by means of angle iron supports 22. In order to circulate the heating liquid through the fixed shelves or platens, a pump 24 is provided which delivers liquid to a heating coil 26 disposed within a tank 28. The tank is filled with a liquid 30 such as ethylene glycol, that is heated by a resistance heating unit 32. Liquid 30 serves to conduct heat from the heating unit to the heating coil.

A temperature gauge 33 is installed in the tank. At the outlet of the heating coil a pressure gauge 34 is fitted, from which a delivery pipe 36 leads to an inlet header 38 connected to the shelves 20 by conduits 39. The header will deliver the heating liquid to each of the fixed shelves or hollow platens.

Leading from the shelves are conduits 41 which open into an outlet header 40 which is connected to a return pipe 42 that directs the liquid back to the pump 24. An expansion tank 44 is provided, the upper end of which is open to the atmosphere. The heating liquid such as ethylene glycol, indicated at 45 in the tank is selected for operation at temperatures in the order of 140°F and the liquid should not be volatile at this temperature. A valve 46 is provided in return pipe 42 for controlling flow of the heating liquid through the shelves 20.

As seen, the product F to be dried is mounted on trays 50, which are supported by the inflatable platens P. The product may be slices of meat, fish or suitably arranged layers of vegetables.

There is also a heating liquid system for the inflatable platens. A pump 52 delivers the heating liquid (turbine oil) there being a pressure gauge 54 for setting the delivery pressure of the pump. The liquid is delivered to an adjustable needle valve 56, which is followed by a second pressure gauge 58. The gauges when read together will give the pressure drop across the needle valve.

The liquid is delivered to a heating coil 60 disposed in a heating tank 28a, and a delivery pipe 62 leading from the coil conducts the heating liquid to an inlet header 64, connected to the inflatable platens P by conduits 65. An outlet header 66 and conduits 67 conduct the liquid from the inflatable platens to a return pipe 68 which returns the liquid to the pump 52.

The platen is formed of fabric-reinforced rubber material. The reinforcement is a nylon fabric and the rubber is a synthetic oil resistant rubber compound, such as Hycar. The walls are joined by seams 84 and between the walls are formed baffles 86 that provide a sinuous passage for the heating liquid as it passes through the platen. The tubes 65 and 67 which connect the platens to the inlet header 64 and outlet header 66 are bonded between the wall parts of the platen.

In operation, the trays 50 are loaded with the product F to be dried which, as illustrated, may be patties of ground beef. The door to the vacuum chamber is closed and the vacuum pump or ejector started. The pressure in the vacuum chamber is brought well below that of the triple point of water, which is somewhat over 4 mm of mercury, so that sublimation can take place.

Heated liquid is then circulated through both the inflatable and the fixed platens.

As the pressure in the vacuum chamber is lowered, the air trapped above the liquid 72 in the tank 70 will apply a steadily increasing pressure to the inflatable platens. This action, if permitted to continue, would bring the mechanical pressure against the product above that of the triple point, and thereby prevent the desired sublimation. In order to provide a precisely controlled subatmospheric pressure in the tank, valve 76 is opened; whereupon, the air trapped above the liquid 72 in the tank is withdrawn by the vacuum pump through bleeder tube 74. As the vacuum pump pulls the vacuum in the vacuum chamber, pressure above the column of liquid 72 in the tank 70 is correspondingly reduced.

This prevents mechanical pressure against the food product from rising above that corresponding to the triple point of water. When the final conditions are reached in the vacuum chamber V, the pressure in the chamber will be very low and in fact may be as small as one hundredth of a millimeter of mercury. The pressure of the air trapped above the liquid 72 in tank 70 is not maintained at this low a value, because the air trapped in the tank supplies part of the force applied to the inflatable platens.

Accordingly, the vacuum gauge 78 is observed, and the valve 76 in the bleeder tube 74 is closed when the pressure trapped above the liquid in the tank 70 is in the order of 1 pound per square inch absolute. This pressure is sufficient to inflate the platens and establish heating contact, without raising the pressure exerted against the ice crystals in the product above the triple point.

The needle valve 56 that controls the circulation of heating liquid in the inflatable platen P is adjusted by observing the gauges 54 and 58, to insure that the pressure of the heating liquid in the hollow platens does not exceed the desired value. The total pressure within the hollow platens is kept below the pressure corresponding to the triple point of water by manipulation of the bleeder valve 76 and the needle valve 56 as described.

Some of the heat reaching the lower surface of the product is applied directly from the fluid circulating in the hollow platens, and some of the heat is supplied from the fluid circulating in the fixed platens or shelves 20 with the inflatable platens P serving as intermediate heat conductors for such heat. If the temperature gradient across the walls of the inflatable platens is steep, the temperature of the tank 28a that heats the liquid circulating in the hollow platens may be adjusted to be somewhat higher than the temperature in the tank 28 that heats the liquid circulating through the fixed shelves.

By establishing direct contact with the top and bottom surfaces of the product, virtually all of the heating is by direct conduction or contact heating and the temperature of the product can be controlled more precisely than is the case where radiant heating is relied upon.

Heat Input into Product at $-10°F$ by Heat Pumping

The freeze drying method developed by *L.A. Hernandez, Jr.; U.S. Patent 3,376,652; April 9, 1968* attains much higher rates of heat transfer by conduction through the frozen product from the heat source to the subliming surface. Thus, freeze drying cycles of about one-third to about one-fourth the time duration of the customary cycles are attained without danger of product thawing or deterioration of the final dried product.

Heat Transfer Mechanisms 43

In accordance with the process, the temperature of the heat source has been discovered to cause optimum conditions when maintained at approximately $-5°$ to $-15°F$. Ice at $-5°F$, for example, has a vapor pressure of about 737 microns Hg. This pressure is below the pressure exerted by a frozen product of approximately one-half inch thickness undergoing drying by sublimation, such pressure being approximately 930 microns, or about 0.018 pound (per square inch), corresponding to the weight of an originally one-half inch thick unfrozen product with a density close to unity.

The sublimation pressure at the initial ice front being about 52 microns, corresponding to about $-50°F$, the net pressure exerted on the effective interface between the product and heat source is about 982 microns, i.e., the sum of 930 and 52 microns, or 245 microns higher than the vapor pressure of the product of 737 microns at such interface. This differential pressure of 245 microns is adequate to keep the frozen product tightly against the heat source surface in order to assure the intimate contact required for optimum heat conduction.

Otherwise, any vapor formation would break the solid state continuity so that the desirably high heat transfer by conduction would no longer take place. Figure 2.6a shows water vapor or subliming pressure, solids gravity-pressure, and their total pressure plotted on pressure versus ice thickness coordinates.

By this chart, the optimum range of subliming temperatures between $-5°$ and $-15°F$ becomes apparent. While the exampled interface temperature of $-5°F$ is enough to allow a theoretical 245 microns of pressure differential to keep the product in intimate contact with the tray at the beginning of the drying cycle, as the product thickness falls to one-fourth inch, theoretically, the summation of the subliming pressure of, say, 208 microns (at $-27.5°F$) plus the product gravity pressure of approximately 465 microns or 0.009 pound per square inch equals only 673 microns which is less than the vapor pressure of 737 microns at the plate-product interface.

At this condition, a separation of the product from the plate would take place. Nevertheless, as the product thickness decreases still further to zero, the combination of the gravity pressure of the solids plus the exponentially increasing pressure at the subliming surface becomes greater than that of the product-plate interface pressure. A theoretical temperature threshold of approximately $-13°F$ would be necessary at the interface to prevent any product-plate separation.

In practice, molecular adhesion of the product to the plate and, to a certain extent, the shear strength of the product slab acting on the side walls of the plate seem to help under these conditions. For various products, it has been found that a product-plate interface of $-5°$ to $-15°F$ is adequate, a practical average being $-10°F$. Heat input into the product at these temperature levels may be accomplished by direct electric resistance heating of the product tray bottom, by circulating heat transfer fluids in passages attached to the product tray bottom, or by the preferred heat pumping method described hereinafter which attains the highest possible overall coefficient of performance.

Under the conditions the process a heat flux of approximately 10 Btu's per hour per square inch (depending upon the water content of the product) can be obtained between the heat source surface and the subliming surface.

44   Freeze Drying Processes for the Food Industry

The process whereby the low temperature heat input and condensing and defrosting cycles are attained is designated as heat pumping. An initial thermal bank is created by freezing the product in the freezing sublimation trays as by direct expansion of a low temperature refrigerant such as, for example, the type commonly known as R-13, in the coils or passages adjacent to the bottom of the trays. Then, under vacuum, this frozen product serves as the heat sink for the same refrigerant in the same coils during a condensing phase wherein the refrigerant is supplied at a high pressure at approximately $-15°$ to $-5°F$.

FIGURE 2.6:   HEAT INPUT INTO PRODUCT AT $-10°F$ BY HEAT PUMPING

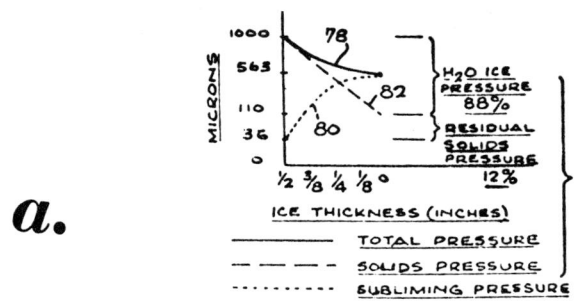

a.

Graph of Pressure versus Ice Thickness

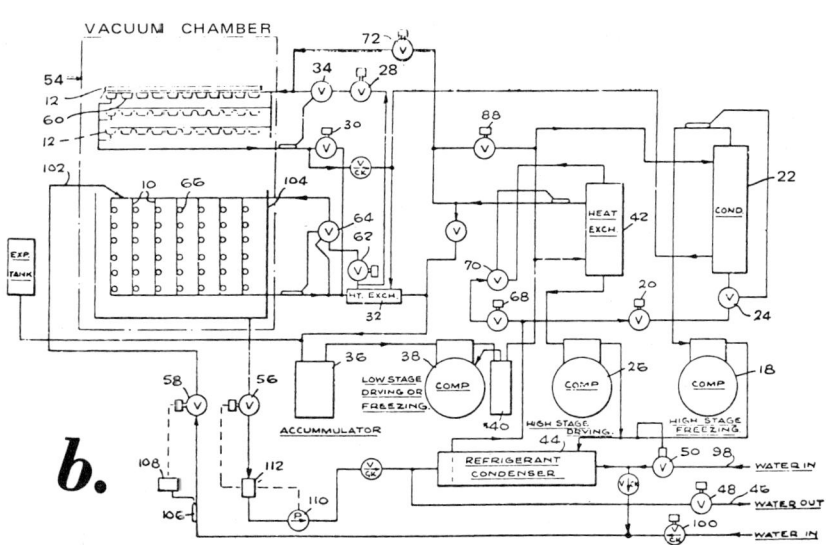

b.

Diagrammatic Illustration of Apparatus

(continued)

FIGURE 2.6: (continued)

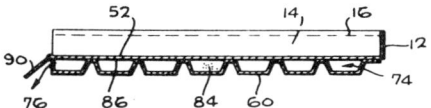

Product Tray and Associated Refrigerated Coil

Source: U.S. Patent 3,376,652

Such low temperature refrigerant, at such low stage, expands in the water vapor condenser coils or plates at a temperature low enough to give an initial condenser surface temperature of approximately −65°F. Accordingly, the heat pumped to the product tray coils, now serving to furnish the necessary heat of sublimation, equals the heat of condensation extracted by the water vapor condenser coils or plates, thereby constituting a balanced low temperature refrigeration cycle, except to the extent of unbalance created by the heat of compression and other minor losses which can be readily counteracted or effectively removed by a cascade system employing a high temperature refrigerant (high stage) such as ammonia, R-12, R-22, R-502, and so forth, thus furnishing properly desuperheated high pressure vapor of the low temperature refrigerant to the product tray coils for its final condensation.

Still further, once the product has been freeze dried, the necessary heat of fusion plus some sensible heat for defrosting the ice at the water vapor condensing plates is supplied by the necessary portion of water used to carry away the heat of condensation of the high temperature (high stage) refrigerant, accomplished by flooding or spraying the thus heat-laden water over the ice bank which has been formed on the water vapor condensing plates and recirculating such water as required. This latter operation is therefore performed simultaneously with the freezing of a new product batch in the manner described hereinabove. Thus, a new cycle commences and advantage is taken of the bank of ice available as a heat sink at both ends of the heat pump cycle.

A complete cycle of operation of apparatus is described with reference to Figure 2.6b. At the start of the initial freezing cycle, as differentiated from subsequent cycles where an ice bank has been built onto the water vapor condensing plates **10**, all valves are in their normally closed positions and all motors and actuating devices are inoperative.

The tray **12** is filled with the liquid product **14** to a desired operating level **16** (generally one-half inch in depth). The high stage compressor **18** and a solenoid valve **20** are turned on to cool the cascade condenser **22** to the operating temperature, preferably between 10° and −10°F depending upon the mechanical design. High stage liquid refrigerant now flows through expansion valve **24**. By means of a switch (not shown) of either a temperature or pressure sensing type at the low stage high pressure inlet to the cascade condenser the low stage compressor **38** is automatically turned on when a predetermined low-stage condensing temperature or pressure (saturated refrigerant vapor under static condition) has been achieved.

Simultaneously, a pair of solenoid valves **28** and **30** are turned on, allowing low-stage liquid refrigerant from the cascade condenser to flow through the liquid-to-suction heat exchanger **32**, the first solenoid valve **28**, and a thermostatic expansion valve **34** where pressure is reduced from about 130 to 190 psig (pounds per square inch gauge) to any low-side pressure which may range from 0 to approximately 50 psig.

The low pressure refrigerant is evaporated by the sensible and latent heat of fusion of the product load and flows through the second solenoid valve **30**, the suction side of the heat exchanger **32**, and the accumulator **36** where any excess saturated liquid is trapped and slowly metered out, and to the low stage compressor **38** where refrigerant vapor is compressed, passed through the oil separator **40** and fed again to the cascade condenser where high pressure super-heated low temperature refrigerant vapor is liquefied to a saturated condition and passed through the liquid-to-suction heat exchanger where this liquid is subcooled by the low temperature (-50° to -90°F) suction vapor. The foregoing cycle is continuously maintained during the product freezing portion of the operation.

The product's sensible and latent heat, the low stage heat of compression, and the transmission heat gains are rejected at the cascade condenser where the high stage freezing compressor **18** pumps this heat onto the high stage refrigerant condenser **44** for transfer to its cooling water which is discharged at **46** through a valve **48**.

The inlet water valve **50** is regulated by the high stage discharge pressure and meters the flow of cooling water to maintain a predetermined constant high stage refrigerant condensing temperature in accordance with the prevailing ambient conditions relative to the cooling water, such ambient conditions including the temperature and cost of the available water as supplied by the local water utility.

Regarding cost, it may be noted that, if water is expensive, then the compressor is used to pump at a higher compression ratio for causing a greater temperature differential through the water, thus using less water but more power for compression to achieve the same effect.

Once the product has been duly frozen to a temperature of approximately -60°F at the product-tray interface **52** as determined, among other means, by a predetermined pressure drop in the low stage suction as may be caused by a small product load, i.e., a small amount of sensible heat consisting of the difference between low-stage evaporating temperature and frozen product temperature, the product freezing cycle is about to be terminated.

After completion of the product freezing portion of the operation, the vacuum chamber **54** is ready to be evacuated for the next portion of the cycle. A pair of solenoid valves **56** and **58** are closed as well as the product filling (in case of liquid product) and vent valves (not shown) and all other apertures so as to isolate the chamber **54** from the atmosphere, and the vacuum pump (not shown) is started. Once the chamber pressure has been reduced to approximately 5 to 10 torrs, solenoid valve 28 is closed to stop the low stage liquid refrigerant flow to the product tray coils 60 and, simultaneously, another solenoid valve 62 is opened to divert such liquid refrigerant to an expansion valve 64 and to the coils 66 of the water vapor condensing plates 10 where the refrigerant is evaporated internally, thus causing any remaining water vapor in the chamber 54 to freeze out on the external surfaces of the water vapor condensing plates 10.

This step prevents water vapor migration to the vacuum pump since the temperature of the condensing plates is lower than that corresponding to the total chamber pressure. As a result, evacuation time is shortened because the vacuum pump does not have to handle water vapor.

This step also serves to precool the condensing plates 10 for the next cycle or portion of the freeze drying operation. A few minutes before the next cycle is initiated, another solenoid valve 68 is opened to permit the feeding of high stage liquid refrigerant to an expansion valve 70 and thus to the heat exchanger (or desuperheater) 42, thus precooling the heat exchanger 42 to its operating temperature.

Evacuation is continued until the chamber pressure has been reduced to a proper level of approximately 500 to 30 microns, the proper vacuum level depending upon various parameters as will be clear to those skilled in the art such as, for example, product thickness, the thermal diffusivity of the product, and other design factors.

As soon as the appropriate vacuum has been achieved in the chamber 54, the low temperature heat pump freeze drying cycle portion of the complete cycle is initiated by turning off the solenoid valve 20 which has been feeding high stage refrigerant to the cascade condenser 22, turning on the solenoid valve 72 to start feeding low stage saturated vapor to the coils 60 of the product trays 12, and turning off solenoid valve 30 to divert the condensed liquid refrigerant to another solenoid valve 62 which is turned on simultaneously so as to feed the condensed liquid refrigerant through the expansion valve 64 to the coils 66 of the water vapor condensing plates 10 wherein the liquid refrigerant evaporates.

All of the foregoing steps are accomplished simultaneously and, at the same time, the high stage drying compressor 26 is turned on and the high stage freezing compressor 18 is turned off. Thus a -5° to -15°F low stage saturated refrigerant vapor is delivered at a corresponding pressure of approximately 150 psig to 123 psig by the heat exchanger 42 to the product tray whereupon such refrigerant vapor is condensed at a temperature which is uniform throughout the entire bottom surface area of the tray 12, thereby affording uniform rates of sublimation of the product.

The condensed low stage refrigerant is now fed in its liquid state into the water vapor condensing plates where it is evaporated at approximately -70°F, resulting in an initial water vapor condensing temperature of about -66°F on the outside surface of the water vapor condensing plates, assuming normal refrigerant film factors. The outer surface of the condensing plates thus being lower in temperature than the surface of the frozen product, the water vapor evolving from the latter is condensed on the surfaces of the former.

As a typical example, with a product tray interface temperature of approximately −10°F and with a temperature difference across the frozen product (i.e., between the exposed surface **16** of the product **14** and the product tray interface at **52**) of approximately 45°F at the start of the cycle for a normal frozen product (i.e., containing approximately 88% water), the sublimation temperature is approximately −55°F saturated, corresponding to about 36 microns of water vapor pressure.

In accordance with the Napierian equations for adiabatic gas flow, a 55% pressure difference results in the maximum possible rate of flow of water vapor as it evolves from the frozen product to the surfaces of the water vapor condensing plates **10** and, accordingly, a condensing plate temperature of about −65.5°F (equivalent to about 19 microns of pressure) achieves such maximum flow rate.

As sublimation proceeds, the temperature difference across the product **14** decreases linearly with the decrease in the remaining thickness of the frozen product, with a correspondingly linear rise in the subliming temperature. Conversely, as the ice on the condenser **10** increases, the temperature difference across such frozen condensate will also increase linearly for all practical purposes since the frozen product and the condensate ice are roughly of the same density in the typical case of the normal product having about an 88% water content.

However, as the fundamental parameter of the process, the corresponding subliming pressure increases exponentially. Thus, while at the beginning of the cycle the subliming pressure is about 36 microns corresponding to about −55°F, the subliming pressure rises to about 563 microns corresponding to about −10°F at the end of the cycle when the last film of product ice is subliming; thus, for an increase in absolute temperature equivalent to only about 10.5%, the subliming pressure increases by more than 1,500%, i.e., over an entire order of magnitude.

Referring again to Figure 2.6a, particularly, it is seen that the total pressure **78** or sum of the subliming pressure **80** plus the product pressure **82** (both frozen and dry product), such additive pressures being effectively at the product tray interface **52**, must be higher than the vapor pressure of water at the product tray interface at any and all times during the freeze drying cycle.

Thus, if the interface vapor pressure is 563 microns, the sum of the subliming pressure and the product gravity pressure must be higher than 563 microns in order to maintain intimate molecular contact between the product and its tray so as to assure true conductive heat transfer between the heat input, surface for the condensing refrigerant and the product subliming surface which, as explained before, is the ice front.

In other words and referring to Figure 2.6c with the foregoing described interface pressure relationships, the only heat transfer mechanism occurring is that of true conduction, and such heat flow occurs in an uninterrupted manner from the refrigerant **84** (which is continuously supplied in a vaporized state under pressure and continuously condenses as its heat is transferred, thereby functioning as a heat source rather than as a refrigerant in the common and usual sense), to the refrigerant coil surface and the bottom of the tray **86**, then through the product tray interface **52** and, finally, through the frozen product **14** to the latter's ice front (at **16** initially) which constitutes the subliming surface.

Heat Transfer Mechanisms

For an average product with an approximately 88% water content, approximately two hours or less is required for the freeze drying portion of the cycle to remove by sublimation all but a trace content of moisture from an initially one-half inch thick frozen product.

Material Swept Downward on Stacked Heating Plates

H.G. Kessler; U.S. Patent 3,460,269; August 12, 1969; assigned to Krauss-Maffei AG, Germany describes a process in which frozen foodstuffs are continuously treated in a vacuum chamber by being moved with constant agitation over successive hot surfaces advantageously to progressively lower levels with consequently lower temperatures and partial pressures.

These surfaces are formed by a stack of heated annular disks with alternately smaller and larger inner and outer diameters, the disks being swept by continuously rotating scraper blades acting radially inwardly in the case of the larger disks and radially outwardly in the case of the smaller disks whereby the goods are alternately pushed over the inner and outer edges of successive disks onto the next lower disk and, finally, onto the floor of the chamber. The process is described with reference to Figure 2.7.

FIGURE 2.7: VACUUM CHAMBER WITH STACKED HEATING PLATES AND CONTINUOUSLY OPERABLE STIRRING MEANS

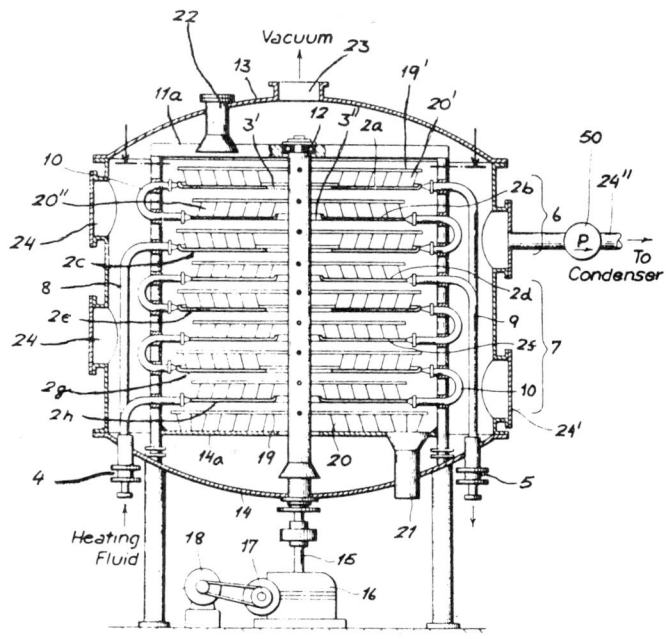

Source: U.S. Patent 3,460,269

A vacuum chamber is supported by posts extending into the interior thereof and terminating at the top in a cross beam **11a** forming an upper journal for a vertical shaft **12** which is also rotatably held in the bottom **14** of the chamber housing and is driven by a motor **18** via a transmission **17**, a speed reducer **16** and a coupling **15** which may include a slipping clutch.

Shaft **12** passes with annular clearance through central apertures **3'**, **3''** in a stack of horizontal annular disks **2a**, **2b**, **2c**, **2d**, **2e**, **2f**, **2g** and **2h** fixedly mounted on the posts through the intermediary of conduits **8**, **9** and **10**. The disks are hollow and receive a heating fluid, such as steam, from an inlet nozzle **4** via conduits **8** terminating at disks **2c** and **2h**; connecting pipes **10** direct this fluid through successively higher disks until it leaves via conduits **9**, attached to disks **2a** and **2d**, for discharge at an outlet nozzle **5**. Thus, the 8 disks are subdivided into two groups **6** and **7** of three and five disks, respectively, with separate circulation; the temperature of these two groups can therefore be individually adjusted.

Counting from above, one notes that the odd-numbered disks **2a**, **2c**, **2e** and **2g** have longer inner and outer diameters whereas the even numbered disks, **2b**, **2d**, **2f** and **2h** have shorter diameters. Shaft **12** carries outrigger arms **19'**, **19''** which support respective sets of scraper blades **20'**, **20''** so arranged as to sweep the goods on the large radius disks, inwardly toward and into their central openings **3'** and the goods on the small radius disks, outwardly toward and over the outer edges thereof.

From the lowest disk **2h** the goods drop onto the floor **14a** which is swept by further scraper blades **20** on outrigger arms **19**, the latter blades being so oriented as to concentrate the dehydrated material at an outlet **21** through which it may leave the chamber.

An inlet **22** for the material to be treated opens onto the top disk **2a**. The lid **13** of chamber **1** also has a central port **23** connectable to a vacuum pump not shown. Side ports **24**, normally closed by covers **24'**, give access to the interior of the chamber for purposes of inspection and repair. One of these side ports is provided with a conduit **24''** leading, via a suction pump **50**, to a condenser (not shown) for the removal of evaporated moisture. The inlet and outlet ports **22**, **21** are provided with gates which can be intermittently opened to a supply source for the goods to be treated and to a receiver for the treated goods.

ARRANGEMENT

Rectangular Container Contacted by Parallel Planar Heating Elements

J. Brouwer and J. Veldstra; U.S. Patent 3,391,466; July 9, 1968; assigned to Unilever NV, Netherlands describe a process for freeze drying a food product comprising the steps of filling package containers having a pair of planar parallel sides with a quantity of the food product, subjecting the product to freeze drying while the containers are in a partly open condition and the pair of planar parallel sides of the containers are firmly contacted by parallel planar heating elements which thereby supply heat to the product.

The heating elements having parallel surfaces may comprise the upstanding ribs of a ribbed tray within which the package container is arranged as a snug fit with the rib surfaces firmly contacting the parallel walls; or the heating elements having parallel surfaces may be formed by a series of plates assembled to simulate a ribbed tray.

A preferred form of the process comprises filling the foodstuff into package containers with at least two plane parallel vertical sides, which after the filling operation remain at least partly open at the top, loading the filled package containers in a vertical position into the heating elements, such as ribbed trays in which they snugly fit, subjecting them to freeze drying conditions, subsequently unloading the package containers from the ribbed trays and treating them in such a way as to make them impervious to water vapor.

One of the advantages of this method of working is that the dosing of the package containers tends to be more accurate, since any deviation from the mean weight during the dosing of the product into the containers before it is dried will be diminished by the drying process; and the method in some cases also offers the advantage that the freeze drying operation is accelerated. The freeze drying in the pack also tends to cause some sintering to occur between the particles, thus diminishing the amount of fine after the transportation of the packages.

The package containers may be made of cardboard, provided with a coating (e.g., of wax or plastic) at the inside to prevent moisture from the foodstuff being absorbed by the cardboard and causing unsightly blots. Also aluminum foil may be used or aluminum foil laminated with plastic. Good results were obtained with 9 micron thick aluminum foil laminated with 22 micron thick polyethylene foil, and with 30 micron thick aluminum foil laminated with 40 micron thick polyvinyl chloride foil.

Example: Deep frozen shrimps with a dry substance content of 25.6% were, after defrosting at 7°C for 20 hours (water loss 18.8%), weighed into cardboard boxes (wax-coated on the inside). The dimensions of these boxes were: height 75 mm; length 96 mm; and width 30 mm. The tops of these boxes had an opening occupying 74% of the total top surface area. In each box were put 114.3 grams of shrimps.

The boxes with shrimps were then placed in aluminum ribbed trays. The distance between the ribs was 30 mm; the height of the space between the ribs was 75 mm. The trays with the boxes were subsequently placed in a freeze drier on heating plates. After closing the freeze drier, vacuum was applied. The product was frozen by evaporation.

As soon as the pressure in the drying room was 0.6 torr (corresponding with a temperature of the ice in the product of −22.5°C), the temperature of the heating plates was increased to supply heat to the product in order to sublimate the ice. This temperature was adjusted in such a way that no defrosting of the product and no burning of the already dry product occurred.

After a total stay in the freeze drier of 16 hours the average moisture content of the shrimps had decreased to 2%. Vacuum was then released by allowing air into the freeze drier. The boxes with dry product were taken from the trays and wrapped in Rayseal film, which was heat-sealed. The quality of the dried product was excellent. The product showed no shrinkage or burning.

Receptacle for Materials and Heating Apparatus Arranged in Vertical Direction

G.A.H. Wehrmann; U.S. Patent 3,728,798; April 24, 1973 is concerned with a freeze drying system characterized by the fact that by arranging the receptacles for the material and the heating apparatus in a vertical direction, the heating devices that extend in the vertical longitudinal plane of the chamber are firmly mounted in the chamber and the receptacles can be separately transported between the stationary heating devices. The vertical arrangement of the product pockets represents a good use of space of the cylindrical vacuum chamber.

The vacuum freeze drying apparatus for particulate food material comprises a cylindrical vacuum chamber, a plurality of fixed, vertical, spaced apart, longitudinally-extending heating coils which are constantly and firmly connected to a common supply pipe and means for guiding a plurality of vertical receptacles containing particulate food material, through the chamber and between the heating coils.

The receptacles are combined, as vertically extending pockets which can be quickly filled with particulate material, into a movable structural unit which is movable between the heating devices in a comb-like manner and which is provided at both sides with rollers that run on rails mounted on the internal side of the vauum chamber. The receptacles are pendently mounted on a portion of the structural unit that moves above the heating devices. The receptacles have ventilator openings in the side walls, the openings being formed by baffle plates on the side walls, the baffle plates being directed outwardly and diagonally upward. The process is described with reference to Figure 2.8.

The system of heating device **2** is stationarily mounted in the cylindrically built vacuum chamber **1** in such a manner that there is provided, in a predetermined distance from each other, a plurality of heating coils, heating plates and the like that are guided in a vertical plane.

The heating devices **3** are connected to a common supply pipe **4** that cross-sectionally extends through the chamber. At the same time the heating devices and/or the cross-sectionally extending supply pipe **4** can be firmly arranged on the brackets **5** of the walls of the chamber. The common feed pipe **6** is passed at a suitable point through the walls of chamber **1**.

The receptacle **7** that receives the material has a plurality of vertically extending pockets **8** that are firmly mounted in an upper portion **9** of the receptacle or form therewith a structural unit. The receptacle is provided at the sides with rollers that run on rails **11** that are supported by brackets **12** mounted on the internal side of the walls of the chamber. By means of the rails the receptacles of the material can be guided with relative accuracy and in suspension.

The pockets **8** that extend in the vertical plane, when the receptacle is in operation in the longitudinal direction of the chamber, in a comb-like manner pass through the intermediate spaces between the stationarily arranged heating devices.

FIGURE 2.8: RECEPTACLES FOR MATERIALS AND HEATING APPARATUS ARRANGED IN VERTICAL DIRECTION

a.

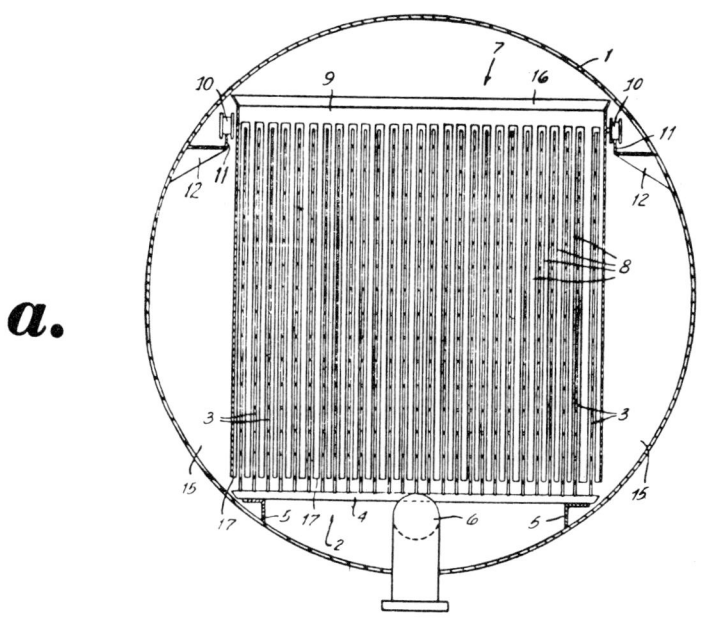

Cross Section of Vacuum Chamber Showing Arrangement of Receptacles and Heating Elements

b.

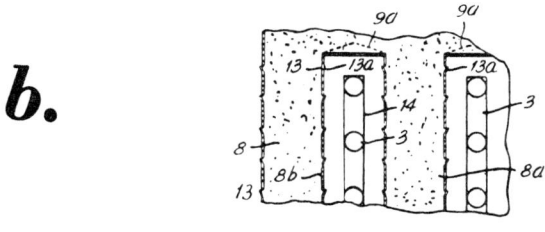

Enlarged Detail of Figure 2.8a

Source: U.S. Patent 3,728,798

In the side walls of the pocket there are ventilator openings **13** that consist of partially punched out blades that are oriented outwardly and diagonally upward. The size of the ventilator openings and of the blades **13a** depends on the size or fineness of the material to be freeze dried.

The heating device, a heating coil for instance, that extends in the vertical plane can be surrounded by a protective covering **14**. Between the covering and the chamber wall **8a** a certain free space is provided so that the chamber wall **8b** cannot rub against the heating device or the protective covering. The receptacle **7** that has the disposable width of the cylindrical chamber **1** can be shut off toward the top by a lid **16** or several adjacent lid portions. But it is possible also to open the receptacle **7** toward the top. The bottoms **17** of each pocket are advantageously made so that they can be downwardly open.

It is possible to provide a bottom lock that can be controlled by tension springs suspended from knee joints. This can be done for instance by means of an apparatus by which all the bottom locks can be opened or closed at once. The same can be said of a lid in the upper side of the receptacle **7**.

The receptacles that have the vertical pockets can be successively introduced in the vacuum chamber. At the end of the cylindrical vacuum chamber that can have a length of many meters, the receptacle is taken out from the chamber with the freeze dried material that is ready, while at the beginning of the chamber a receptacle already filled is introduced into the chamber, the pockets gripping in a comb-like manner between the heating devices.

Chamber Containing Material to Be Dried in Stacked Arrangement over Cold Trap

J.L. Dantoni; U.S. Patent 3,574,950; April 13, 1971 provides an apparatus in which several chambers with material to be freeze dried may be stacked in communication with each other so that all may be processed at once and all use the same cold trap and source of vacuum. The process is described with reference to Figure 2.9.

10 represents a cabinet which encloses a refrigeration plant or system **11** with the usual condenser portion **12** located in one portion **13** of the cabinet and the evaporator or cooling portion **14** located in a second portion **15** of the cabinet. The cooling portion is in the form of a coil wound around the exterior or interior of a well-like receptacle or cooling trap **16** which is open through the top of the cabinet and is there provided with a slightly raised turned over rim **17**.

The cold trap **16** is provided with a drain **18** and cutoff valve **19** and a second pipe **20** extends upwardly through the bottom thereof which, when the device is in use, is connected to a source of vacuum, not shown. The pipe is suitably valved at **21**. The well-like cold trap **16** with its surrounding cooling coil **14** is enclosed in insulating material **22** in the portion **15** of the cabinet to isolate these parts from the outside temperature.

The condenser portion of the refrigeration system is located in the cabinet portion **13** adjacent the insulated evaporator section thereof. The condenser, gives off heat and to dissipate this heat a fan **23** is mounted in the side of the cabinet to force air into the cabinet portion and out through the top opening **24** thereof.

A hood, **25**, is located on the cabinet above the opening **24** and is provided with inclined louvered panels **26** and **27**, the louvers of which may be opened or closed by means of a lever **2**. Chamber members are provided for receiving the materials to be subjected to the freeze drying process.

FIGURE 2.9: CHAMBERS CONTAINING MATERIAL TO BE DRIED IN STACKED ARRANGEMENT OVER COLD TRAP

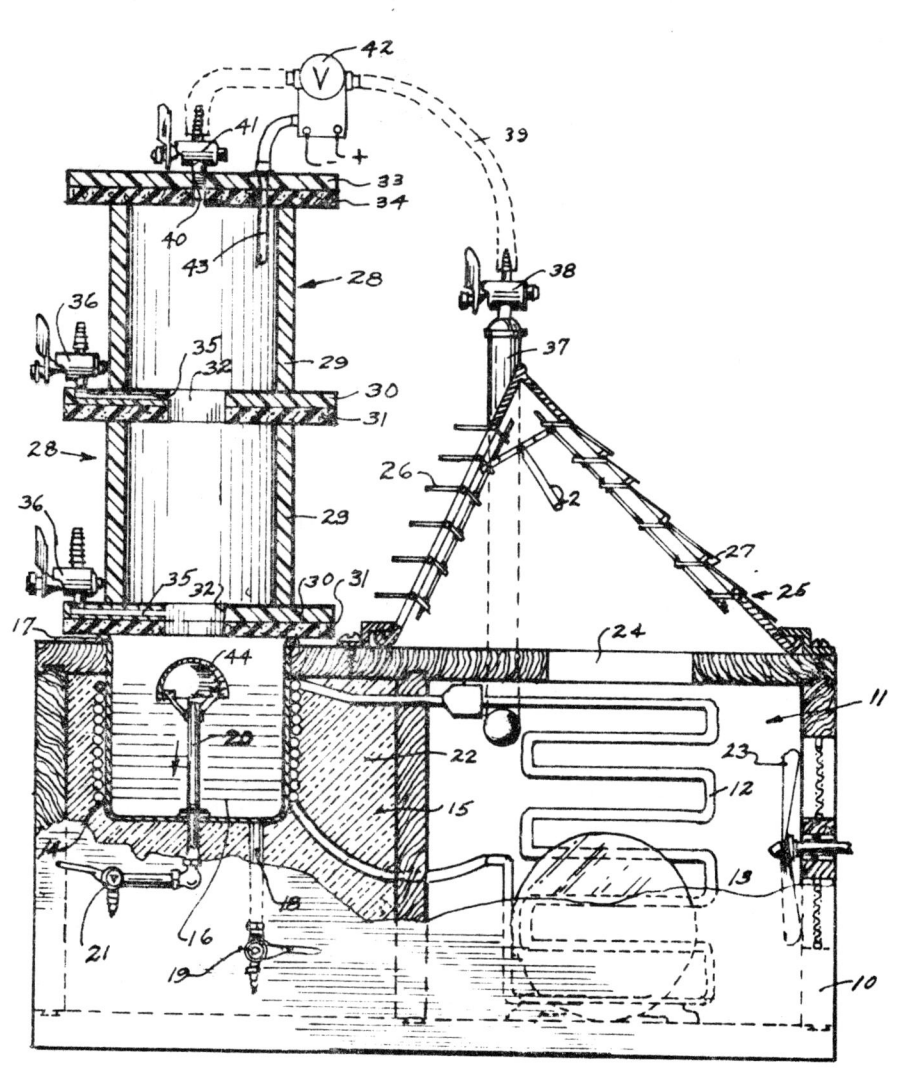

Source: U.S. Patent 3,574,950

Each chamber consists of a sidewall portion **29** preferably of transparent plastic such as Lucite with a bottom wall **30** also of plastic such as Bakelite sealed to the bottom edges of the sidewalls and a relatively thick resilient gasket **31** fixed and sealed to the lower surface of the bottom wall. An opening **32** of fair size through the bottom wall and gasket affords communication therethrough.

A chamber may be placed directly on the rim 17 of the cold trap and one or more chambers may be placed or stacked one above the other on the first chamber, the lowermost in sealing engagement with the rim of the cold trap and all the other chambers in sealing engagement with the one below. The uppermost chamber is closed at its top with a closure plate 33 having a sealing gasket 34 attached to its under surface similar to the gaskets 31 on the bottoms of the chambers.

Each chamber is provided with a port 35 extending outwardly through the bottom thereof from the opening 32 to the outside and is there provided with a stopcock 36 to which a hose may be attached.

The condenser section of the refrigeration system is located adjacent the evaporator section and cold trap. This places the hood 25 with its inclined louvers 26 facing the chamber or chambers 28 positioned over the cold trap whereby the heat from the condenser section may either be exhausted through the louvers 26 and directed against the chambers to assist in the evaporation and sublimation of the moisture contained in the materials in the chambers and it may also be regulated by the lever 2 to exhaust through the louvers 27 or through both sets of louvers to regulate the amount of heat thus supplied to the chambers. In addition the amount of heat thus supplied to the chambers through regulation of the louvers may be accomplished automatically by thermostatic control if desired.

Provision is also made to supply heated air directly to the interior of the chambers through the pipe 37 leading from the interior of the condenser section of the cabinet through a control valve 38 and hose 39 to the port 40 in the top plate 33 closing the uppermost chamber. For convenience in closing off the port 40 when interior heating is not used a valve 41 may be used.

The operation of the apparatus is as follows: With the refrigeration machinery in operation one or several of the chambers 28 containing the prefrozen materials to be lyophilized are placed over the cold trap in stacked relation and a cover plate 33 is placed over the uppermost chamber. The pipe 20 through its valve 21 is then connected to a source of vacuum.

This will subject the interiors of the chambers to a very low pressure and will cause the frozen materials to begin losing moisture by sublimation. The moisture will be drawn into the cold trap and there condensed and frozen on the low temperature walls of the cold trap. This process will continue until all the moisture has been extracted from the materials being processed, after which the source of vacuum may be cut off and the dried materials removed from the chambers.

It may be desirable under certain circumstances to heat the chambers during the drying process to speed evaporation. This may be accomplished by opening the louvers 26 to direct the heated air from the refrigeration condenser toward and against the chambers or heated air may be introduced directly into the chambers through the pipes 37 and 39 by opening the valves 38 and 41. The sensor 43 acting on thermostatic valve 42 may then be used to control the amount of heat to suit the particular conditions. In addition the freeze drying process may be effected in an atmosphere of inert gas such as nitrogen or sterile air by connecting a source of supply of such gases through the valved ports 35 and 36 in the chambers.

Vacuum pipe **20** is provided with a deflector or hood **44** for the purpose of preventing entry into the vacuum pipe of moisture and ice crystals and thus avoid clogging and damage to the vacuum system. After completion of the freeze drying run the refrigeration system is shutdown and the frozen moisture in the cold well allowed to melt whereupon it may be drained off through pipe **18** by opening the valve **19**. In a modification the chamber is equipped with a plurality of outlets to which bottles or vents may be attached for processing.

OTHER

Sensing Glow Discharge in Microwave Heating

When using radio-frequency or microwave energy it is important to keep the rate of application of energy below the point at which the moist frozen material reaches the melting point of the included liquid. If this precaution is not taken and some of the included material condenses on the substance being dried, this liquid will absorb radio-frequency energy and be heated, together with the surrounding region of dry material which may be burned or otherwise damaged.

This is particularly true when the freeze drying process has reduced the moisture content to a relatively low value and the applied radio-frequency or microwave energy is relatively great in proportion to the remaining moisture content in the material being processed. At this portion of the freeze drying cycle, the microwave energy density and the molecular vapor concentration surrounding the material or food being processed approaches critical values and glow discharge or the development of an intensely blue colored atmosphere occurs in the evacuated container. This discharge is undesirable because, when foods are being processed, it contributes a characteristic spark discharge type of flavor to the foods whenever they are exposed to an appreciable amount of discharge.

In a process developed by *D.A. Copson; U.S. Patent 3,020,645; February 13, 1962; assigned to Raytheon Company* a photoelectric cell, mounted on an evacuated microwave oven is used to sense for the presence of the development of glow discharge whenever the microwave energy density exceeds the density required to obtain a desirable rate of sublimation. When glow discharge is sensed, the photoelectric cell provides a responsive signal which is amplified and used as a control signal to reduce the microwave energy input to a power level below that required to melt moist frozen material within the container.

In a further modification of the process the control signal is used to increase the pumping capacity of the evacuating means to draw off the vaporized component of the sublimation process at a faster rate so as to reduce the molecular vapor concentration in the container. The process is described with reference to Figure 2.10.

Numeral **10** designates a conductive cavity or metallic oven having a door **11** to permit the insertion of food material and a source of microwave energy, a magnetron **12**, coupled to the oven through a coaxial transmission line **13** of the waveguide type. Frequencies which are particularly significant for this purpose are those in the microwave range (between 300 megacycles per second and

30,000 megacycles per second). Material **14** in the frozen state to be dried is inserted into and hermetically supported on the nonconductive tray **15**. Cavity **10** together with transmission line **13** may be referred to as a propagated electromagnetic wave guiding structure since this structure is used for guiding and directing the propagated electromagnetic microwave energy to the material.

Magnetron **12** is energized from a suitable power supply source **18** in series with a well-known voltage regulator **20** and, when so energized, delivers microwave energy having a predetermined wavelength to coaxial transmission line **13**.

FIGURE 2.10: SENSING GLOW DISCHARGE IN MICROWAVE HEATING

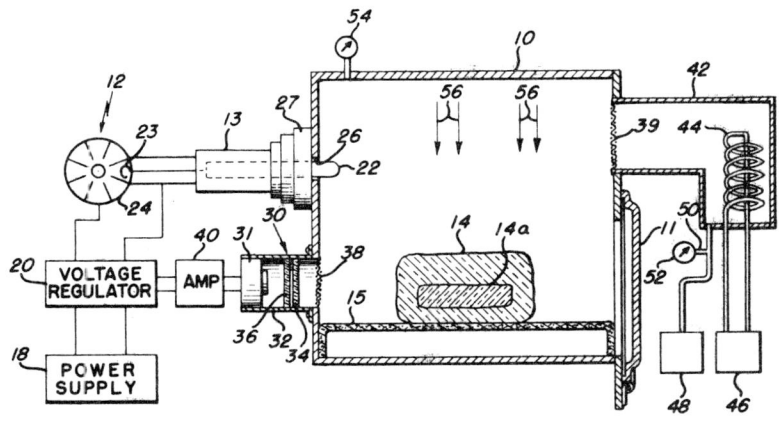

Source: U.S. Patent 3,020,645

In order to sense the development of glow discharge in container **10** a photoelectric sensing mechanism **30** comprises a photoelectric cell **31** hermetically sealed in a metallic tube **32** which, in turn, is hermetically sealed to the wall of the container.

Red and yellow color filters **34** and **36**, respectively, are mounted in tube **32**, in front of the face of the photoelectric cell. In this manner, only the light having a characteristic blue glow will pass the light filters and become sensed by the photoelectric cell. The wall of the container is provided with a screen **38** formed with openings of a diameter small with respect to the wavelength of the microwave energy, to prevent the propagation of microwave energy into the photoelectric cell **31** while permitting the passage of light into the filters.

When glow discharge is sensed by the photoelectric cell, a control signal is fed to a conventional amplifier **40**. The output of the amplifier is the amplified control signal which is fed to the voltage regulator **20** to reduce the input power to magnetron **12** and, thereby to reduce the microwave energy input into container **10**.

On the opposite side of the container is a metal conductor **42** hermetically sealed to the wall of the container and provided with a cooling coil **44** for removing the sublimated vapor from the container. The cooling coil is connected to a conventional compressor **46** adapted to pump cooling fluid through the coil.

In operation, the material **14**, such as meat, is supported by the tray of dielectric material **15**, such as glass, affording a minimum of impedance to the passage of microwave energy. As microwave energy, represented by the arrows **56**, penetrates the frozen material, the ice or other component to be vaporized, evaporates without first becoming a liquid, that is, it sublimates due to the absorbtion of microwave energy-producing heat.

This sublimation produces a region of dry material, represented by the section **14** and leaves a core of frozen undried material represented by the dotted section **14a**. The dried material offers no appreciable impedance to the microwave energy, thus permitting the sublimation of the moist material to continue at a substantially uniform rate.

However, as the freeze-drying cycle progresses, the load placed upon the source of microwave energy by the decreasing volume of the moist material becomes smaller, and the microwave energy density and molecular vapor concentration surrounding the dried material approach critical values. When this condition occurs, the development of glow discharge occurs throughout the atmosphere in the evacuated container.

The occurrence of glowing is sensed by the photoelectric cell **30**, which produces a control signal to decrease the microwave energy density and to prevent the spark discharge type of flavor. In this manner, the microwave energy can be applied at a considerably more rapid rate without danger of damaging the dried region. By preventing the development of glow discharge, the microwave energy is also prevented from being applied fast enough to raise the temperature of any portion of the material above the melting point of the vaporizable portion of the material.

Flow of Refrigerant Through Internal Passages in Both Freezing and Sublimation Steps

According to *P.B. Mason; U.S. Patent 3,281,956; November 1, 1966; assigned to Mitchell Engineering Ltd., England* the fluid to be dried is frozen in a layer of predetermined thickness onto the external surface of a structure having an internal passage by passing a refrigerant through the passage and then subjecting the frozen matter to such conditions of reduced pressure that sublimation of at least one constituent of the frozen matter occurs, while supplying heat to the frozen matter by the maintenance of a flow of refrigerant through the passage at a refrigerant temperature controlled to be below the fusion point of the frozen matter at least until substantial drying has occurred. Until drying has proceeded to a substantial extent, the refrigerant temperature although below the fusion point is controlled to provide heat to compensate for the sublimation effect.

By maintaining the flow of refrigerant through internal passages during both the freezing and the sublimation steps there is no tendency for the frozen matter to separate from the structure and so lose contact with the surface from which it gains the heat required for sublimation. The energy required to effect freeze drying may be substantially reduced, since the energy for the sublimation step is a by-product of the refrigeration plant.

The apparatus shown in Figures 2.11a-d comprises a drying chamber **10** having in it a plurality of downwardly extending structures **11** on which the fluid to be dried is frozen and then subsequently sublimated.

FIGURE 2.11: FLOW OF REFRIGERANT THROUGH INTERNAL PASSAGES IN BOTH FREEZING AND SUBLIMATION STEPS

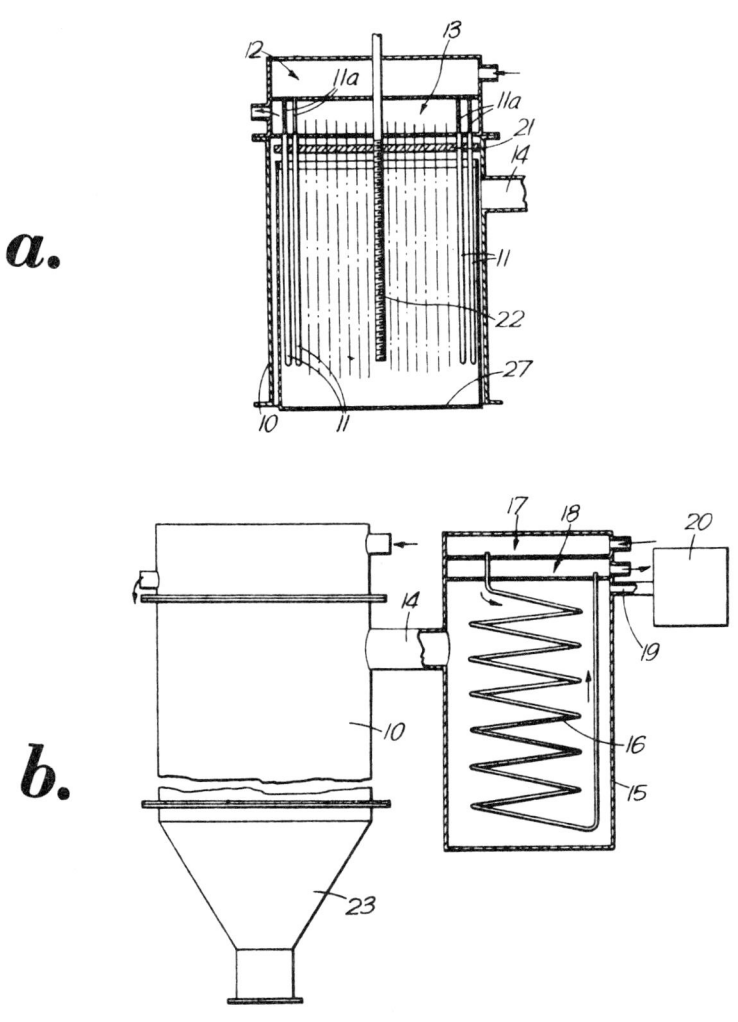

(continued)

Heat Transfer Mechanisms 61

FIGURE 2.11: (continued)

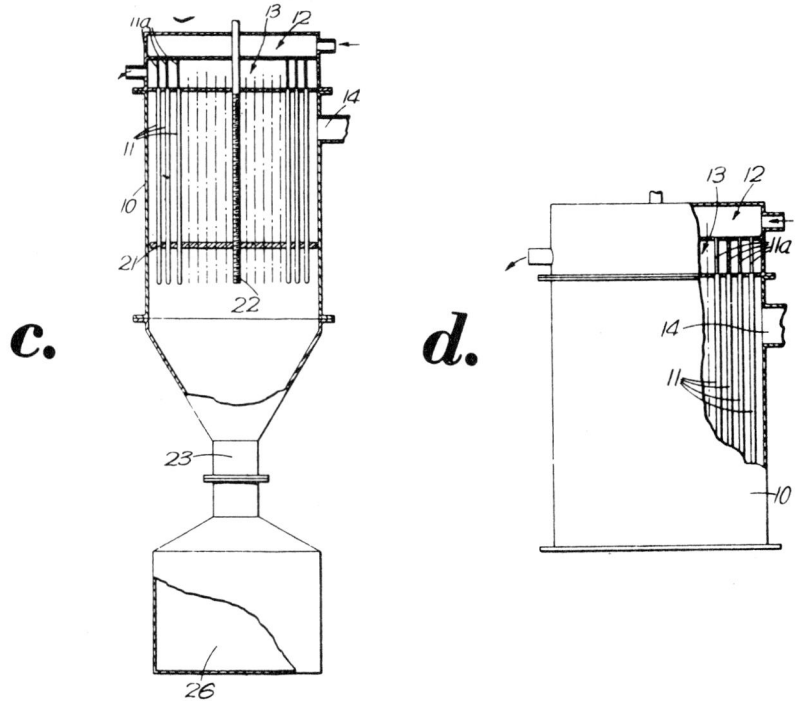

(a)-(d) Successive Steps of Process

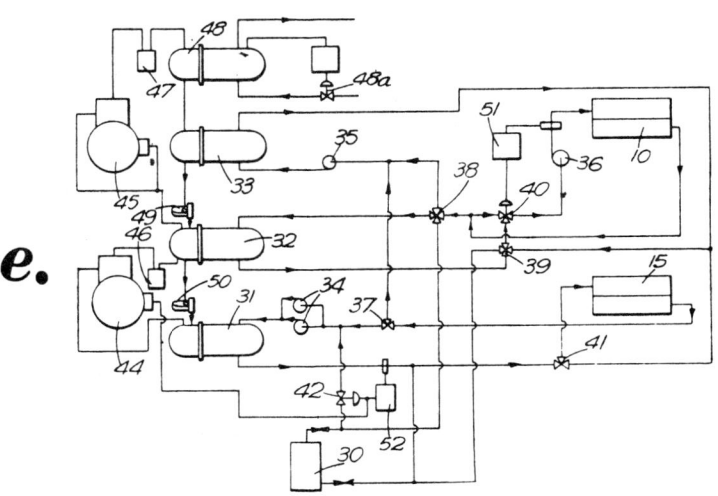

(e) Circuit Arrangement for Refrigerant

Source: U.S. Patent 3,281,956

The structures are tubes closed at their lower ends and each has extending within it an inner tube **11a** leading from a refrigerant inlet header **12** to the lower ends of the structures. The upper ends of the tube structures open into a refrigerant outlet header **13**.

The drying chamber **10** is connected by a duct **14** to a condenser chamber **15** containing a vapor condenser **16** leading from a refrigerant inlet header **17** to an outlet header **18**. The vapor condenser may be of any convenient construction and for instance may be a tube structure like the structure **11**, or may be a coil. The chamber **15** is connected by duct **19** to a vacuum pump **20**.

A scraper **21** is mounted on a central threaded rotating shaft **22** and on rotation of the shaft the scraper which is in the form of a perforated plate through which the tube structures **11** extend, travels up or down according to the direction of rotation of the shaft. The drying chamber has an associated removable bottom closure **23** which has at its lower end a sealing valve structure **24** of any conventional kind through which dried material can be discharged into containers **26**.

In use the fluid to be dried is first frozen to a predetermined thickness, so that for instance for 1 inch diameter tubes the thickness of the frozen layer is up to one-fourth inch. This is done for instance by immersing the tubes **11** in a tank **27** of the fluid (Figure 2.11a) and, while still passing a refrigerant through the structures from the inlet header **12** down the tubes **11a** and up to the tubes **11** to the outlet header **13**. The wall of the drying chamber may be detached from the headers during this step. The scraper **21** is at the top of the tube structure **11**. The tank **27** is now removed.

The closure **23** is now fitted (Figure 2.11b) and, while still passing a refrigerant through the tube structures to keep the fluid frozen, the pressure within the chamber **10** is reduced to say 0.1 mm Hg by means of the vacuum pump. Since the material is kept in the frozen state, there is no need that this step be effected rapidly.

Sublimation of the frozen material now commences and the vapor so produced passes into the condenser chamber **15** and is condensed on the vapor condenser **16** through which refrigerant is being passed. The heat required for sublimation is derived from the refrigerant passing through the internal passages of the structures **11**. This refrigerant is clearly cooled during this step and refrigerant leaving the header **13** may be employed directly or indirectly for effecting condensation of the vapor on the vapor condenser **16**.

After an appropriate drying time, the dried matter is removed from the structures **11** by traversing the scraper plate **21** downwardly along them by rotating shaft **22** (Figure 2.11c). This may be done after bringing the pressure in chamber **10** back to atmospheric pressure. Figure 2.11e shows a circuit arrangement for the refrigerant. The operation of the two-heat-transfer-medium-circuit can be summed up as follows.

During the freezing steps, cold brine at $-25°$ to $-40°F$ is circulated as follows: from the tank **30** through valve **39**, throttle **40**, pump **36** and the structures, such as **11**, in the drying cabinet **10** and then back to the tank **30**; from the cooler **31** through valve **41**, the vapor condenser **16** in chamber **15**, valve **37**, pumps **34**, back to the cooler **31**.

When the freezing is complete and after closure of the cabinet **10** and reduction of the pressure therein, the valves **38** and **39** are closed so as to stop the freezing step and to initiate the drying step. During the major parts of the drying step valves **38** and **39** are set so that brine at about 25°F or below can flow from the heater **32**, through valve **39**, throttle valve **40** which controls the brine temperature by means of the temperature controller **51**, pump **36**, the structure covered with frozen material in the cabinet **10**, and then back to the brine heater **32**. The vapor condenser **15** remains connected to the cooler **31**.

During the final portion of the drying step, the valves **39** are set to allow brine at up to about 86°F to flow through to valve **40** and thus to the structure such as **11** in the cabinet **10**. Under these conditions, the valve **42** is opened by the temperature controller **52** so allowing some of the cold brine from the cooler **31** to be bypassed from the condenser **15** to flow through the cold brine tank **30** to recool the brine therein. This is possible because only small amounts of vapor are reaching condenser **15** which is therefore operating at low load.

During removal of the product, the vapor condenser **15** is defrosted by adjusting valves **41** and **37** so that brine from heater **33** can be circulated through the condenser coil by pump **35**.

MAINTAINING OPTIMUM CONDITIONS FOR SUBLIMATION

USE OF INERT GAS

Introduction of Inert Dry Gas to Heating Zone

B. Kan; U.S. Patent 3,263,335; August 2, 1966; assigned to United Fruit Company describes a process in which an inert dry gas is injected into the drying chamber while heating the product. Heat transfer within the dry outer layers of the product is improved. The dry gas also prevents the formation of local concentrations of water vapor in the chamber. The preferred gases for this purpose are hydrogen or helium, which have high conductivities, but present minimum impedance to the flow of water vapor, because of their high rates of interdiffusion with water vapor.

Referring to Figure 3.1, the apparatus comprises mesh shelves **10** for holding frozen particles **12** of food to be dried, such as shrimps or peach slices or chunks of meat. Radiant heating panels **14** are disposed between the shelves for heating the product. Such heating carried out at low pressures sublimes water vapor from the ice within the frozen food product. The water vapor is condensed at refrigeration coils **16** to remove it from the system.

In using the apparatus for drying shrimp, the temperature of the heating panels **14** is maintained at about 300°F for the initial portion of the drying cycle and then lowered to about 150°F to avoid scorching dry outer layers of food product. The condenser temperature is maintained at about -35°F. This provides a sufficiently rapid condensation to remove the evolving water vapor before its partial pressure can rise above the critical limit of 4 mm Hg. Lower condenser temperatures would unduly raise refrigeration costs in proportion to the benefits gained thereby.

The drying and refrigeration zones are enclosed in a common vacuum chamber **18** evacuated by a rotary gas ballast pump **20** via a valve **22**. After filling the shelves **10** with frozen food product **12** a chamber door (not shown) is closed, and the chamber is evacuated to a residual air pressure, dependent upon con-

Maintaining Optimum Conditions for Sublimation

denser temperatures of less than 0.2 mm Hg absolute (torr) by pump **20** in about five minutes. The valve **22** is then closed to completely isolate the chamber. Heaters **14** are operated and a very rapid evolution of water vapor from the food product begins.

In accordance with the process a charge of dry gas is injected into the chamber via valve **24** during an early part of the cycle. The dry gas, which may be hydrogen or helium, remains in the chamber throughout most of the drying cycle and is handled in the manner described below to shorten the cycle time necessary to achieve a given amount of drying.

The principal benefits realized are that the total pressure within the food product and in the zone between the heaters and food is raised and that local concentrations of water vapor are avoided. The pressure increase, together with the high thermal conductivity of hydrogen or helium, substantially raises the amount of heat supplied to the food at a given radiator temperature.

Thus, the partial pressure of dry gas may be as high as 10 mm Hg while maintaining the partial pressure of water vapor below four mm Hg, depending on operating temperatures and food product. In high density shelf packing practice, it is preferred to maintain a backfilled chamber pressure of 7 torrs after a substantial outer layer of food product has formed. During the early part of the cycle, a pressure of 1 to 4 torrs is preferred if no shrinkage is tolerable and 4 to 12 torrs if slight shrinkage is tolerable, under the temperature conditions described above for a preferred embodiment.

A principal obstacle to realizing the benefits of the dry gas is that the water vapor sweeps some of the dry gas before it into the refrigeration zone. The dry gas cannot condense on the coils **16** or otherwise interfere with the condensation process; but it is useless in the region of coils **16**. The dry gas must be maintained in the heating zone to realize the above benefits.

FIGURE 3.1: INTRODUCTION OF INERT DRY GAS TO HEATING ZONE

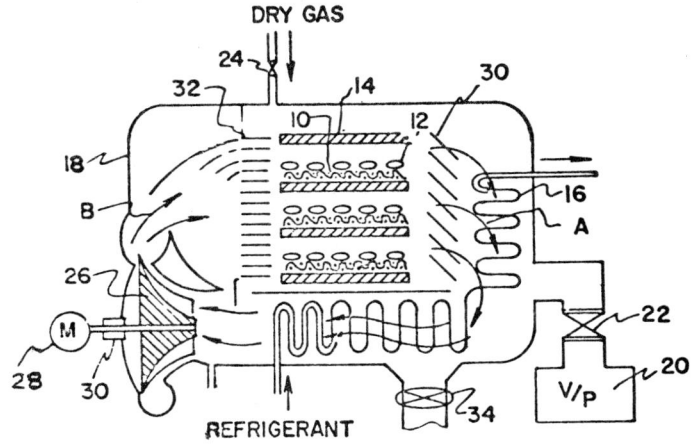

Source: U.S. Patent 3,263,335

A fan **26** is provided to accomplish this purpose. The fan is of the centrifugal flow type to provide high circulation rates with sufficient head to overcome pressure drops due to losses in chamber **18** at low cost. A motor **28** drives the fan via a conventional rotary seal **30**. Axial flow fans are preferred in those commercial production units where flow losses are low. Centrifugal flow fans are preferred for commercial production units having high flow losses. Small size positive displacement blowers are preferred for laboratory size freeze dryers.

Louvers **30** and **32** are provided to smooth out the flow of water vapor and dry gas between the heating and condensing zones and to serve as heat reflecting barriers between these zones. In the limited space afforded by the common chamber **18** it would be less desirable to have the coils **16** directly exposed to radiant heaters **14**. Louvers **32** smooth flow from the fan to the heating zone and are dimensioned to uniformly distribute the helium flow over all the shelves **10**. Ice collected on coils **16** and on the chamber walls is cleaned out via drain valve **34** between drying cycles.

Periodic Pulsing of Dry Inert Gas Through Drying Chamber

In a process developed by *J.G. de Buhr; U.S. Patent 3,262,212; July 26, 1966; assigned to United Fruit Company* a dry and inert gas is periodically pulsed into the chamber during the process. This causes a temporary rise in heat transfer from the heaters to the product as well as within the product. At the same time the water vapor diffuses well through the dry gas so that its partial pressure does not build up rapidly. The dry gas pressure is not raised so high that it presents a substantial impedance to water vapor travel. If this happened, a water vapor equilibrium exceeding the melting point of ice would be reached. The pulsing of dry gas is repeated cyclically throughout the drying cycle. Use of this technique permits a reduction in drying time in excess of 30%.

An apparatus for the drying phase is shown in Figure 3.2a. Within vacuum drying chamber **10** with a door **12**, are arranged heating platens **14** alternating with mesh shelves **16**. Frozen product **18** is placed on the shelves. The chamber is evacuated by a vacuum pump **20**, typically of the rotary gas ballast type, via passage **22**. A refrigerated condenser **24** is placed in the passage in series with the pump and chamber to selectively remove water vapor sublimed from the frozen product. An air operated slide valve **50** is provided for quickly isolating the vacuum pump from the system.

In the apparatus arrangement provided by the process, there is provided a source **52** of dry gas with a valve **54** for selectively admitting the gas to the chamber and a refill valve **56**. A control **58** automatically schedules the opening and closing of valves **50** and **54** in accord with a predetermined time schedule and pressure in the drying chamber **10**, as measured by a pressure gauge **60**.

Valve **150** cuts the chamber off from the condenser. This can be operated from time to time by control **58** to make indirect temperature measurements by the technique described in U.S. Patent 2,944,132. The measurement function of valve **150** does not form part of this process. The essence of the process is to avoid presenting a substantial impedance to removal of water vapor. This is generally done by closing valve **50** while admitting the helium or other dry gas while valve **150** remains open.

Maintaining Optimum Conditions for Sublimation

FIGURE 3.2: PERIODIC PULSING OF DRY INERT GAS THROUGH DRYING CHAMBER

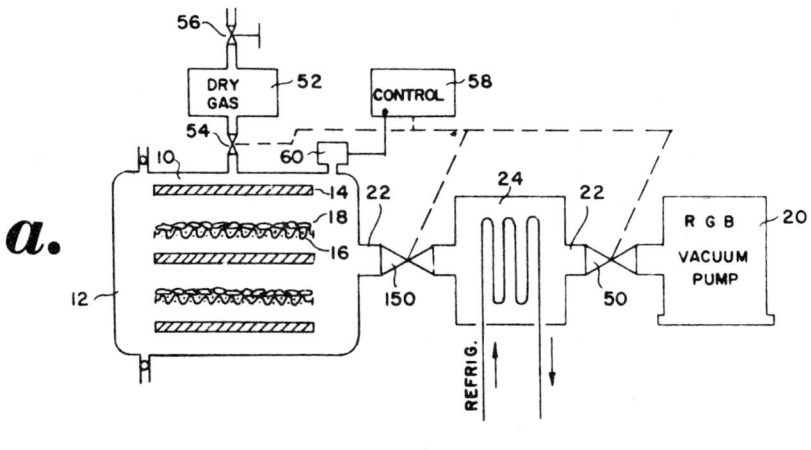

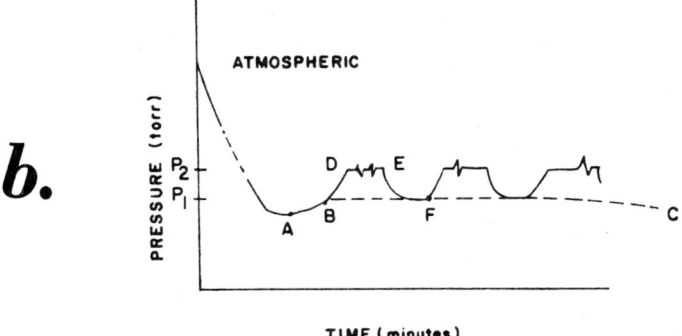

Source: U.S. Patent 3,262,212

Referring to Figure 3.2b, the operation of the above apparatus is shown on a pressure-time trace. The total pressure in chamber **10** is initially at atmospheric as the product is loaded and the door **12** is closed; valves **150** and **50** are open, valve **54** is closed and pump **20** is off. The heaters **14** are turned off. Refrigerant circulated through condenser **24** maintains it at a temperature of about $-35°F$.

The pump is switched on and the chamber is evacuated to a pressure on the order of a few mm Hg absolute in less than 10 minutes. At point **A**, the heaters **14** are turned on to produce a heater temperature of about $300°F$. The product is spaced about one inch from each of its associated heaters. Water vapor is sublimed from the product. After a transient instability, this results in an equilibrium, at pressure P_1, and the total pressure would approximate the dashed line

B—C if no further process steps were applied. However, at point B, the control 58 is put into effect. The control closes valve 50 to isolate the pump from the chamber and opens valve 54 to admit the dry gas to the chamber. The condenser continues to remove water vapor, but the total pressure rises due to the admission of dry gas without any means for pumping it.

The pressure rise continues until a desired P_2 is reached at point D. Then the pressure gauge 60 signals the control to limit the pressure rise. This is accomplished by partially opening the valve 50. Pressure P_2 is maintained for a few minutes. However, it should be understood that the process is really a dynamic one, as indicated by the wavering line D—E. At point E the control 58 completely closes valve 54 and opens valve 50 to pump the chamber down again until point F where the cycle is repeated. The dry gas is preferably hydrogen or helium.

The refrigerated condenser would be operated so that the pressure P_1 would be below 4 mm Hg absolute, where water vapor is the volatile constituent to be removed. The use of dry gas, such as helium or hydrogen, in accord with the pulsing technique allows a P_2, intermittently maintained, on the order of 8 mm Hg absolute. It is preferred to make the dry gas pulsing a regular cycle.

Frozen shrimp were dried in several series of runs. Each series was done in accord with the pressure and temperature cycles indicated in the table below. The refrigerated condenser was maintained at -60°F for all the runs of the 7 series. The heating platens were spaced about one inch from the shrimp to be dried which varied in weight between about 10 to 15 grams each. Shrimp were treated singly and in arrangements where surrounded by four other shrimp. As an aid in reading the table series No. 5 is described as an example.

In the runs of this series, the equilibrium pressure P_1 in the chamber after turning on the heaters was 0.03 torr. This pressure was held for three hours. Then valve 50 was closed and valve 54 was opened to admit nitrogen to the system. The total pressure rose quickly to 2 torrs and this was held for 5 minutes, followed by a 5 minute pumpdown to get the pressure back to 0.3 torr, followed by a second pulse of nitrogen in the same fashion. The pulsing of nitrogen was thus continued at a rate of 6 pulses per hour for the last 3½ hours of the cycle. After the runs, the shrimp were baked overnight and the weight loss noted as a measure of residual moisture. Mean values for the several runs of series 5 are noted below as 1.04% for single shrimp and 5.05% for surrounding shrimp. In all the series of runs, the temperature of the heaters 14 was maintained at 300°F for the first 1½ hours and at 150°F for the remaining 5 hours of the cycle.

Series	Total Pressure Cycle	Gas Introduced	Results—Mean Residual Moisture (%)	
			Single Shrimp	Surrounded Shrimp
#1*	0.03 torr for 6½ hr	None	2.82	9.4
#2	0.03 torr for 3 hr; gas added to produce 2 torrs, held for 3½ hr (helium held at 8 torrs for 3½ hr)	Nitrogen	2.12	8.47
#3		Argon	1.95	6.53
#4		Helium	1.33	2.38
#5	0.03 torr for 3 hr; then pulsing** to 2 torrs by addition of gas (8 torrs for helium) for 3½ hr	Nitrogen	1.04	5.05
#6		Argon	1.14	4.27
#7		Helium	0.45	0.41

*Control
**Rate of pulsing was 6 per hr (each pulse comprising 5 min at elevated pressure; 5 min pumpdown—5/5).

Series 5, 6 and 7 were in accord with the process. In each of these cases, the residual moisture was less than that obtainable in control series 1. Holding a partial pressure of dry gas, as in series 2 to 4, provides less improvement than the pulsing technique of the process.

Alternate Pulsing of the Two Drying Chambers with Inert Gas

A substantial reduction in cycle times for freeze drying is achieved by the techniques described in U.S. Patent 3,262,212 above. In accord with such techniques a dry inert gas is periodically pulsed into the drying chamber whereby the total pressure in the chamber is periodically raised and lowered. *B. Kan; U.S. Patent 3,255,534; June 14, 1966; assigned to United Fruit Company* describes an improved apparatus for carrying out the process.

Referring to Figure 3.3a there is shown a freeze drying apparatus comprising a drying chamber A connected to a vacuum pump **10** via a passage **30A**. A refrigerated condenser A is connected to the passage in a series with the drying chamber to reduce the vapor tension of ice in the product to be dried. Similarly, a drying chamber B is connected to the pump via a passage **30B** and condenser B. Chamber A is provided with a source **12** of preferably hydrogen or helium or from among the noble gases. A fuller description of the method of periodically pulsing a drying chamber with a dry inert gas is found in U.S. Patent 3,262,212.

A Roots blower **16**, or other high speed rotary blower, is provided for circulating gas between the drying chambers A and B. It operates in only one direction and is connected to the chambers A and B via their respective condensers and the valve system **20**. The valve system comprises a first valve **21** controlling inlet flow from condenser A, a second valve **22** controlling outlet flow to condenser B, a third valve **23** controlling outlet flow to condenser A and a fourth valve **24** controlling inlet flow from condenser B. The valves are actuated by air operated servomechanisms controlled by automatic controller **18** which sequences the operation of these four valves. The controller also controls valves **14A** and **14B** in response to signals received from pressure gauges **26** and **28**.

Referring to Figure 3.3b, the operation of the apparatus is shown on a pressure-time trace. The variation of total pressure in chamber A is shown by the solid curve A and the variation of total pressure in chamber B is shown by the dashed curve B. Initially both chambers are at atmospheric pressure, blower **16** is idle and valves **21–24** are closed.

Valves **14A** and **14B** are open and operation of pump **10** begins. This pulls the pressure in both chambers down to point **1** which is on the order of about 1 mm Hg, absolute. Then helium is admitted to dryer A and valve **14A** is closed under manual control. Meanwhile control **18** is switched on so that further operations are automatically controlled. The total pressure in chamber A rises until a pressure indicated at point **2** in excess of 4 mm Hg, and preferably 8 mm Hg, is reached. Then pressure gauge **26** signals the controller to partially open valve **14A** to prevent any further rise. In general, it will not be necessary for such a signal to be transmitted to the controller since a limited amount of helium is used to charge the dryer A.

The primary cause of pressure rising above point **2** will be excessive evolution of water vapor from the frozen product. This pressure is maintained for a limited time which may vary from as low as 30 seconds to as high as 30 minutes, depending on the nature of the frozen product and the equipment selected. At point 3 representing the end of this limited time, the controller shuts valve **14B** and opens valves **21** and **22**. The high speed blower **16** removes gas from chamber A to chamber B rapidly via the condensers.

The routing of the gas via the condensers assures the removal of a substantial portion of water vapor so that the greater portion of gas shuttled from one chamber to the other is helium. Thus the total pressure in chamber B is rapidly raised to that at point **4** by the addition of dry gas while the total pressure in chamber A is correspondingly decreased to that at point **5**. At point **4–5** of time, valves **21** and **22** are closed and the elevated pressure is maintained in chamber B. Overpressure surges, if any, are eliminated by valve **14A** or **14B** in response to signals from gauges **26** or **28**, respectively.

FIGURE 3.3: ALTERNATE PULSING OF THE TWO DRYING CHAMBERS WITH INERT GAS

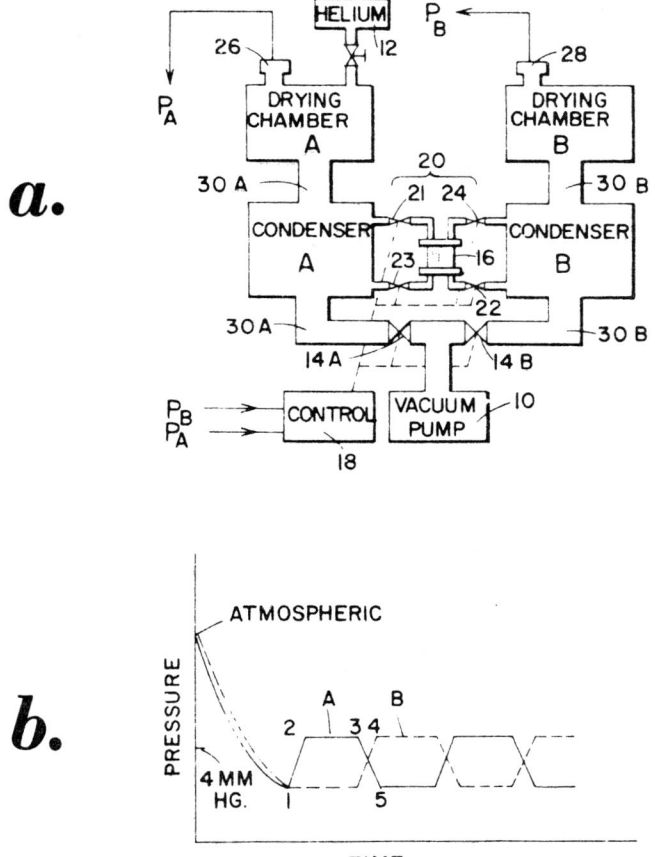

Source: U.S. Patent 3,255,534

Maintaining Optimum Conditions for Sublimation 71

The alternate pulsing of chambers A and B is continued in cyclic fashion throughout the drying cycle. The helium is conserved by this technique. At intermittent times a little helium should be added to chamber A to make up for losses due to intermittent opening of valves 14A and 14B due to overpressure surges detected at one of the pressure gauges 26 or 28.

Addition of Helium to Drying Atmosphere

J.C. Harper; U.S. Patent 3,271,873; September 13, 1966 discovered that the addition of certain light gases such as hydrogen or helium to the drying atmosphere results in a substantial increase in the drying rate.

Figure 3.4 illustrates this finding as applied to a freeze drying process using a refrigerated condenser. In freeze drying materials, such as food, the material is first subjected to an ambient atmosphere having a temperature low enough to turn the moisture within the material from a liquid state to a solid state more commonly known as ice. By maintaining the ambient pressure of the freeze atmosphere at a moderate vacuum pressure such as 5 to 10 mm Hg during freezing, a substantial portion of trapped air can be removed from the material. In addition, the vacuum has a tendency to evaporate a portion of excess moisture and results in increased efficiency of freezing. However, it is not necessary to freeze the material in a vacuum and the control valve 12 to the vacuum pump can be closed off.

The frozen material can then be sealed in an air-tight drying chamber to vaporize the frozen moisture during a drying cycle. First, the ambient pressure within the drying chamber is reduced to an appropriately low level, on the order of 1.0 mm of Hg, and external heat is applied to the material by means of direct radiation or heated plates.

During the initial part of this drying cycle, sublimation or vaporization of the frozen moisture occurs at a very rapid rate and, by nature, starts in the outer layer and progresses inwardly toward the center of the material. This results in the sublimed portion of the material forming a dry porous shell about the still frozen core. Since the thermal conductivity of this shell is a combination of the thermal conductivities of the dried material and whatever gas is left in the pores, it becomes an effective insulator and hinders the externally applied heat of sublimation from reaching the still frozen core. Thus, the insulation reduces the drying rate to an undesirably low rate.

In order to overcome this problem, the chamber is further evacuated to a low pressure, preferably 0.1 mm Hg or less, in order to remove as much air as possible. A light gas having a relatively high coefficient of thermal conductivity and high coefficient of diffusion for water vapor is then allowed to flow into the drying chamber until a desired gas pressure is attained. As a result of this procedure, the air in the pore spaces of the shell is substantially replaced with the light gas, thereby providing for improved conduction of heat and diffusion of water vapor. Usually the desired gas pressure will lie in a range from about 0.5 mm Hg to 10 mm Hg, although pressures both below this range and above it as high as atmospheric pressure can be used.

One gas which has been found to be especially effective is helium. This gas is characterized by a coefficient of thermal conductivity approximately six times

FIGURE 3.4: ADDITION OF HELIUM TO DRYING ATMOSPHERE

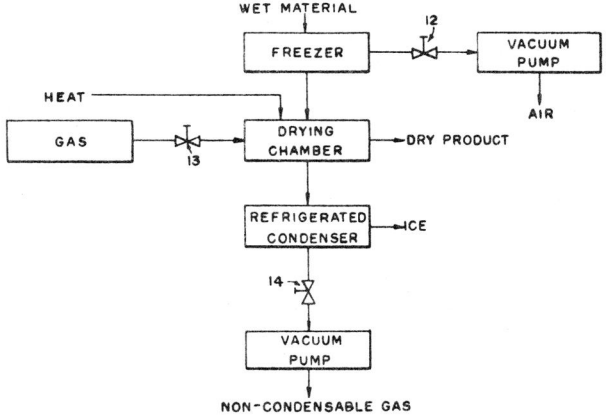

Source: U.S. Patent 3,271,873

as great as that of air and a coefficient of water vapor diffusion approximately four times as great as that of air. As a result of these two combined characteristics it can be shown that the drying rate for the pure helium atmosphere, as compared with an all air atmosphere, will be increased almost by a factor of three.

In order to decrease operating costs by reducing the amount of helium used, the helium gas can be combined with air if desired, the amount of helium added to the air should not be less than 10% helium and 90% air at the lower limit.

Another gas having the desired characteristics which could be used is hydrogen. Hydrogen, however, has the disadvantage of creating an explosive atmosphere. In addition, helium, hydrogen and air could be intermixed within the above stated limits with regard to the helium gas.

The above discussed drying by light gas is fully capable of being used in either intermittent or continuous processes. When an intermittent, or batch, process is used, the frozen material is first placed in the vacuum sealed drying chamber, and the air or gas within the drying chamber brought to the previously described desired level for the initial drying period by means of the vacuum pump. After the initial period is completed, which might be from one-half hour to one hour or longer, the vacuum is brought to the highest possible value, i.e., lowest pressure by fully opening vacuum pump valve **14**.

Valve **14** is thereafter closed and gas flow control valve **13** opened for a long enough time to permit light gas to flow into the drying chamber until the desired gas pressure is reached. At the same time, the heat input must be increased by means of a suitable source in order to supply the necessary heat for the increased drying rate. After a period of time, the pressure in the chamber will rise as a result of air leaking into the chamber and occluded air released from the frozen material. This air will impede the movement of water vapor through the dry porous shell and will cause the drying rate to decrease. In order to restore the drying rate, valve **14** is again opened until the chamber reaches the previously discussed high vacuum; thereafter, vacuum pump control valve **14** is again closed and gas flow control valve **13** again opened to admit more light gas. These alternating steps of evacuating the chamber and adding light gas are continued until the drying process is completed, after which the chamber is opened and the dry product removed.

Another method of operation, which avoids the alternate operation of valves **13** and **14**, is to set both of these valves in fixed positions, thus allowing light gas to flow in at a constant rate and maintaining a constant pressure in the drying chamber. With this method of operation, the gas in the chamber is a constant mixture of air and light gas. The composition of the gas and the total pressure in the chamber can be controlled at any desired values by proper positioning of the valves **13** and **14**.

The chamber must be open to the refrigerated condenser to exhaust the air polluted gas from the drying chamber. The moisture laden gas is directed into contact with refrigerating coils or plates to condense any condensible material. This results in ice forming which is thereafter tapped from the condenser as waste. The vacuum pump is also connected to the refrigerated condenser and is operable to exhaust any noncondensible gas thereby maintaining the high vacuum when vacuum control valve **14** is open.

Use of Inert Gas During Terminal Portion of Drying

E. Thuse; U.S. Patent 3,299,525; January 24, 1967; assigned to FMC Corporation found that it is possible to shorten the drying cycle by circulating an inert or nonoxidizing carrier gas through the drying chamber, after a large proportion of the water vapor has been removed from the product by conventional evacuation and vapor condensation methods. The process is described with reference to Figure 3.5.

The product is introduced and removed from the drying chamber **10** on a car **12**. The car is provided with heated shelves **14**, through which a heated liquid is circulated for supplying the heat of sublimation to the product. A condenser plate assembly **38** is mounted on each side of the car bearing the product laden trays.

For removal of air and other noncondensable gases initially in the chamber, as well as for making up for any leakage of such gases into the chamber as may occur during the drying cycle the system is provided with line **46** connected to a vacuum pump (not shown).

A feature of the system is the use of an inert or nonoxidizing carrier gas during a terminal portion of the drying cycle, for assisting in conducting heat to the ice core of the product particles, and for sweeping away the additionally sublimed water vapor.

The carrier gas, which may be helium, is supplied in a pressure vessel **50**, which carries a gas valve **52** that controls a gas admission line **54** leading into the system. A carrier gas circulating conduit **56** connects between the lower and upper

FIGURE 3.5: USE OF INERT GAS DURING TERMINAL PORTION OF DRYING

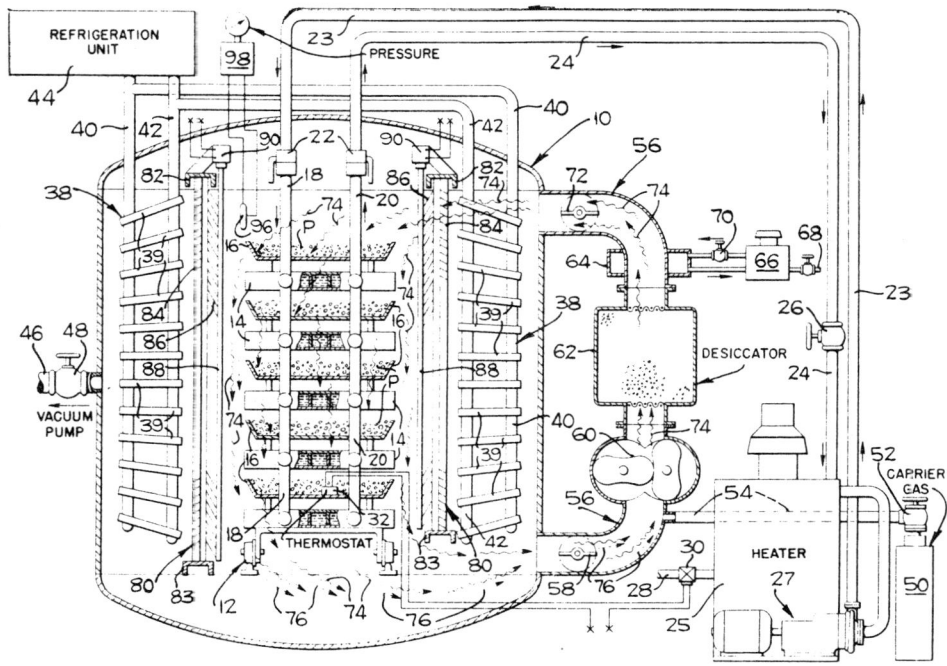

Source: U.S. Patent 3,299,525

portions of the drying chamber **10**. This conduit has an inlet butterfly valve **58**, and the gas is circulated by a positive displacement blower **60**. After leaving the blower, the carrier gas is directed through a desiccator **62** filled with a water absorbent agent.

The carrier gas may be heated by a heater, in the form of a heating chamber **64** surrounding the carrier gas circulating conduit **56**. Usually, however, the carrier gas picks up sufficient heat from the heated shelves **14**.

The end of the carrier gas circulating conduit that connects to the upper portion of the drying chamber **10** is also provided with a butterfly valve **72**. The wavy arrows **74** indicate the flow of carrier gas through the system, when the carrier circulating system is in operation.

Means are provided to isolate the condenser assemblies **38** from receiving radiant heat from the heated shelves **14**, and for discouraging the flow of carrier gas into the condensing zones, when the carrier gas is being circulated through the chamber. In order to accomplish this, a louver assembly indicated generally at **80** is provided between each condenser assembly **38** and the car with its trays and shelves within the vacuum chamber. Two sets of louvers are employed, a set of fixed louvers **84** and a set of adjustable louvers **86**.

Maintaining Optimum Conditions for Sublimation 75

In operation, the car **12** will have been removed from the drying chamber and its trays **16** will have been loaded with the frozen product **P** and placed on the car. The condenser plates **39** will have been flushed clear of ice from the previous cycle. The car is pushed into the drying chamber and the heating liquid connections **22** are made. The butterfly valves **58** and **72** in the carrier gas circuit will be closed. The door (not shown) to the drying chamber is closed and sealed, and the vacuum pump connected to line **46** is started up with the vacuum valve **48** open.

With the vacuum pump in operation, the noncondensable gases in the vacuum chamber are soon exhausted and the pressure drops to well below the triple point. Heated liquid is now circulated through the shelves **14**. As soon as the pressure thus drops, water sublimes from the frozen product, this sublimed water vapor passes freely through the adjustable louvers **86** and the fixed louvers **84**, to the condensing plates **39**. Here the water vapor is at once frozen to form ice particles, deposited on the condenser plates. This freeze drying by sublimation continues and the ice core of the product **P** gradually is reduced in size and recedes from the surface of the product particles. This recession of the ice core leaves a surrounding volume of dried cellular material, that acts as a good heat insulator.

Toward the end of the drying cycle, the ice core has become so small and the insulating volume so large, that the effectiveness of heat radiation in penetrating to the ice core is reduced, thereby prolonging the drying time unnecessarily. When this condition occurs, which is approximately after 90% of the water vapor has been sublimed from the product, the adjustable louvers **86** are adjusted to their substantially closed position. Also, the butterfly valves **58** and **72** are opened and the blower **60** is started. Carrier gas such as helium is admitted by the opening of the gas valve **52**, and circulation of the carrier gas begins. The admission of the carrier gas continues until the pressure gauge unit **98** indicates the desired pressure, which may be very low, in the order of 10 millimeters of mercury.

A good operating zone is at a pressure in the order of 2 to 30 millimeters of mercury. This makes an effective compromise of the conflicting requirements as to heat transfer into the ice core and diffusion of the water vapor through the dried shell of product from the ice core.

The carrier gas, sweeps through the shell of dried material and contacts the ice core, thereby giving up heat to the ice core and assisting in further sublimation of water vapor. The carrier gas picks up heat from the shelves **14**, which continue to receive heating liquid. The water vapor is picked up and entrained by the carrier gas, and is carried down past the inlet valve **58** through the blower **60** and into the desiccator **62**. Here the water vapor is removed from the carrier gas, and the dry carrier gas is recirculated through the outlet valve **72**, back into the drying chamber and across the product and shelves **14**.

Use of Flash Evaporator for Readily Condensable Heat Carrier Heptane

In this process developed by *J.H. Blake, J.P. Pelmulder and E. Thuse; U.S. Patent 3,382,584; May 14, 1968; assigned to FMC Corporation* a readily condensable, heat carrier vapor (such as heptane) makes at least two passes through a layer or bed of product moving through a vacuum drying chamber. The carrier vapor

is immiscible with water. It is slightly superheated initially and picks up heat
between the passes, and thus supplies the heat of sublimation to the product.
The carrier gas and entrained water vapor are both condensed, the condensates
separated, and the carrier fluid recirculated into the chamber. The process is
described with reference to Figure 3.6 which shows a preferred drying process.

Carrier Fluid Circuit: Starting with the liquid carrier C falling into basin 80 in the
drying chamber, a mixture of the carrier and the diluted brine B drains freely
through pipe 82 into a decanter 120. The carrier C and the immiscible brine B
separate by gravity, and the carrier if it is heptane, will form a layer on top of the
brine. The carrier is withdrawn through a jointed or flexible pipe 122 supported
by a float 123 and connected to a pipe 124. The carrier is withdrawn by means
of a constant flow rate pump 126. This pump is made to deliver the carrier at a
constant rate by variable speed drive 127 controlled by an orifice flow sensor unit
128 of conventional design. In order to prevent the decanter 120 from becoming
airbound an air bleed line 129 leads from the decanter to the drying chamber adjacent the inlet for the vacuum line, so that noncondensable gases in the decanter
will be withdrawn through the vacuum evacuation system of the drying chamber.

The carrier liquid C passes from the pump 126 to a filter 130, which separates
any brine that remains entrained with the carrier. This filter may be a fuel-type
filter having a hydrophobic element therein that permits only organic liquid to
flow readily through the filter element.

The purified carrier liquid then passes through a line 134 to a heat exchanger
154, which operates as a heater for the carrier (heptane) and a cooler for the
refrigerant (ammonia). The preheated carrier liquid then passes through line
138 to a flash heptane vaporizer 140. Here, the liquid heptane is both flash
vaporized and superheated slightly by a low pressure steam source LPS.

The liquid carrier entering the vaporizer in line 138 passes through coils within
a steam chest 143 of the vaporizer. The heat transfer surface of the coil with
the given steam heat source is more than sufficient to evaporate all of the carrier
liquid at the temperature and flow rate desired. The coefficient of heat transfer
in the vaporizer is relatively high when the carrier is in its liquid phase and hence
the carrier is readily vaporized. The carrier vapor is superheated, but not excessively, when it leaves the flash vaporizer 140 and the carrier vapor reenters the
drying chamber by means of the line 110, previously described. Automatic controls are provided for the carrier vaporization step.

Assuming that 2,000 pounds of water per hour will be evaporated from the
product and assuming that the carrier is n-heptane 56,000 pounds of carrier per
hour will be circulated through the carrier circuit of the system. Evaporated
refrigerant (ammonia) leaves the drying chamber condenser coils (which serve
as the evaporator in the refrigeration system) through a line 66 and enters the
low pressure receiver 148 of the refrigeration system.

The gaseous refrigerant is drawn into the compresser through a line 150, and
the hot refrigerant is supplied by the compresser through a line 152 and preferably through a water cooled condenser 153 to a heat exchanger 154, which
serves as a heptane (carrier) heater and an ammonia condenser.

Maintaining Optimum Conditions for Sublimation

FIGURE 3.6: USE OF FLASH EVAPORATOR FOR READILY CONDENSABLE HEAT CARRIER HEPTANE

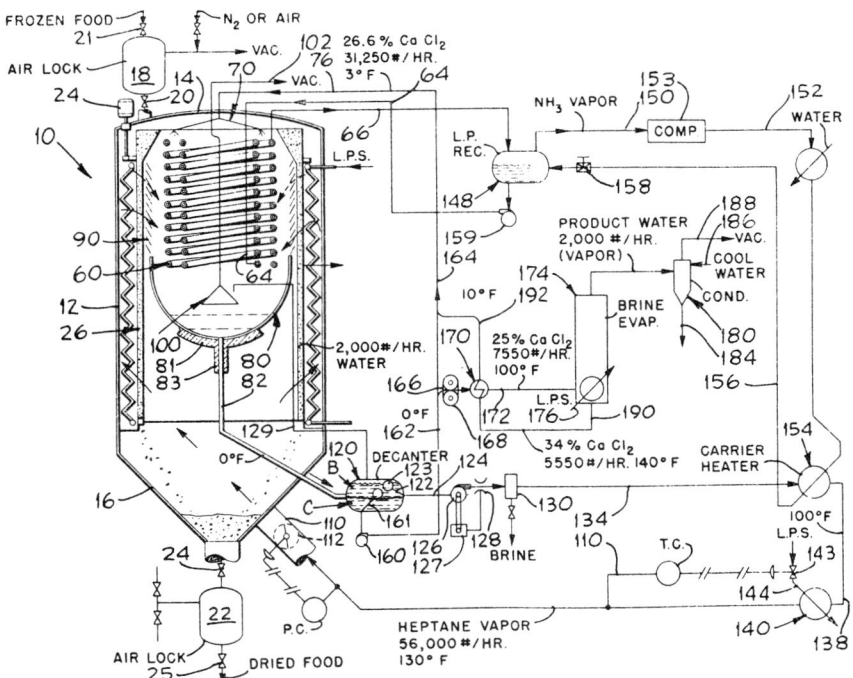

Source: U.S. Patent 3,382,584

The liquefied refrigerant leaves the heat exchanger **154** through a line **156** and is conducted to an expansion valve **158** back to the low pressure receiver **148**. The liquid refrigerant is pumped from the receiver by a pump **159** which returns it to the line **64** forming the inlet line for the drying chamber condenser coil **62**. The liquid refrigerant entering the condenser coil **62** in the drying chamber will be at a temperature of approximately 25°F and the refrigeration system has sufficient tonnage to condense the carrier and water vapor that reaches the condenser **60** in the drying chamber.

Brine Circuit: The brine sprayed onto the condenser **60** is diluted with water vapor sublimed from the product and this solution drains from the gas barrier basin **80** by means of pipe **82** and flows into the decanter **120**. Here the brine B separates from the carrier C by gravity, and the brine pump **160** takes suction below a bulb **161** which floats at the brine carrier interface and the pump circulates the brine via a delivery line **162** which has a main branch **164** for returning the brine to the inlet pipe **76** for the brine spray nozzles.

Part of the brine from the decanter **120** is diverted into a line **166** by means of a constant delivery pump **168** and flows into a heat exchanger **170** of a conventional design. The diluted brine drawn off in branch line **166** leaves the heat exchanger **170** via a line **172** and enters a brine evaporator **174**. Here the brine is heated by a low pressure steam LPS from a line **176** in a conventional manner at a temperature sufficient to evaporate 2,000 pounds per hour of product water. This is the amount of water vapor that has been sublimed from the product in the drying chamber, frozen on the condenser **60**, and washed into the brine from the condenser.

The brine evaporator is operated under a vacuum using a barometric condenser **180**, the product water leaves the evaporator **174** via a line **182** leading to the barometric condenser **180**. Condensed water vapor flows out of a leg **184** of the barometric condenser. Cooling water enters the barometric condenser **180** through an inlet line **186**, and a vacuum connection from the vacuum pump is made to the condenser by a line **188**.

Returning to the brine evaporator **174**, concentrated brine leaves the evaporator by a line **190** which passes through the heat exchanger **170** and hence is cooled by the cold incoming brine in line **168**. This cooled, concentrated brine, which will be at approximately 10°F, leaves the heat exchanger **170** through a line **192** which connects with the main line **164** for return to the spray nozzles **172**.

In the example being given the brine entering the spray nozzle **172** will be a 26.6% calcium chloride solution and will be flowing at the rate of 31,250 lb/hr at a temperature of about 3°F. The brine diverted in the concentrating line **166** will be diluted to a 25% calcium chloride solution, and will be delivered to the brine evaporator at a rate of 174 lb/hr and at about 100°F.

The concentrated brine leaving the brine evaporator **174** in line **190** will have a composition of about 34% calcium chloride and will leave at a flow rate of 5,550 lb/hr at about 140°F. As mentioned, after passing through the heat exchanger **170**, this concentrated brine rejoins the main stream in pipe **164** at a temperature of 10°F, so that since main steam initially in pipe **164** is at a temperature of about 0°F the temperature of the brine entering the spray nozzle **72** in the drying chamber is about 3°F.

Gas Circulated Through Bed of Frozen Particles Mixed with Molecular Sieve

C.J. King, III and J.P. Clark, III; U.S. Patent 3,453,741; July 8, 1969; assigned to U.S. Secretary of Agriculture describe a process for freeze-drying wherein pieces of food or other material to be dried are frozen and placed in proximity to molecular sieve granules. A gas, preferably of low molecular weight, such as helium, is circulated through the system whereby to scavenge water vapor formed by sublimation of ice, to transport the water vapor to the molecular sieve for adsorption thereby, and to transfer heat generated by this adsorption to the material under treatment to supply heat required for sublimation.

The substances commonly known as molecular sieves are aluminosilicates or zeolites, the crystals of which contain minute pores. In fact, the cross sections of these pores are of molecular dimensions. As the molecular sieve, one may employ natural zeolites or synthetic zeolites. The system is operated at a moderate vacuum, a pressure of about 10 to 100, preferably 25 to 75 mm Hg absolute. It is preferred that the gas stream entering the system be at least at room temperature (25°C) and usually it is at about 50° to 60°C.

Maintaining Optimum Conditions for Sublimation

A typical application of the process is explained in connection with Figure 3.7 shown on the following page. Within dehydration chamber 1, provided with a removable lid 2 is a foraminous platform 3, made of wire cloth, perforated sheet metal, or the like, which supports bed 4, a mixture of molecular sieve granules and frozen pieces of food to be dried. A removable screen 5 is disposed over a bed 4 to prevent particles from entering the piping system during operation.

For circulation of gas through the bed there are provided pipes 6, 7, 8, 9, and 10 and blower 11. Interposed in pipe 10 is heat exchanger 12 whereby the stream of gas can be maintained at a predetermined temperature, by applying heating or cooling as necessary. Also connected into the gas recirculation system are vacuum pump 13 and gas source 14.

The gas recirculation system also includes means whereby the gas stream can be directed upwardly or downwardly through the bed. This means includes two-way valves 16 and 17. The broken line designated by numeral 18 represents a linkage between valves 16 and 17, permitting their operation together, either manually or by a timer mechanism.

In the position shown, the gas stream is impelled by blower 11 through heat exchanger 12 into pipe 7 and then into the base of chamber 1. The return flow is via pipes 6 and 8 back to the blower. By suitable rotation of valves 16 and 17, the flow is reversed, i.e., from the blower, through heat exchanger 12, into pipe 6 and then into the upper portion of chamber 1. The return flow is via pipes 7 and 8 to the blower.

In starting a run, frozen pieces of food or other material to be dried are mixed with molecular sieve granules, using at least enough of the latter to take up the amount of water to be removed from the food. The mixture is deposited on platform 3. Screen 5 is placed over the bed and top 2 is put in place to seal chamber 1 from the atmosphere.

Vacuum pump 13 is actuated for a period long enough to substantially exhaust the chamber and the appurtenances communicating therewith. The pump is then shut off and remains off during the run. Valve 15 is opened to invest the system with gas, preferably, a low molecular weight gas such as helium, from pressure source 14. When the pressure of gas in the system reaches a level of about 25 to 75 mm Hg absolute, valve 15 is closed and kept closed during the run. Blower 11 is now turned on to circulate gas through bed 4, circulation being continued until the material is dried to the desired degree.

Also during the run, the positions of valves 16 and 17 are periodically changed to alternately direct the gas upwardly and downwardly; this ensures a uniform dehydration effect. The blower is operated at such a speed that the gas flows through bed 4 at a rate which will effectuate the desired transfer of heat and scavenging of water vapor. However, the flow is not so high as to effect a fluidization of the material in the bed which remains in a static condition whereby attrition is avoided and at the same time rapid and uniform dehydration are achieved through the coaction of the circulating gas and the molecular sieve.

During the operation, heat exchanger 12 is activated to provide heating or cooling as necessary to maintain the gas stream at a predetermined temperature.

FIGURE 3.7: GAS CIRCULATED THROUGH BED OF FROZEN PARTICLES MIXED WITH MOLECULAR SIEVE

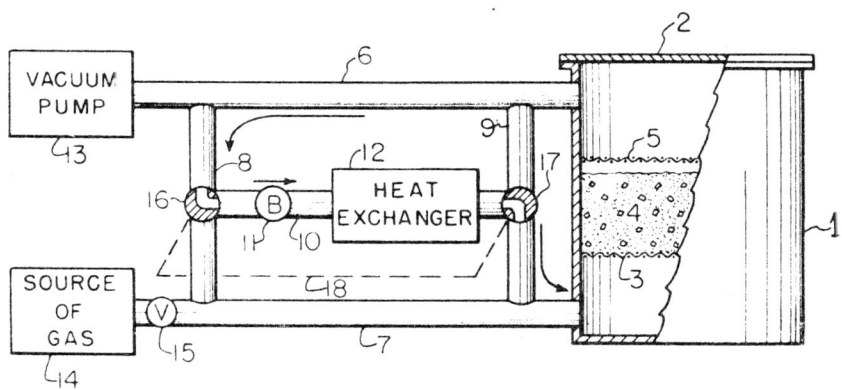

Source: U.S. Patent 3,453,741

Generally, it is preferred to have the gas stream at as high a temperature as possible without thawing the frozen material or causing it to shrink or become sticky. After the material has been dried to the desired extent, the chamber is opened and the bed is removed and subjected to a screening or equivalent operation to separate the pieces of dried material from the molecular sieve granules.

PRESSURE MODIFICATIONS

Increasing Pressure After Predetermined Portion of Cycle Is Completed

W.H. Hamilton, U.S. Patent 3,230,633; January 25, 1966; assigned to Pennsalt Chemicals Corporation found that the drying cycle may be substantially reduced by increasing the pressure within the chamber after a predetermined portion of the cycle has been completed. The remainder of the drying cycle may be substantially reduced by increasing the pressure in the chamber when the rate of moisture loss starts to decrease. Preferably, the increased pressure is effected by passing a gas through the housing.

By introducing gas into the housing to increase the pressure in the chamber, and by operating the vacuum pump in a throttled condition, the articles will be dried by convection as well as radiation during the remainder of the drying cycle. While the gas increases the vapor pressure on the article, this effect is offset by the increase in the vapor loss due to convection drying. Thus, articles dry faster. The process is described with reference to Figure 3.8.

In apparatus **10**, the drying chamber within the housing **12** may be evacuated through the conduit **14**, which is in open communication with the drying chamber and a condenser **18**, via valve **16**. A vacuum pump **22** has its inlet side in communication with the outlet side of the condenser. A solenoid operated valve **20** is disposed intermediate the condenser and the pump for throttling the pump.

An electrically heated shelf **24** (only one shown) is supported within the drying chamber. The electrically conductive coating of the shelf will be connected to terminals **30** and **32** by wires **26** and **28**. Wire **28** is preferably provided with a rheostat **34**.

A perforated tray **38** for the articles **40** to be dried is supported by the shelf **24**. A spacer **36** of mesh-type electrically nonconductive material is provided so that the articles within the tray **38** will be heated by radiant heat from above and below.

One of the articles **40** will have a resistor embedded therein. The resistor will be connected to a weight recorder and controller **48** by wires **42** and **44**. An amplifier **46** will be connected to one of the wires, such as wire **44**. The weight recorder and controller **48** will be connected to a source of potential having terminals **50** and **52**.

The weight recorder and controller may be connected to a temperature recorder and controller **54** by wires as illustrated. The temperature recorder and controller may be connected to the rheostat **34** by wires **56** and **58**. A conduit **60** has one end in communication with the drying chamber in the housing **12**. The other end of the conduit is connected to a source of gas **61** which may be a tank of pressurized inert gas such as nitrogen.

A solenoid operated pressure regulator type valve **62** is disposed within the conduit. The valve is normally closed and will be opened at a predetermined point as will be made clear. The solenoid operator for the valve **62** is connected to the weight recorder and controller by wires **64** and **66**. The solenoid operator for the throttling valve **20** is connected to the weight recorder and controller by wires **68** and **70**.

The operation of the apparatus is explained in conjunction with the graph illustrated in Figure 3.8b which is based on test results wherein cooked whole shrimp was freeze dried with the shrimp being placed on end in a wire tray having an area of approximately 0.5 square feet. The shrimp was stacked in a layer approximately two inches thick. The original weight of the shrimp was 1,084 grams.

The frozen shrimp were disposed within trays and supported by the shelves within the drying chamber in spaced relation therewith so that the shrimp were heated by radiant heat from above and below. The shelf temperature at the beginning of the heat cycle was approximately 150°C and the pressure within the drying chamber was maintained at approximately 2,400 microns. As indicated by the graph of the product temperature, the temperature of the product at the beginning of the cycle was below freezing.

As seen from the graph of the moisture loss, the largest portion of the moisture in the shrimp is sublimed during the first half of the cycle. Thus 90% of the moisture is removed from the shrimp in the first 6½ hours. During a normal cycle, the next 5% of the moisture would require about 9½ hours to remove the same. At about this point in the cycle, the pressure within the chamber was doubled from 2,400 to 4,800 microns and maintained at the latter pressure for the remainder of the cycle. Likewise, the shelf temperature was reduced from about 150° to 65°C. At this point, it will be noted that the product temperature decreased substantially and the rate of moisture removal increased. Accordingly, the abovemen-

FIGURE 3.8 INCREASING PRESSURE AFTER PREDETERMINED PORTION OF CYCLE IS COMPLETED

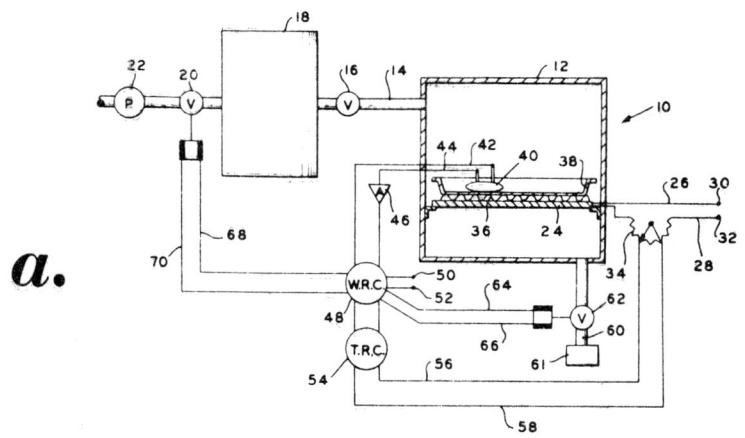

a.

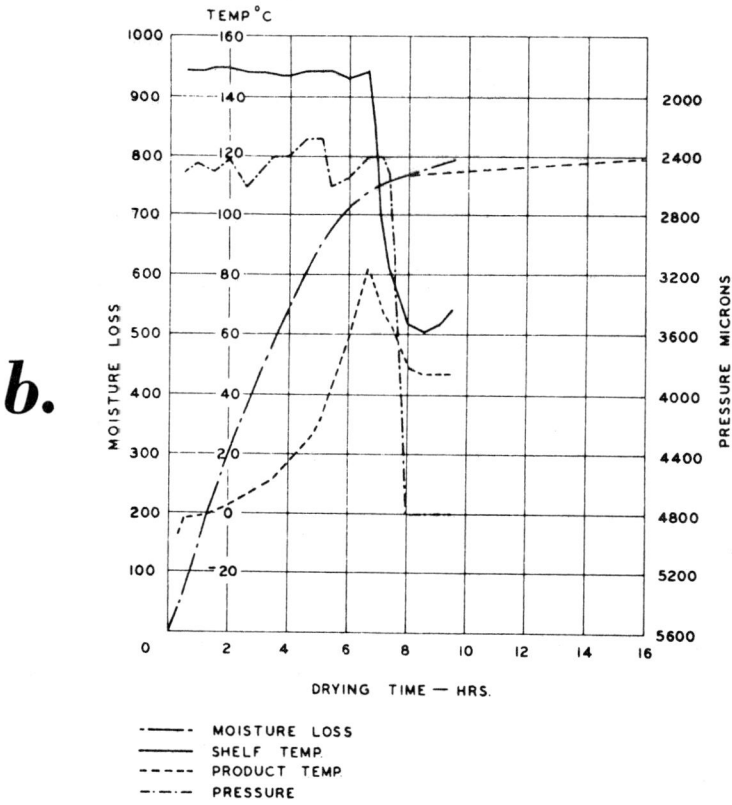

b.

- — · — MOISTURE LOSS
- ——— SHELF TEMP
- – – – – PRODUCT TEMP
- — · · — PRESSURE

Source: U.S. Patent 3,230,633

tioned changes in the drying cycle enable the shrimp to be dried to a point where only 5% of the moisture remained therein within 9½ hours as compared with 16 hours during a normal cycle.

As the moisture remaining within the shrimp constantly changes, the electrical resistance likewise changes. When the electrical resistance reaches a predetermined point, the resistance coupled to the weight recorder and controller **48** through the wires **42** and **44** enables the controller **48** to open valve **62**, vary the rheostat **34** through the temperature controller **54**, and throttle the valve **20**. The opening of the valve **62** enables a gas from source **61** to flow into the drying chamber.

The entry of gas into the drying chamber increases the pressure therein. The gas introduced into the drying chamber is removed at the same rate that it enters the chamber by the pump **22**. Thus, gas is flowing through the drying chamber and heating the shrimp by convection. The shelf temperature need not be reduced. However, such reduction in the shelf temperature prevents overheating of the shrimp, and prevents boiling off fats and other nutrients in the article being freeze dried. Merely throttling the valve **20** will increase the pressure within the drying chamber. However, such throttling of the valve is preferably performed in conjunction with the introduction of gas into the drying chamber.

Pressure Modification via Introduction of Incondensable Gases to Achieve Temperature Control

L.M.A. Rieutord; U.S. Patent 3,192,643; July 6, 1965; assigned to Societe d'Utilisation Scientifique et Industrielle du Froid Usifroid, France provides an apparatus for regulating freeze drying operations in which the substance to be treated is placed in an enclosure in which it is possible to create a vacuum and on a support whose temperature can be varied, the water vapor leaving the substance being eliminated by condensing on refrigerator coils located in the enclosure. The apparatus is such that a continuous transfer of the water vapor from the substance to the trap is maintained. The total pressure prevailing within the enclosure is modified as a function of the variations in a given property of the substance, this pressure modification being achieved by introducing in the enclosure controlled amounts of incondensable gases (such as air or nitrogen).

The process is carried out in the apparatus of Figure 3.9 in the following manner. The substance **S** placed in the containers **5** is previously frozen. A container containing the substance is thereafter placed on the plate **P** within drying chamber **C**. The thermometric device or devices **T** placed in the substance in one or several chosen containers, is or are then connected to the circuit of the galvanometer **28**. The refrigerator **M** was started up so as to bring the condenser **G** to the required temperature, which is generally about −50° to −60°C.

The galvanometer is so adjusted that the contacts **63** and **64** separate when the temperature of the substance reaches the maximum chosen value, t, which varies with the type of substance to be treated and the chosen sublimation rate. The pump **V** is then actuated so as to create a vacuum in the drying chamber **C** and the conductors **37** and **65** are connected to the AC supply.

As the temperature of the substance is lower than the extreme chosen temperature t, the frame of the galvanometer urges the index **62** in the direction for

FIGURE 3.9: PRESSURE MODIFICATION TO ACHIEVE TEMPERATURE CONTROL

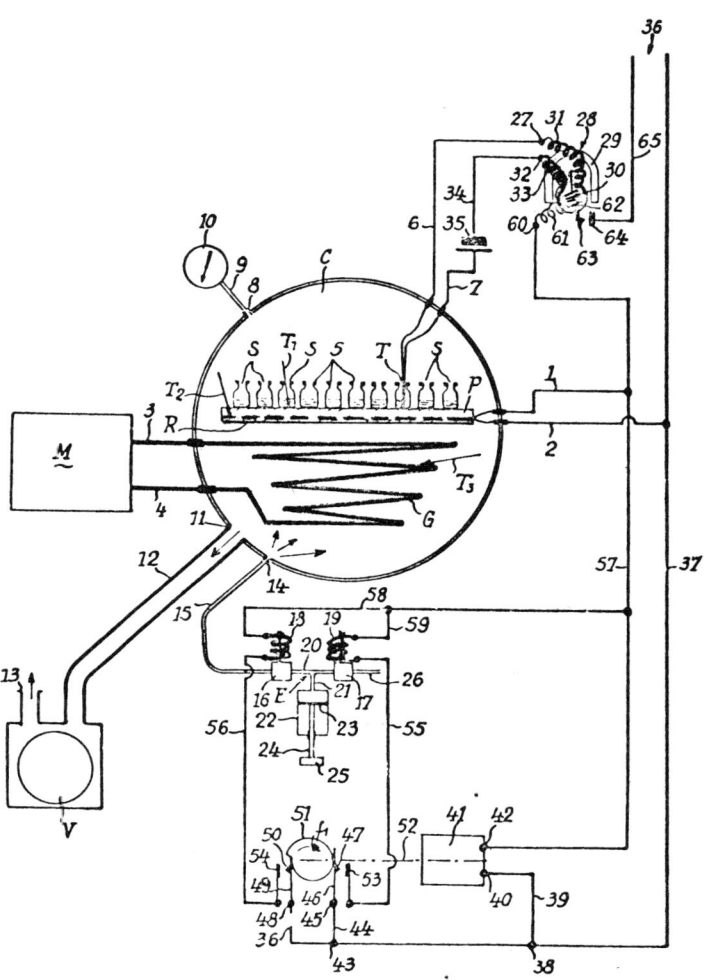

Source: U.S. Patent 3,192,643

applying the contact **63** against the contact **64**. The actuating circuit is therefore closed, the motor **41** is supplied with current by the conductors **37** and **57** and rotates the cam **51** in the direction of arrow f_1 and the rise in the cam brings the contact **47** against the fixed contact **53**, the contacts **50** and **54** being then separated and the valve **16** closed. The current then passes through the solenoid **19** of the valve **17** (by way of conductors **37** and **44**, strip **46**, contacts **47** and **53**, conductor **55**, solenoid **19**, conductors **59, 58, 57, 61**, index **62**, contacts **63, 64,** and conductor **65**) and this solenoid opens the valve and allows the air coming from the tube **26** to fill the variable volume chamber **E**. The volume of this chamber, namely the volume available for the gases, is chosen according to needs by acting on the regulating knob **25** which determines the position of the piston **23** in the cylinder **22**.

As the cam continues to rotate, the strip **46** drops down the step in the cam and the contact **47** moves away from the contact **53** and causes closure of the valve **17**. At this moment the strip **49** is acted upon by the rise or boss of the cam and comes into contact with the contact **54** so that the current is then carried by the solenoid **18** which causes the valve **16** to open. The gas contained in the chamber E passes through the pipe **15** and enters the chamber C by way of the aperture **14**. This arrival of a certain amount of gas increases the total pressure in the chamber C but it will be clear that the transfer of the steam from the substance to the trap constituted by the condenser G has not been modified.

When the strip **49** in turn drops down the step in the cam **51**, the contacts **50** and **54** are separated and the positions shown in the diagram are resumed. The profile of the cam is so designed that at least one of the two valves **16** or **17** is always closed so as to avoid any accidental supply of gas to the chamber C. The motor **41** continues to operate and the procedure of admission of the incondensable gases in the chamber C is reproduced in a regular manner. By this process it is possible to supply the chamber C with puffs of gas of given volume and, by modifying this volume in regulating the volume of the chamber E, it is possible to more or less act on the total pressure prevailing in the chamber C.

If it is desired to introduce an incondensable gas other than air (for example nitrogen), the tube **26** is connected to a suitable source of this gas. While the incondensable gases are admitted into the chamber the resistance of the plate is supplied conductors **57** and **37**. This resistance therefore raises the temperature of the plate simultaneously with the introduction of incondensable gases. As the supply of the gases raises the pressure prevailing in the chamber, the temperature of the substance rises owing to the improvement in the heat transfer from the plate to the substance due to increased convection phenomena. The simultaneous rise in the temperature in the plate P therefore increases this effect by an additional supply of heat and causes an even more rapid rise in the temperature of the substance.

While the actuating device is supplied with energy the temperature of the substance rises. When it reaches the extreme fixed value t the intensity of the current supplied by the battery **35** in the galvanometer circuit reaches the value at which the frame **30** and the index **62** are driven in such direction that the contacts **63** and **64** are separated. At this moment the circuit of the actuating device is opened and current no longer passes through the motor **41** nor in the valves **16** and **17** and in the resistance R. Consequently incondensable gas is no longer supplied, the pressure drops in the chamber C under the effect of the pump V and the thermal exchanges between the substance and the plate are slowed down.

The sublimation, in continuing, takes the heat necessary thereto from the substance, the temperature of which drops. This temperature drop is still more accelerated if the supply of heat to the plate P is stopped. When the temperature of the substance is once more lower than the value t, the frame of the galvanometer is driven in the opposite direction and returns the contacts **63** and **64** to the position in which they bear against one another. This alternating procedure occurs in this way during the entire sublimation operation which is therefore perfectly regulated and controlled throughout the operation.

Pressure Controlled by Temperature of Material Being Dried

Investigations have shown that the quantity of heat supplied to the material to be dried for a given, constant temperature of the heating elements or heatable supports depends to a very high degree upon the pressure inside the drying chamber, i.e., on the number of molecules per unit volume. The larger the number of molecules present per unit volume, the larger is the quantity of heat supplied to the material to be dried from the heating elements or heatable supports by direct transport of heat (convection).

K. Neumann; U.S. Patent 3,077,036; February 12, 1963; assigned to Leybold-Hochvakuum-Anlagen GmbH, Germany provides a method and apparatus for expediting the drying process, in which the pressure in the vacuum drying chamber is controlled as a function of the temperature of the material being dried, in such a manner that each batch is maintained at substantially a predetermined optimum drying temperature which is so selected that when the material is kept at such temperature, the heat flow to the material corresponds to the cold due to evaporation which is produced by the sublimation, which cold results at the maximum permissible temperature of the frozen material.

The pressure in the drying chamber is no longer sought to be kept as low as possible, which result was heretofore obtained only by economically impractical means; instead, this pressure is allowed to be so high, depending on the temperature of material to be dried, that the heat transfer due to the convection contributes to the total heat flow to the ice-containing material. Another way of expressing this would be as follows: the sublimation of the ice from the material to be dried is controlled by the pressure in the vacuum chamber in such a manner that the cold due to evaporation is not greater than the heat supplied by the heating device. In this way, the frozen material will not be cooled unnecessarily, but remains at its maximum permissible upper temperature. In this way, the drying time is approximately half of that required in heretofore known freeze drying processes.

The effective cross sectional area of the conduit means which place the drying chamber and the condenser in communication with each other is varied, such as by a suitable throttling valve. The change in the surface temperature thus acts in the known manner to influence the partial pressure of the water vapor.

During the drying process the temperature difference between the surface of the condenser and the ice-containing core changes constantly, because during the progressive drying the thickness of the layer of already dry material continuously increases and thus offers a greater resistance to diffusing, so that a greater temperature differential must be maintained if the material to be dried is to be maintained at the optimum drying temperature.

By controlling the throttling valve a low partial pressure due to the water vapor can be obtained, but due to the throttling effect, this pressure is not equalized with that prevailing in the drying chamber so that in each of the two chambers there is a certain pressure differential of the partial pressure of the water vapor, depending on the settling of the throttling valve.

FIGURE 3.10: PRESSURE CONTROLLED BY TEMPERATURE OF MATERIAL BEING DRIED

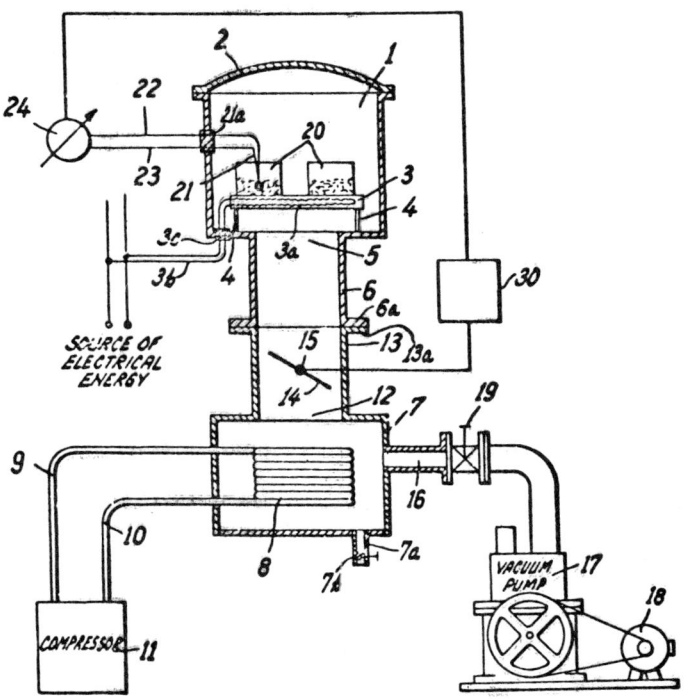

Source: U.S. Patent 3,077,036

The drying apparatus comprises means forming an hermetic enclosure, a condenser arranged in the enclosure and occupying part of the space therewithin, and means for varying the cross sectional area between the part of the space occupied by the condenser and the remainder of the space within the enclosure. Such apparatus is shown in Figure 3.10.

The drying apparatus comprises a vacuum drying chamber **1** having a removable cover **2**, within which drying chamber there is arranged an electrically heated support plate **3** resting on legs **4**. The drying chamber is formed with a circular opening **5** which is the mouth of a downwardly directed connecting conduit **6** having a flanged end **6b**.

The drying apparatus further includes a chamber **7** which houses a condenser coil **8** whose supply and discharge tubes pass through the walls of the chamber **7** and communicate with a compressor **11**. The chamber **7** is formed with a circular opening **12** which is the mouth of an upwardly directed connecting conduit **13**, the latter having a flanged end **13a** connected to the flanged end **6a** of the downwardly directed conduit **6**, thus placing the drying chamber **1** and the condenser chamber **7** in communication with each other to form a single hermetic enclosure.

Within the conduit 13 there is arranged an adjustable butterfly or throttling valve 14 which is mounted on an axle 15 that extends exteriorly of the conduit 13 through an appropriate sealing (not shown). Thus, upon rotation of the axle, the cross sectional area of the conduit which places the chambers 1 and 7 in communication with each other may be varied. A suitable actuating element (not shown) may be attached to the axle for this purpose.

The condenser chamber 7 also has a connecting stud 16 which places this chamber in communication with a vacuum pump 17 driven by a motor 18, a closure valve 19 being included for closing the communication between the chamber 7 and the vacuum pump.

The support plate 3 supports the containers 20 which hold the material to be dried. In order to measure the temperature of the ice-containing center or core part of the material, there is provided a thermocouple 21 whose leads 22 and 23 pass through the wall of the chamber 1 by way of a sealing bead 21a and are connected to a measuring instrument 24. The instrument is connected to the axle 15 of the butterfly valve 14 by way of an automatic control mechanism 30 so that the position of the valve is regulated automatically in response to the temperature of the ice-containing core of the material being dried.

Adjusting Feed Rate to Maintain Constant Pressure Within Drying Chamber

J.-P. Bouldoires and J. Bally; U.S. Patent 3,882,610; May 13, 1975; assigned to Societe d'Assistance Technique Pour Produits, Nestle SA, Switzerland provide a freeze-drying process, comprising continuously feeding frozen product into a freeze drying chamber and withdrawing dried product therefrom, in which the pressure is measured at at least one point within the chamber and the rate of feed of frozen product into the chamber is continuously adjusted to maintain the measured pressure substantially constant. The process is described with reference to Figure 3.11.

The apparatus comprises a vacuum drying chamber 1 (means for creating a vacuum are not shown) into which the frozen product is introduced by a feeding device 2, which comprises an inlet air-lock 3 and a hopper 4 supplying a feed screw 5 driven by a motor 6. Frozen product 7 is delivered by the feeding screw onto a heated vibrating conveyor 8. Dried product is delivered by the conveyor into a discharge hopper 9 which terminates into an outlet airlock 10.

FIGURE 3.11: ADJUSTING FEED RATE

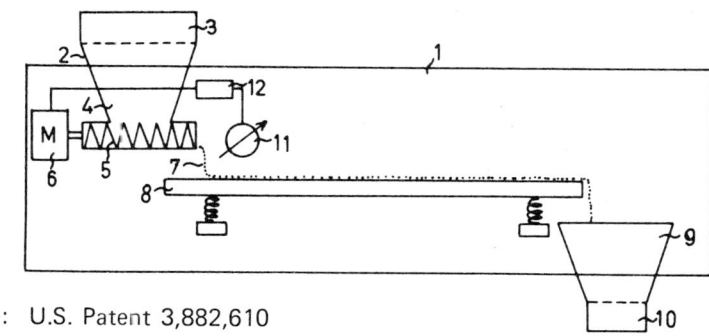

Source: U.S. Patent 3,882,610

The total pressure in the drying chamber 1 is measured by a diaphragm gauge 11 located close to the point of entry of the product into the chamber. The gauge emits a signal which is used for varying the speed of rotation of the feed screw by way of an electronic control means 12 which regulates the speed of rotation of the motor 6 driving the screw.

Preferably, the control means includes a governor for limiting the maximum speed of rotation of the motor so that the screw is not rotated at too high a speed in case of a product supply failure, which may be caused by bridging in the hopper. Rotation of the screw at an excessive speed could lead to a surge of a large amount of product when the supply is restored.

In operation, the speed and temperature of the conveyor are maintained constant, at optimum values determined by preliminary trails. The vacuum source, including condensers and pumping means for noncondensables, are operated conventionally. The condenser temperature is chosen having regard to the nature of the product.

The optimum value at which the pressure is held substantially constant, that is the reference pressure p_c, is determined as a function of the nature of the product, the solvent, and of the desired final moisture content. This reference pressure p_c as well as an upper limiting value of the speed of rotation of the motor, is set in the controller 12 before the start of the drying operation.

A diaphragm gauge is preferred for measuring the pressure, because such a gauge functions independently of the nature of the gas and thus always measures the total pressure. Moreover, such a gauge gives a linear response as a function of the total pressure, since the diaphragm is not displaced by more than a small amount (a few microns) and its response time is extremely short (of the order of 0.01 second).

For example, it has been observed with the control system shown in Figure 3.11 that the speed of rotation of the feed screw oscillates periodically about an approximately constant average speed, corresponding to the reference pressure, and that these oscillations coincide, but in the opposite sense, with the regular pressure oscillations caused by the fact that the feed screw delivers the product in packets.

A recording made during a drying operation with the feed screw rotating at 4 revolutions per minute shows, for example, that the pressure measured by the control gauge 11, a diaphragm gauge, oscillated at the frequency of 4 cycles per minute between 0.20 and 0.23 torr, with the pressure determined by a safety gauge of the thermocouple type, remained constant: Recordings of the speed of rotation of the feed screw, as determined by a dynamo tachometer connected to the shaft of the screw, also show that the speed varies, but inversely, with the pressure variations determined by the diaphragm gauge.

OTHER METHODS

Determining Temperature of Ice Within Material Being Dried

In freeze drying the actual drying operation is carried out in two stages. First the ice which has formed within the material is sublimed; this causes the formation of a dry outer zone of increasing thickness surrounding a progressively shrinking core of ice. When the ice core completely disappears, the rough drying stage is complete. The fine drying stage then serves to remove the water bound by absorption in order to give the desired residual moisture content. The latter will vary according to the kind of material being dried. For a freeze drying process to succeed, it is important that the temperature be kept within certain limits.

According to the physical laws involved in the drying process the speed and with it the economy of the drying operation increases with the temperature maintained within the material being dried. In other words, the greater the temperature-dependent difference of vapor-pressure between the material being dried and the surface of the ice condenser can be made, the more rapidly will the drying be completed. Now, if it is desired to obtain as high as possible a difference of pressure between the material and the surface of the ice condenser, the temperature must be maintained at the maximum permissible temperature which the inner core of the material, which is still embedded in ice can withstand without sustaining any damage. This requires an accurate knowledge of the temperature actually within the parts of the material still enclosed by the ice. The already dry outer marginal zones require far less consideration, since the temperatures tolerable in these zones are in any case considerably higher.

The process of *K. Neumann; U.S. Patent 2,994,132; August 1, 1961* permits the temperature of the ice in a substance being freeze dried to be measured with accuracy during the drying process. This is accomplished in a freeze drying apparatus by successively shutting-off the drying chamber for a short period of less than one minute, preferably for from 2 to 10 seconds, from the means for removing or condensing the water vapor and from the vacuum pump, the shutting-off being so effective that the flow of water vapor to the condenser is largely stopped, or preferably so to an extent of at least 90%; furthermore by measuring the pressure in the drying chamber by means known per se at the end of the period of shutting-off, this period being extended to at least the end of the nonlinear rise in pressure; and finally, by determining the temperature of the ice in the material being dried from the pressure measured at the end of the shutting-off period on the basis of the vapor pressure curve of ice.

In the case of the pressure gauge used, there exists the secondary requirement that its time of response must be short compared to the period of shutting-off. An ionization manometer of the type utilizing a radioactive exciting substance may be used with advantage as a gauge.

The apparatus is schematically represented in Figure 3.12. Within the vacuum chamber **1** there is contained the material **2** to be dried to which heat from a suitable means of heating **3** can be supplied. The material to be dried may be introduced into the vacuum chamber through an opening which can be closed by a cover **4**. Within the vacuum chamber there is also provided a pressure gauge **5** of the type having a short time of response which is connected to a suitable pressure indicator and to a means of power supply **6**. A chamber **7** adjoining the vacuum chamber contains an ice condenser which may, for example, be in the form of a cooling coil **8**. Chambers **1** and **7** may be separated from each other by a rapidly closing valve, for example, a valve disk **9** which can be shifted by a rod **10** by means of an electromagnetically, hydraulically, pneumatically or otherwise operated actuator **11** of known type. In the closed state (as indicated by broken lines), the disk **9'** bears against the valve seat **12**. Chamber **7** is connected through a pipe **13** with a vacuum system **14** of known type.

The automatic actuation of the apparatus and control of the drying procedure described above may be effected by means of a programming apparatus **15** which receives the measured data from the pressure indicator **6** and causes actuation of the valve **9, 10, 11**. The controls of the heating etc., are indicated schematically by the arrows **16** and **17**.

The apparatus makes it possible to dry a material under optimum conditions of heat supply with the highest possible speed, and at the same time to ensure that the substance being dried is protected against excessive heating.

FIGURE 3.12: DETERMINING TEMPERATURE OF ICE WITHIN MATERIAL BEING DRIED

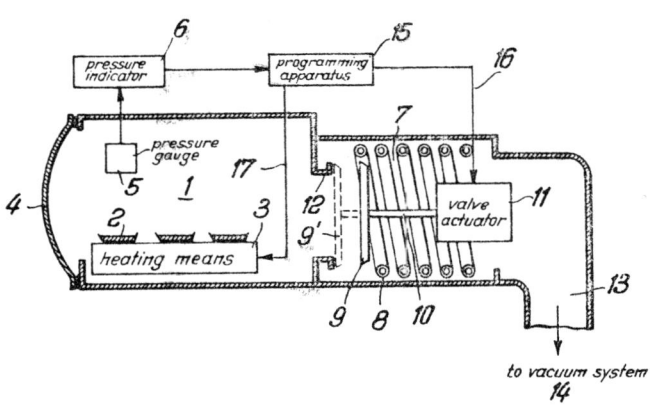

Source: U.S. Patent 2,994,132

Throttling Flow of Water Vapor from Container

H.P. Thompson; U.S. Patent 3,318,012; May 9, 1967; assigned to Vickers Limited, England provides a method of freeze drying material, including the steps of placing in a container frozen material to be dried, subjecting the frozen material to vacuum, supplying heat to the material in the container while maintaining the material under vacuum to effect drying of the material in a manner such that the water contained in the material is drawn off directly from the ice state to the water vapor state, and adjusting the extent to which the flow of this water vapor from the container is throttled thereby to vary the thermal conductivity of the water vapor.

The drying apparatus shown in Figure 3.13 includes a container in the form of a tray **1** having a rectangular base **2** formed by a fine mesh of expanded or perforated metal, or a wire mesh. The edges **3** of the mesh base are secured to bent-inward bases **4** of rigid upstanding imperforate flanges **5** which form the sides of the tray, all four flanges being of the same height. The bases of all four flanges rest on a lower heater plate **6** mounted in a vacuum-tight chamber **9**. An upper heater plate **7** also mounted in the chamber is disposed above the tray. The upper edges **5A** of the flanges of the tray terminate beneath the upper heater plate **7** and this plate extends beyond all these edges.

The heater plates **6, 7** are disposed in a vacuum-tight chamber **9**, each supported by a pair of links **10**, respectively pivotally connected to the ends of the plate. The two rods **10** supporting adjacent ends of the plates are pivotally connected to one or the other of a pair of spaced ball-crank levers **11**, as shown, which levers are respectively pivotally mounted on spaced supports **12** within the chamber.

The heater plates hang freely in spaced relation from the lower arms of the ball crank levers with the tray resting on plate **6**. The upwardly-extending arms of levers **11** are interconnected by a link **13** and the arm of one lever is also connected, by a link **14** to the piston rod **17** of a double-acting hydraulic cylinder and piston assembly **15** having a piston **16** to which the piston rod is connected. The assembly is connected to a hydraulic fluid pump **18** via a hydraulic fluid flow control unit **19**.

The heater plates are thus mounted on the movable supports formed by the links **10** for movement, by the linkages made up of the parts **11, 13, 14** and **17**, in the up and down direction, that is, the plates are movable towards and away from one another by assembly **15** so that the width of the gap between the upper edges of the flanges **5A** and the lower surface of the upper heater plate **7** can be varied, to in turn vary the throttling of the vapor flow from the container **1**. The plates are electrically heated, for example by electrical heating elements **20** embedded in the plates.

In operation of the apparatus the tray **1** is removed from the vacuum chamber and frozen material **8** to be freeze dried, for example a frozen foodstuff such as a frozen leafy vegetable, is placed in the tray **1**, the amount of material **8** placed therein being such that the upper edges **5A** of the flanges **5** of the tray **1** are above the upper surface of the material **8**. The tray **1** is then placed in the chamber **9**, the heater plates **6, 7** set so that the gap between the upper edges **5A** of the flanges **5** and the lower surface of the upper heater plate **7** is of a desired width, the door **9B** closed, and the vacuum pump **30** set in operation to evacuate the chamber to a very low pressure (e.g., 1 torr or below).

This results in the temperature of the frozen material **8** dropping well below the freezing point of the material **8**. When a desired minimum temperature of the material **8** has been reached, this being measured by the temperature measuring devices **27, 28**, the low pressure in the chamber is maintained and the heater plates **6, 7** are heated to a desired temperature.

The heater plates **6, 7** radiate heat onto the material **8** in the tray **1**, heat from the lower heater plate **6** passing through the mesh base **2** of the tray **1**, and sublimation takes place, the water present in the material **8** being driven off directly from the ice state to the water vapor state without passing through the liquid phase. This water vapor passes into vacuum-tight compartment **9A** to condense to ice on the condenser plates **31** that are in this vacuum-tight

FIGURE 3.13: THROTTLING FLOW OF WATER VAPOR FROM CONTAINER

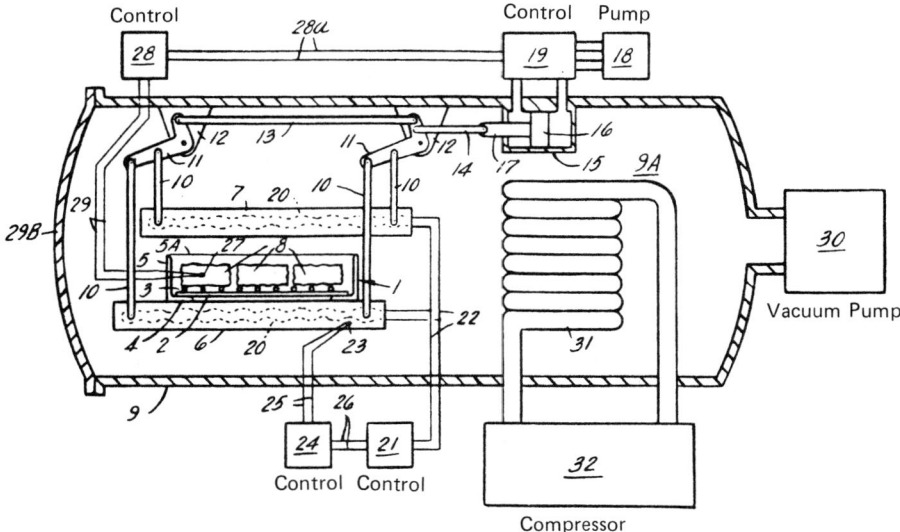

Source: U.S. Patent 3,318,012

compartment. The water vapor passes only through the gap between the upper edges **5A** of the imperforate flanges **5** and the lower surface of the upper heater plate **7**, the mesh base **2** being closed by the lower heater plate **6**. In addition to the radiant heat received by the material **8** being dried, the water vapor surrounding the material **8** acts as a heat transfer medium transferring heat from the heater plates **6, 7**, to the material **8**, the thermal conductivity of the water vapor surrounding the material **8** being dependent on the vapor pressure with the tray **1**.

As drying proceeds, the ice surface of the frozen material **8** recedes and a dry layer of material is formed around the still frozen material. This dry layer, which gradually increases in thickness, is of poor thermal conductivity and it will be appreciated that the resistance offered by the dry layer to the transfer of heat to the still frozen material by conduction through the dry layer of material increases as the dry layer becomes thicker. To offset the effect, on drying, of this increasing resistance, the heater plates **6, 7** are moved to reduce the width of the gap between the flanges **5A** and the upper heater plate **7**, thereby throttling the flow of water vapor through this gap which has the effect that the vapor pressure within the tray increases whereby the thermal conductivity of the water vapor surrounding the material being dried also increases.

The thermal conductivity of the water vapor in the container may be varied by adjusting the extent to which the flow of water vapor from the container is throttled. The temperature of the material being dried in the container is measured, and the heating of the material and the throttling effect on the vapor are varied in accordance with these temperature measurements, thereby to control the temperature and heating of the material in the container.

Removal of Thin Surface Layer as It Dries with Scraping Means

The process of *L.A. Hernandez, Jr.; U.S. Patent 3,448,527; June 10, 1969* relates to a system and apparatus employing heat transfer by means of direct heat radiation to the quick-frozen product and involving successive removal of thin dried layers of the product to effectively minimize the occurrence of heat transfer by conduction, minimize the period of time during which the dried product is subjected to heat, minimize the degree of heat to which the dried product is subjected, and increase the effective sublimation rate so that the total time involved in the freeze drying process is greatly reduced for improved efficiency and economy. The operation is described with reference to Figure 3.14. A liquid product, such as fresh orange juice or coffee extract is supplied via the conduits **22** to each of the freezing-sublimation trays **14** until each is full to the level permitted by the exit port **48**. Upon completion of the freezing step, the refrigeration system **16** is turned off and, when the appropriate vacuum has been achieved, the hot fluid supply **26** is turned on for continuously supplying heat energy to the heat radiators **24** and **142**.

As heat is radiated from the heat radiators **24** onto the exposed top surface **156** of the frozen product **148**, the ice sublimes therefrom and its vapors are driven, by difference in vapor pressure, toward the condensing plates **30** for collection thereon in the form of ice, later to be melted during defrosting for collection by the drain pan **32**.

As a relatively thin and horizontally uniform portion or layer at the upper surface **156** of the frozen product becomes dried or dehydrated in accordance with the sublimation occasioned by the received heat flux, the scraper blade driving mechanism is actuated for continually driving the scraper blade **66**. As the edge **70** of the scraper blade traverses the tray **14** in intimate contact with the upper surface **156**, the dried product layer is scraped and removed thereby in substantially powder or crystal form, such powder or crystals being indicated at **158** in Figure 3.14a during the middle of the scraping step, and is dumped out the open end of the tray **14** for reception by the collector bucket **118**. Thus, a new ice front is exposed as an effectively new top surface of the frozen product for receiving the heat flux from the heat radiators **24** directly in the form of radiated heat rather than conducted heat through the previously dried (and now removed) thin layer.

The foregoing steps of continuous heat radiation and continual successive scraping are maintained until the entire frozen product has been dried with the trays **14** and removed incrementally by the scraping action into the bucket **118**. Due to the presence of water vapor within the vacuum chamber **10** during the foregoing sublimation step, the dried product within the bucket undergoes final drying and is maintained in the dry state and prevented from water vapor absorption and/or adsorption by the final heat radiator **142**. When all of the dried product resides in the bucket, the vacuum may be broken, the door **124** opened by any convenient manual or automatic mechanical means, and the dried product is dumped from the bucket into the hopper **126** and the product handling system **128**.

The scraper means are illustrated in Figure 3.14c. A scraper blade driving mechanism is provided for driving the scraper blade **66** along a generally rectangular path from right to left while scraping the frozen product, then upwardly for a

FIGURE 3.14: REMOVAL OF THIN SURFACE LAYER AS IT DRIES WITH SCRAPING MEANS

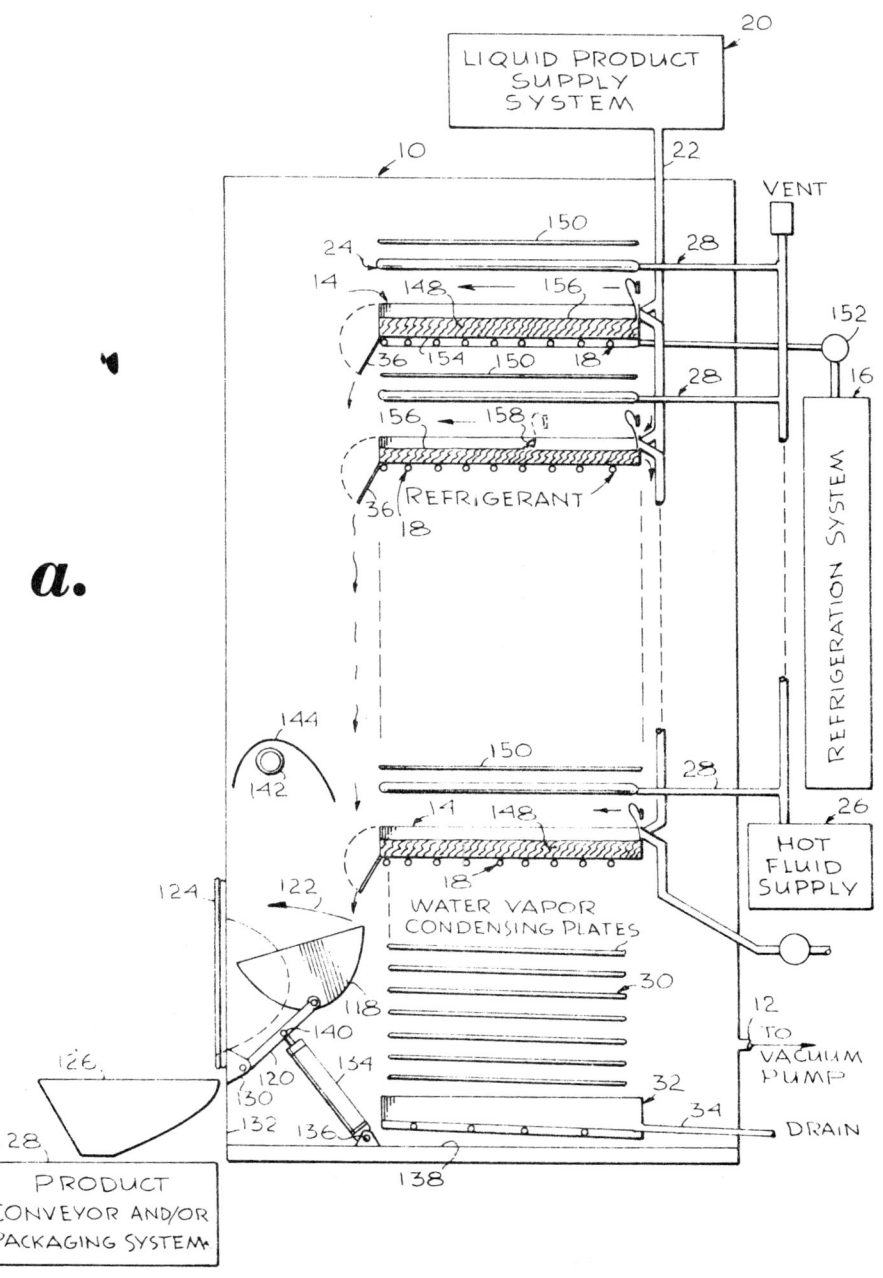

a.

Side Elevational View of Apparatus

(continued)

FIGURE 3.14: (continued)

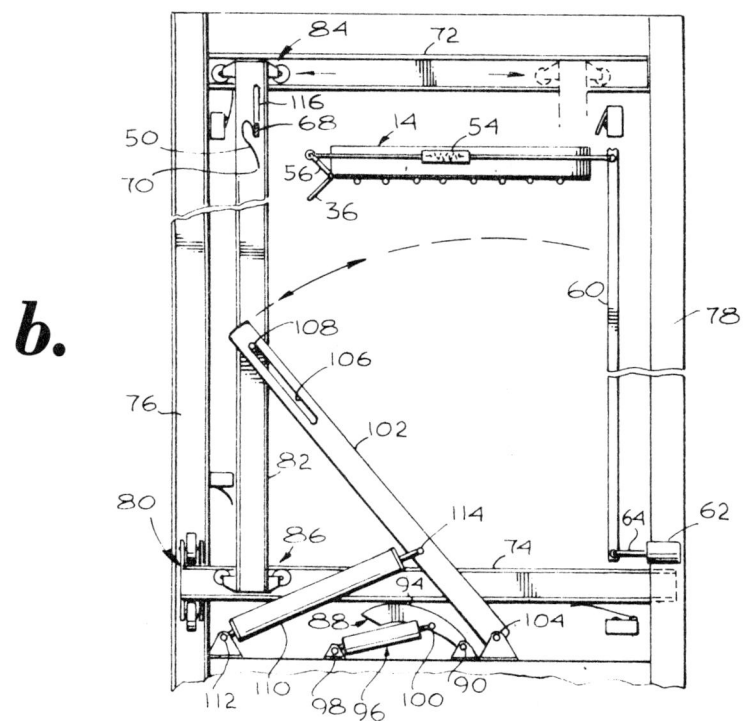

b.

Scraper Means

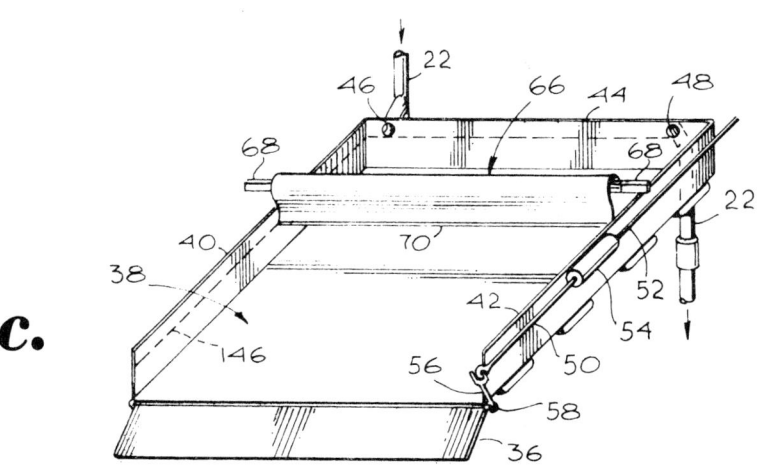

c.

Freezing-Sublimation Tray and Scraper Means

Source: U.S. Patent 3,448,527

slight distance to assure avoidance of return scraping, then from left to right for nonscraping return of the scraper blade to its starting position, and then downwardly until contact is made by the blade's bottom edge **70** with the frozen product for commencement of the scraping motion from right to left again. Only the mechanism on the near side of the chamber **10** is illustrated.

A pair of horizontal frame members or tracks **72** and **74** are provided to guide and support horizontal movement and provide vertical lift, each being mounted at its opposite ends in a corresponding pair of vertical tracks **76** and **78** as by vertical travel support rollers, one such mounting being indicated generally at **80**, whereby the horizontal frame tracks are movable upwardly and downwardly together. A vertical support pillar member **82** is mounted at its opposite ends onto each of the horizontal frame members by horizontal travel support rollers **84** and **86**, respectively, whereby the vertical support member **82** is guided in its horizontal movement back and forth by the horizontal track members and also moves vertically with such track members when the latter are vertically driven.

Vertical movement of elements **72, 74** and **82** is accomplished by a crank cam **88**, one end of which is pivotally mounted at **90** to a stationary element of the chamber, a curved cam surface **94** at the other end portion being in slidable abutment with the lower horizontal support track **74** whereby pivotal movement of the lower frame track and, consequently, the remainder of the aforedescribed frame and support members, such pivotal movement being caused by pressure actuation of hydraulic cylinder **96**, one end of which is pivotally connected at **98** to a stationary element, with the piston connecting rod pivotally connected to the crank cam **88** at a fulcrum point **100**.

Horizontal movement of the vertical support member **82** is accomplished by a lever **102**, one end of which is pivotally mounted at **104**, the other end of which is provided with a lost-motion slot **106** in slidable and pivotal relationship to a pin **108** projecting laterally from the vertical support member **82**. One end of a hydraulic cylinder **110** is pivotally mounted at **112** and its piston connecting rod is pivotally connected at **114** to the lever **102** for arcuate driving of such lever whereby the support pillar **82** is moved back and forth along the horizontal track members.

The ends of the transverse rod **68** are mounted within a vertical slot **116** in the vertical support member **82** for carrying the slot permitting slidable movement of the rod in the vertical direction to accommodate changes in thickness in the frozen product during the scraping portion of the process cycle.

Continuous Removal of Fully Dried Layer

U. Hackenberg; U.S. Patent 3,234,658; February 15, 1966 provides a freeze drying method and apparatus which reduce drying time, increase heat utilization efficiency and prevent damage to the freeze dried product caused by application of excessive heat. During the freeze drying process an external fully dried layer portion of the frozen product core is continually removed. The installation is described with reference to Figure 3.15.

In the operation of this device the cylindrical surface of material drum **16** is provided with an inner ice layer **27** and an outer frozen core **28** composed of a layer of the material to be freeze dried. These frozen layers can be applied

FIGURE 3.15: CONTINUOUS REMOVAL OF FULLY DRIED LAYER

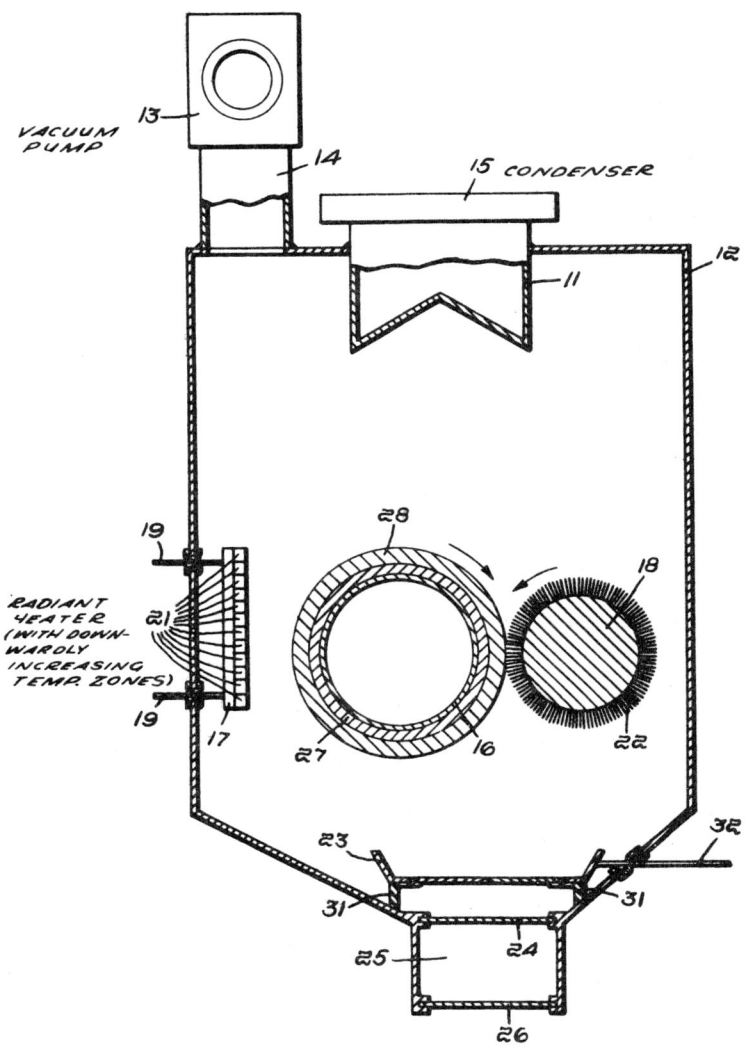

Source: U.S. Patent 3,234,658

with the material drum positioned either within the vacuum chamber **12** or while it is removed therefrom. Then with the vacuum chamber pressure reduced by the vacuum pump device **13**, the water vapor condenser **11** cooled by the refrigeration device **15**, and the plural zoned radiant heat source **17** energized via feedthroughs **19**, the frozen material covered drum **16** and scraping device **18** are energized to provide rotation thereof.

Maintaining Optimum Conditions for Sublimation

As a given exterior segment of the frozen material core **28** passes by the radiant heat source **21** its water content will be removed by sublimation. The water vapor resulting therefrom will move upward in the chamber **12** in response to the pressure gradient created by the water vapor condenser **11** and vacuum pump device **13**. However, upon reaching the scraping device the surface temperature of the material core **28** will have decreased to approximately the equilibrium temperature appropriate to the steam pressure in the vacuum chamber **12**. Therefore during the scraping operation there is produced little or no water vapor which would carry the dry loosened product toward the pumping components **11** and **13**.

Upon contact with the scraping device the external dry product layer on the material core **28** will be removed by the brush-like covering **22**. As a result of the counterrotational directions of the material drum **16** and the scraping cylinder device **18** (which is preferably adapted for faster rotation than the product core **16**) the dry loosened product particles will be compelled downward tangentially to the product core and scraping device and into the product receptacle **23**.

After the product receptacle has been filled in this way it can be automatically emptied through the opened valve **24** into the airlock chamber **25**. The upper vacuum valve **24** will then again be closed to maintain vacuum within the vacuum chamber **12** and the vacuum valve **26** opened to allow removal of the dried product from the airlock chamber.

The purpose of the ice layer **27** in this operation is to prevent damage to the material drum by the scraping device upon removal of the entire frozen material layer **28**.

To provide a greater compelling force for the detached product particles the product receptacle **23** can be insulated from the vacuum chamber by insulating supports **31** and electrically energized via an electrical feedthrough **32** which passes through and is insulated from the wall of the vacuum chamber. The electric field produced thereby will provide an attractive force on the neutrally charged product particles toward the energized product receptacle.

The removal of the dried surface layer on the material core **28** continually exposes a new frozen surface to the applied drying heat. The problems of inefficient heating, heat damage and water vapor flow restriction are substantially eliminated. Furthermore the fact that the drying material surface is moving by the heat source allows for much more uniformity in the heating of localized surface areas.

Proper heating is further enhanced by the provision of the decreasing temperature zones **21**. The frozen surface which has been freshly exposed by the scraping device initially moves into the first zone of radiant heating device **17** which has been adjusted for maximum permissible heat input. As the dry layer begins to form on the product surface, this maximum permissible temperature is reduced because of the lower heat capacity of the dried surface layer. However, as this dry layer is forming it will be passing with the rotating material drum **16** into zones of properly selected lower temperatures.

DEICING

USE OF TUBULAR INTERNAL FREEZE DRYING CONDENSER

Baffle Arrangement for Separation of Condensable and Noncondensable Vapors

The process of *W.B. Ludwig, B.E. Elerath and J.F. Ewald, Jr.; U.S. Patent 3,443,324; May 13, 1969; assigned to General Foods Corporation* relates to a tubular internal condenser for a vacuum freeze drying chamber. The condenser consists of two sets of tubes mounted on the sides of the cylindrical chamber. All of the tubes mounted on one side of a chamber are referred to as a set. The tubes within each row are uniformly spaced in a vertical direction. Tubes in adjacent rows are staggered to provide a tortuous path of travel for vapors within the chamber. The sets of tubes are mounted in tube supports which are shaped to make close contact with the wall of the chamber. The sets of tubes are spaced to provide room for a product load between the sets. The spacing of the tubes should be sufficient to allow for a uniform ice build-up around the tubes.

Adequate protection for the vacuum pumps used in a freeze drying system is provided by minimizing the amount of water vapor which bypasses the condenser and flows with the noncondensable gases to the vacuum pump. This is accomplished by creating a tortuous path of travel for the vapors past the condenser tubes and by mounting the tubes in support plates that are in intimate contact with the chamber wall, thus preventing condensable vapors from bypassing the condenser.

The separation of condensable and noncondensable vapors is further improved by placing a baffle plate between two rows of tubes such that a secondary or clean-up zone is created. The primary condensing zone encompasses the bulk of the condensing tubes in the set. The secondary zone includes those tubes between the wall of the chamber and the baffle plate. The outlet for noncondensable vapors must now be located in the wall of the chamber forming part of the secondary condensing zone. The process is described with reference to Figure 4.1.

Deicing

Figure 4.1a shows a cylindrical vacuum chamber 1 from the front with the door (not shown) open. The front edge of the chamber 2 is a flat, circular flange surface against which the door can be seated to seal the vacuum chamber. Within the chamber are two sets 3 of condenser tubes 4 mounted in tube supports 5 on either side of the chamber. The sets are spaced so that a cart containing product can be moved into the chamber on the tracks 6 located at the bottom of the chamber. The refrigerant inlet line 7 is connected to an internal refrigerant inlet header 8 through which refrigerant flows to the inlet risers. Refrigerant from the condenser tubes passes through the outlet risers 9 which are connected by a refrigerant outlet header 10.

Figure 4.1b shows one set 3 of condenser tubes 4 showing three tube supports 5a, 5b, 5c. Tube support 5a shows how the supports are mounted flush against the chamber wall. A baffle plate 12 is shown mounted vertically between the outer two vertical rows 11 of condenser tubes. The baffle plate extends vertically down from the wall of the chamber past half of the tubes in a vertical row and the plate extends longitudinally to make contact with tube supports 5b and 5c. The tubes between tube supports 5b and 5c behind the baffle plate are in the secondary condensing zone and a duct 13 for removal of noncondensables enters the chamber with an opening through the chamber wall above the secondary condensing zone. At the rear of the chamber 1, the condenser tubes 4 are shown passing into a refrigeration outlet riser 9. The outlet riser is connected to the refrigerant outlet line 17 by an outlet header 10. The refrigerant

FIGURE 4.1: BAFFLE ARRANGEMENT FOR SEPARATION OF CONDENSABLE AND NONCONDENSABLE VAPORS

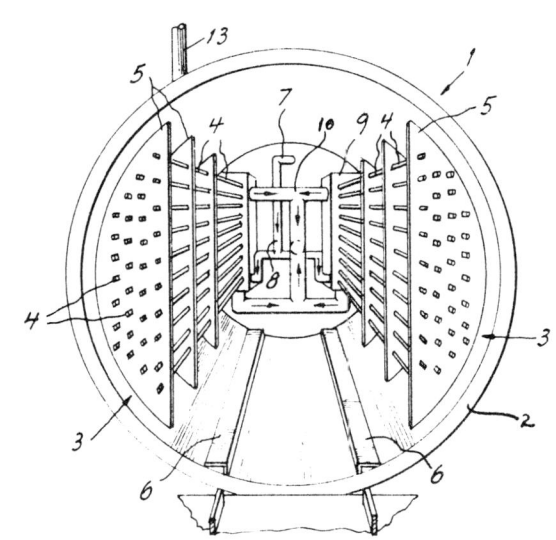

(a) End Perspective View of Vacuum Chamber   (continued)

FIGURE 4.1: (continued)

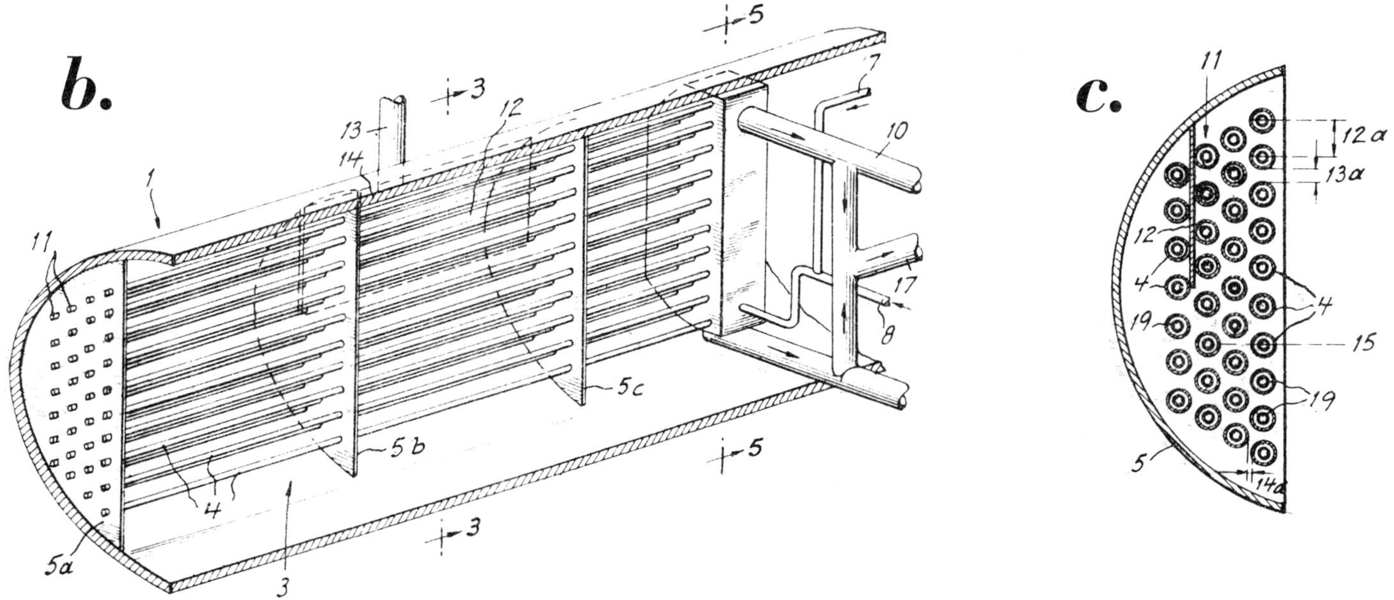

(b) Longitudinal sectional view of vacuum chamber showing one set of condenser tubes and connection to refrigeration piping
(c) End view of tube support

Source: U.S. Patent 3,443,324

inlet line **7** is shown connecting to the inlet header **8** which passes into the outlet riser **9**.

Figure 4.1c shows the arrangement of condenser tubes **4** in vertical rows **11**, the center to center distance **12a** between tubes within a row and the vertical distance **13a** between the outer wall of two adjacent tubes being uniform throughout the row. The rows of tubes are laterally spaced and adjacent rows are staggered such that a horizontal line through the center of a tube in one row **15** will pass through the vertical space **13a** between two tubes in an adjacent row.

The lateral spacing between rows can be uniform throughout the set or the lateral spacing between rows can diminish toward the wall of the chamber. In all cases, the spaces must be of sufficient magnitude to accommodate a desired build-up of ice around the condenser tubes. An ice build-up of about ½" is a preferred thickness for design of the spacing.

Example: In a cylindrical chamber, two sets of condenser tubes were mounted as shown in Figure 4.1b. The condensing tubes were 1" schedule 40 pipe arranged in a set as shown in Figure 4.1c where **12a** is 3". There were 7 rows of tubes and the lateral spacing between the inner rows including the center row is $2^{19}/_{32}$" and the lateral spacing between the outer rows including the center row is $1\frac{7}{8}$". The feed tubes have an internal diameter of $\frac{1}{8}$". Condenser tube supports were spaced at intervals of $3\frac{1}{2}$' and a baffle plate was inserted between the outer two rows of condenser tubes extending down from the wall of the chamber past half of the condenser tubes in the adjacent rows and extending longitudinally between two tube supports. A batch of particulate frozen coffee extract was placed in the drying chamber, the chamber door was closed, the chamber evacuated to a pressure of less than 1 mm of mercury, and heat applied to the frozen extract under controlled conditions in order to vacuum freeze dry the coffee solids.

The condenser efficiently handled water vapor at condensation rates in excess of 700 pounds of water per hour. Ice build-up on the condenser tubes was uniform and the total ice load was in excess of 2,000 pounds. The oil in the vacuum pumps showed no sign of water contamination.

The condenser was used in successive runs for a period of one month. Between drying runs, the pressure in the chamber was increased to atmospheric pressure, the chamber door was opened, the freeze dried product was removed from the chamber and the condenser was defrosted.

After one month, no plugged condenser tubes were encountered. Ice build-up on all condenser tubes was uniform from run to run. No water contamination of the vacuum pump oil was encountered and use of auxiliary equipment for stripping water out of the oil was discontinued.

Discharge End of Tube Elevated from Horizontal

The process developed by *R.F. Bardsley, H.S. Bower and S. Katz; U.S. Patent 3,543,411; December 1, 1970; assigned to General Foods Corporation* is concerned with increasing the condensing capacity of the condenser tubes described above in U.S. Patent 3,443,324.

It was discovered that if the condenser tubes were installed on an angle such that the discharge end of the tube was elevated with respect to the inlet end of the tube the efficiency and capacity of the tubes were significantly increased. This improvement was shown by the fact that the ice build-up around the tube was more uniform and by the fact that for a given flow rate of refrigerant the condensing load could be significantly increased while maintaining a uniform ice build-up around the tubes.

Inclining the tubes as shown in Figure 4.2 is one method of accomplishing the desired goal. Other methods involve inserting a weir at the discharge end of the tubes or shaping the tubes such that there is a rise in a vertical direction at the discharge end of the tubes. Each of the methods and variations thereof will have the effect of increasing the liquid level in a condenser tube and may collectively be referred to as liquid level pipe dams.

FIGURE 4.2: DISCHARGE END OF TUBE ELEVATED FROM HORIZONTAL

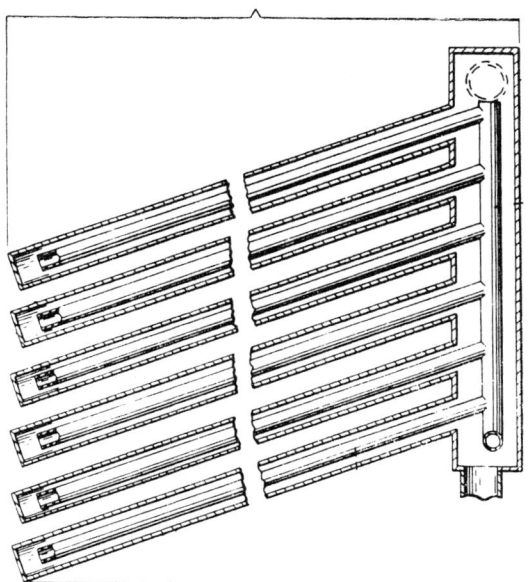

Source: U.S. Patent 3,543,411

The liquid level pipe dam consists of mounting the tubes on an incline such that the discharge end of the tubes is elevated from the horizontal, the elevation being at least one tube diameter and the elevation being restricted such that the condenser tube does not form more than a 45° angle with the horizontal. A preferred incline from the horizontal is from 1 to 10 tube diameters.

Example: In a cylindrical chamber, two sets of condenser tubes were mounted as shown in Figure 4.1b. The condensing tubes were 1" stainless steel schedule

10 pipe arranged in 7 vertical rows. The lateral spacing between the inner rows including the center row was 2¹⁹⁄₃₂" and the lateral spacing between the outer rows including the center row was 1⅞". The tubes within a row were uniformly spaced at center to center distance of 3". The refrigerant inlet and outlet risers were arranged as shown and the feed tubes had an internal diameter of ⅛". The condenser tube supports were spaced at intervals of 3½ feet and a baffle plate was inserted between the outer two rows of condenser tubes extending down from the wall of the chamber past half of the condenser tubes in the adjacent rows and extending longitudinally between two tube supports.

The tubes were installed on an incline such that the discharge ends of the tubes were 1¾" higher than the inlet ends of the tubes. A batch of particulate frozen coffee extract was placed in the drying chamber, the chamber door was closed, the chamber evacuated to a pressure of less than 1 mm of mercury and heat was applied to the frozen extract under controlled conditions in order to vacuum freeze dry the coffee solids.

The condenser efficiently handled water vapor condensation rates in excess of 700 pounds of water per hour. The surface temperature of the condenser tubes was essentially uniform across the entire tube and ice build-up on the tubes was uniform.

OTHER TECHNIQUES

Deicing in Tetra-2-Ethylhexyl Silicate Condensing Medium

E. Thuse; U.S. Patent 3,132,929; May 12, 1964; assigned to FMC Corporation provides a freeze drying process where the ice crystals frozen from the sublimated water vapor during the drying process are continuously and automatically removed from the system before they can reduce the heat transfer efficiency of the condensing medium.

The water vapor is frozen into ice crystals by contact with a film or stream of low temperature condensing liquid falling through the vacuum chamber adjacent the frozen articles. This liquid or condensing medium has a freezing point well below that of water, and the condensing liquid is insoluble in water and is introduced at a temperature substantially below freezing. Thus the ice crystals become entrained with the liquid, and this entrainment permits removal of the slurry of condensing liquid and ice crystals from the drying chamber by an ordinary pump.

Figure 4.3 is a diagram of the system employing a condensing liquid that is not soluble in, and that is immiscible with water. The system includes a vacuum drying chamber **10** made to withstand atmospheric pressure, which can be evacuated by a vacuum pump **11**. Mounted within the drying chamber is a series of hollow article support platforms **12** for receiving metal trays **13** containing the articles **14** to be dried. The usual access door **16** is provided for loading and unloading the drying chamber. The article support platforms are hollow and are connected to a reservoir **17** containing a heating liquid **18** such as water (at 110° to 150°F) which is heated by resistance heating coils **19**. The heated water is directed from a discharge line **21** to a pump **22** that circulates the hot water upwardly through the hollow article supports and back to the reservoir by means of a discharge line **23**.

FIGURE 4.3: DEICING IN TETRA-2-ETHYLHEXYL SILICATE

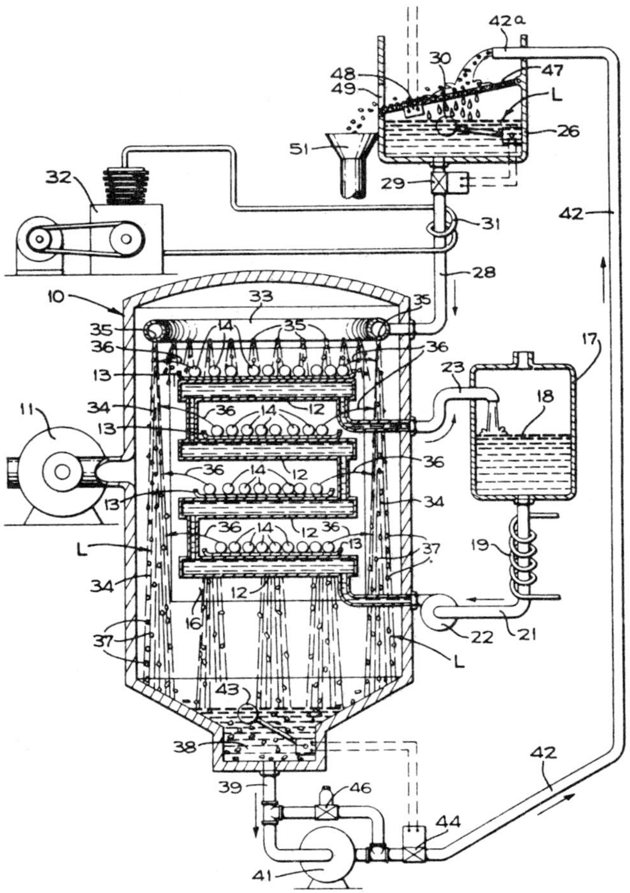

Source: U.S. Patent 3,132,929

A reservoir **26** is provided for the liquid condensing medium **L**, which is insoluble in water, nonvolatile, odorless, and nontoxic, the preferred liquid being tetra-2-ethylhexyl silicate. Other suitable nonsoluble liquids are di-2-ethylbutyl adipate and di-2-ethylbutyl azelate. The specific gravity of these liquids is about the same as that of ice. The condensing liquid leaves the reservoir by means of discharge line **28** under control of an electrically controlled flow control and float assembly **30**. When the liquid level in the reservoir falls to a certain minimum height, the float of the liquid level control assembly causes valve **29** to close by means of the electric circuit, preventing further depletion of the tank. The purpose of the flow control system is to ensure that there is a supply of liquid in the reservoir **26** to provide a gas seal for the drying chamber so that the vacuum pump will not draw in atmospheric air.

The condensing liquid **L** is brought down to a temperature of $-40°F$ by cooling coils **31** surrounding the line **28** and connected to a compressor and refrigerating unit 32. A manifold 33 is mounted in the upper portion of the drying chamber and is formed to direct a stream or sheet 34 of cooling liquid downwardly around the periphery of the article supporting platforms. The manifold has discharge openings 35 spaced about its periphery and, as indicated by arrows 36, water vapor sublimed from the articles 14 finds its way directly to the stream or sheet 34 of condensing cooling liquid L, whereupon the water vapor is frozen into ice crystals 37.

The slurry of cooling liquid and ice crystals drops to the sump 38 of the drying chamber, and is discharged by line 39 connected to a recirculating pump 41. The pump forces the slurry through a line 42 back into the liquid reservoir 26. In order to ensure that there is a supply of liquid in the sump, to provide a vacuum seal at this connection a liquid level float assembly 43 is provided which controls an electrically operated flow control valve 44 in the discharge line 42. When the flow control valve is closed to prevent emptying of the sump 38, the pump merely recirculates the condensing liquid through a spring-loaded bypass valve 46.

The articles are frozen before insertion into the drying chamber and have an initial temperature of about +10°F. The heat supplied from the hot water 18 and conducted through the platforms 12 and the trays 13, soon sublimes the ice in the articles directly into water vapor. The pressure in the drying chamber 10 will be low enough (under 4.5 mm of Hg) so that the operation takes place below the triple point, at which there is no liquid phase, but the ice in the articles sublimes directly into water vapor. Thus the articles are never wetted.

Means are provided at the condensing liquid reservoir 26 to separate or screen the ice particles 37 from the condensing liquid L. Such means are in the form of a screen 47 interposed between the discharge nozzle 42a of the condensing liquid recirculating line 42 and the bottom of the reservoir. The screen may be vibrated by electromagnetic vibrator 48, so that the ice particles 37 are discharged through an opening 49 in the reservoir 26 and collected by chute 51, whereupon they are melted and disposed of as water.

Condensing Water Vapor in Low Temperature Condensing Spray

The work of *D. Eolkin; U.S. Patent 3,210,861; October 12, 1965;* assigned to *Gerber Products Company* relates to a process and apparatus for removing large volumes of water vapor from a dehydration area by condensing the water vapor in a shower of low temperature condensing spray which causes at least a portion of the vapor to solidify.

The process is described with reference to Figure 4.4. From drying chamber 10, an exhaust outlet 11, having condensation chamber 12 therein, leads to a vacuum pump 14. A spray nozzle 15 at the top of chamber 12 permits the introduction of a finely divided mist of material which will fall to the bottom of the chamber by means of gravity. The material is then collected and withdrawn through an outlet 16 by means of a pump 17.

For a commercial food freezing and dehydration system, a condenser oil serves adequately as the carrier material to spray through the chamber 12. Condenser oils have low vapor pressure at the temperature of condensation in the chamber, are not miscible with water and present a very low solubility characteristic to

FIGURE 4.4: CONDENSING WATER VAPOR IN LOW TEMPERATURE CONDENSING SPRAY

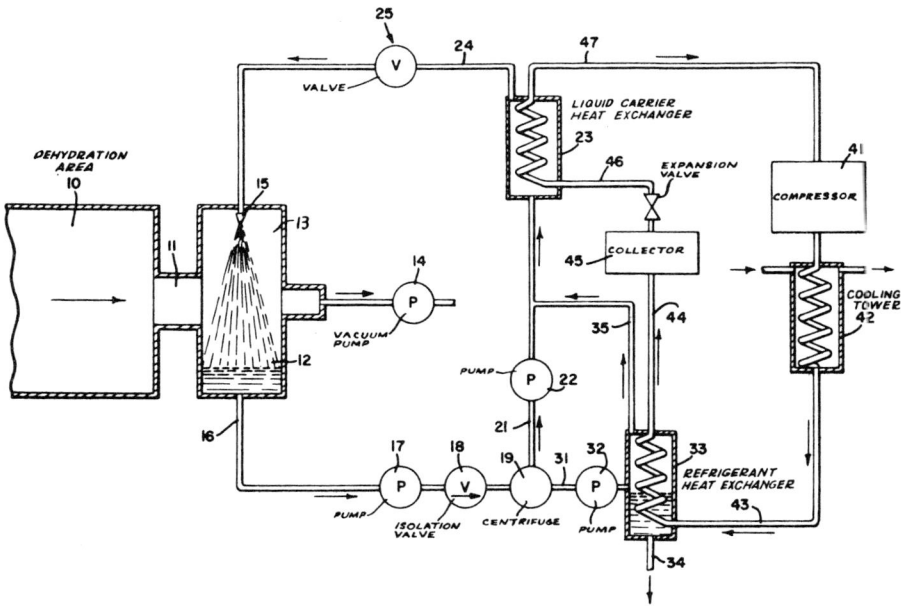

Source: U.S. Patent 3,210,861

water. They have a low viscosity at condensing temperature and are very stable. Pump **17** is provided in flow communication with the chamber **12** to pump oil and ice conglomeration through outlet **16**. The conglomeration is pumped through a vacuum isolation valve **18** to a centrifuge **19**. At the centrifuge station, the bulk of the clean condensing oil is separated from the water and the oil is then taken off in line **21** by pump **22** to flow into condenser oil heat exchanger **23**. During the passage from chamber **12** through the centrifuge to the heat exchanger for recycling back to chamber **12** through line **24** the condenser oil shows little or no temperature increase.

The remaining conglomeration of oil and ice consisting of only a minor portion of oil is in a sufficiently fluid state as to render it pumpable and it passes from the centrifuge through line **31** and pump **32** to the refrigerant cooling heat exchanger **33**. At station **33** the ice melts and separation of oil and water occurs with the water being dumped from the system through drain **34**. The separated condenser oil is then passed through line **35** to rejoin the major portion of the clean cold condenser oil in line **21** for recycling through heat exchanger **23** for cooling to the proper point before recycling to chamber **12** through line **24** and vacuum isolation valve **25**.

The refrigerant system comprises a compressor **41**, a cooling tower **42**, and line **43** which leads to the coil of heat exchanger **33**. A refrigerant line **44** leads from the coil of heat exchanger **33** to a collector or sump **45**. A take-off line **46** and the associated expansion valve are connected to the collector **45** and the coil of heat exchanger **23**. A return flow line **47** connects the coil of heat exchanger **23** back to the compressor **41** for recycling and recooling of the refrigerant. The ice and oil fall to the bottom of the chamber **12** where they are intimately mixed but not dissolved in one another. The exceedingly fine nature of the solidified ice from the water vapor enables the material conglomeration at the bottom of the chamber to flow as a pumpable fluid. Separation of the oil and water by means of the centrifuge station results in recovery of a major amount of the oil by mechanical means leaving only a minor part of the oil intermixed with the ice. There is very little heat gain, if any, in the major part of the oil thus recirculated to the refrigerant heat exchanger **23** and recycled to the chamber **13**, and thus little energy requirement is made upon the refrigerant system including compressor.

The minor quantity of oil separated for further processing is subjected to the refrigerant at heat exchanger **33** at which point the water is dumped out and the oil is recirculated to the heat exchanger **23** with resultant small loss in the lower temperature value of the refrigerant material. The amount of oil so treated is relatively small and thus requires a correspondingly small amount of energy from the refrigerant system. The two-stage treatment of the condenser oil enables a balance to be preserved in the system which contributes to its efficiency.

Vacuum Dryer with Vapor Absorbent Means in Unitary Construction

R.G. Gidlow; U.S. Patent 3,311,991; April 4, 1967; assigned to The Pillsbury Company combines a vacuum dryer with vapor absorbent means in a unitary construction. The system promotes more efficient drying by minimizing travel distance for the water vapor given off by the material being dried. This is accomplished by cascading a refrigerated absorbent liquid over the walls of the vacuum chamber in which the materials to be dried are placed. This obviates the need for the usual large vapor piping and eliminates the frictional resistance caused by it. The absorbent liquid may be water miscible or water immiscible. It is concentrated by separation of absorbed moisture and continuously recirculated through the system.

The process is described with reference to Figure 4.5 which is adapted to the use of a water-miscible absorbent liquid. Housing **10** is provided with a hinged wall or door panel **11** for providing access to the drying chamber within the housing. The door fits tightly with respect to the housing by means of appropriate sealing means to render the interior drying chamber gastight. The drying chamber is largely occupied by a plurality of spaced parallel horizontal shelves **12** adapted to support trays bearing the material to be dried. The shelves are positioned with their edges spaced inwardly from the vertical walls of the housing.

The shelves are preferably of hollow double-walled construction or provided with integral tubular coils to permit heating by circulation of a heating fluid. An inlet **13** is provided for introduction of hot water or a heated antifreeze solution for circulation to heat the shelves and trays. The material to be dried may also be heated by the use of resistance heating elements in the shelves, by radiant heating means disposed below the next shelf above, and the like. The drying housing is provided with an opening **14** through which the drying chamber is connected by means of a suitable conduit **15** to a vacuum source, such as vacuum pump (not shown). Opening **14** is spaced above the floor of the chamber to prevent the condensing liquid from being drawn into the vacuum source.

FIGURE 4.5: VACUUM DRYER WITH VAPOR ABSORBENT MEANS IN UNITARY CONSTRUCTION

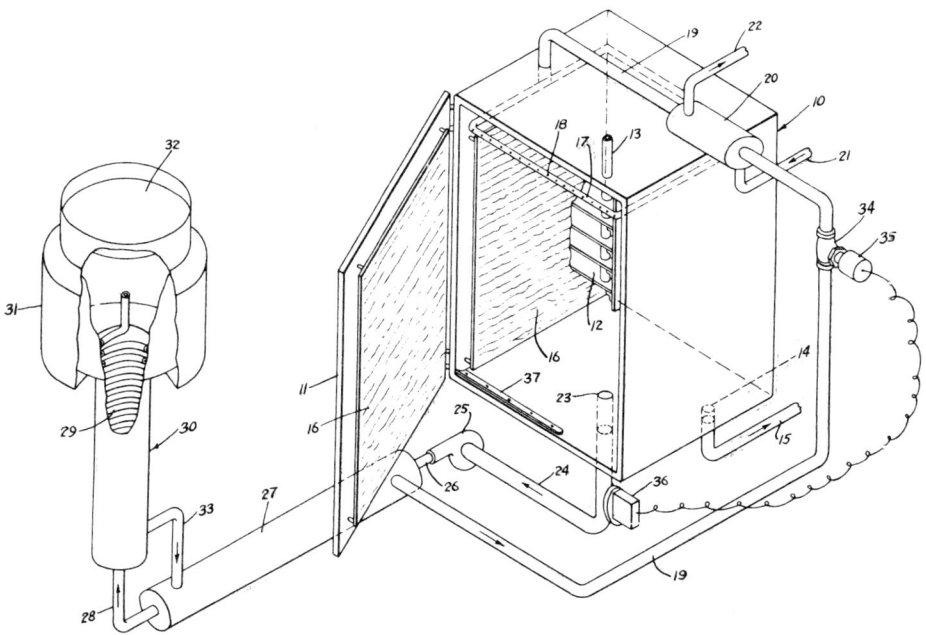

Source: U.S. Patent 3,311,991

Each of the inside vertical walls of the chamber housing is provided with a condenser plate or panel **16**. These condenser plates may be integral with the housing walls or are spaced inwardly from and parallel to the housing walls. The condenser plates are preferably formed from a heat conductive sheet material and are preferably immune from corrosive attack by the absorbent liquid cascaded over them. If desired, the condenser panels or the walls of housing **10** may be of double-walled construction for circulation of a refrigerating fluid through the panels or walls, or cooling coils may be disposed in the space between the inner housing wall and the condenser plates in contact with the plates.

The condensing surfaces are desirably corrugated to increase the effective surface area. In some instances the condenser panels may be formed from mesh material or perforated sheet material where increased liquid surface area is desired. Louvers or equivalent open shielding means may be disposed between the edges of the shelves and the condensing surfaces in order to avoid any possibility of contamination of the material being dried by contact with the condensing agent. Such shielding means must merely deflect the condensing liquid away from the material being dried without materially affecting the passage of vapor toward the condensing surfaces.

A distributor tube **17** is positioned within housing **10** at the top of the drying chamber. The distributor tube **17** is in the form of a loop having a geometrical shape corresponding generally to the floor plan of the drying chamber (rectangular). The distributor tube is positioned so that one leg of the distributor tube is disposed adjacent the top edge of each of the condenser plates. The distributor tube is provided with a plurality of openings **18** or a continuous narrow slot for distributing absorbent liquid for cascading over the condenser plates. The distributor tube is supplied with absorbent liquid from a liquid supply tube **19** which passes through the housing wall and connects with the distributor tube. The supply tube passes through a chilling unit **20** to cool the absorbent liquid to the desired low temperature before introduction to the drying chamber. The chilling unit is supplied through an inlet pipe **21** with a liquefied refrigerating gas. As is well known, cooling takes place as a result of evaporation of the liquefied gas which is then discharged through outlet **22** to a compressor unit (not shown) for reliquefication and reintroduction into the chilling unit.

The refrigerated absorbent liquid discharged through the holes **18** of the distributor tube **17** cascades down over the surfaces of the condenser plates **16** where it absorbs water vapor given off by the material undergoing dehydration and collects in the bottom of the drying chamber where it is discharged through sump **23** in the floor of the chamber to an outlet conduit **24**.

There is shown means for separating absorbed moisture from the absorbent liquid by evaporation. Outlet tube **24** is connected to a pump **25** which forces the absorbent liquid and absorbed water vapor through a conduit **26** to a heat exchanger **27** where the cooled liquid passes in heat exchanging relationship with warmer reconcentrated absorbent liquid being returned to the chilling unit and drying chamber to absorb further moisture. From the heat exchanger the absorbent liquid passes through conduit **28** and into a coil **29** which is contained in an evaporator unit, indicated generally at **30**.

The evaporator is in the form of a vertical column. The upper portion of the column is surrounded by a steam jacket **31** and the top **32** of the column is open to the atmosphere. The top end of coil **29** is open and discharges into the evaporator column. The condensed moisture absorbed by the water-miscible condensing agent dilutes it. The diluted absorbent liquid is heated to its normal boiling point in its passage through the heat exchanger and evaporator and the absorbed moisture is discharged as vapor through the open top of the evaporator to the atmosphere. Venting means (not shown) are preferably provided to carry off the freed vapors.

Evaporation of moisture from the diluted mixture has the effect of reconcentrating the absorbent liquid and raising its boiling point. The concentrated absorbent liquid freed from the absorbed moisture is drawn from the bottom of the evaporator column through conduit **33** and passed in heat exchanging relationship through heat exchanger **27** with the chilled solution from the drying chamber. This partially cooled absorbent liquid is then passed through supply tube **19** for passage through the chilling unit **20** and reintroduction into the drying chamber.

The hydrostatic head in the evaporator column and reduced pressure in the drying chamber forces the return of absorbent liquid to the chiller and drying chamber, or alternatively, additional pump means may be employed. Flow of absorbent liquid into the drying chamber is controlled by motorized valve **34**. Valve **34** is operated by a motor **35** actuated in response to a level detector unit **36** positioned on the outlet conduit **24** below sump **23**. In order to prevent an accumulation of absorbent liquid in the bottom of the drying chamber, further introduction of liquid into the drying chamber is halted when the level in the outlet tube is higher than a predetermined level. At the same time, in order to insure maintenance of reduced pressure in the drying chamber the liquid in the outlet tube is not permitted to fall below a predetermined level. The valve may also be controlled by determining the level of liquid in the bottom of the drying chamber with a float, or with simple contact points using the brine condensing agent as a conducting medium. A drip guard 37 is disposed on the bottom of the drying chamber adjacent the door opening to minimize the possibility of liquid dripping from the chamber when the door is opened.

Example: Freeze Drying Process Utilizing a Water-Miscible Liquid Condensing Agent — Frozen potato flakes spread in a thin layer on suitable trays are introduced to the drying chamber at an initial temperature of about $-10°$ F. The chamber is closed and vacuum is applied to reduce the pressure to less than about 1.2 mm. A lithium bromide brine of 55% concentration is used as the condensing fluid. When the desired reduced pressure condition is reached, brine is introduced into the drying chamber to cascade down over the condenser panels.

The brine is introduced at an initial temperature between about $15°$ and $19°$F and at a rate of flow of about $1/12$ gallon (315 ml) per hour for each square inch of exposed condenser plate surface. At the same time the shelves supporting the trays of frozen potatoes are heated from their initial low temperature to about $120°$F during the course of the drying cycle. This heating causes sublimation of ice crystals within the product being dried. With this material and under these conditions, the rate of evaporation is such that the temperature of the condensing fluid rises about $2°$F in the course of its passage through the drying chamber.

In the course of its downstream passage through the heat exchanger unit, the temperature of the brine, diluted due to the absorbed moisture contained therein, is raised to or just under the boiling point of the brine. This heated material is then passed into the evaporator column, the top portion of which is maintained at a temperature of about $289°$F at which temperature the water vapor is released to the atmosphere. The removing of moisture from the brine has the effect of increasing the concentration of the brine and raising its boiling point. This material is then returned to the drying chamber for recirculation.

In the course of its return passage through the heat exchanging unit in countercurrent flow with the diluted brine from the drying chamber, the temperature of the reconcentrated brine is lowered to within about $3°$ to $5°$F of the temperature at which it is desired to reinject it into the drying chamber. The brine is passed through the chilling unit to reduce its temperature to the $15°$ to $19°$F range desired for introduction to the drying chamber. Upon completion of the drying cycle, the circulation of condensing fluid is stopped, the vacuum is relieved to permit the drying chamber to return to normal pressure and the dried product is removed.

Freeze Dryer Utilizing an Expendable Refrigerant

The process developed by *D.S. Fraser; U.S. Patent 3,564,727; February 23, 1971; assigned to The Virtis Company, Inc.* relates to a freeze dryer of the chamber type which may also be included with connections for manifold-type drying as well as chamber drying. The design includes a refrigeration system which utilizes an expendable refrigerant located outside the freezing chamber. An exchange system of closed circuit design is used to transfer the heat from the expendable refrigerant to the interior of the vacuum chamber. This system may also be utilized to maintain the products in a frozen state or freeze dry products while positioned on the shelf.

FIGURE 4.6: FREEZE DRYER UTILIZING AN EXPENDABLE REFRIGERANT

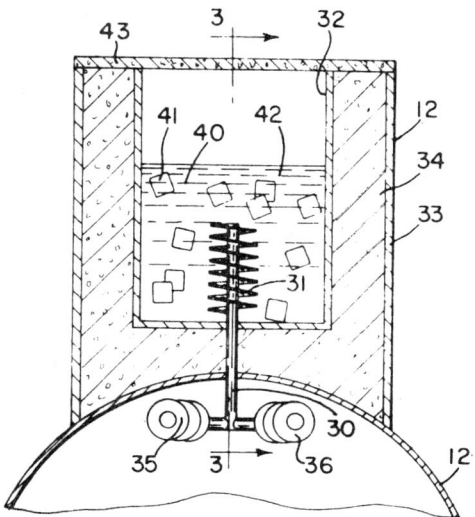

Source: U.S. Patent 3,564,727

The freeze drying arrangement includes a vacuum chamber and refrigeration system which is disposed on top of the vacuum chamber. The chamber includes a bottom shelf on which is positioned a tray having bottles of material to be freeze dried. A stoppering plate is located above the bottles. The front of the chamber is closed off by a transparent door of plastic having metal deposited thereon. This permits the operator to observe the freeze drying process while reflecting heat from outside sources and maintaining the heat level in the chamber fairly constant. External heat is reflected while full viewing of the specimens being dried is available. The total reflectance should be of the order of about 75%. The chamber is provided with suitable connections (not shown) for attachment to a vacuum pump capable of evacuating the interior of the chamber.

The expendable refrigeration unit **12** is of box-like form and disposed on the top of the vacuum chamber. As seen in Figure 4.6, the refrigeration unit includes a closed tube **30** having heat exchange means **31** located centrally of a receptacle **32** which is located within a housing **33**. An insulating foam **34** is formed in the space between the receptacle **32** and the housing **33**, both of which may be rectangular as shown, or of any other suitable shape. The closed tube **30** extends within the vacuum chamber **12** and communicates with a generally X-shaped condenser having a plurality of condensing fins at the respective ends. Obviously, the condenser may be of any desired configuration. The heat exchange surfaces are formed by a plurality of metallic fins which are joined to the closed tube **30** by suitable means such as soldering, welding, or the like, to provide for optimum heat transfer.

The heat exchange means **31** is surrounded in the well **32** by an expendable refrigerant **40** which may include dry ice chunks **41** immersed in an alcohol or acetone bath **42**. Inasmuch as the well **32** is well insulated from the outer housing **33** by a layer of foam several inches thick and covered by a suitable cover **43**, the expendable refrigerant will last for approximately 24 hours.

The tube **30** forms a closed refrigeration circuit and is filled with a suitable refrigerant such as Freon 13 which boils at about $-115°F$. Gases which would serve as a refrigerant in this system are nitrous oxide, refrigerant 503, ethane, ethylene and possibly others. The pressure range would vary between about 200 to 700 psi, depending upon the refrigerant (300 psi for Freon 13).

The refrigeration cycle which occurs is as follows. As the refrigerant is cooled at the upper end of the tube by the expendable refrigerant, it goes to a liquid state, draining down to the bottom of the tube and into the condensing area of the heat exchangers. As the heat exchange means in the chamber **12** is warmed through loss of heat during the freeze drying or refrigeration of the material to be freeze dried, the refrigerant is heated to above its boiling point and assumes a gas form, rising in the tube **30** to the area of the heat exchange means **31**. There the refrigerant is again cooled and the cycle repeats itself.

MONITORING THE FREEZE DRYING CYCLE

MONITORING ELECTRICAL PROPERTIES

Monitoring Drying by Measuring Variation in Resistivity

L.P. Rey; U.S. Patent 3,078,586; February 26, 1963; assigned to Centre National de la Recherche Scientifique, France has established that an extremely convenient and reliable means of directly ascertaining and monitoring the gradual modifications in the structure of the substances being processed, is provided by indicating the variations of electrical characteristics, the variations of which are a direct reflection of the structural modifications in the substance. One such magnitude that can be advantageously used is electrical resistivity of the substance, which resistivity rises to high values following the drying or desiccation step, and drops sharply as the structure of the substance changes from a totally frozen form to a more fluid state.

Since the electrical resistivity of a large number of the substances to which lyophilizing processes are applicable all lie within a relatively narrow range it becomes possible in a great number of cases to use substantially identical laws of variation for the resistivity of the sample with time without extensive preliminary testing of each individual substance.

If the selected characteristic is resistivity, a DC voltage source may conveniently be used and the circuit would measure the DC resistance of the sample. If the selected characteristic is the dielectric constant, the electrodes provided in the sample container would be in the form of capacitor plates and an AC voltage source would be used so as to measure the capacitance of the sample.

In carrying out the process, a sample of the substance being processed and subjected to the same conditions as those to which the substance as a whole is being exposed may have a constant voltage applied to it through suitable electrode means, and the variations in electrical impedance, i.e., resistance or capacitance, of the sample (due to the structural changes in the substance), may be measured by means of any suitable measuring circuit. The process is described

116   Freeze Drying Processes for the Food Industry

with reference to Figure 5.1 in which a freeze drying apparatus **1** comprises a sealed enclosure containing an intermediate tray-like support **2** on which is positioned a container **3** having a sample **4** placed therein.

FIGURE 5.1: MONITORING DRYING BY MEASURING VARIATIONS IN RESISTIVITY

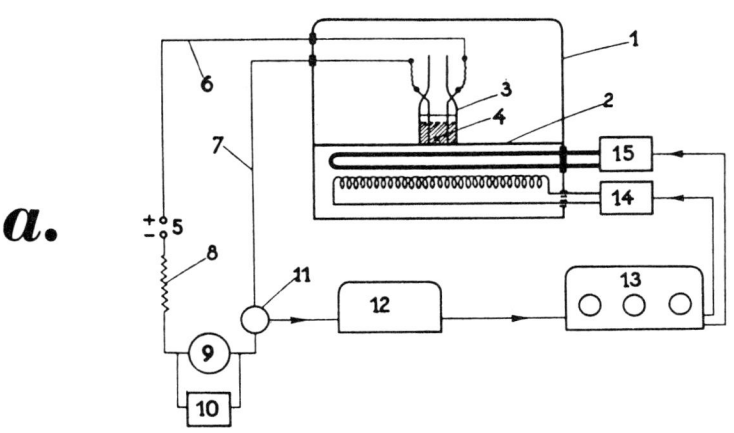

a.

Freeze Drying Apparatus Showing Sample Bottle

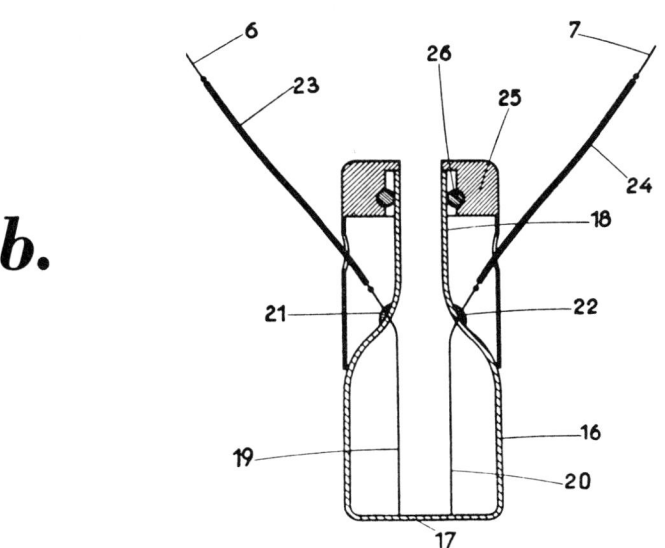

b.

Sample Container for Measuring Resistivity

Source: U.S. Patent 3,078,586

The enclosure 1 is connected to suitable evacuating means, capable of maintaining in the enclosure a degree of vacuum, substantially in the range from 10^{-2} to 10^{-4} mm Hg, or other suitable range depending on the temperatures used.

The sample flask (see Figure 5.1b) is a glass flask 16 having a flat bottom 17 and a cylindrical neck 18 of relatively narrow elongated form. Extending vertically in the body of the flask are a pair of spaced wires 19, 20 made of platinum or other suitable conductive and inert metal which are sealed to the bottom 17 at their lower ends, and are passed out through opposite sides of the flask wall by means of fused joints 21 and 22 for connection with the pair of insulated conductors 23, 24 the outer terminals of which are adapted for connection with an external circuit.

A protective outer insulating cover 25 is fitted around the neck of the flask and the electrode sealing means by way of an annular gasket 26. The external circuit is shown herein as comprising a source of stable DC voltage 5 having one end connected through lead 6 with one of the electrodes, 19, and its other terminal connected by way of a limiting resistor 8 to a suitable impedance, e.g., an indicator instrument 9 in parallel with which a recording device may be connected as indicated at 10.

The opposite terminal of the indicator is connected with one input terminal of a sensitive electric signal generator or detector device 11 having its other input connected through lead 7 with the other electrode wire 20. The output from generator 11 may be applied to the input of a switching control unit 13 adapted to operate both control circuits 14 and 15 of the heating and cooling means respectively in response to the electric signal applied to its input. For example, with a signal of one polarity applied to control switching unit 13, the unit may act to establish a first circuit condition in which an energizing circuit is completed at 14 for the heating resistance.

Another energizing circuit for operating an electrovalve feeding cooling fluid from circuit 15 to the cooling tube in the enclosure is closed whereby the space within the enclosure is heated. In the event of a signal of the opposite polarity applied to switching unit 13 the reverse conditions would be established and the space in enclosure 1 would be cooled.

If desired, a programming unit 12 may be further connected in the circuit ahead of the switching unit 13 and may serve to modify the operation of the switching control unit in accordance with some preestablished program recorded therein.

In operation, sample 4 is placed in the flask 16, and may either be previously cooled to a desired low temperature before introduction into the enclosure 1 or the cooling may be effected within the enclosure as a result of the vacuum therein attaining a sufficiently high value to cause sublimation of ice within the sample and consequent drop in temperature of the sample.

In the freeze dried sample the resistivity is high, and it is generally possible so to adjust the control system that the switching unit 13 will energize heating circuit 14 to apply heat to the sample at the start of the process. As the temperature of the sample rises, its resistivity decreases. This decrease in the resistance in one arm of the circuit including the source 5, limiting resistor 8, indicator means 9–10, in its opposite arm, is manifested as a differential signal of a predetermined polarity generated by the device **11** and applied to the switching control unit **13**, which thereupon may cut off the heating means and (if necessary) energize the cooling means for cooling the enclosure **1**. By such feedback regulation the temperature of the tray **2** in the enclosure **1** is continually controlled so that the resistivity of the sample **4** may be held to a substantially constant preset value.

Monitoring Dielectric Properties Without Disturbing Drying Conditions

With a view to improving the yield of freeze drying operations, and in particular to regulate the heat energy supplied to the product at an optimal level without producing localized melting, efforts have been made to develop methods and means for measuring physical characteristics related to the state of the treated product during freeze drying. One of these methods, consisting of measuring the dielectric characteristics of the product—these are closely related to the amount of ice present—permits observation of the evolution of drying (that is the progressive removal of the ice), avoids melting, enables the heat energy supplied to be adjusted at an optimal level and also gives an accurate indication of the moment when all the ice crystals have disappeared, so that the end of the sublimation period may be detected.

J.-P. Bouldoires; U.S. Patent 3,811,199; May 21, 1974; assigned to Societe D'Assistance Technique Pour Produits Nestle SA, Switzerland provides a freeze drying apparatus that is particularly simple to operate and which permits monitoring of the dielectric characteristics of the product to be treated without disturbing the drying conditions.

The freeze drying apparatus comprises an impervious chamber, means for maintaining a reduced pressure in the chamber, at least one tray defining a receptacle for product to be dried and including a plurality of electrically insulated parallel metallic partitions dividing the receptacle into compartments, means for supplying heat to the product to be dried, and at least one device, for measuring the capacitance and/or the associated resistance of a condenser, connected to at least two of the partitions so as to measure the capacity and/or the associated resistance of a condenser the electrodes of which are constituted by the partitions.

As shown in Figures 5.2a and 5.2b, the freeze drying apparatus comprises an impervious chamber **1** communicating with a condenser **2** and a vacuum pump **3**. The chamber **1** contains a series of superimposed trays **4**, only the upper tray being visible. Each tray comprises a body **5** made of an electrically nonconducting thermosetting plastic material defining a receptacle in the form of a right-angled parallelepiped and consisting of a bottom **6** and a rectangular frame **7**.

The bottom is traversed by two series of tubes, **8** and **9** respectively, these tubes being made of an electrically insulating material, and each series of tubes is connected to a manifold **10** and **11** respectively. These manifolds are connected to a circuit **12** for circulating heated glycol which comprises a heat exchanger **13** for heating the glycol and a circulating pump **14**. The free ends of the tubes protrude upwards from the inner side of the bottom **6**, and form outlets evenly spaced on two straight paths parallel to the short sides of the frame **7**. The tray **4** also comprises partitions **15** consisting of hollow metallic members in the form of parallelepipeds the cavity of which is in communication with the two series of tubes, and each partition constitutes a circulation element for glycol between two tubes one each from the series **8** and **9**. The partitions of one of the trays are electrically connected in parallel to the measuring device **16** located outside the chamber, by leads **17** and **18** which pass through an impervious passage **19** in the wall of the chamber **1**.

On the other hand, in each compartment defined by the partitions **15** is placed a screen **20** made of insulating material for facilitating the passage and removal of the water vapor emitted from the product during freeze drying.

The partitions **15**, which serve to heat the product, also act as the electrodes of adjacent plate condensers connected in parallel to the measuring device **16**.

FIGURE 5.2: MONITORING DIELECTRIC PROPERTIES WITHOUT DISTURBING DRYING CONDITIONS

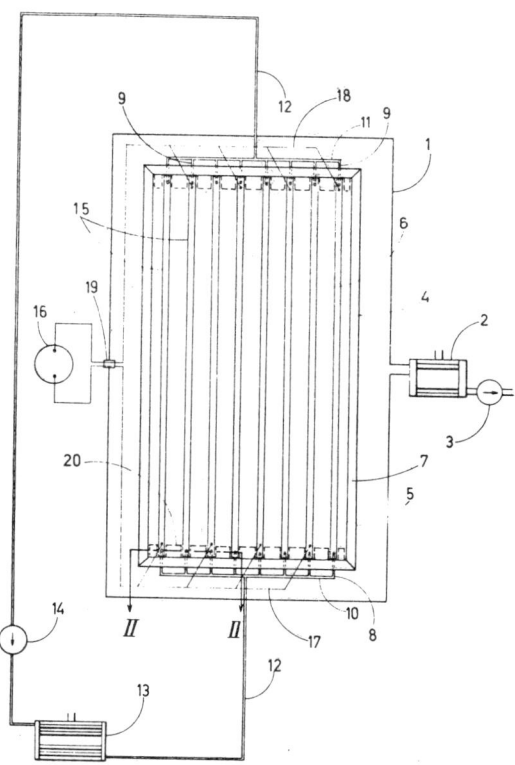

a.

Plan View of Apparatus

(continued)

FIGURE 5.2: (continued)

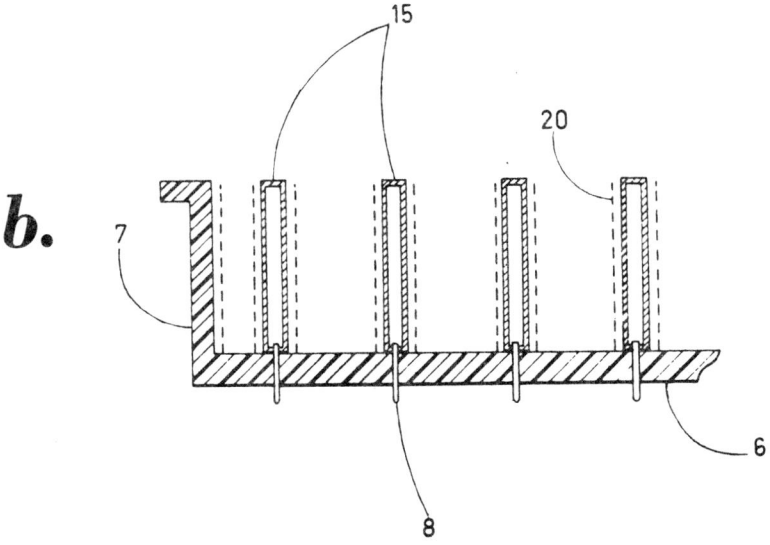

Partial Section of Tray Along II–II of Figure 5.2a

Source: U.S. Patent 3,811,199

MEASURING WEIGHT LOSS

Use of Food Product Weight as Parameter in Controlling Dehydration

G.R. Cox; U.S. Patent 3,178,829; April 20, 1965; assigned to J.P. Devine Manufacturing Co. is concerned with control of freeze drying operations. There is provided a control that continuously compares the actual state or degree of dehydration with the instantaneous desired state or degree of dehydration, to thus determine whether the actual dehydration processing is accurately following the scheduled optimum rate of dehydration.

Frozen food products placed in the processing apparatus will be subjected to a subatmospheric pressure and heat causing sublimation of the moisture therefrom. The heat is supplied by a controlled system which includes a closed loop, time rate of dehydration schedule that is effected by comparing a desired food weight loss schedule to the actual food weight loss. The heat supply is further controlled to maintain a sufficiently low processing pressure, especially during early stages of the processing and to maintain a sufficiently low food temperature, especially during latter stages of the processing. As a matter of safety, the heat supply is automatically limited to protect heater components from overtemperature conditions.

The processing is monitored by a degree of dehydration indicator that has a compacted scale range during a major portion of the processing and has an automatically selected, expanded scale range of increased sensitivity during terminal stages of the processing. An end point indicator is automatically actuated to provide a readily discernible signal indicating that the processing has been completed.

To maintain proper sublimation conditions, the pressure ambient to the food products is maintained below a predetermined value, usually in the neighborhood of 0.5 to 1.5 mm Hg absolute pressure by temporarily discontinuing the heat supply. As the operation progresses, the rate of drying becomes much slower and the temperature of the products tends to rise. When the temperature reaches a maximum allowable value, the heat is temporarily discontinued to avoid cooking the food.

As the products reach the latter stages of dehydration, for example, 10% of their original weight, the food product weight indicator automatically changes its scale to a scale having about ten times greater accuracy; for example, the starting full scale may have a range of about 0 to 50 pounds, whereas the scale during the terminal stages may have a range of about 0 to 5 pounds.

Thus, during the latter stages, the degree of dehydration of the food products may be more accurately monitored. Moisture will continue to be driven from the food products until their weight reaches a value predetermined to be the final desired weight, at which time a lamp will light signaling the end of the drying cycle. The food products are now ready to be packaged in an inert atmosphere such as nitrogen for storing or shipping.

The process is described with reference to Figure 5.3. Vacuum chamber **11**, is provided with heating means **13**, and vacuum creating system **20**.

The vacuum chamber **11** is shown provided with control sensing means or elements for measuring the parameters used in the control system. One of these sensing means is a load cell or dehydration sensing means **30** which is connected through suitable means, such as a link or chain **31**, to at least one of the food product carrying trays **T** within the chamber to support the tray in a manner such that a known percentage of its weight is imposed upon the load cell. The load cell **30** consists essentially of a strain gauge **30a** or other means that will vary a measurable electrical property as a function of the force applied thereto.

Also within the chamber is a pressure-responsive means, such as an evacuated bellows **32**, that will indicate by its degree of contraction, the absolute pressure within the chamber. A temperature-responsive device, such as a bimetallic strip **33**, is located in the chamber to accurately and continuously sense the food product temperature.

Each of the sensing means is provided with electrical signal transmitting means, lines or conduits, for relaying the information sensed in the chamber to the control system or computer. For simplicity, the various sensing means have been shown in position in the control system and their positions within the chamber are indicated only schematically by broken line circles.

The control system comprises an output switch **34** that controls the supply of electric current from a power source **35** to a heating coil **36** or other suitable electric heating means within fluid or oil heater **15**. The output switch is normally held in an open position by a compression spring **38**, and is actuated from

122 Freeze Drying Processes for the Food Industry

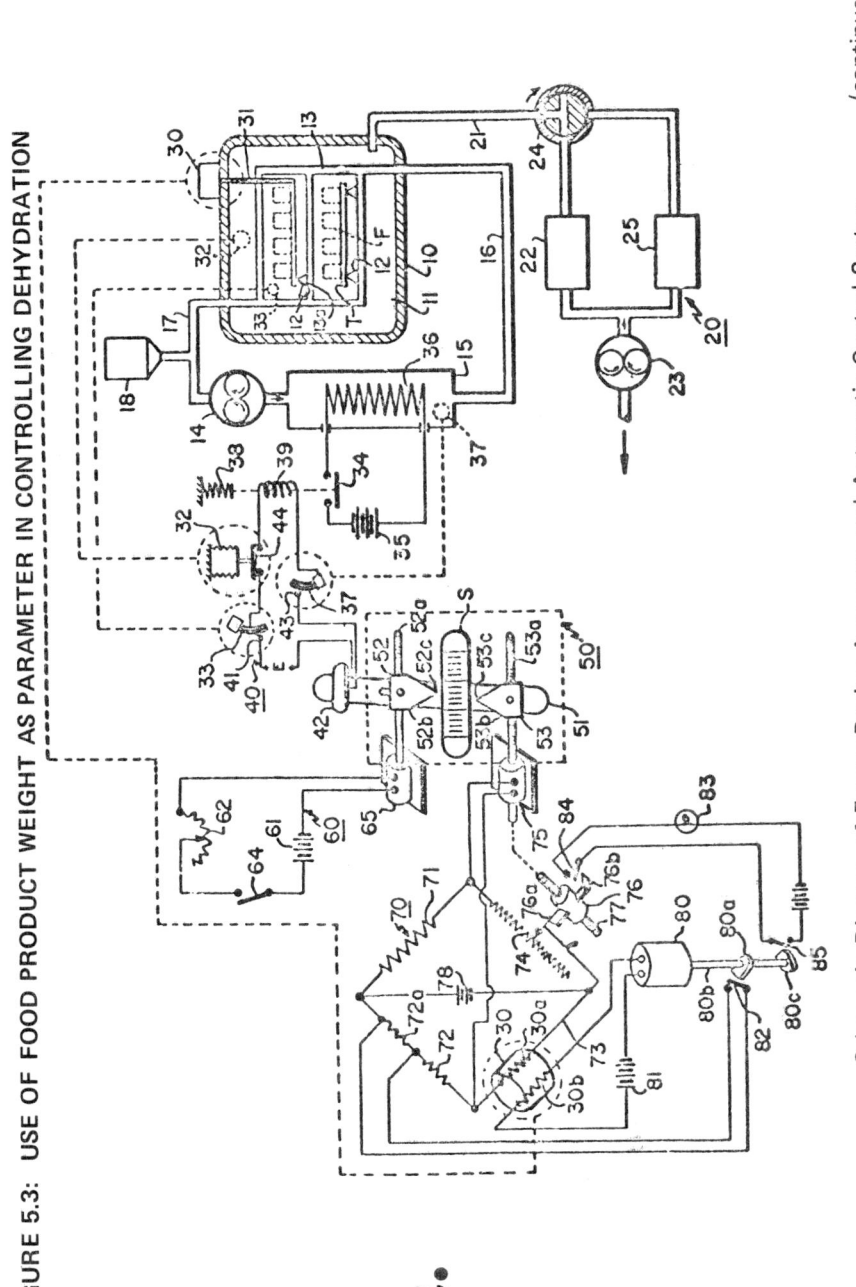

FIGURE 5.3: USE OF FOOD PRODUCT WEIGHT AS PARAMETER IN CONTROLLING DEHYDRATION

a. Schematic Diagram of Freeze Drying Apparatus and Automatic Control System

(continued)

FIGURE 5.3: (continued)

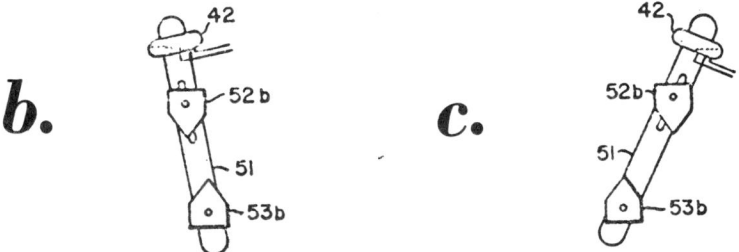

b.   **c.**

(b)(c) Enlarged View of Controller Elements Showing Two Operative Positions

Source: U.S. Patent 3,178,829

its normal position by a solenoid **39** that is energized by a controlled electrical circuit **40**. The controlled electrical circuit includes a suitable current source E and four control switches **41, 42, 43**, and **44**. The control switch **41** is actuated in response to the food product temperature, as sensed by the temperature-responsive means or device **33**. The switch **41** is normally closed and will open when the food product temperature exceeds a predetermined maximum limit. Such an over-termperature condition will usually occur, if at all, during terminal stages of the processing.

The mercury control switch **42** is the main or primary control element of the system and functions to detect the difference or error between the actual instantaneous degree of dehydration and a scheduled instantaneous degree of dehydration.

The controlled circuit **40** may also include an oil temperature limit switch **43**. This switch will be normally closed and will be actuated to an open position by any suitable temperature responsive means, such as bimetallic strip **37**, upon an over-temperature condition of the oil in the heater **15**. Finally, the circuit **40** includes a chamber pressure responsive switch **44** that is actuated by the evacuated bellows **32** upon an over-pressure condition within the chamber **11**.

It will be apparent that the solenoid **39** will be energized by the circuit **40** to close the switch **34**, only when the control switches **41, 42, 43**, and **44** are all in a closed position. Thus, an over-temperature condition of the food product, as sensed by means **33**, an instantaneous over-schedule dehydration, as sensed by the mercury switch **42**, an over-temperature of the heater oil, as sensed by means **37**, or an over-pressure condition in chamber **11**, as sensed by means **32**, will open the control circuit **40** to deenergize solenoid **39**, open switch **34**, and temporarily discontinue the heat supply to oil heater **15**.

A differential sensing means or comparing unit **50** is provided for actuating the primary control switch **42** to maintain the actual dehydration processing in step with the desired or scheduled dehydration processing, at least during a major portion of the operation. The differential sensing means preferably includes a floating lever controller or other mechanical differential means **51** upon which the mercury switch **42** is fixedly supported.

Two separate means or screw jacks **52** and **53** are provided for linearly-displacing opposite ends of the floating lever **51**. The floating lever controller receives a first or time scheduling signal adjacent its upper portion from the screw jack **52** that includes a rotatable threaded lead screw **52a** and a follower member or part **52b**.

The lower end portion of the floating lever controller receives a signal that is a function of the actual instantaneous degree of dehydration or food product weight from the screw jack **53** that includes a rotatable, threaded, lead screw **53a** and a weight indicating output follower member or part **53b**. The follower members **52b** and **53b** may be conveniently located adjacent a calibrated scale and provided with pointers **52c** and **53c**, respectively, to visually indicate the instantaneous, scheduled, and actual dehydration values.

The lead screw **52a** is rotated by a time schedule drive motor **65** that turns at a rate that is a function of its current supply to move the follower **52b** as a preset function of time. The current supplied to the motor **65** is provided by a scheduling circuit **60** that includes an electrical power source **61** and a current-varying rheostat **62**. The scheduled rate will be selected according to the nature of the particular food being processed. A starting switch **64** is provided for initiating the process by beginning the schedule motor drive.

Lead screw **53a** is rotated by a motor **75** that receives current from a computing or weight sensing bridge circuit **70**. The bridge circuit is of the self-rebalancing type and basically includes a first leg **71** having a substantially fixed resistance, a second leg **72** having one of two fixed resistances, a third leg **73** having a resistance **30a** provided by a load cell **30** and varied as a function of the actual food product weight, a fourth leg **74** having a resistance that is varied by a follow-up follower or arm **76** of a follow-up lead screw **77** that is driven by a motor **75**, and a power source **78**.

It will be seen that upon variation of the resistance **30a** in the load cell **30** due to a change in food product weight, an error will develop across the bridge **70** to cause the motor **75** to rotate. This rotation will cause the lead screw **77** to drive follower **76** and reposition its associated contact arm or brush **76a** to reduce the resistance of the fourth leg **74** of the bridge to eliminate the error. When the motor has turned an amount sufficient to eliminate the error, both the follow-up follower **76** and the output follower **53b** will have changed their positions by an amount that is a direct function of the change in food product weight.

The operation of the device thus far described is as follows: Upon closure of the starting switch **64** (which may be of a push button type), schedule motor **65** is actuated to rotate lead screw **52a** that drives follower **52b** to displace the upper portion of the floating lever controller **51** to the right. This displacement causes closure of the control switch **42** to supply energizing current to the solenoid **39**, causing the switch **34** to be pulled upwardly, close its contacts, and complete the heating circuit. The heat supplied drives off moisture from the food products by the process of sublimation and the moisture thus driven off is removed by the subatmospheric pressure creating system **20**.

As the moisture is driven from the food products, the food product weight decreases, thus decreasing the resistance of element **30a** within the load cell **30**. The decrease in resistance **30a** causes an upset in the weight-sensing bridge circuit

70. This upset causes the resistance in the fourth leg **74** of the bridge circuit **70** to also be reduced by movement of the motor **75**, its lead screw **77**, and follower **76** that carries the brush **76a**. The output follower **53b** is displaced by this movement of the motor **75** to displace the lower end portion of the floating lever controller **51** to the right, thus tending to restore the lever to a vertical orientation. If at any time, the actual dehydration exceeds that called for by the predetermined schedule, lever **51** will rotate counterclockwise to open switch **42**, and interrupt the heat supply.

If during the early stages of dehydration, the moisture is driven-off at a rate that cannot be handled by the subatmospheric pressure creating system **20**, thus causing the pressure to rise within the chamber **11** to a point (usually about 0.5 to 1.5 mm Hg absolute) near which sublimation without melting can no longer occur, the pressure responsive means **32** will open its associated switch **44** to deenergize solenoid **39**, open switch **34** and interrupt the heat supply.

During the latter stages of dehydration, there is a tendency for the food product to overheat. To prevent food damage from such a tendency, the temperature-responsive switch **41** in the controlled circuit **40** will open upon an over-temperature condition of the food product F, as sensed by means **33**, to terminate current flow to the solenoid **39** and allow the spring **38** to open switch **34** and temporarily discontinue the heat supply.

The process also provides means for indicating the instantaneous weight of the food products and for increasing the accuracy of this indication near the critical end point of the processing. To this end, a scale **S** is conveniently positioned alongside pointer **53c** of the weight-responsive output follower **53b**. The pointer **53c** of the follower **53b** will visually indicate the instantaneous food product weight, as determined by the load cell **30** and the bridge **70**. Also a pointer **52c** cooperates with the scale S to visually indicate, at least during the early portions of the process, how the actual weight is following the scheduled desired weight change.

It is desirable to provide means for expanding the scale and increasing the sensitivity of the weight indicator during the terminal stages of the process. For example, the scale could go from 1 to 50 pounds during early processing stages and from 1 to 5 pounds during the terminal stages.

Automatic Control in Response to Weight Loss

M.E. Fuentevilla; U.S. Patent 3,176,408; April 6, 1965; assigned to Pennsalt Chemicals Corporation has developed a method and apparatus in which the freeze drying cycle is automatically controlled in response to the loss of weight of the articles being freeze dried. When using the method, it is not necessary for an operator to set any controls as to the timing of the cycle nor periodically inspect the articles being freeze dried so as to ascertain the status of the articles. The process is described with reference to Figure 5.4 which shows freeze drying apparatus **10** comprising a housing **12** having a plurality of hollow shelves such as **14** disposed one above the other. The interior of the shelf **14** is connected to an inlet conduit **16** and an outlet conduit **18** for circulating heating medium. The conduits are connected to a pump **20** which continuously circulates the medium through the shelf. A heat exchanger **22** is disposed in the conduit **16** between the pump and the shelf.

FIGURE 5.4: AUTOMATIC CONTROL IN RESPONSE TO WEIGHT LOSS

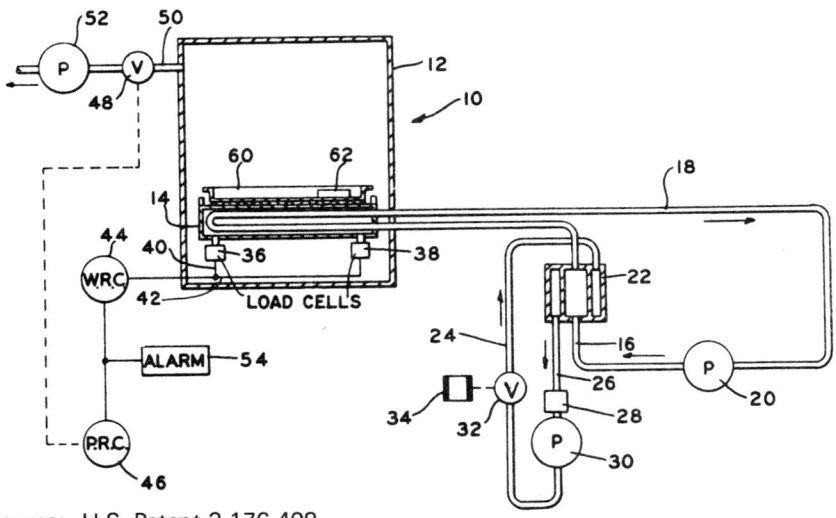

Source: U.S. Patent 3,176,408

The temperature of the heat exchanger is regulated by steam which is circulated therethrough by means of the inlet conduit **24** and the outlet conduit **26**. The outlet conduit is connected to a steam generator **28** having a pump **30**. The outlet side of the pump **30** is connected to the inlet conduit **24**. A valve **32** having a solenoid operator **34** is disposed within the conduit **24**.

The shelf **14** is supported by load cells **36** and **38**, which are commercially available devices which generate an electric current in response to pressure. Such load cells are extremely accurate and are not affected by the temperatures and pressures within the housing **12**. The load cells are connected to the weight recorder and controller **44** by wires **40** and **42**. The weight recorder and controller is a commercially available device which permanently records the electrical current generated by the load cells on a graph which may be interpreted in terms of weight. In addition, the weight recorder and controller relays the current received from the load cells to a controller **46**.

The controller **46** regulates the flow of motive fluid to the diaphragm on a valve **48** in the conduit **50**. The conduit is connected to the interior of the housing **12** and is provided with a vacuum pump **52** for evacuating the interior of the housing. An alarm **54** is connected to the weight controller and recorder **44** and is responsive to a minimal amount of current.

A tray **60** of prefrozen articles **62** which are to be freeze dried is disposed within the housing **12** on the shelf **14**. The door (not shown) providing access to the interior of the housing **12** is provided with a seal means so that the housing is hermetically sealed in the closed disposition of the door. The circulating fluid within the shelf heats the articles by radiation. All surfaces on the shelf and the inner surface of the housing are colored black while the inner surface on the door is colored white.

As the articles are heated by radiation, they are subject to a vacuum within the housing effected by the pump **52**. The combination of heat and the vacuum causes the ice within the frozen articles to sublime with the resultant vapor being removed by the pump. As the ice within the articles sublimes, the weight supported by the load cells **36** and **38** decreases.

After a predetermined period of time, the weight of the articles will become stabilized thereby indicating that substantially all of the ice is removed therefrom by sublimation.

During the entire cycle, the controller **44** is recording the change in the current generated by the load cells and relaying the same to alarm **54** and controller **46**. In response to a minimum current indicating that substantially all ice has been removed from the articles, the alarm is activated and controller **46** is activated so as to close the valve **48**. Thus, it will be seen that the freeze drying cycle is controlled so as to stop the cycle as soon as substantially all of the ice has been removed from the articles.

It has been determined that there is an optimum rate of change of weight of the articles **62** with respect to time. It is possible to utilize a controller **46** which is of the type which is capable of receiving a signal from the controller **44** so that the valve **48** is constantly varied during the freeze drying process.

Recording Apparatus for Measuring Weight Loss

R.D. Cole, R.R. Gidner and A.W. Reichert; U.S. Patent 3,280,471; Oct. 25, 1966; assigned to American Sterilizer Company describe a supporting arrangement for measuring the moisture removal from a product in a freeze dry process. This is capable of operating accurately during the rapid changes of temperature encountered.

In the structure disclosed, a load cell is contained in the small water jacket chamber attached immediately above the freeze dry chamber with a rod made of a low heat conductivity material passing into the freeze dry chamber. Water is passed through the water jacketed chamber. This water jacket maintains the load cell at a constant temperature and shields it from radiant conduction, and convection heat since the heat which might otherwise pass through a load cell is absorbed and carried away by the water in the jacket. The low conductivity rod conducts little or no heat from the chamber to the load cell.

The system also effectively eliminates all mechanical vibrations transmitted to the load suspended in the freeze dry chamber. These vibrations are picked up and transmitted to the indicator or recording instrument and result in variable electrical signals. These variable signals ordinarily appear on the load chart as a wide line which is difficult to interpret. By connecting an electrical condenser across the output electrical lines, the electrical signal from the load cell is averaged and the system then performs as if no mechanical vibrations were present.

The process is described with reference to Figure 5.5 which shows a controlled freeze dry chamber, capable of being reduced to a minimum temperature of approximately −50°C. The chamber is provided with suitable refrigeration equipment, evacuating equipment, etc.

A rack **11** supports the trays **12** which contain the product to be freeze dried.

The rack is supported by the low conductivity rod **18** which passes through a vacuum seal **19** at the top of the pressure and temperature controlled freeze dry chamber.

Supported on top of the freeze dry chamber **10** is the water containing chamber **13** which contains an inlet pipe **30** and an outlet pipe **31** for circulating water **21** through the chamber. The water chamber has the hydraulic cylinder **22** supported on it and this cylinder has a piston **23** reciprocably supported therein. The piston in the cylinder may be actuated by air admitted below the piston in line **24** and exhausted from above the piston from line **25**. Thus, the piston can be moved up and down to lift the rack **11** so that the rack is supported by the low conductivity rod **18** through the load cell **14** and the piston rod **26**. The load cell is disposed in water jacketed chamber **41**.

The load cell is of the conventional type of strain gauge load cells which may be connected through suitable electrical wires **27** and **28** to the recording instrument **17**.

The recording instrument is of the type of strip chart recorders or other continuous recording or indicating instruments, and will provide a line **29** on the chart **20** or indication on an indicating type meter.

Due to mechanical vibration of the chamber, the line will be wide as at **40** as indicated by the dotted line. It has been discovered that this line will be of reduced width when a capacitor **16** is connected across the lines **27** and **28**. If an indicating type instrument is used, the capacitor will reduce the fluctuations of the indicating pointer.

FIGURE 5.5: RECORDING APPARATUS FOR MEASURING WEIGHT LOSS

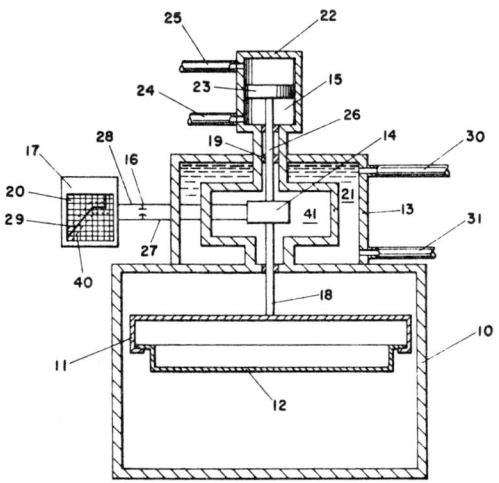

Source: U.S. Patent 3,280,471

In carrying out the process the product to be dried is placed on the tray **12** on the rack **11**. Air pressure or other fluid is admitted in the line **24** which will lift the rack through the low conductivity rod **18**, load cell **14**, and piston rod **26** so that the weight of the rack, product, and tray is supported on the load cell. The strain gauge type load cell **14** will thus transmit a signal to the recorder **17**. Then as the goods are dried, the recorder will indicate a line such as the inclined portion of the line **29** and when the moisture has been removed from the goods, this line will flatten out. At the time that the line flattens out, the goods are dry.

OTHER MONITORS

Controlling Temperature of Heating Means

In a process developed by *R.A.J. Ridge; U.S. Patent 3,343,273; Sept. 26, 1967; assigned to Vickers-Armstrongs (Engineers) Limited, England* there is provided control apparatus in equipment for freeze drying liquid, semiliquid or granular material, the equipment comprising a vacuum-tight chamber, means for evacuating the chamber, and heating means for supplying heat to frozen material in the chamber, and the control apparatus comprising means for controlling the temperature of the heating means, during a freeze drying operation, such that, after an initial heating-up of the material, the heat supplied by the heating means to frozen material being dried in the evacuated chamber maintains the temperature of a free surface of the material being dried at a substantially constant predetermined value. The process is described with reference to Figure 5.6.

With frozen material **7** that is to be dried in the trays **2**, and with the chamber **1** evacuated to a very low pressure (e.g., 0.15 mm Hg absolute), the equipment is operated as follows. Heat of sublimation is supplied to the frozen material from the heater plates **3**, the water present in the material as ice being driven off as vapor without passing through the liquid phase. The pump is run continuously to maintain the vacuum in the chamber **1** and draw off the water vapor which condenses to ice on the condenser plates.

As drying proceeds the temperature of the surface **8** of the material **7** in each tray **2** tends to rise and the control apparatus in the equipment acts as described later to cause the mean temperature of the heater plates **3** to drop as drying proceeds and to cause the mean temperature of the surfaces of the material in all the trays to be maintained at a constant desired level. The control apparatus operates as follows. When the equipment is not in use the moving contacts **18**, **21** of the controllers **12**, **14** respectively make contact on the contacts **19** and **22**. The contact arms of the controllers **12**, **14** are set, the arm of the controller **12** to the maximum mean temperature which it is desired the heater plates **3** should reach, and the arm of the controller **14** to the maximum temperature which it is desired the mean temperature of the surfaces of the material in all the trays should reach.

Upon switching on the equipment by means of a switch not shown, the light sources of the controllers are illuminated and the time delay relay **27** is energized.

The light from these sources falls on the phototransistors causing a current to

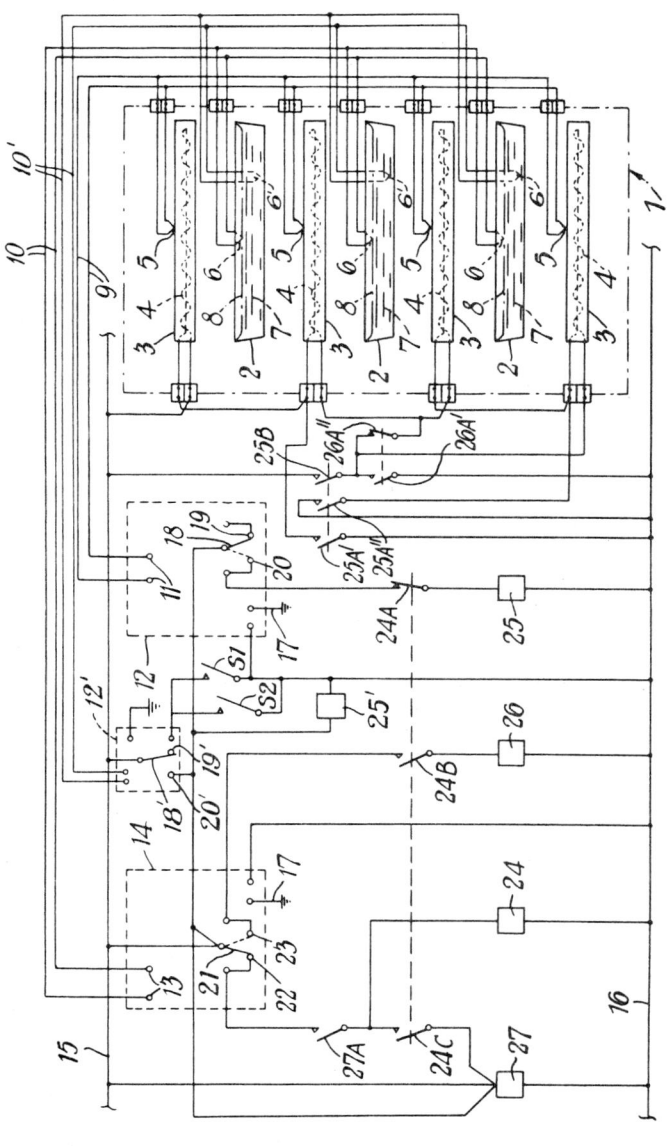

FIGURE 5.6: CONTROLLING TEMPERATURE OF HEATING MEANS

Source: U.S. Patent 3,343,273

Monitoring the Freeze Drying Cycle

flow and thereby causing the relays of the controllers to be energized. The time taken for the relays of the controllers to operate produces a delay, i.e. a "warming-up" time. During this warming-up time the contact **27A** is open. The time delay relay **27** is set to close the contact **27A** after a period of time sufficient for the controllers **12, 14** to warm up has elapsed, for example, approximately 10 seconds. When the controllers have warmed-up their relays become energized and their moving contacts **18, 21** are actuated by the relays to close, respectively, on the contacts **20** and **23**.

With the moving contacts in these respective positions, the relays **24** and **26** are deenergized while the relay **25** is energized and closes its contacts **25A', 25A"** and **25B**. The elements **4** are therefore connected to the supply lines **15, 16** in parallel and the heater plates **3** heat up.

As the plates heat up the signal passed to the controller **12** from the thermocouples **5** causes the indicator pointer of the controller to move towards the control arm of the controller, the control arm it will be recalled having been set to the maximum mean temperature (e.g., 140°C) which it is desired the plates should reach. As the indicator pointer approaches the control arm the resistance/capacitance circuit becomes operative so that the relay **25** is alternately deenergized and energized.

The contacts **25A', 25", 25B** therefore alternately open and close so that the rate at which the temperature of the heater plates rises slows down as the plates approach their maximum mean temperature. When the heater plates reach this maximum mean temperature the shutter carried by the indicator pointer cuts off the light beam falling on the phototransistor mounted on the control arm and the relay of the controller **12** becomes continuously deenergized so that the moving contact **18** moves from the contact **20** to remain on the contact **19** and the supply of current to the elements **4** is cut off.

Since the current supply to the elements is cut off the plates cool down and the signal passed to the controller **12** from the thermocouples **5** causes the indicator pointer to move away from the control arm. Light therefore again passes from the light source to the phototransistor, the relay of the controller is reenergized, and the moving contact **18** moves back onto the contact **20**. The relay **25** therefore becomes energized again and closes its contacts so that current is again supplied to the elements **4**. The heater plates are thus maintained at their maximum mean temperature, the resistance/capacitance circuit tending to prevent undue fluctuations in the temperature of the plates.

While the heater plates are maintained at their maximum mean temperature the temperature at the surface **8** of the material **7** in each tray **2** rises and the signal passed to the controller **14** from the thermocouples **6** causes the indicator pointer of the controller to move towards the control arm of the controller.

As previously mentioned this control arm is set to the temperature (e.g., 60°C) at which it is desired the mean temperature of the surfaces **8** of the material **7** in all the trays **2** should be maintained during drying. This temperature is chosen, in dependence upon the characteristics of the material, such that the heat applied to the material to maintain this temperature is high enough satisfactorily to dry the material without being so high as to burn the material or to thaw to liquid any of the water present as ice in the material.

When the surfaces reach this mean temperature the shutter carried by the indicator pointer of the controller **14** cuts off the beam of light falling upon the transistor carried by the control arm of this controller and the relay of the controller **14** is deenergized. The moving contact **21** therefore moves onto the contact **22** and since the contact **27A** of the relay **27** is closed the relay **24** becomes energized.

The contact **24A** therefore opens and the contacts **24B** and **24C** close. Due to the closure of the contact **24C** the relay **24** becomes selfenergizing. Since the contact **24A** is open the relay **25** becomes deenergized and its contacts **25A'**, **25A"**, **25B** open. The current supply to the elements **4** is therefore cut off and the heater plates begin to cool. Since the plates begin to cool the mean temperature of the surfaces of the material in the trays also drops and the indicator pointer of the controller **14** moves away from the control arm of this controller so that light again falls on the phototransistor and the relay of the controller **14** becomes energized to move the moving contact **21** onto the contact **23**.

Contact **24B** is closed, so the relay **26** is energized and closes its contact **26A'**, contact **26A"** opening. Due to the closing of the contact **26A'** and the opening of the contact **26A"** the elements **4** are connected, in series, to the supply lines **15, 16** and the temperature of the heater plates begins to rise again. The rise in the temperature of the heater plates results in a rise in the mean temperature of the surfaces of the material in the trays so that the relay of the controller **14** again becomes deenergized and the supply of current to the elements is again cut off. The controller **14** continues to control the temperature of the heater plates, as drying continues, so that the mean temperature of the surfaces remain constant. The controller **12** has no further controlling effect on the temperature of the plates.

When the material is sufficiently dry the equipment is switched off. This switching off can be effected by a timer set for a desired length of time.

Determination of Practical End Point of Drying

G.M. Illich, Jr.; U.S. Patent 3,259,991; July 12, 1966; assigned to Abbott Laboratories provides a method and apparatus for accurately determining the practical end point of a high vacuum drying operation, such as freeze drying, where water vapor is transported through a vacuum atmosphere.

The method consists of observing the pressure indications of two separate and dissimilar electrical vacuum pressure measuring devices placed in the vacuum chamber. By proper selection and utilization of control means in conjunction with the vacuum pressure measuring devices, the drying operation can be made automatic, permitting it to proceed unattended.

Two gauges are utilized; one an ionization type gauge of the kind described in U.S. Patent 2,497,213, and the other a thermocouple type gauge. It is essential that the two gauges be calibrated accurately for a vacuum atmosphere of air. The operation of each type of gauge is based upon different physical properties used in measuring the absolute pressure. Low molecular weight vapors such as water with a molecular weight of 18, as compared with air having an average molecular weight of 29, cause an ionization type gauge to read lower than for an equivalent pressure of air. Conversely, low molecular weight vapors have higher thermal conductivities and cause thermocouple type gauges to read higher than for an equivalent pressure of air. During the early part of a high vacuum drying operation, the two gauges differ in readings, the ionization type gauge reading lower than the thermocouple type gauge. The ratio of the readings may be as low as 1 to 2 if the vacuum system has minute air leakage. When all of the ice has been sublimated and essentially only air remains in the vacuum atmosphere, the gauges will have corresponding readings since they are originally calibrated accurately for a vacuum atmosphere of air.

Thus, the end point of the drying operation, at which time all of the ice has been sublimated, can be determined accurately and the drying operation to remove the water bound by adsorption can be commenced immediately in order to obtain the desired residual moisture content. If desired, automatic control of the drying operation can be effected by automatic actuation of a vacuum control valve and a programmed heating cycle. The method is described with reference to Figure 5.7.

Within the vacuum drying chamber **10** on support **13** is placed a container **11** in which is confined the material **12** to be dried. Heat can be supplied to the material **12** by a circulating fluid heating system **14**.

FIGURE 5.7: DETERMINATION OF PRACTICAL END POINT OF DRYING

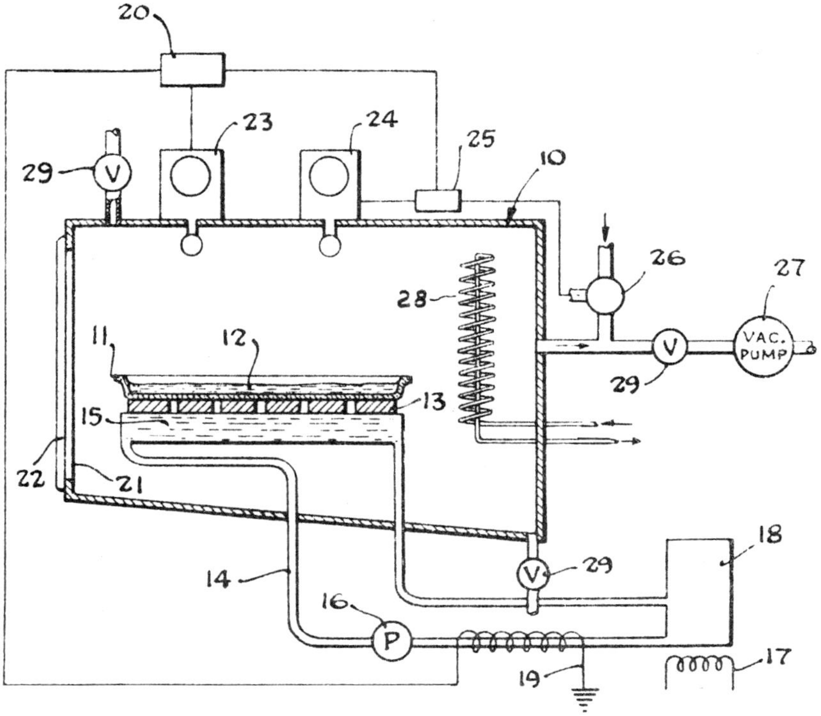

Source: U.S. Patent 3,259,991

The material to be dried is introduced into the vacuum drying chamber through opening **21** closed by a door **22**. Within the vacuum drying chamber there is provided a thermocouple type vacuum pressure gauge **23** and an ionization type vacuum pressure gauge **24**, both of which are initially calibrated for a vacuum atmosphere of air. A pressure sensing control and transmitting device **20** is employed in conjunction with the thermocouple type gauge to actuate the heating means **19** at the desired point in the drying cycle.

Likewise, a pressure sensing control and transmitting device **25** is employed in conjunction with the ionization type gauge **24** to actuate a vacuum pressure control valve **26** located at the vacuum side of the suction pump **27**. The function of this pressure sensing control and transmitting device is to operate the vacuum control valve in order to maintain the desired operating pressure. When the thermocouple type gauge reaches the set point, at which time the major portion of the moisture has been removed from the material being dried, the pressure sensing control and transmitting device **20** will deactivate the pressure sensing control and transmitting device **25**, utilized in conjunction with the ionization gauge **24**, to actuate the vacuum pressure control valve **26**.

In this way, the pressure control valve which may be a vacuum control air bleed for example, is closed at the appropriate time to thereby decrease the pressure in the vacuum drying chamber and aid in removing the small amount of adsorbed moisture which remains in the material to be dried. When drying temperature-sensitive materials such as enzymes or proteins, the drying temperature can be maintained constant throughout the run.

In such a case, the thermocouple gauge **23** and pressure sensing and control device **20** can be used solely to reduce the pressure within the drying chamber at the desired point in the drying cycle, in the manner described.

Part II.
Equipment for Freeze Drying

CONTAINERS

EXTENDED SURFACE TRAYS

Ribbed Tray for Freeze Drying

W.H. Hamilton and D.E. D'Alessandro; U.S. Patent 3,247,602; April 26, 1966; assigned to Pennsalt Chemicals Corporation describe the manufacture of a ribbed tray for drying purposes which eliminates the need for welding the ribs to the bottom of the tray. There is a substantial clearance between the end of each rib and the end of the tray for simple cleaning of the tray. The manufacture of the tray is described with reference to Figure 6.1. A member is extruded from standard metal stock by any of the standard methods of extruding metal. The member (not shown) consists of a base member from which extend perpendicularly side walls and ribs. The side walls are parallel to each other and to the ribs. The ribs have a height slightly less than the height of the side walls. In the center of the bottom portion there are drilled holes between the ribs for more efficient heat transfer.

The next step in the manufacture of the freeze drying tray (shown in Figure 6.1a) is the milling of the side walls **28** and **30** and the ribs **32** along a line **38**. The side walls **28** and **30** and the ribs **32** are milled leaving a flap portion **36** of the bottom wall **26**. A second milling operation forms the third step in the manufacture of the freeze drying tray. This second milling step includes milling the still exposed ends of the ribs **32** along a line slightly inwardly and parallel to the line **38**.

The next step in the manufacture of the freeze drying tray is the folding of flap **36** along fold line **42** spaced from the mill line **38**. This can best be seen with respect to Figure 6.1b. The folding of the flap **36** along line **42** forms clearance spaces **44** (not shown) and **46** between the flap **36** and the ends of side walls **28** and **30**. In one instance, the clearance spaces **44** and **46** were approximately one-quarter inch wide. In use, a food specimen can be placed between the ribs **32**. It is to be understood that the holes have a cross-sectional area less than the cross-sectional area of the specimen F to be freeze dried.

FIGURE 6.1: RIBBED TRAY FOR FREEZE DRYING

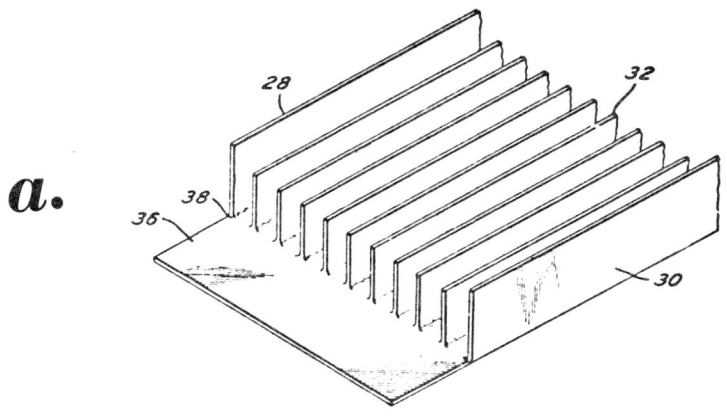

View of Extrusion After Second Step of Process

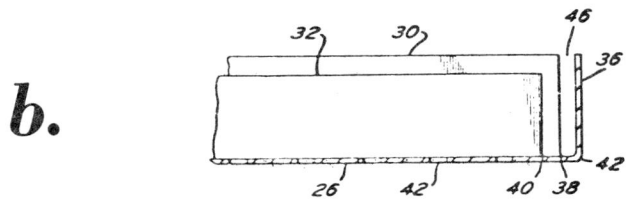

Cross-Sectional View of Tray

Source: U.S. Patent 3,247,602

The ribs **32** have a height slightly less than the height of the side walls **28** and **30** so as to prevent overflowing of the freeze drying tray. Clearance has been provided between the side walls **28** and **30** and the end wall **36** so that the tray may be easily cleaned and has no corners. The ribs **32** are provided with an even greater clearance to the end wall **36** so that the entire unit may be easily cleaned and reused.

High Thermal Conductivity Tray

The process of *P.F. Waltrich; U.S. Patent 3,545,097; December 8, 1970; assigned to Pennwalt Corporation* relates to an extended surface ribbed tray, especially for freeze drying of products, comprising a high thermal conductivity plastic composition molded or extruded with a high thermal conductivity filler and including an appropriate pigment for radiant or conductive heating. This finned or ribbed tray which is molded of a plastic composition is lighter in weight, easier to handle, and less expensive in cost than a comparable extruded aluminum tray while at the same time having comparable thermal conductivity and emissivity characteristics. The tray is fabricated by preferably molding or extruding it from a thermally conductive plastic composition. A preferred molding compound is a polycarbonate resin. The polycarbonate resins offer superior heat resistance and thermal conductivity and provide excellent dimensional stability and high corrosion and impact resistance in the temperature range between −65°F to +300°F, the range encompassed by freeze drying. Also, the polycarbonates have a specific gravity less than half that of aluminum.

In order to improve the thermal conductivity characteristic of the polycarbonate thermoplastic resin itself, it is desirable to incorporate a thermally conductive filler, such as graphite, copper, or aluminum powder, thereto in amounts of from between 10 to 40 parts per weight based on 100 parts of resin. A satisfactory composition is 25 parts per weight of graphite to 100 parts of resin.

It is also desirable to incorporate a dark colored pigment, such as ferrous oxide or carbon black, 1 to 3% by weight, in order to bring the emissivity characteristics down to a coefficient of 0.9. Aging and light stabilizers, molding lubricants and plasticizers may be added in appropriate quantities to suit the particular application or molding equipment. Additives, such as glass fiber, may also be incorporated for reinforcement. An example of a polycarbonate composition which will satisfy the nondegassing and thermal requirements of vacuum freeze drying, as well as the resistance to chemical attack and impact is:

	Parts by Weight
Polycarbonate resin	100
Graphite filler	25
Carbon black pigment	2
Metal stearate (alkali and alkaline earth)	0.5
p-Phenylenediamine	0.5

VENTING ELEMENTS TO FACILITATE BULK DRYING IN CONTAINERS

Wedge Shaped Venting Element

H. Eilenberg; U.S. Patent 3,529,362; September 22, 1970; assigned to Leybold-Heraeus-Verwaltung GmbH, Germany is concerned with facilitating the drying of bulk materials in a freeze drying container. The process proposes to provide a support means having an element which extends along the length of a container used in freeze drying operations. The support means element is water-vapor-permeable and serves as a venting element for water vapor drawn from the drying material.

The venting element has a wedge shaped cross section and is so disposed in the container that the narrower end portion of the wedge shaped cross section rests on the bottom of the container and the wider portion of the wedge shaped cross section extends vertically at least to the level of the upper surface of the material to be dried. By thus positioning the venting member within the material to be dried, openings in the venting element can advantageously be placed at that level where the ice-containing core of the material being dried is disposed during the freeze drying operation. Moreover, the wedge shaped configuration of the venting element advantageously provides a means for reaching water vapors trapped in the upper reaches (with reference to the bottom or base of the container) of the material being dried. The sloping sides of the venting element facilitate the collecting of water from the relatively thick material at the bottom of the container. Thus, the venting element, in effect, provides a means or channel by which water vapor is drawn from the material being dried. The angle between the sides of the wedge shaped venting member can be relatively narrow so that the volume, hence the load capacity, of the container is not unnecessarily reduced.

In another form the venting element is a good heat conductor and is in heat-conducting contact with the bottom and/or wall surfaces of the container. In addition to drawing water vapor from the material being dried, heat can be transferred to the material to facilitate the drying operation. In such an arrangement, the venting element is formed as a perforated or sieve-like material with good heat-conducting property. The use of suitable sintered, screened, or sponge-like materials is also possible.

In another specification, the venting element is constructed as a separable element so that it may be removed as a unit from the container, as desired. It has also been found advantageous to provide partitioned portions within the container and a venting element within each portion, venting elements in adjacent partitioned portions being connected to one another.

Referring now to Figure 6.2, the venting elements **2** are centrally disposed with respect to the space between partition members **3** within the container **1**. The venting elements **2** are so arranged that the narrower end **4** of each element is at the bottom of the container. The wider end **5** is vertically disposed with respect to the bottom of the container so that it extends above the top surface of the material in the container being freeze dried.

FIGURE 6.2: WEDGE SHAPED VENTING ELEMENT

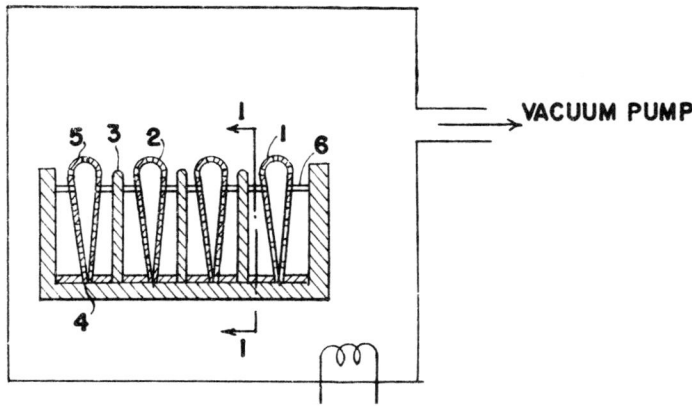

Source: U.S. Patent 3,529,362

Containers

The venting elements are connected to each other by spacers **6** which are used to hold the venting elements in position when the material to be dried is poured into the container.

Venting Channel Member in Form of a Partition

This process of *H. Eilenberg; U.S. Patent 3,590,496; July 6, 1971; assigned to Leybold-Heraeus-Verwaltung GmbH, Germany* involves the use of a water-vapor-permeable channel member provided in an insertable container for a freeze drying apparatus. The channel member is designed to provide a venting channel with respect to the surface of the material being dried in such container. The channel member is in the form of a partition disposed in the fill area of the container and extends along the entire length of a given horizontal inner surface of the container.

Specific forms of these partitions are disclosed. Some are described with reference to Figure 6.3. Referring to Figure 6.3a the partitions **20**, **21** and **22** are provided on an insert **36** which is provided with perforations and which is inserted into the container **1** and pressed to the bottom surface of container **1**. The partitions **20**, **21** and **22** are spaced from the intermediate wall members and container sidewalls, thus forming a space between such walls and the partitions. Moreover, the partitions are connected to extend over the intermediate sidewalls. The space thus defined provides a venting space in which vapors from the materials being freeze dried are collected.

Referring to Figure 6.3b partitions **23** and **24** having corrugated exterior surfaces are provided. Two corrugated surfaces or walls are provided and a plurality of pocket-type venting spaces or channels **37** which extend perpendicularly with respect to the bottom of container **1** is provided between these walls. The venting channels thus formed provide a considerable improvement in the gas and water vapor exchange within the container, during the freeze drying process. The corrugated surface can be zigzag shaped or sinusoidal shaped.

Referring to Figure 6.3c partitions **25**, **26** and **27** are permanently connected to the bottom of container **1** by a metallic bond.

FIGURE 6.3: VENTING CHANNEL MEMBER IN FORM OF A PARTITION

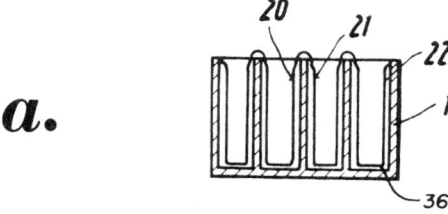

a.

Cross-Sectional View of Insertable Partition

(continued)

FIGURE 6.3: (continued)

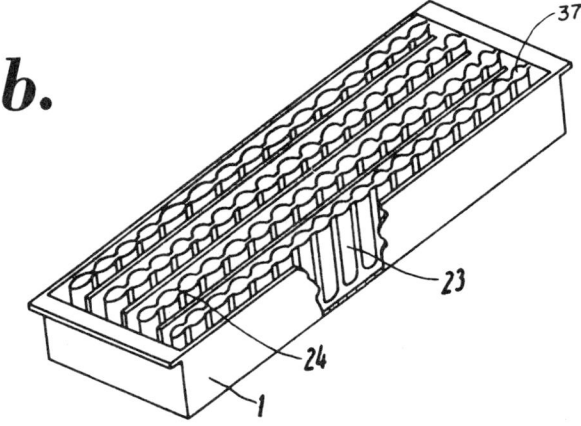

Perspective View—Venting Channel of Corrugated Material

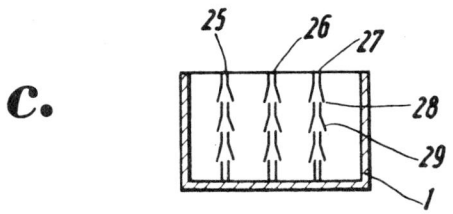

Cross-Sectional View of Permanently Connected Partition

Source: U.S. Patent 3,590,496

The partitions are double walled and are provided with slit-type openings **28** positioned opposite louvered covers **29**, which serve to direct gas or water vapor into the interior of channel members **25, 26** and **27** while they reduce the chances of the undesirable entry of the material being freeze dried into the space defined between the double walls of the partition. The permanently connected partitions provide excellent heat conduction and the advantages flowing therefrom are realized by the drying process.

Container with Vapor Venting Passages for Drying Granulated Material

H. Eilenberg and F.-J. Schmitz; U.S. Patent 3,401,468; September 17, 1968; assigned to Leybold-Anlagen Holding AG, Switzerland provide a container for the drying of granulated material which provides additional vent passages for the removal of vapor and other gases from the product. The container comprises an open topped tray-like member having bottom and side walls, and the combination therewith of a vapor permeable liner disposed adjacent to one or more of the walls, the liners being spaced from the walls to define vent chambers or passages for receiving vapor from the product and passing the vapor to the exterior of the container.

Such containers may be larger and deeper since vapor may be removed from the product beds through vent passages adjacent to the bottom or side walls of the container, as well as from the normally exposed upper surfaces of the beds. Through their use larger beds may be dried with acceptable efficiency, or smaller beds may be dried with much improved efficiency.

The process is illustrated with reference to Figure 6.4. The container includes a tray **1**, which is fabricated from a good heat conductive material such as aluminum, by an extrusion process, and consists of a bottom wall **12**, side walls **7**, and end walls **8**.

FIGURE 6.4: CONTAINER WITH VAPOR VENTING PASSAGES FOR GRANULATED MATERIAL

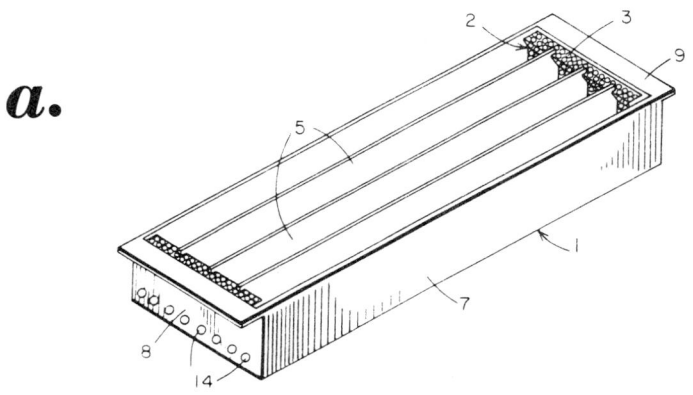

Perspective View of Product Container

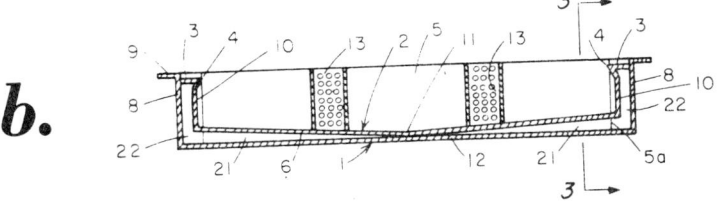

Longitudinal Section Taken in a Vertical Plane Through
One of the Compartments of Figure 6.4a

(continued)

FIGURE 6.4: (continued)

c.

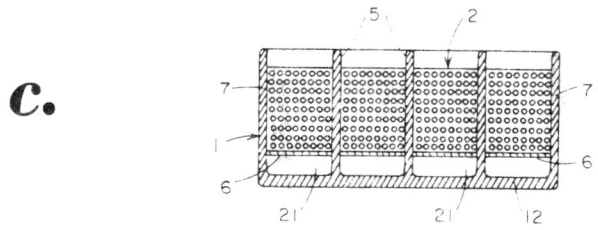

Transverse Section Along Line 3-3 of Figure 6.4b

Source: U.S. Patent 3,401,468

The end walls are bent outwardly to form flanges 9 which lie in the plane defined by the top of the tray. The tray also includes a plurality of longitudinal ribs 5 which form partitions dividing the tray into four longitudinal compartments. These partitions do not extend to the end walls, but terminate short of the end walls as indicated by the terminal edge 5a in Figure 6.4b.

The container also includes a liner 2 fabricated from a good heat conducting material, such as aluminum or other good heat conducting metal. The liner may take the form of a metallic sheet having generally uniformly distributed perforations or slit openings which will permit the passage of vapor while retaining the granulated material; or it may take the form of a wire mesh screen. The liner consists of four parallel strip portions, adapted to be received in the four compartments of the tray 1, and common connecting end portions. The strip portions of the liner form generally horizontal bottom walls 6; and the common end portions of the liner are bent upwardly to form generally vertical end walls 10 and then they are bent outwardly to form generally horizontal support portions 3.

The width of the support portions which is the longitudinal dimension in relation to the length of the tray is somewhat greater than the distance between the ends 5a of the partitions 5 and of the tray end walls 8. A relatively sharp ridge 4 is provided where the support portions of the liner meet the end walls, and this relatively sharp ridge is provided so that the support portion 3 might be wedged into the space between the tray end wall 8 and partitions 5 for the purpose of supporting the ends of the liner 2. The height of the liner end walls 10 is less than the height of the tray 8 so that the bottom wall 6 of the liner is supported in spaced relation to the bottom wall 12 of the tray, at the ends of the tray.

The liner bottom wall 6, while generally horizontal, is actually bent to define a very shallow V in longitudinal section as seen in Figure 6.4b, the V being defined by a bend 11 intermediate the ends of the strips defining the bottom walls 6. When the container is filled with granulated product to a level near the top of the container, the granulated product will be supported by the liner 2 so that the vent spaces will be maintained underneath the bed of granulated product and at the ends thereof. When the container is now placed in a vacuum chamber, the vapor within the product may pass either upwardly to the surrounding atmosphere at the upper surface of the bed or may pass downwardly through the vapor permeable bottom wall **6** of the liner **2**, and then pass to the exterior of the container through the vent spaces or passages **21** and **22**. It may be seen that the water vapor, or other gas, need pass through only half the distance withwithin the bed, which would otherwise be required were it not for the liner **2** and defined vent spaces.

In Figure 6.4b there is shown another form of vent passage which consists of tubular vent members **13** supported on the bottom wall **6** of the liner and extending upwardly to a point adjacent to the top of the container. These tubular vent members **13** may merely define vents for the horizontal vent spaces **21**, and in that sense may be fabricated of a solid material; or the vent members **13** may be formed of a vapor permeable material similar to that of which the liner **2** is fabricated. When constructed of a vapor permeable material, of course, the vapor from the product bed may pass laterally to the tubular vent members **13**.

OTHER EQUIPMENT

Assembly of Stacked Containers Rotatable over 90° Arc

A. Baer and M. Tribout; U.S. Patent 3,281,954; November 1, 1966; assigned to Compagnie Francaise Thomson-Houston, France provide a stack of tray-like containers for freeze drying.

Provision is made for bodily rotating an assembly of stacked, parallel spaced containers between two positions over an arc of 90°, whereby in the first position the tray-like containers extend horizontally and are stacked vertically for being easily and quickly filled with metered charges of material through cascading overflow down the containers of the stack, while in the second position the containers extend vertically and are spaced horizontally for being easily and quickly discharged of their contents into underlying receiver means, and flushed clean if desired.

The process is illustrated with reference to Figure 6.5, for the freeze drying of milk. The frame **121** supported a stack of thirty trays **116** each 1.50 m^2 in plan contour and 22 mm deep spaced 50 mm apart. The stack was thus about 1.50 m high, providing a generally cubic assembly. The vacuum vessel **202** had a cylindrical body section 2.50 m diameter and 3 m high. The frustoconical bottom funnel **135** was 0.75 m deep while the domed top had a vertical chord of 0.20 m, so that the total capacity of the vacuum enclosure was 16 m^3.

The overflow nozzles of all the trays were adjusted so that the trays were all charged to a depth of 15 mm. This advantageously small depth of the liquid charge to be subjected to freeze drying is made possible by the relatively very large total surface area simultaneously exposed to vacuum in the stack of thirty trays, i.e., $30 \times 1.5^2 = 67.5$ m^2. The total charge present in the trays per treating cycle is somewhat greater than 1 m^3 or 1 metric ton. The skim milk was stored in overhead storage containers (not shown) maintained at a temperature of about +4°C and provided with agitators for homogenization, and was delivered through pipe **129** by gravity. The fat content of the skim milk was 10 gpl. At the time of charging, the interior of vessel **202** was held at a pressure of 6 torrs, which is approximately the vapor pressure of milk at 4°C. The charging step proceeded with the milk cascading down the stacked trays so as to fill each tray with an accurately metered charge 15 mm deep. The excess milk from the bottom tray was discharged through means not shown. After the trays **116** were filled with the prescribed charge of skim milk, cooling liquid at a temperature of –30°C was circulated through the piping **128** and the associated flow passages of the trays.

The temperature control liquid used was a 1/1 water/glycol mixture and was cooled to the indicated temperature in a separate and conventional cooling plant not shown. The charge of skim milk quickly froze and after about one hour circulation of the cooling liquid the milk reached a temperature of –25°C. At this point the flow of cooling liquid was cut off and the vacuum pump **160** was connected to apply a vacuum of 0.1 torr through condenser **133** to the interior of vessel **202** by means of the valving shown.

FIGURE 6.5: ASSEMBLY OF STACKED CONTAINERS ROTATABLE OVER 90° ARC

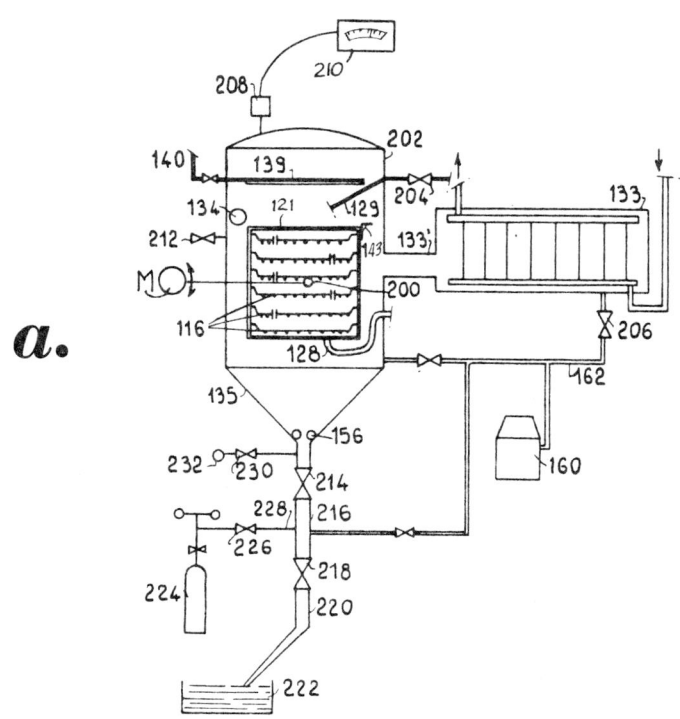

Freeze Drying Installation Shown in Upright Position
(continued)

FIGURE 6.5: (continued)

b.

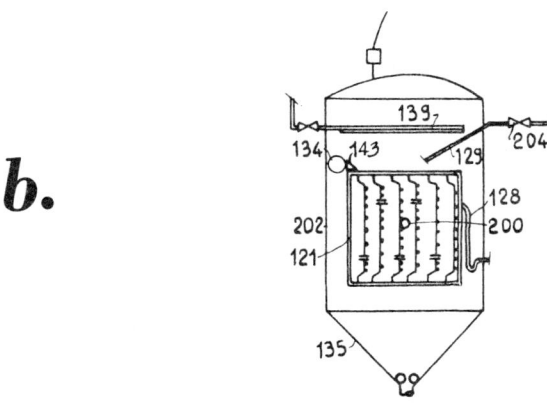

Tray Assembly of Figure 6.5a Shown in
Pivoted Position

c.

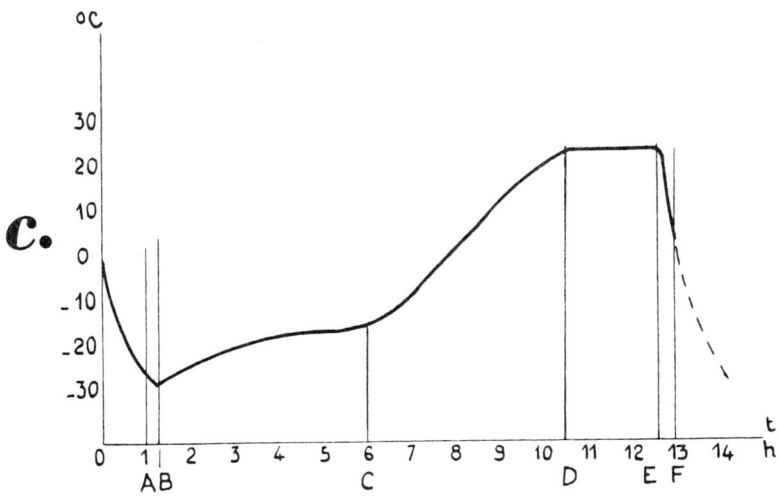

Typical Freeze Drying Cycle

Source: U.S. Patent 3,281,954

As will be noted from the temperature/time graph (Figure 6.5c), where the instant of vacuum application is indicated as the point A on the time axis, the temperature of the milk thereupon dropped a few additional degrees to about −27°C due to the action of the vacuum.

At this point, point B on the graph, the frozen mass can be considered completely free from any liquid particles therein, and the freeze drying stage proper, or sublimation stage, of the process is commenced. The flow and temperature of the temperature control liquid was controlled so as to provide in the charge of solid milk a temperature cycle in accordance with the indicated graph. From point B to point C about five hours later, the temperature was increased slowly from -27°C to about -15°C, then from point C to point D (about 4½ hr) the rate of temperature rise in the charge was increased until the temperature reached about +25°C, and the temperature was then held stationary at this value for about two hours (point E).

During the controlled application of heat under vacuum, commencing at point B, the charge of frozen milk exposed to vacuum in the trays **116** is subjected to sublimation, that is, the particles of ice therein are directly converted to vapor without passing through the liquid state, and the vapor is discharged through pipe **133'** to condenser **133** where it recondenses. The ultimate stage of the cycle conducted at relatively high constant temperature (to point E) is a final evaporation stage and is found necessary to remove the moisture down to a final residual value of 2% water content, leaving the skim milk in the form of a dry powder in the trays. The flow of temperature control liquid in piping **128** was then cut off.

Frame **121** was rotated about shaft **200** to the tipped position. Most of the milk powder dropped into the bottom of funnel **135**, and vibrator **134** was operated for a short time to remove residual powder clinging to the trays. Owing to the presence of vacuum in the enclosure the powder particles drop without creating flying dust. Lock inlet valve **214** was then opened and crusher device **156** was operated to pass the milk powder into lock chamber **216** in the form of a free-flowing powder product.

When the full charge of milk powder had been collected in the lock chamber, nitrogen valve **216** was opened to fill the lock chamber with nitrogen at atmospheric pressure, while avoiding a danger of oxidation. Lock outlet valve **218** was finally opened to discharge the product to receiver **222** for subsequent processing. After the milk product has been discharged out of the vessel **202**, valve **140** is operated to deliver washing liquid through pipe **139** and flush the trays **116** in their vertical tipped condition. First however, the vent valve **212** is opened to put some air into the vessel, to a pressure of about 24 torrs which is the vapor pressure of water at 25°C, the temperature of the washing water used.

The washing water which has streamed down the trays **116** collects in funnel **135** and is discharged by way of valve **230**, opened at this time, and pump **232**. The frame **121** can be rotated back to its upright processing position and the described cycle can be repeated. It will be noted from the graph that the freeze drying cycle described has a duration of about 13 hours, and that the freeze drying unit shown therefore has a production rate of about 1.85 metric tons per 24 hour day.

Housing Means for Drying in Sealable Containers

U. Hackenberg; U.S. Patent 3,286,365; November 22, 1966; assigned to Leybold Anlagen Holding AG, Switzerland provides a freeze drying apparatus which greatly reduces the requirements for product and container handling and in addition permits efficient transfer of heat to the drying product during the freeze drying process without danger of product heat damage. There is provided a freeze drying installation which comprises heating plates for furnishing conduction heating to the product being dried, a product housing adapted to shield the product from direct heat radiation and to permit escape of sublimating water vapor, and an additional product container adapted for placement within the product housing thereby allowing use of the additional container as the final packaging for the dried product.

The product housing is constructed of metal walls having a thickness of at least 1 mm so as to exhibit good heat conduction properties. In addition the product housing can contain optically dense water vapor escape openings which shield the product from direct heat radiation.

Referring now to Figure 6.6 there is shown a vacuum chamber wall **11** which encloses a plurality of vertically ribbed heating plates **12**. The vertical central column **13** supports the horizontal bars **14** to which are attached the horizontal angle bars **15**. The top wall extensions **16** of the rectangular product housing **17** provide shoulders **18** which are demountably supported by the horizontal angle portions of the angle bars.

FIGURE 6.6: HOUSING MEANS FOR DRYING IN SEALABLE CONTAINERS

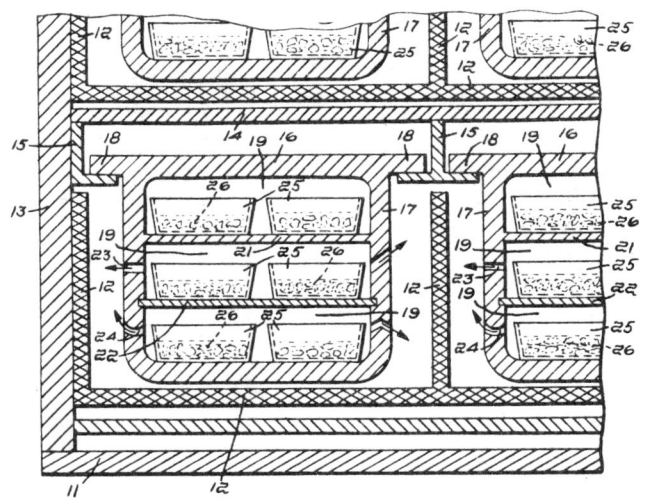

Cross Sectional View of Segment of Apparatus

Source: U.S. Patent 3,286,365

The product housing is divided into vertical sections **19** by dividers which can be either an integral part **21** of the housing or removable parts **22** thereof. Water vapor escape openings **23** are cut in the side wall of the housing and preferably these openings can be made optically dense by providing a directional change as shown in the openings **24**. Supported in the vertical sections **19** by the bottom wall of housing **17** and dividers **21, 22** are the tray containers **25** filled with a product **26** to be dried. The containers are passed into the housing through the open ends thereof.

During operation of the apparatus the containers **25** are filled with a frozen product **26** to be dried and placed in the housing **17**. After the loaded housing has been positioned within the vacuum chamber, the chamber is evacuated to a pressure of about 10^{-1} mm Hg and the heater plates **12** are energized. The top, bottom and side walls of the housing shield the product from direct heat radiation from the heating plates while the open end walls and openings **23, 24** permit escape of sublimating water vapor from the frozen product.

The walls of the housing are preferably composed of metal and have a thickness of at least 1 mm (preferably about 2 mm) so as to provide a good heat conduction path to the containers and product. Upon completion of the freeze drying process, the containers still filled with the dried product can be sealed in a gastight manner by suitable sealing equipment (not shown). This sealing can be performed in an inert atmosphere if so desired. The completely packaged product is then ready for commercial distribution. The housing assemblies are immediately available for reuse without any cleaning requirement since they have not been in direct contact with the product itself.

Drying in Small Bottles with Freedom from Contamination

M. Seligman; U.S. Patent 3,286,366; November 22, 1966; provides a method of freeze drying in small containers, while maintaining the product free from contamination.

Referring to Figure 6.7 there is provided a freeze dryer **10** having a cover **11**. A vacuum source **12** (not shown) is connected to the freeze dryer so that it may be evacuated when the cover has been positioned in a sealed relationship as illustrated. The freeze dryer is provided with a cooling coil and condenser **8** having a refrigeration system connected thereto. A fan **9** is provided to move the atmosphere within the container for cooling and also there is provided a heating element **7** connected to a switch, a potential supply and a variable resistor control. The freeze dryer is also provided with a movable platen **14** that is retained in the raised position as illustrated by a plurality of springs **15** and restricted in its upward movement by a plurality of stop nuts **13**.

The platen is also provided with an expandable air cylinder **16** such as the bellows illustrated. The upper limit of the bellows is retained by a structural element **17** which is affixed to the container and may be further retained by a plurality of bolts **18** by providing a plurality of nuts **19** to position element **17**. An air supply **20** is connected to the bellows through port **21** using valve **22**. In use a plurality of bottles **25** may be stacked on the bottom of the container **10**. The bottles are cooled to freeze the contents by circulating the atmosphere surrounding the bottles, by utilizing a condenser and refrigerant coil **8** within the freeze drying chamber and providing a fan **9** behind the refrigerant coil. The bottles are subject to the lowering of their temperature to a predetermined point (just about freezing temperature of the contents of the bottle). This may be noted by a thermometer **6**.

FIGURE 6.7: DRYING IN SMALL BOTTLES WITH FREEDOM FROM CONTAMINATION

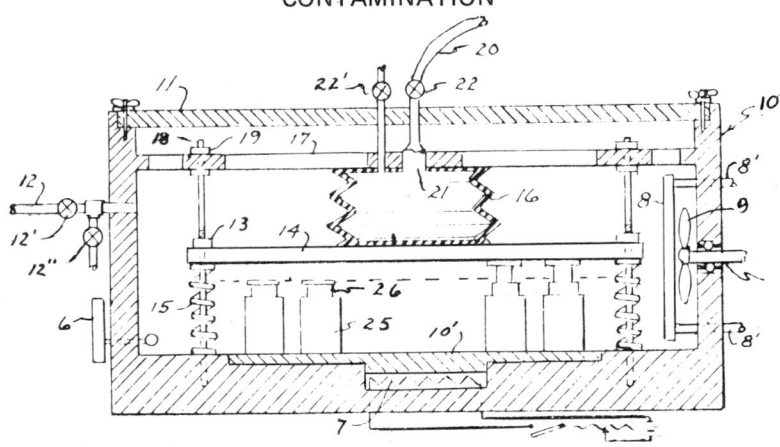

Source: U.S. Patent 3,286,366

With the initial freeze of the contents of bottles **25**, the platen is lowered to the dotted position and retained in this position. The vacuum source **12** is turned on (valve **12'**) to partially evacuate the air within the freeze dryer. As the pressure within container **10** drops to or approaches a difference of one atmosphere (between outside and inside) the platen is raised by suddenly releasing the pressure within the expandable cylinder **16**. The air pressure within the bottles being greater than the pressure surrounding the bottles causes the stoppers **26** to pop.

The platen **14** may be in a predetermined position above the bottles to thus prevent the stoppers from popping to the extent that they leave the neck of the bottle. Rather the stopper pops sufficiently to allow the exposure of the contents of the bottles to the conditions existing within the freeze dryer. After the contents of the bottles have all been dried, that is, after the water vapor has been released from the frozen content to the degree desired, valve **22** may be actuated or opened permitting the air pressure from **20** to expand bellows **16** moving the platen **14** downward to reinsert the stoppers.

When the stoppers have been sealed in the containers **25** the vacuum valve **12'** may be closed. The vacuum is then released by a release valve **12"**. A release valve **22'** may then be opened releasing pressure from the bellows **16** and allowing springs **15** to raise the platen **14** to the position illustrated. The cover **11** may then be removed and the containers may be removed from the vacuum chamber.

APPARATUS MODIFICATIONS

COMBINATION FREEZER-FREEZE DRYER

Double Walled Container

In a process developed by *G. Seffinga; U.S. Patent 3,298,108; January 17, 1967; assigned to SEC NV, Seffinga Engineering Co., Netherlands* the drier is so constructed that it serves also as the freezing apparatus, while the heat of solidification is removed by a circulating gas. With reference to Figure 7.1, the apparatus comprises a number of plates, expanded sheets, or the like, on which the product **2** is placed in layers. If the heat of sublimation is supplied by radiation, a radiation plate **1** is fitted parallel to the layer of product, on either side of it.

The apparatus is further provided with an air cooler-condenser **3**, which corresponds to the evaporator of the cooling plant. The air cooler-condenser consists of two parts, between which is arranged a fan **4**. Between the system of cooler **3** with fan **4** and the space in which the product is present a radiation screen **5** has been placed.

During the freezing process, which usually takes place at atmospheric pressure, a gas, such as air, is circulated by the fan(s) through the air cooler-condenser and over and in contact with the product. In the condenser, the gas is cooled; as this gas flows over and in contact with the product, heat of solidification is removed from the latter. After the product has been frozen, the fans are disconnected and the apparatus is evacuated, after which sublimation heat is supplied to the frozen product. The water vapor coming from the product condenses on the air cooler-condenser, which is cooled by the cooling plant also during the drying process.

The arrangement of the fan **4** between the two parts of the air cooler-condenser involves the advantage that the fan does not form a resistance to the water vapor flowing to the condenser. In consequence of the arrangement of the air cooler-condenser with the radiation screen **5**, heat losses from the space in which the product is present to the condenser are avoided, while this arrangement at the same time reduces the flow resistance during the freezing and drying processes to a minimum. The body of the container, which may have a diameter of 350 cm and a length of 10 m, has been constructed with double walls, for the reasons which follow.

During the freezing process the apparatus must be cooled and during the drying process it must be heated. The cooling and the heating capacity respectively required for this must be kept as small as possible. If the body were to consist of a single wall, in view of the reduced pressure prevailing in the container during the freeze drying process this wall would have to be made of fairly thick sheet steel, e.g., 10 mm thick. This sheet would then have to be cooled and heated alternately, which requires a large amount of energy. By providing two walls, viz an inner wall **6** and an outer wall **7**, with a space **8** in between, it is possible to set up in the space a pressure which is equal to or lower than the pressure prevailing in the container during the freeze drying process.

FIGURE 7.1: COMBINED FREEZER-DRIER IN DOUBLE WALLED CONTAINER

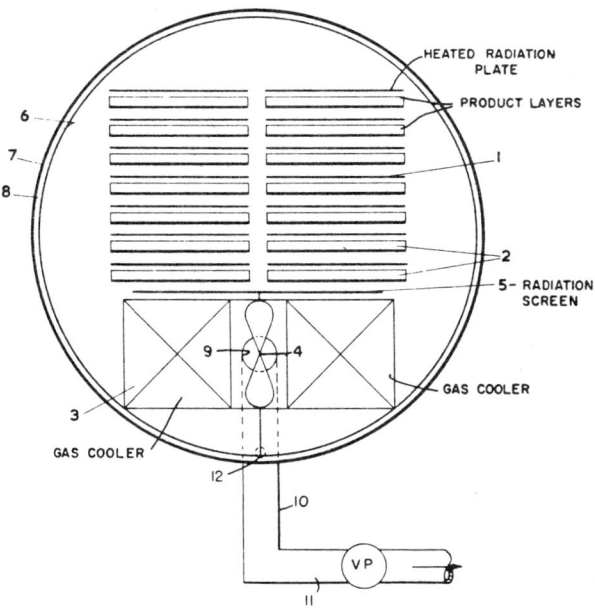

Source: U.S. Patent 3,298,108

The outer wall will then be subject to external over-pressure, and will therefore have to be constructed to be thick, but the inner wall will only be exposed to internal over-pressure, in view of which this wall can be made of much thinner sheet steel, e.g., 2 mm thick. When the space is used at the same time to insulate the container, it is only the thinner inner wall that has to be cooled and heated respectively, not the thicker outer wall. For maintaining the vacuum required during the freeze-drying an outlet **9** leading from the space inside the container is connected with a vacuum pipe **10** having an evacuating device **VP** in **11**. The vacuum pipe may also communicate, as through a connection **12**, with the space between the container walls.

Feed Frozen on Continuous Belt

R.M. Stinchfield; U.S. Patent 3,218,731; Nov. 23, 1965; assigned to Arthur D. Little, Inc. describes a method of introducing feed material continuously to the vacuum chamber of a freeze dryer which utilizes a continuous surface for support and conveyance of the material during drying. The primary object is to feed material continuously to a freeze dryer and, moreover, to prefreeze the material to the desired operating temperature before exposure of the material to the vacuum of the chamber. It is desirable to positively exclude air from entering the vacuum chamber when the feed material is liquid, and to minimize air introduction when the feed material is solid.

The liquid feed material is introduced through an extended flat housing disposed along a short portion of the top of the dryer belt or other continuous surface forming therewith a freezing chamber. The feed is frozen on the belt or surface while in the housing and before being subjected to the vacuum of the vacuum chamber, even though the belt upon which it is deposited is contained within the vacuum chamber. Refrigeration means are positioned beneath the belt and opposing the extended flat housing. The housing is provided with top and three sides, but no bottom. However, in use, it is open only at the end toward which the belt is traveling because the three sides are in a sliding seal contact with the belt.

The liquid feed is frozen prior to emerging through the open end of the housing, at which point it acts as a pressure seal to prevent air or excess feed from entering the vacuum chamber. An added advantage is that the feed material has intimate contact with the belt by being frozen to it and will thus have better heat transfer characteristics in the subsequent drying. The process is described with reference to Figure 7.2.

Vacuum chamber **11** is provided with a continuous belt **15** moving over roll **17** which is mounted on axes **19** and driven in any appropriate manner. The belt is of good heat conductivity, e.g., of aluminum or stainless steel. A similar roll (not shown) is located at the other end of the chamber, together with suitable means (scraper blade) for removing the dried product. This dried product may be taken out of the vacuum chamber through any suitable air locks. Adjacent the end of the belt near roll **17** is located refrigerating means **21**, provided with inlet **23** and an outlet for introducing and removing refrigerating fluid **27**. On the opposite side of the belt is the extended flat housing **31**, forming with the belt a freezing chamber. This housing is fed through conduit **33** projecting through the wall of the vacuum chamber.

Thermocouples **35—35** may be provided along the top of housing **31**. The feed enters in the form of liquid **37** and becomes frozen as it travels along the belt and through the housing toward the exit end **40**. This frozen product is represented by the numeral **39**. The refrigerating means **21** consists generally of a bottom plate **41** and sides **43** which confine the refrigerating fluid **27** against the bottom of the belt **15**. The sides **43** may be provided with low-friction edges **45** (Figure 7.2c). These low-friction edges may also constitute the entire side wall of the refrigerating means **21** (not shown).

FIGURE 7.2: FEED FROZEN ON CONTINUOUS BELT

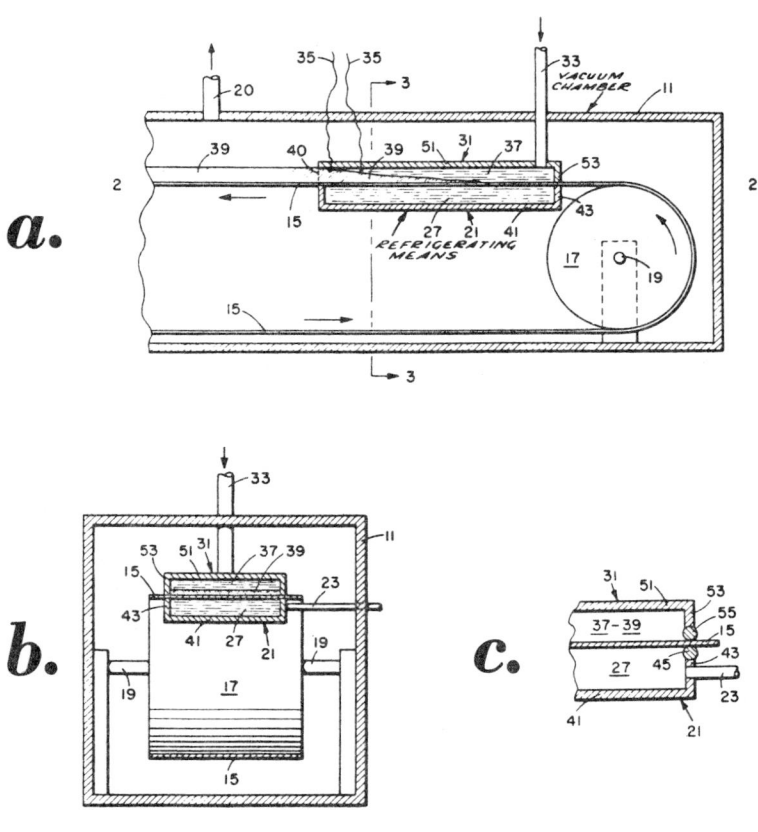

(a) Sectional Side View of Apparatus
(b) End Sectional View Along 3–3 of Figure 7.2a
(c) Details of Housing and Refrigerating Means

Source: U.S. Patent 3,218,731

The housing **31** consists of a top plate **51** and sides **53** which extend all around except at exit end **40**. As in the case of the refrigerating means, these sides may have low friction edges **55** which may constitute part of the sides as shown in Figure 7.2c or the entire sides (not shown). The top may be made of transparent material in order to observe the progress of the liquid and frozen product through the housing; however, this is not necessary and the thermocouples **35** can be used instead to determine the condition of the feed material as it progresses through the housing. Although this process is especially useful for the freeze drying of

Isolating Valve Mechanism in Shield Between Heater and Condenser

The object of this process of *U. Hackenberg, H. Kamps and H. Rink; U.S. Patent 3,270,434; September 6, 1966; assigned to Leybold Anlagen Holding AG, Switzerland* is to provide a freeze drying installation which exhibits the economy and advantages of a condenser element positioned directly in the drying chamber but which also provides efficient, compatible use of the installation's condenser and vacuum pumps as well as allowing the regulation of conductance between the drying chamber and the condenser elements. The freeze drying installation provides a closed gas circulation path between a condenser element and the material being processed in a drying chamber and includes a valve mechanism permitting gas isolation of the condenser element from the material being dried.

The isolating valve mechanism comprises a pair of perpendicularly aligned, parallel plates each possessing gas circulation openings which are not in perpendicular alignment so as to be closed upon surface contact of the parallel plates. The isolating valve mechanism is also adapted to provide a radiant heat shield between material heating elements and the condenser device. The process is described with reference to Figure 7.3.

In the operation of the freeze drying installation, the closed evacuable chamber 11 is first sterilized by a quantity of steam directed into the chamber through the lower aperture 12. The steam is supplied by a conventional steam producer (not shown) connected to the connecting flange 61. The steam supply device is then sealed from the chamber by a suitable valve mechanism (not shown). With the chamber interior sterilized the cover plates 15 and 31 are opened and the heating plate assembly 62 removed. A plurality of open flasks 63 filled with a material 64 to be freeze dried, such as for example a pharmaceutical product, are positioned on the lower heating plate 51. The heating plate assembly is then lowered by the carrying handle 58 into the chamber internal portion 27 and attached with the holding nut 53. The cover plate 31 is then replaced by the cover plate 15 rotated into alignment with the top flange 14 and sealed thereto.

The freezing of the material 64 is begun with the housing 26 in its lowest position A. A supply of cooling fluid is circulated through the condenser coil 38 from a conventional cooling medium source (not shown) connected to the inlet 39 and outlet 41. Rotation of the fan 24 is started by a suitable driver (not shown) connected to the rotary feed-through 23. The rotating fan causes air in the chamber 11 to circulate in a closed gas circulation path which passes the condenser coil 38 and the internal chamber portion 27 containing the product 64.

This closed path extends from the interior of auxiliary housing 42 which encloses the condenser 38, through the apertures 45 in lower circular plate 32, through the apertures 46 and 47 in circular valve plate 43, through the internal chamber portion 27, out of the housing 26 through the rotating fan 24, and down the external annular portion 28 to again enter the open end of the auxiliary housing 42. This closed circulation path provided by the divider housing 26 greatly reduces the time required to freeze the product **64** within the evacuable chamber. On completion of the freezing process the fan **24** is de-energized and the housing **26** raised to its intermediate position **B** by upward movement of the central column **33**. In this position the housing will engage an annular sealing gasket **65** attached to the annular flange **16** of cover plate **15**. A suitable vacuum pump (not shown) connected to the rotatable evacuation tube **21** by a flexible vacuum line and vacuum valve (both not shown) is then energized to reduce the pressure in the internal chamber portion **27** to a desired operating pressure of, for example, 10^{-1} millimeters of mercury.

FIGURE 7.3: ISOLATING VALVE MECHANISM IN SHIELD BETWEEN HEATER AND CONDENSER

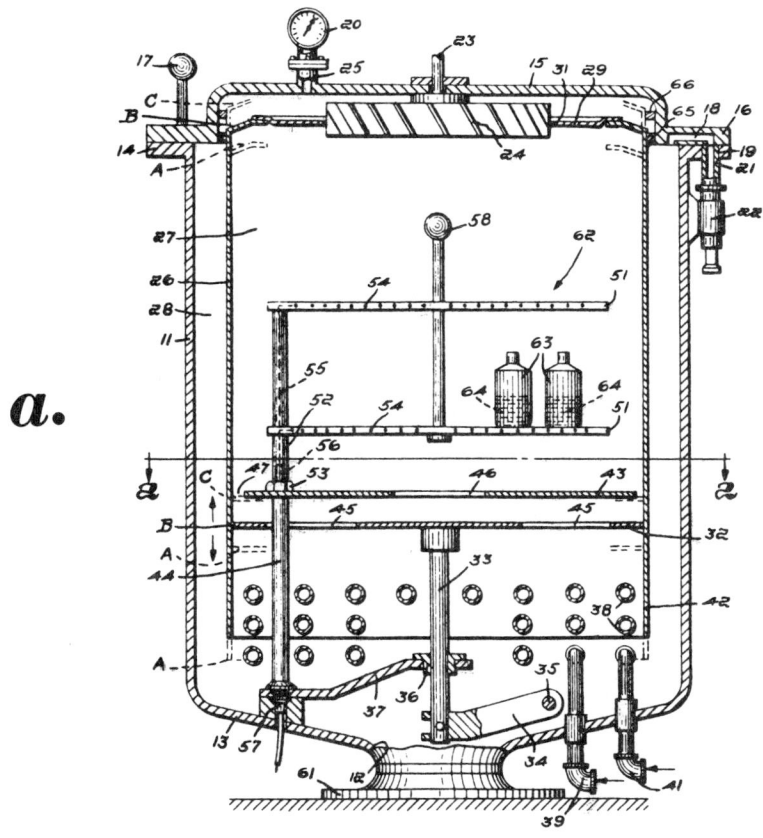

a.

Longitudinal Cross Section Through Freeze Drying Apparatus

(continued)

FIGURE 7.3: (continued)

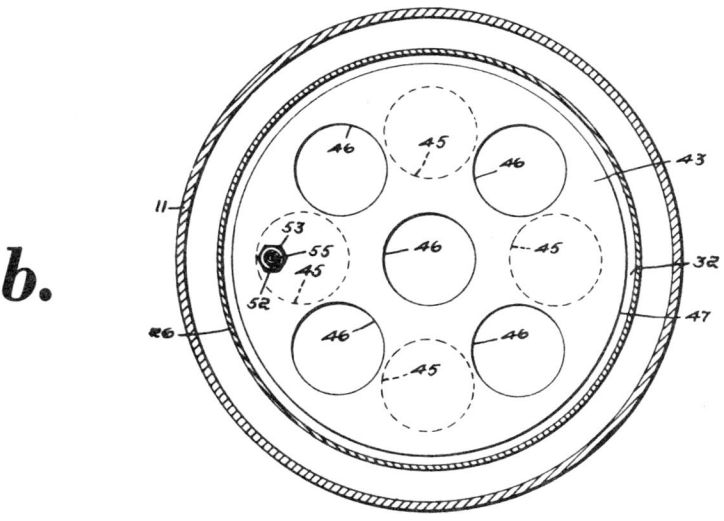

b.

Horizontal Section of Figure 7.3a

Source: U.S. Patent 3,270,434

During this drying portion of the freeze drying process, water vapor sublimating from the frozen product **64** in the well-known manner will be drawn from the internal chamber portion **27** through apertures **45, 46,** and **47** to be condensed on the condenser coil **38**. Noncondensable gases mixed with the water vapor are pumped out of the open lower end of auxiliary housing **42**, up the external annular portion **28**, and out of hollow channel **18** and tubes **21** by the attached vacuum pump. Since all paths between the upper end of housing **26** and the vacuum line **21** have been closed by the sealing gasket **65**, all removed gas must pass the condenser element **38** before reaching the evacuation line **21**. This is a very desirable arrangement since substantially all condensable gas in the mixture will be condensed on the condenser coil **38** thereby substantially reducing the gas load on the connected vacuum pump.

When a degree of product dryness test is desired, the housing is raised to its top position **C** forming an additional seal between the housing **26** and the annular gasket **66**. In this position the circular plate **32** is in surface contact with the valve plate **43** thereby sealing the openings **45, 46,** and **47** and isolating the internal chamber portion **27** from the condenser and vacuum pumps.

The resulting pressure rise in the housing **26** will be indicated by the pressure gauge **20** which is in gas communication therewith through the fan **24**. This pressure rise indication can then be used to determine the stage of the drying process. After the pressure rise has been noted, the housing is returned to position **B** and the drying process continues in the manner described above.

Vertical Tube Dryer for Freeze Drying Liquids

The process developed by *G. Seffinga; U.S. Patent 3,264,745; August 9, 1966; assigned to SEC NV, Seffinga Engineering Company, Netherlands* makes it possible for the freezing and the freeze drying operations to take place in a single apparatus. The tubes of a bank of tubes are filled with the material to be treated, freezing medium is circulated about the tubes until a layer has frozen onto the inner walls of the tubes, the nonfrozen material is removed, heating medium is circulated about the tubes while the layer frozen in them is subjected to vacuum for the sublimation of the ice from the frozen layer of material until the ice has completely sublimed, the interior of the tubes at the same time being connected with a discharge device for the water vapor, and finally the material is removed from the tubes.

Such a tube drier presents the advantage that the layer of product has much less tendency to be detached from the tube than from a dish, for instance. Owing to the improved thermal contact between product and wall the drying time can be reduced and a lower temperature of the heating medium will suffice. For a given capacity the tube drier as an apparatus is simpler and less expensive than a plant in which freezing takes place separately and drying dishes are used. Moreover, with such a drier it is possible without much difficulty to subject large quantities of material under sterile nonoxidative conditions to freeze drying.

The vertical tube drier according to Figures 7.4a and 7.4b comprises a sheath **1** enclosing tubes **2** which end at the top and the bottom respectively in spaces **3** and **4**, respectively, which are formed by an upper lid **5** and a lower lid **6** respectively. The upper space **3** moreover communicates with a source of vacuum through a discharge pipe **7** for the water vapor formed. Connected to the lower lid which is attached to a sheath **1** at **8** by means of a hinge, is a conduit **9** with a shut-off valve **10**.

The compartment or space **11** within the sheath not occupied by the tubes **2** is shut off both at the top and at the bottom from the spaces **3** and **4** respectively. Encircling this space **11** both at the top and at the bottom are annular channels **12** and **13** respectively, which communicate with the space. These channels are furnished with connections **14** and **15** respectively.

The material to be frozen and dried, such as a liquid, is introduced into the drier until the tubes **2** are completely filled with it. Subsequently via connection **14** and the channel **12** freezing medium is admitted round the tubes and in the space **11** and discharged via channel **13** and connection **15**.

After a sufficiently thick layer of the product has frozen onto the inside of the tubes, the supply of the freezing medium is stopped, whereupon via the conduit **9** and shut-off valve **10** the liquid still present is drawn off. Via connections **15** and channel **13** heating medium is subsequently admitted round the tubes in the space **11** and discharged via channel **12** and connection **14**.

At the same time the discharge pipe **7** is connected to a vacuum forming discharge device for the water vapor. When the ice from the frozen product has completely sublimed, the supply of the heating medium is stopped and the connection with the discharge device for the water vapor is interrupted. Then, the lid **6** is turned back about the hinge **8** and the product is removed from the tubes.

FIGURE 7.4: VERTICAL TUBE DRYER FOR FREEZE DRYING LIQUIDS

a.

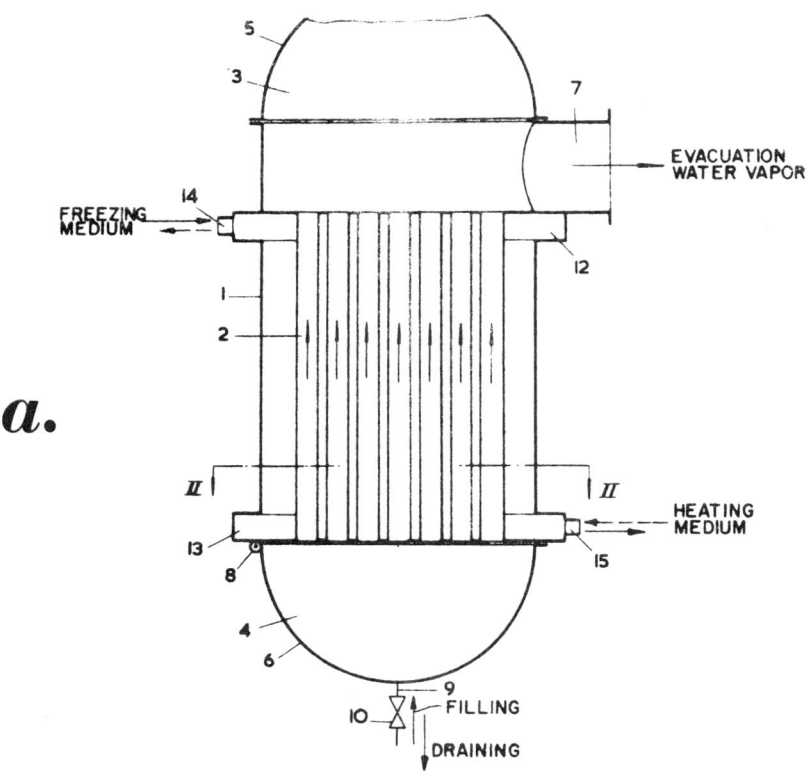

b.

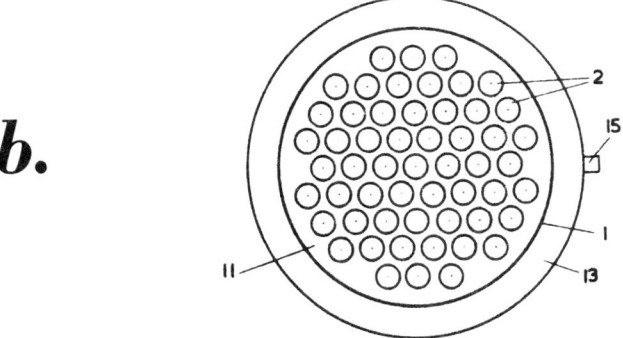

Source: U.S. Patent 3,264,745

VIBRATORY INSTALLATIONS

Vibration by Means Outside Vessel

R.C. Bonteil; U.S. Patent 3,733,716; May 22, 1973; assigned to Matsushita Electric Works, Ltd., Japan describes a lyophilization installation comprising a vacuum vessel having a plate rigidly fixed inside the vessel and upon which, in operation, a vibration bed is produced and particles of a product to be lyophilized are dried. Vibration means are provided and connected to the vessel for vibrating the vessel as a whole. The vacuum vessel is constructed to define, above the plate, an increasing flow cross section for vapors liberated from the particles of the product. The process is described with reference to Figure 7.5.

FIGURE 7.5: VIBRATION BY MEANS OUTSIDE VESSEL

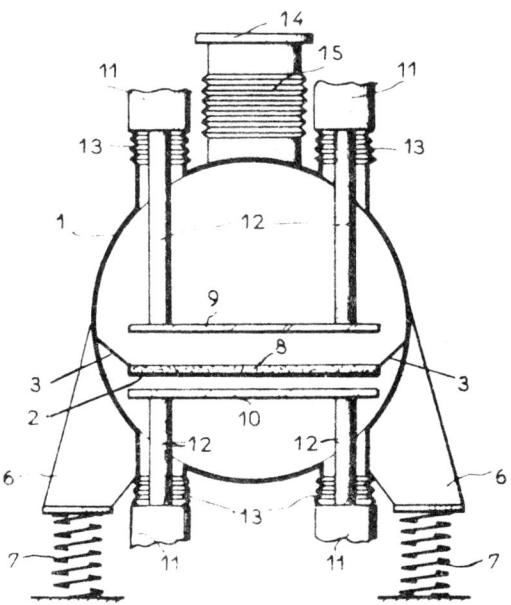

Source: U.S. Patent 3,733,716

The installation comprises a vacuum vessel **1** of cylindrical shape and horizontally disposed. Inside the vessel there is a horizontal plate **2** adapted to support the particles of the product. The plate is disposed in the bottom part of the vessel and is rigidly connected to the side walls thereof by inclined ribs or rims **3**. The vessel rests on the ground by four feet **6** provided with elastic suspension elements **7** and is vibrated by external means (not shown). The vibrations to which the vessel is subjected are such that the particles of the product deposited on the plate **2** are transported from upstream (i.e., adjacent the inlet) to downstream (i.e., adjacent the outlet) in the form of a vibrating or fluidized bed **8**. Inside the vessel on each side of the plate there is disposed an upper heating plate **9** and a lower heating plate **10**. These two heating plates are rigidly connected to fixed supports **11** by uprights **12** which pass through the walls of the vessel in flexible connections **13** to ensure airtightness of the vessel. The water vapor, freed from the particles of the product during its drying, is discharged through a top opening **14** also provided with a flexible connection **15**, passing to a steam trap or condenser (not shown). In the course of their transport from upstream to downstream on the plate **2**, the particles of the product are dried by the heating plates **9** and **10** and the water vapor freed from them passes to the top opening **14**.

The entire vessel **1** is vibrated. The particles of the product liable to be deposited on the walls of the vessel fall back automatically into the vibrating bed **8** as a result of the vibration of the vessel and with the aid of the inclined rims **3**. The yield of this lyophilization installation is, therefore, improved and, in addition, as the inside walls of the vessel remain clean, the vessel does not need to be cleaned as frequently as the vessels of conventional lyophilization installations.

Since the plate **2** is disposed in the bottom part of the vessel **1**, this arrangement being made possible by the fact that the vessel has no vibration means disposed inside it, the flow cross section for water vapor liberated from the particles of the product first increases above the vibrating bed **8** and then decreases towards the top opening **14**. Consequently, the particles of the product which may be suspended in the flow of water vapor and run the risk of being entrained in the flow of water vapor to the trap undergo an abrupt drop in speed and tend to return to the vibration bed or contact the inside walls of the vessel by the action of gravity.

Vibrating Housing Within Vacuum Chamber

H. Eilenberg and F.-J. Schmitz; U.S. Patent 3,364,591; January 23, 1968; assigned to Leybold-Anlagen Holding AG, Switzerland provide freeze drying apparatus which will substantially reduce processing costs by providing maximum utilization of vacuum chamber volume and efficient heat transfer and vapor removal during the drying process.

There is provided a freeze drying apparatus having a housing with bottom, top and side wall portions which substantially isolate the interior volume of the housing from the interior of an enclosing evacuable chamber. The apparatus includes a vibrator device for producing, within the housing, vibratory motion of contained frozen material particles which are retained by the isolating top and side wall portions of the housing. By preventing escape of the agitated particles, the top and side wall housing portions permit the vibration and efficient drying of relatively thick layers of frozen particles in addition to providing desirable additional surface area for heat transferring contact therewith.

Referring to Figures 7.6a and 7.6b, there is shown within the vacuum chamber **11** the housing **12** having bottom wall **13**, top wall **14** and side walls **15**. The housing defines a volume **16** in which the frozen particles **17** are to be dried. Pressed into the top wall are a plurality of indentures **22** in each of which is formed an aperture **23**. As shown, the apertures lie in planes which are perpendicular to the bottom wall **13**. Also formed in the top wall **14**, at one end of the housing **12**, is the housing inlet opening **25**; while the housing outlet opening

Apparatus Modifications

26 is formed in the bottom wall 13 at the opposite end of the housing 12. The mechanical vibrator 27, of a conventional type, is attached to the bottom wall 13 adjacent the housing inlet 25. Mechanical vibration of the entire housing is produced by actuation of the vibrator 27, which is also adapted to permit vertical positioning of the inlet end of the housing.

The material feeder lock 31 provides communication between the supply hopper 32, and the vacuum chamber inlet 33, while connected to the vacuum chamber outlet 34 is the material discharge lock 35. Connected by valve 36 between the vacuum pump 37 and the vacuum chamber 11 is the refrigerated condenser 38. The hollow heating coil 41 is adapted for connection to a suitable source of heating fluid (not shown), so as to heat the encircled housing 12 which is preferably made of a good heat conducting material such as aluminum.

FIGURE 7.6: VIBRATING HOUSING WITHIN VACUUM CHAMBER

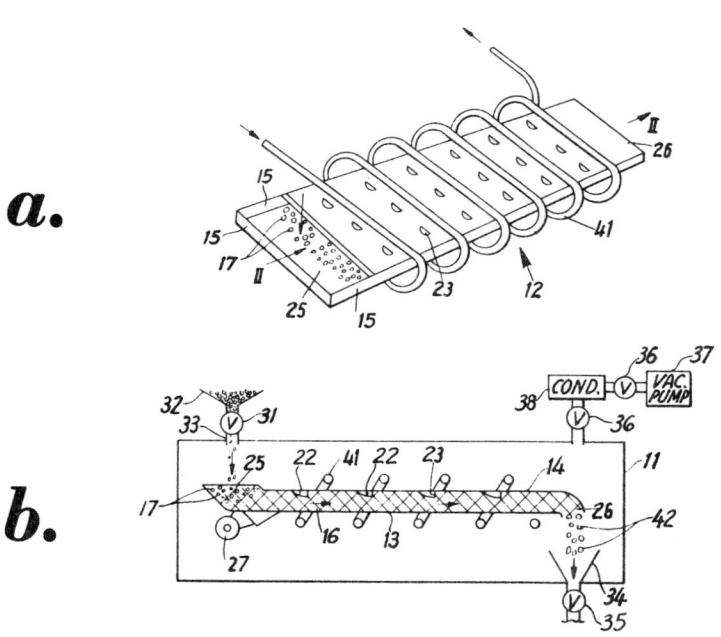

(a) Schematic View of Vibrating Housing
(b) Schematic Cross Section Along Line II—II of Figure 7.6a

Source: U.S. Patent 3,364,591

Typical operation occurs in the following manner: the inlet end of the housing is raised relative to the outlet end, with the vibrator mechanism 27, so as to provide a desired inclination for the bottom wall 13. The desired degree of inclination which causes linear movement of the frozen particles 17 from the housing inlet 25 to the housing outlet 26 will vary in dependence upon the material being processed. Opening of the valves 36 and actuation of the vacuum pump

37 will then effect evacuation of the chamber 11. Frozen substance particles obtained, for example, by grinding of frozen substance blocks, by spray freezing processes, etc. are supplied from the supply hopper 32 into the vacuum chamber through the feeding lock 31. The introduced frozen particles 17 pass through the chamber inlet 33 into the housing through the adjacent housing inlet 25. Within the housing 12 the particles are continuously agitated by the vibrating walls thereof so as to come into periodic contact with both the bottom wall 13 and the top wall 14, as they travel from the inlet 25 toward the outlet 26 under the influence of the vibratory motion and the inclination of the housing. Upon reaching the housing outlet 26, the completely dried particles 42 fall through the adjacent chamber outlet 34 and are removed through the discharge lock 35.

Because the housing side walls and top wall prevent escape from the housing of the vibrating particles, relatively thick substance layers of, for example, 5 to 10 times the individual particle thickness can be transmitted through the housing. In addition, the side wall portions and top wall provide additional heated surfaces for contact with the vibrating particles thereby substantially increasing the heat transfer capabilities of the apparatus.

The apertured indentures 22 in the top wall 14 provide the necessary conductance paths for the escape of water vapor sublimating from the frozen particles 17. Furthermore, they are uniquely suited for this particular application in that they do not provide obstruction free paths between the inside and outside of the housing in a direction perpendicular to the bottom wall 13. Since this is the primary direction of movement of the vibrating particles, substantially no particles will escape through the apertures 23. The heating coils 41 are so positioned relative to the apertured indentures 22 as to provide no direct paths between the surfaces of the heating coils 41 and the interior of the housing. Thus, the coils 41 may be operated at relatively high temperatures without danger of burning the particles with direct radiant heat.

Supplying Continuous Apparatus with Material at Proper Rate

H. Eilenberg; U.S. Patent 3,513,559; May 26, 1970; assigned to Leybold-Heraeus-Verwaltung GmbH, Germany is concerned with an apparatus in which frozen granulate may be brought into the vibratory freeze drying installation in sufficient quantity so that the full capacity of the drying assembly may be fully utilized. This is achieved by the provision of a vertically displaceable helical stirring mechanism in a funnel-shaped storage hopper, which stirring mechanism has a sealing plug fixed to it whose position within a discharge opening in the storage hopper can be adjusted to vary the discharge opening cross section. In addition, the discharge opening of the storage hopper is cooled by a cooling element which surrounds it. The cooled discharge opening extends into the upper portion of the vacuum lock without a heat-conducting connection.

Raising or lowering the sealing plug regulates the amount of granulate which flows from the storage hopper into the vibratory freeze drying apparatus. The cooling of the discharge opening and the fact that there is no heat-conducting connection between this opening and the vacuum lock prevents heat transfer from the vacuum lock to the discharge opening of the storage hopper. Even in small quantities such heat transfer could partially thaw the frozen granulate and cause an undesirable agglomeration of the individual granules. When the vacuum lock is closed, it is possible to vent the storage hopper for the purpose of adding further quantities of granulate without influencing the subatmospheric pressure in the drying chamber. Accordingly, continuous feeding of the drying chamber is possible. The process is described with reference to Figure 7.7.

FIGURE 7.7: SUPPLYING CONTINUOUS APPARATUS WITH MATERIAL AT PROPER RATE

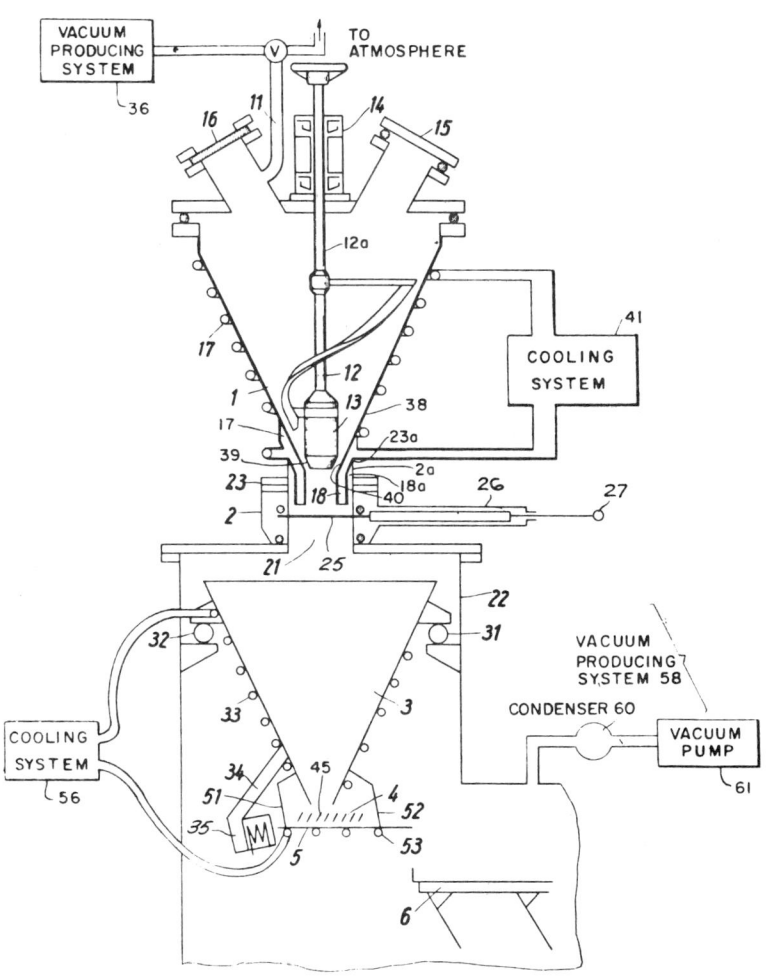

Source: U.S. Patent 3,513,559

The material to be dried is supplied in the form of frozen granulates through the opening **15** into the storage hopper **1**. Previously, the storage hopper will have been sealed off from the dosaging hopper **3** by closed position and vented by appropriate manipulation of the two-way valve. After the storage hopper is filled, the opening 15 is locked to make the hopper airtight. Valve V is then turned to connect the storage hopper to the vacuum-producing system 36 and the storage hopper is evacuated to the same subatmospheric pressure as exists in the vacuum drying chamber 22. For example, if coffee is being treated, the pressure will be reduced to about 500 μ, although this pressure may be different for other products. While hopper 1 is being filled, shaft 12a is slowly rotated to stir the granulate. Stirring is continued as the granulate falls into the hopper 3. The vacuum lock 2 is then opened and the helical stirring mechansim 12 is vertically displaced, and the sealing plug 13 is moved to open discharge opening 40 to the desired size. Although the sealing plug is shown as being large enough to seal the opening, it can also be smaller than the opening if desired.

The granulate then passes in a continuous flow from the storage hopper 1 to the dosaging hopper 3. Preferably, the passage opening is set to a size such that a larger amount is fed into the dosaging hopper than is discharged at its outlet 45. The dosaging hopper is continuously moved by drive mechanism 34 and the material passes through the shear grate 4 and over the vibratory distributor 5. At the same time, the moving means 51, 52 for the vibratory distributor 5 is in operation, moving it in a direction opposite that of the dosaging hopper. The vibratory distributor 5 spreads the material onto the drying beds 6.

The material in the storage hopper is continuously maintained at a desired temperature by the cooling coil 17. The material which is located within the dosage hopper 3, or is being distributed by the vibratory distributing mechanism 5, is maintained at a desired temperature by the cooling coils 33 and 53. The surface temperature of the dosaging hopper and the vibratory distributor are maintained at a temperature approximately 5°C higher than that of the surface of the condenser 60. This prevents the cooled dosaging hopper and the vibratory distributor, due to their arrangement in the vacuum drying chamber, from inadvertently acting as condensers for the water vapor evaporating from the material being dried during the drying process.

Multistage Vibrating Conveyor

J.L. Mercer and L.A. Rowell; U.S. Patent 3,648,379; March 14, 1972 describe a system for the continuous freezing and freeze drying of solids-containing aqueous liquids to obtain a freeze dried product. The feed liquid is frozen as a thin sheet on a continuously moving belt and is broken to form discrete pieces which are further reduced in size prior to freeze drying. The frozen particles are moved on a chilled multistage vibratory conveyor in an evacuated chamber wherein refrigerated condensers and radiant energy sources cooperate with the conveyor to sublimate the ice content of the frozen particles to form a freeze dried product. The vibrating conveyor is constructed to minimize transfer of vibratory forces as well as the effects of changes in length of the conveying sections as a result of temperature change.

The system operates continuously to receive the liquid feed, to discharge the freeze dried product, and to separate and remove ice collecting on the condensers. The system functions to accomplish the freezing and freeze drying operations in the relatively short period of from 40 to 110 minutes. The process is described with reference to Figure 7.8.

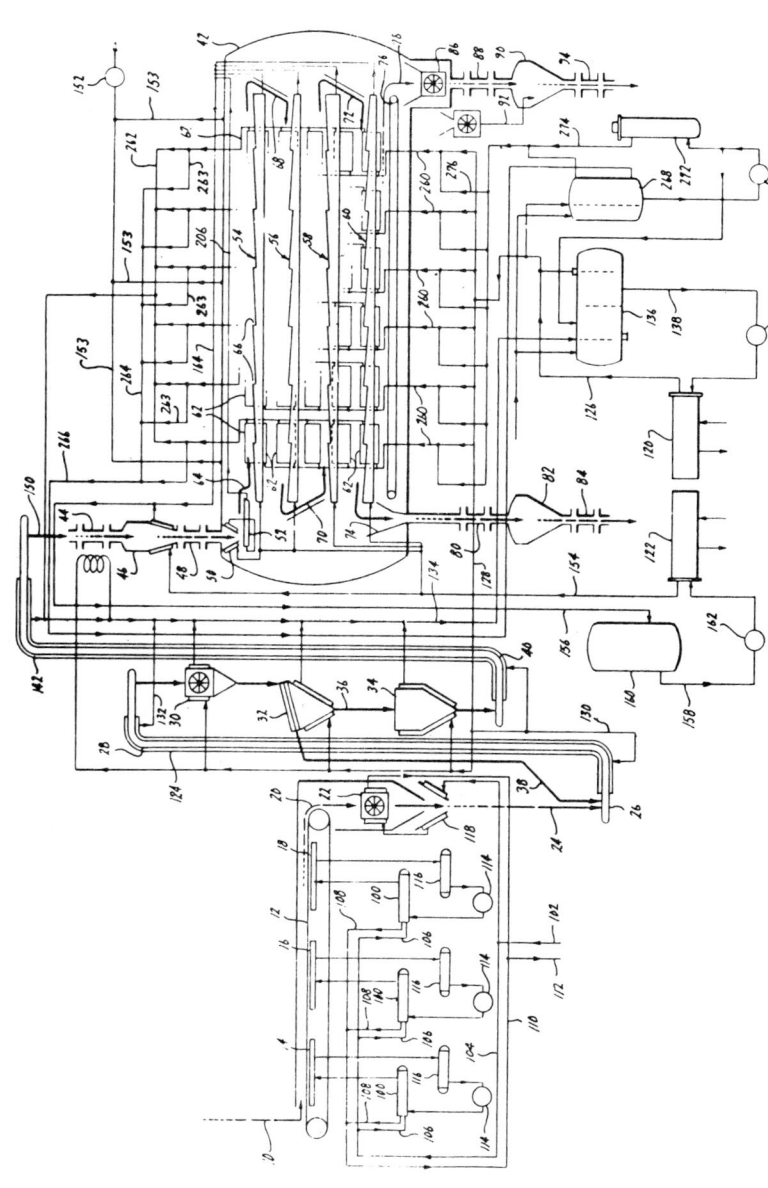

FIGURE 7.8: MULTISTAGE VIBRATORY CONVEYER

Source: U.S. Patent 3,648,379

166   Freeze Drying Processes for the Food Industry

The liquid to be freeze dried, such as an aqueous extract of coffee, is introduced at **10** to a continuously moving contact freezer represented by the endless belt **12**. The liquid is chilled and frozen in 8 to 15 minutes as a thin continuous layer on the surface of the belt at –40°F by circulation of a refrigerant liquid in contact with the bottom of the belt, for example, by trough or spray means **14, 16** and **18**. The resulting continuous layer of frozen material fractures and falls off the end of the belt (arrow **20**) into a breaker mechanism **22** where the fractured pieces of frozen material are reduced in size for further conveyance in the system.

The broken pieces of frozen material subsequently pass into a conveyor mechanism **26**, as represented by the arrow **24**. The conveyor is chilled by refrigerated jacket means **28**, and functions to convey the frozen feed material in the form of discrete individual pieces to comminuting means **30**, where the individual pieces are further reduced in size to produce individual particles generally of a size less than about four mesh. The particles leaving the comminutor are subjected to classification in the screening device **32** so that only particles of a size less than about four mesh pass to the refrigerated hopper **34**, as represented by the arrow **36**. Oversized particles, that is, greater than about four mesh, are recycled to the inlet of the conveyor **26** (arrow **38**) for further processing and reduction in size in the comminutor. Appropriately sized particles in the hopper **34** pass to a second refrigerated conveyor **40** which leads to the freeze drying operation.

The freeze drying is carried out in an evacuated substantially air-free chamber **42**. To maintain desired conditions within the chamber, frozen particles discharged from the conveyor **40** initially pass through a first vapor lock **44** to a refrigerated feed hopper **46**. Thereafter, at appropriate intervals, the frozen material is fed through a second vapor lock **48** to an internal feed hopper **50**, within the drying chamber **42**, from which the frozen particles are fed to a controllable vibrating feeder **52** at a regulated, uniform rate to vibratory conveyor means within the freeze drying chamber. As illustrated, the vibratory conveyor means comprises a series of vertically spaced vibrated conveying decks **54, 56, 58** and **60**. The individual conveying decks are cooled to maintain desired characteristics of the particles as they are freeze dried, while the particles are simultaneously subjected to radiant energy to sublimate the ice content to vapor form.

The vibratory conveyor means is bounded on each side by a series of condensers schematically represented at **62**, which are maintained at a suitably low temperature (i.e., –50° to –100°F) by suitable refrigerating media such as liquid cryogens. In general, the condensers function as high-speed pumps and maintain a low pressure by causing water vapor and other condensable gases to condense and freeze on the cold surfaces of the condenser plate.

The frozen particles fed to the top conveying deck (arrow **64**) move with a bouncing or dancing motion to the right of the chamber, and are periodically rotated and turned over by the vibrating action of the conveyor and also by steps **66** provided at spaced positions on the upper conveyor surface. Upon reaching the end of the conveyor **54**, the particles fall and are deflected onto the subadjacent conveying deck **56** (arrow **68**) where the conveying action is repeated until, upon reaching the end of the conveying deck **56**, the particles again fall and are deflected onto the deck **58** below (arrow **70**). The particles continue along the vibrating pathway provided by the deck **58** until they fall again (arrow **72**) to the subadjacent conveyor **60**, where the vibrating conveyance is repeated until the particles eventually fall into the product hopper **74**. Throughout the progression along the vibrating pathway of the conveyors **54, 56, 58** and **60**, the ice content of the frozen particles is effectively sublimated and removed by transfer to the condenser plates **62**. The latter are periodically deiced by surges of warm refrigerating means with the result that the ice falls to the bottom of the chamber where it is collected on a belt conveyor **76** for removal from the system (arrow **78**).

The removal of a freeze dried product, and the separated ice, is effected by means which maintain the desired freeze drying conditions within the chamber **42**, and without upsetting the equilibrium of the system. Thus the dried product in hopper **74** can be fed through a first vapor lock **80** to a collection hopper **82**, from where it may be discharged through a second vapor lock **84** to dump bins, drums or other suitable means for collecting or storing the product. In like fashion, the ice periodically separated from the condensers, after first being reduced in size in the breaker means **86**, passes through a first vapor lock **88** to a collection hopper **90**. Preferably, the ice in the hopper **90** is melted by contact at atmospheric pressure with hot water or other suitable liquid introduced through the line **92**, and is discharged through a further vapor lock **94** as a liquid waste.

From an operating standpoint, it has been found that effective freeze drying is accomplished on vibratory conveyor decks when the rate of material movement is relatively slow, but the amount of bouncing or dancing on the screen remains relatively rapid. For example, in the vibratory system illustrated this frequency might be of the order of 240 cycles per minute. It has been additionally found that a desired rapid rate of conveyance together with a very rapid bouncing or dancing of the particles can be obtained at the natural vibration frequency of the spring support system which, might be about 360 cycles per minute.

In other words, at 240 cycles per minute, the particles can be held at any desired position on the vibratory pathway with virtually no forward movement whereas at the higher cycling rate of 360 cycles per minute, a forward progress of the order of 10 feet per minute is achieved. By alternating between the lower and the higher vibration frequency, say between 240 and 360 cycles per minute, virtually any rate of movement up to 10 feet per minute can be obtained, to achieve freeze drying cycles of 20 minutes or longer, as may be desired.

Accordingly, in conjunction with an initial freezing cycle of 8 to 15 minutes, together with 1 to 3 minutes for conveyance and classification, a total period for freezing and freeze drying a liquid product in the described equipment is about 30 to 40 minutes, at the low side, up to longer periods, as may be feasible. For practical reasons, a freeze drying cycle in excess of about 150 minutes is generally too long to achieve an economical and efficient use of the equipment.

Vibrated Support plus Filter to Retain Fine Particles

W. Rothmayr; U.S. Patent 3,465,452; September 9, 1969; assigned to Afico SA, Switzerland describes cryodesiccating apparatus which has a vibrated support for the material to be dried and at least one filter for catching fine particles removed when the frozen matter in the material is sublimed. The process is described with reference to Figure 7.9. The apparatus comprises a cryodesiccation chamber **1** connected to a battery of condensers **2**. This chamber is also in contact via the condensers with a pumping installation (not shown) capable of maintaining the interior of the enclosure formed by the walls of the chamber at a reduced pressure of the order of 0.1 mm of mercury, and a refrigerating installation (also not shown) designed to maintain, especially in the condensers, a temperature lower than the final solidification temperature of a liquid.

FIGURE 7.9: VIBRATED SUPPORT TO RETAIN FINE PARTICLES

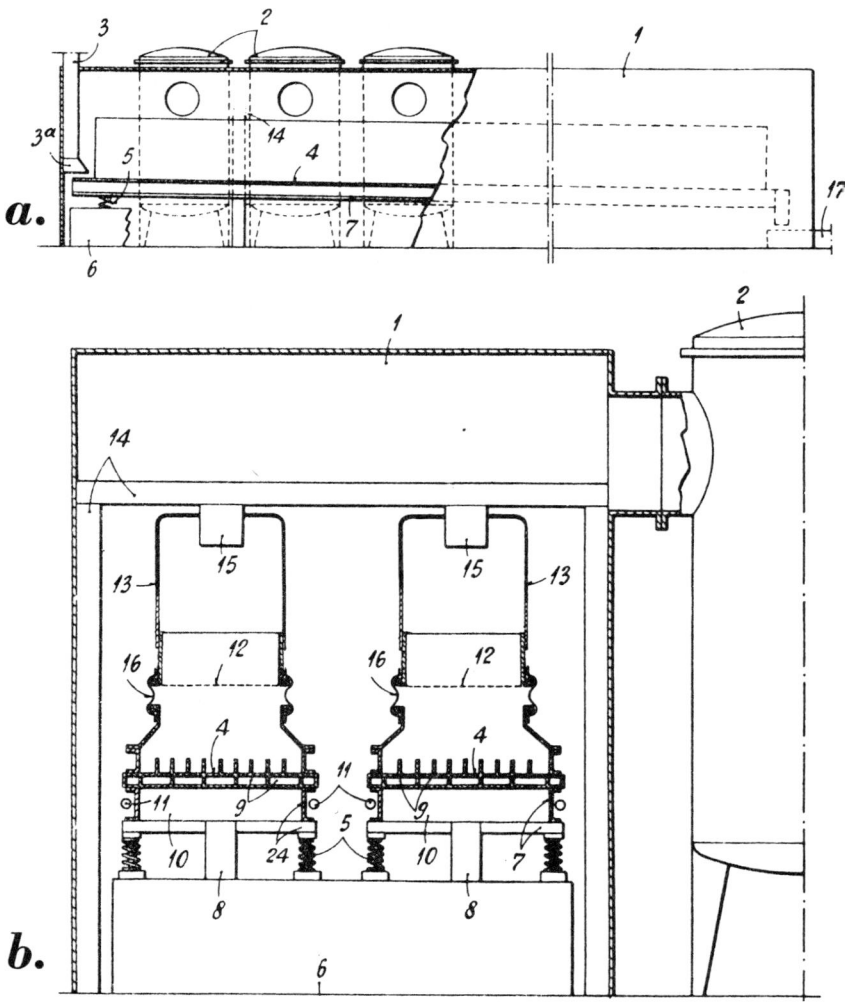

Source: U.S. Patent 3,465,452

Chamber **1** is connected to two drying lines placed side by side. It has at one end at least one air lock **3** prolonged by a duct **3a** opening on to the adjacent portion of the end of desiccation tray **4**. The bottom of duct **3a** is preferably provided with a device for distributing the dry product, previously frozen and

Apparatus Modifications

divided, on the end part of the tray 4. The latter is mounted on a mounting 7 which rests, by means of elastic members 5, on a rigid stand 6. The mounting is subjected, by means of vibrator device 8 of known construction, to vibrations of a relatively high frequency but small amplitude which cause a slow movement of the product from one end of the cryodesiccation tray to the other while constantly altering the orientation of the particles because of the leveling of the latter. The rate of displacement of the product is regulated by the requirement that, when the particles reach the end of the tray, the drying of the product is practically complete.

The rapid sublimation of the frozen liquid implies a progressive heating of the product on the tray. To this effect, the latter is ribbed in order to form longitudinal passages 9 which cover the major part of the useful surface of the tray, ribs in which is circulated heating fluid such as steam, mineral oil etc. In another form of execution (not shown) of the apparatus, heating is assured by an electric device integral with or separated from the tray, for example by an infrared heater across part of the latter. As shown in the drawing, the mounting 7 comprises a chamber 10 in which can be mounted the heating means.

As the apparatus is operated at relatively low temperatures, the heating of the tray leads to temperature differences between the latter and its mounting. The differential expansion of these two elements causes the development of considerable mechanical stress, sufficient to cause rupture of one or other of the elements. In order to avoid this difficulty the mounting has a suitable heating means which allows it to be maintained at a temperature similar to that of the tray which it supports. The heating means is constituted by tubes 11 in which are circulated the same heating fluid as that which passes through passages 9. Preferably, at least one part of the mounting is thermally isolated and the heating of the different elements can be assured equally well from the exterior such as from the inside of chamber 10.

In order to avoid loss of a portion of the product, especially the fine particles which are carried to the condensers by the sublimed vapors, the apparatus has at least one filter disposed in the zone above the plate 4. In the modification shown, the filter consists of a sieve 12 (flat or zigzag), which can be formed as a single or divided piece along the length of several elements mounted end to end. Preferably this sieve is permanently fixed in a rigid chassis 13 in the form of an arch, of which the upper part has the openings necessary to evacuate towards the condensers 2 the filtered sublimation vapors.

The chassis and sieve are not integral with the tray and its support, in order to limit the vibrating mass. The chassis 13 is suspended, by means of devices 15, from a rigid structure 14 integral with the walls of the chamber 1. These devices, of a construction known in the mechanical art, electromagnetic or pneumatic for example, are designed to vibrate the above described chassis at a suitable frequency, the shaking being sufficient to free the sieve 12 of particles held on its surface.

Furthermore, the chassis is joined to the tray-mounting assembly 4–7 by a flexible seal 16 adapted to guarantee, with all the latitude imposed by the relative movements of the elements connected to it, an impervious sealing of the chamber between the tray and the filter element. On reaching the end of the tray 4, the lyophilized product falls through an air-lock 17 where it can be collected directly or transported to a conditioning installation.

AVOIDING LOSS OF FINES DURING DRYING

Deflecting Plates in Drying Chamber for Removal of Entrained Particles

It has been found in instances where the products to be dried are shredded or ground there is associated with the comminuting operation a creation of fine product particles, these particles being of a size of about fifty mesh or smaller. In the course of the drying operation, these fine particles are entrained in the stream of water vapor evolved during the rapid dehydration of water-laden material. High velocity vapor can entrain fine particles and carry them to the condenser sections where they are then lost either in the ice or in the course of ice removal. In both the continuous and static product drying systems where many thousands of pounds of dried product may be processed, any loss of product is a substantial financial factor to be considered in the drying operation.

H.J. Togashi and J.L. Mercer; U.S. Patent 3,276,139; October 4, 1966; assigned to Cryo-Maid, Inc. describe a system whereby minimization of particle loss is achieved at a nominal expense. A series of deflecting plates are inserted between the dehydrating zone and condenser sections to permit the removal of product particles entrained in the vapor. It has been determined that the vapor, which travels rapidly to the surfaces of the various condenser sections, is forced to change direction with the insertion of deflecting plates. The inclusion of the deflecting plates does not impede the conductance of the water vapor to the surfaces of the condenser sections. Because of the vapor deflection caused by the plates, the product particles are unable to maintain the same velocity as the water vapor and therefore drop out of the vapor stream and, as a result, can be collected and returned to the total dried product. The process is described with reference to Figure 7.10.

The drying chamber contains one or more stacks of trays **21** bounded on each side by many individual condenser sections **20**, maintained preferably below -50°F, which function as high-speed pumps, maintaining a low pressure by causing water vapor and other condensable gases to freeze on the surfaces of the respective sections. Each condenser section **20** is provided with connections leading to a source of refrigerating medium. One or more conduits **26** are provided for connection to an evacuating system (not shown) capable of rapidly reducing the pressure in the chamber at the beginning of the drying operation.

Trays **21** are agitated by vibrators **30** in order to impart a dancing or multidirectional motion to the frozen particles. As the particles move along on vibrators **30**, they pass under a source of radiant energy, such as heated plates **24**. Upon heating, sublimation occurs and the frozen water contained in the particles to be dried turns to vapor which rapidly travels to the surfaces of condenser sections **20** where it accumulates on the surfaces in the form of ice. Cooling means **29**, suitably in the form of conduits or coils embedded within or attached to the underside of trays **21**, are provided for cooling the trays so that the frozen particles will not thaw and agglomerate or stick to the trays. Upon completion of the drying operation, the dried particles are collected in hopper **31**.

The vapor evolved from the product carries with it a small percentage of the finely ground particles. To prevent ultimate loss of these particles, directional deflection plates **41** are inserted between the dehydrating zone and sections **20**. For good results the vapor should travel a tortuous path which is optically tight, i.e., the path should be such that no particle could travel in a straight line from the vapor generation zone adjacent the particles being dried to the condenser sections without impinging on at least one baffle. For optimum results, it is preferred that the pathway in which the particles travel contain at least three bends each of substantially 90°.

FIGURE 7.10: DRYING CHAMBER ILLUSTRATING DEFLECTING PLATES USED FOR ENTRAINED PARTICLE REMOVAL

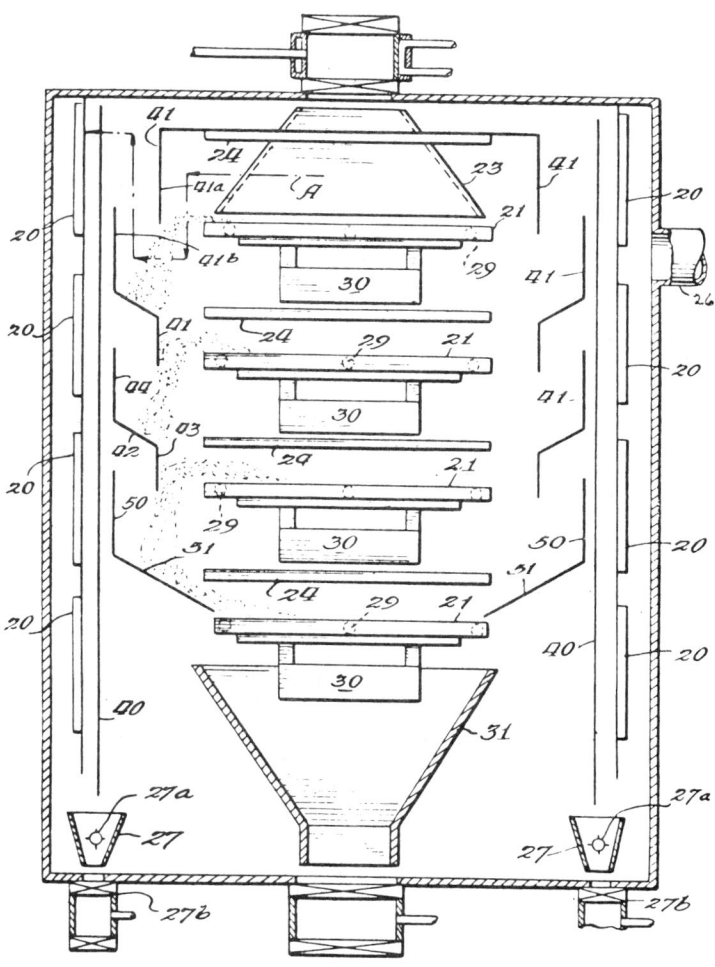

Source: U.S. Patent 3,276,139

It will be observed from arrow **A** that the vapor in its course of travel from the dehydrating zone to sections **20** will be deflected off plates **41a** and **41b**. The purpose of providing a tortuous route for the vapor to travel is to afford an opportunity for the entrained product particles to drop out of the vapor stream. The product particles which are eliminated from this stream gradually drop to deflector plate **50** where leg **51** is extended to end above particle tray **21** located at the bottom of chamber **25**. Here the recaptured fine particles are blended with the products on this particular tray. In the bottom tray only a small percentage of water is removed from the products so that no particle entrainment occurs at this point due to a lack of vapor. If desired, the recaptured fine particles could be dropped directly into hopper **30**.

Example: A test was conducted where a coffee extract containing 27.1% solids was processed with no deflecting plates positioned between sections **20** and the dehydrating zone. Entrained product particles lost in the ice and ice removal system were found to be about 2.77% of the total theoretical dry yield.

When the test run was made with one deflecting plate designed to create a vapor path having an upwardly extending portion and two portions disposed angularly to the upwardly extending portion, the entrained particles lost in the ice and ice removal system were about 1.25% of the total theoretical dry yield.

When the process was operated with a deflecting plate design such as that shown, the entrained particles lost in the ice and ice removal system amounted to about 0.35% of the total theoretical dry yield.

Filter Tray Assembly for Confining Material During Freeze Drying

The tray described by *C.E. Bender, T.N. Thompson and D.S. Fraser; U.S. Patent 3,488,860; January 13, 1970; assigned to The Virtis Company, Inc.* includes a cover which is slidably interlocked with the bottom portion of the tray. The cover is provided with a pair of coarse filter screens, at least one of which is removable to permit the placement of a coarse paper filter therebetween, forming a filter sandwich for use in confining the product or material to be sublimated. The process is described with reference to Figure 7.11.

FIGURE 7.11: FILTER TRAY ASSEMBLY FOR CONFINING MATERIALS DURING FREEZE DRYING

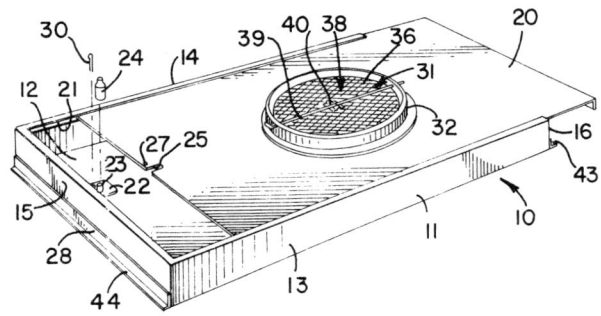

Source: U.S. Patent 3,488,860

Apparatus Modifications

A filter tray assembly **10** includes a receptacle **11** having a bottom wall **12**, side walls **13** and **14**, and end walls **15** and **16** and is rectangular in shape. A cover **20** of generally rectangular form is slidably received in guide channels formed in the opposite side walls. A similar channel is provided in the end wall with all the guide channels formed by joining an angle strip **21** along the wall immediately below the integral inturned flange formed at the upper end of each wall. The end wall includes an inwardly projecting pedestal **22**, located above the level of the cover **20** and supporting an upwardly extending port **23** communicating with the interior of the tray. The port is closed at its outer end by means of an elastomeric valve **24** which permits access to the interior of the tray. A U-shaped slot **25** is formed in the inner end of the cover **20** to permit the cover to be completely closed without cutting off communication with the interior of the chamber through the port **23**.

When the filter cover **20** is completely closed, an aperture **27** in the cover is aligned with an aperture **28** in the pedestal. A cotter pin **30** or similar lock may be inserted into the aligned apertures to lock the cover in place, preventing inadvertent displacement during handling. The central portion of the cover **20** is provided with a filter assembly **31**, consisting of an annular housing **32** which is welded or secured by equivalent means to the cover **20**. The housing surrounds an opening which is of lesser dimension, forming a ledge radially inwardly of the inner periphery of the housing **32**. A filter screen of coarse mesh rest on the ledge to form the lower half of the filter sandwich. A second screen **36** overlies a coarse paper filter disc and a locking means indicated generally at **38** clamps the sandwich in operative position.

The locking means **38** may consist of a stiff wire **39** with a pressure distribution pad **40** at the central portion. One end of the wire is bent to assure retention of the wire with the cover during handling. It is possible for the wire to be moved toward its bent end to release the free or unbent opposite end from engagement with the housing **32**, permitting it to be elevated, the coarse screen **36** removed, and the coarse paper filter replaced.

When a material is enclosed in the tray and the cover closed, the port **23** is closed off by a rubber stopper which is provided with a pierced self-closing aperture. The pierced aperture permits the insertion of a heat-detecting probe, such as a thermistor, which will permit detection of the temperature in the interior of the tray throughout the sublimating process. The overall simplified construction of the tray permits the components of the tray assembly **10** to be easily disassembled for sterilization, washing or the like.

At opposite ends of the tray are provided a pair of outwardly projecting flanges **43** and **44** which, in the preferred form, are coextensive with the bottom of the tray. These flanges cooperate with clamps of known type in the sublimating modules or chamber to permit the bottom of the pan to be tightly pressed against the shelf on which it is located. In this manner, good heat transferring contact between the pan and shelf is assured.

Use of Filter Which Functions as Heating and Agitating Means

The known devices for freeze drying pulverulent materials have substantial drawbacks because they do not permit a quick drying to be obtained by efficient utilization of the heat energy which can be transferred from the heating members to the powder since the high speeds of evaporation have the effect that the steam carries the powder out of the vacuum container to such an extent that usual filters placed in front of steam condensers become clogged and the pressure consequently rises in the vacuum container.

A. Thale; U.S. Patent 3,319,343; May 16, 1967; assigned to A/S Niro Atomizer, Denmark overcomes this drawback by employing inside the vacuum container a filter for the pulverulent material which filter defines a space for containing the material. The simplest and cheapest form of the apparatus is one in which the filter has the function of acting as a heating member and the function of maintaining the material in agitation whereby the filter is the sole member which serves the purpose of maintaining the material in agitation and of heating it. The process is described with reference to Figure 7.12.

FIGURE 7.12: USE OF FILTER WHICH FUNCTIONS AS HEATING AND AGITATING MEANS

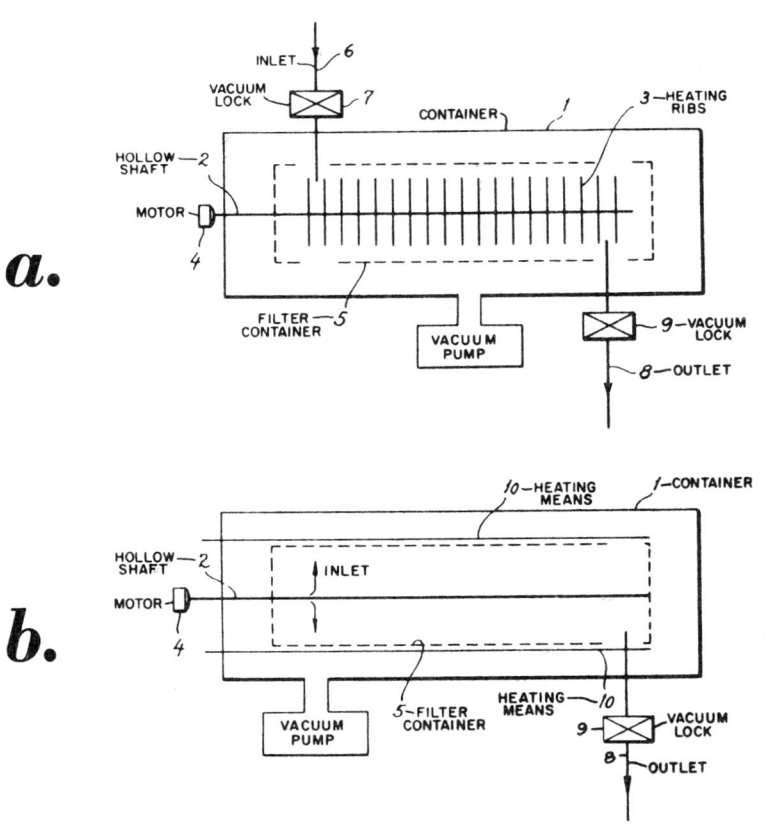

Diagrammatic Side Views
   (a) Material Heated and Agitated in Filter
   (b) Filter Sole Agitating Means; Heating Outside Filter

Source: U.S. Patent 3,319,343

In Figure 7.12a, the apparatus comprises a horizontal cylindrical vacuum container **1** in which is axially mounted a heating member consisting of a hollow shaft **2** in which a heating medium may be introduced and which is provided with ribs **3**. The heating member is rotated by means of a motor **4** situated outside the container. The heating member is completely surrounded by a filter container **5** of cylindrical shape, and the pulverulent frozen material is introduced inside this filter container through a conduit **6** which in case of continuous operation is provided with a vacuum lock **7**. The dried material is discharged through a conduit **8** which is likewise provided with a vacuum lock **9** for continuous operation. The filter container may be arranged for rotation about its axis by means of the motor so that both the filter container and heating member serve as members for maintaining the pulverulent material in agitation.

The vacuum container is connected to well-known means which have not been shown, for establishing and maintaining vacuum and for removal of the generated steam, e.g., a vacuum pump and a condenser.

In the form shown in Figure 7.12b, the apparatus also consists of a horizontal cylindrical vacuum container **1** inside which is mounted a smaller cylindrical filter container **5** arranged to be rotated about its shaft **2** by means of a motor **4**. Feeding and discharge of the pulverulent material takes place through the shaft **2** which is hollow. In this modification, the filter container is the sole member for maintaining the pulverulent material in agitation. Heating is effected by means of heating members **10** situated outside the filter container but close to it so that the effect is substantially the same as if the filter itself were a heating member.

OTHER INNOVATIONS

Providing for Thermal Expansion of Shelves

M.E. Fuentevilla; U.S. Patent 3,146,077; August 25, 1964; assigned to Pennsalt Chemicals Corporation describes a freeze drying apparatus having means providing for thermal expansion of shelves on which articles to be freeze dried are supported, means for reducing the pressure to which the articles are subjected and means for removing moisture which is evolved during the freeze drying process.

The condenser means includes a pair of condensers alternately connected to the enclosure. In this manner, the efficiency of the freeze drying apparatus is increased since one of the condensers will be in communication with the enclosure while the other condenser is connected to a means for melting the ice accumulated on the condenser coils. In addition, a means is provided to reduce the de-icing time for a condenser by generating latent heat which precools the condenser prior to the flow of a coolant medium such as brine through the coils of the condenser. Figures 7.13a and 7.13b illustrate the housing in which the articles to be freeze dried are supported during the freeze drying cycle.

The rectangular housing **12** is provided with insulation **14** and a selectively operable door **15** connected by a pair of hinges. Standards **16, 18, 20, 22** are disposed in pairs upright within the housing **12** along the side walls. The standards are provided with a plurality of spaced notches. A hollow shelf is supported by the standards by the notches on such standards. Thus, shelves **24** and

26 are each supported in separate notches at spaced points along the length of the standards 16, 18, 20, 22. The shelves are identical. The shelf is hollow so that a heat exchange medium may be circulated therethrough via an inlet conduit 28 and an outlet conduit 30. A manifold 32 is in communication with the inlet conduit for each of the shelves.

FIGURE 7.13: FREEZE DRYING APPARATUS

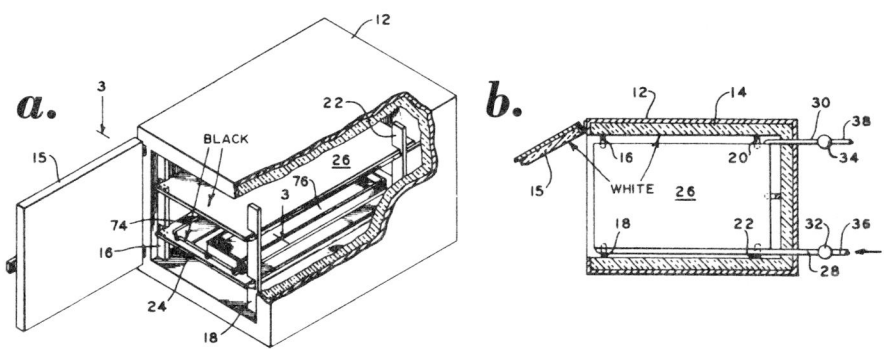

(a) Perspective View of Housing
(b) Transverse Sectional View Along 3–3 of Figure 7.13a

Source: U.S. Patent 3,146,077

A manifold 34 is connected to the outlet conduit for each shelf. A conduit 36 is in communication with the manifold 32 and extends therefrom to the outlet side of a pump. A conduit 38 extends from the manifold 34 to the inlet side of the pump. A heat exchanger is provided in the conduit 36. A pair of condensers are provided for the housing 12. A heat exchange medium is circulated through the shelves so as to raise the temperature thereof.

Pans 76 having articles in a frozen state, to be freeze dried, are supported in spaced relationship to the shelves by means of wire trays 74. After the pans 76 have been disposed on the trays 74, the door 15 is closed so that the housing 12 is hermetically sealed. The combination of the radiant heat from the shelves and the reduced pressure in the housing 12 causes the ice in the frozen articles to sublime.

The moisture from the sublimation of the ice within the frozen articles is condensed by one of the condensers on the cooling coils therein. After a predetermined period of time, the coils in that condenser accumulate ice so that the condenser becomes inefficient. The housing 12 is then arranged via valves in conduits to be in communication with the vacuum pump through the alternate condenser. During the freeze drying process, the shelves 24 and 26 have been heated by the medium flowing therethrough. The cutout notches on the standards 16–22 enable the shelves to expand radially outwardly toward the side walls of the housing 12 and toward the door 15. Care must be taken so that the articles being freeze dried do not become overheated. The articles being freeze dried are heated by radiant heat. Difficulties encountered in regard to the heat supply may be overcome by using black shelves, black trays and black pans. By black is meant the thermodynamic effect of a black body which thermodynamically is an absorptive surface of high emissivity. Shelves, trays and pans may attain their black color by being coated with urethane, baked phenol-formaldehyde resin, etc.

In addition, the interface of the door 15 is preferably white. By using a heat supply in this manner, the heating is uniform and the length of time of the heating may be reduced. Tests showed that one sample required thirteen and one-half hours for the material being dried to reach a temperature of 14°C. When the sample was placed in a housing 12 having black shelves, black trays and black pans with the door coated white, it took only seven hours for the dried articles to reach a temperature of 14°C. Thus, the drying time is reduced almost fifty percent.

Helical Drying Bed

The process of *G.-W. Oetjen and H. Eilenberg; U.S. Patent 3,574,951; April 13, 1971; assigned to Leybold-Heraeus-Verwaltung, GmbH Germany* relates to an apparatus in which frozen granulates are continuously moved through the freeze drying apparatus on a helical drying bed. The process is described with reference to Figure 7.14.

At the start of the operation the product inlet 22 is opened and the product is introduced into the first cooled hopper 10. When a sufficient amount of material has been introduced, the product inlet is sealed and, via opening 24, which is connected to vacuum producing apparatus, a vacuum is produced in the first cooled hopper. When the vacuum in the first cooled hopper is the same as that of the drying chamber 12, the gate valve 26 is opened and at about the same time the stirrer 28 is again moved (it is also moved as the hopper is being filled) so that the granular material which is frozen will pass through the gate valve 26 which has been opened and into the inner hopper 14 inside chamber 12.

The granular material is then moved along vibrating dispenser 16 to the lowest end of the helical bed 18. The material travels up the helical bed to the top and then falls down through a central opening. By the time the particles are falling, they have been fully dried, and the material may be collected in a collecting hopper 30 which is also inside the vacuum chamber 12. When a sufficient amount of material has been collected in collecting hopper 30, the discharge lock 20 via vacuum connection 32, which is connected to a vacuum pump, is evacuated. During this time, the gate valve 34 between hopper 30 and discharge lock 20 is closed as is gate valve 36 which is the product outlet. After the vacuum in the discharge lock is the same as that in vacuum chamber 12, gate valve 34 is opened and the material passes into the discharge lock.

When a sufficient amount of material has passed into the discharge lock, the gate valve 34 is closed and gate valve 36 is opened at which time the discharge lock is vented and the vacuum is lost, and at the same time the product discharges from the outlet valve 36.

FIGURE 7.14: SCHEMATIC VIEW OF HELICAL DRYING BED

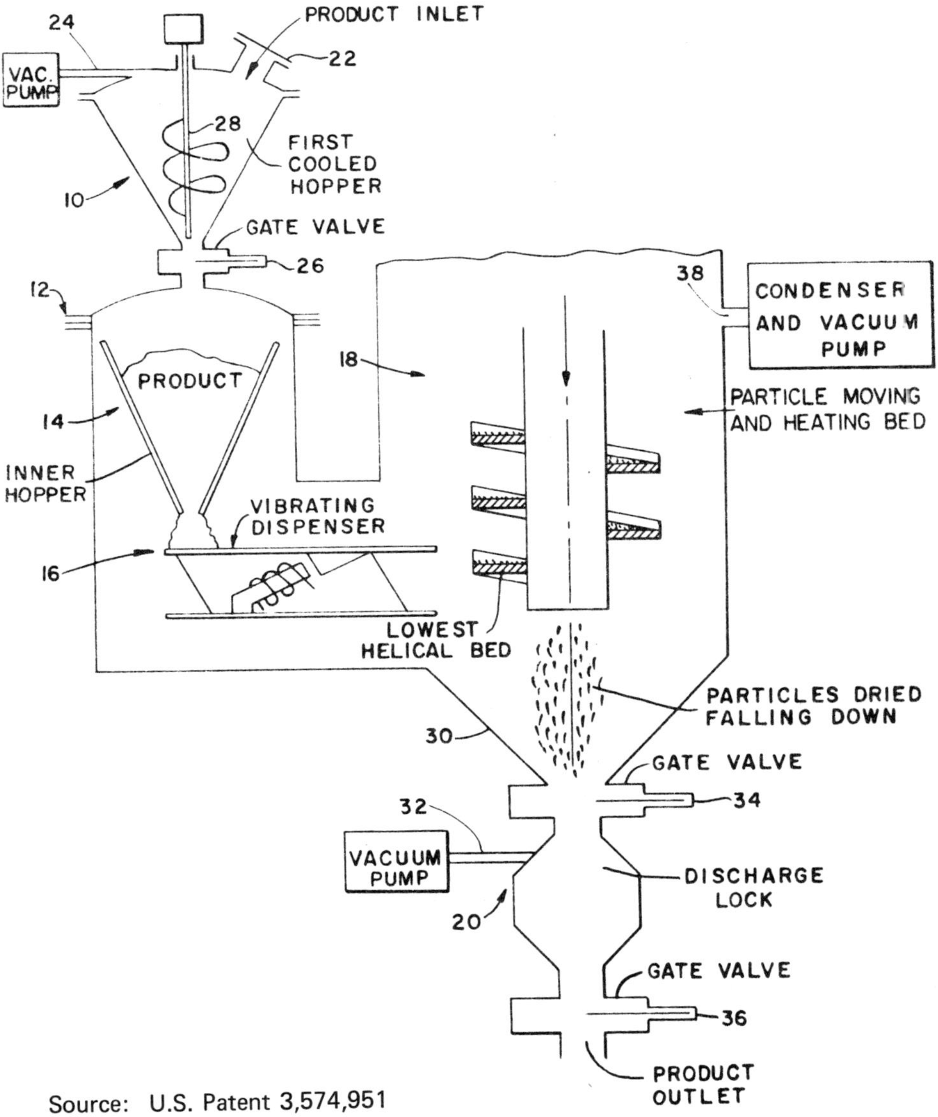

Source: U.S. Patent 3,574,951

Cooling Chamber Separated from Heating Chamber by Partition

The process developed by *J. Lorentzen; U.S. Patent 3,382,586; May 14, 1968; assigned to A/S Atlas, Denmark* relates to an apparatus for freeze drying products containing water in vacuum, the apparatus being of the type that has a cabinet with built-in cooling surfaces for freezing the water vapor out as ice, and means to thaw the formed ice off. The characteristic features of the apparatus are that the cooling surfaces are separated from the rest of the cabinet by

Apparatus Modifications

means of partition walls which have an aperture with a closing member that can be opened and closed. In connection with the thawing of the ice off the cooling surfaces, there only needs to appear a relatively small pressure in the chamber bounded by the partition walls, e.g., 10 to 20 torrs, and as a still smaller pressure exists in the rest of the cabinet, e.g., 1 torr, the stresss on the partition walls will be slight so that the walls may be made with small strength. As there only exists a small pressure difference it is not necessary either to have a large degree of tightness, and the construction can therefore be made relatively cheaply. The apparatus is described with reference to Figure 7.15.

FIGURE 7.15: COOLING CHAMBER SEPARATED FROM HEATING CHAMBER BY PARTITION

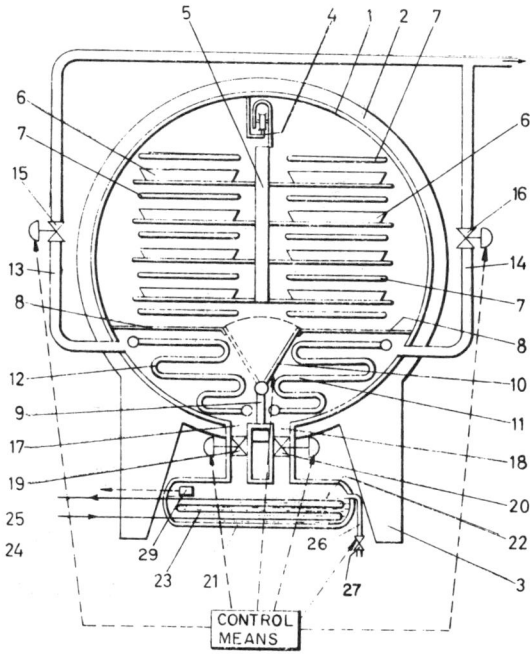

Source: U.S. Patent 3,382,586

A cylindrical freeze drying cabinet is made of steel plate **1** with reinforcement rings **2**. In the bottom of the cabinet there are cooling surfaces **11** and **12** in the form of pipes in which a cold cooling medium can be circulated. The cooling surfaces are separated from the rest of the cabinet and from each other by means of partition walls **8** and **9** and by means of a closing member in the shape of a hinged valve **10** that may be set to close an aperture in one or the other of the chambers formed by the partition walls. These chambers can be connected to an evacuation plant by means of pipes **13** and **14**, in which closing members **15** and **16** are arranged. Pipes **17** and **18** are connected to the bottom of the two chambers, closing valves **19** and **20** being arranged in the pipes, the latter forming a connection to a water vapor generator in the shape of a container **21** that is filled with water up to a level **22**. The water in this container can receive heat from pipe **23**, through which a medium warmer than the water can be circulated by means of the connection pipes **24** and **25**. A temperature control member **29** is provided inside the container **21** to maintain temperature of the water below a certain maximum value which is determined by the strength of the partition walls. The container has a spillover pipe **26** with a valve **27** through which the water above the level **22** can be removed in the pause between the freeze drying periods, atmospheric pressure existing in the cabinet and parts connected thereto in the pause, that is, while a new batch of material to be freeze dried is being inserted in the apparatus.

The freeze drying process takes place in the following manner. Fresh goods in the trays **6** are placed in position in the cabinet between the heating plates **7**. The cabinet is hermetically sealed and is evacuated, through pipe **13** with valve **15** open, until the pressure has been reduced to approximately 1 torr. The necessary conditions for freeze drying the goods are then present. Heating medium is then supplied to the heating plates **7** and cooling medium is supplied to the cooling surfaces **12**. The closing valve **10** is in its right hand position, shown with full lines. The valve **19** is closed. So as to avoid rupture of the partition walls evacuation of the space around the cooling surfaces **11** must be provided for. This can possibly be done by not closing the valve **10** tightly or by keeping the valve **16** open during the first evacuation period in each freeze drying process. As a result, the stress on the partition walls **8** and **9** will be reduced so that the walls may have substantially smaller strength than the steel plate **1**.

The goods in the trays **6** emit water vapor which is precipitated as ice on the cooling surfaces **12**. When the ice thickness on the cooling surfaces **12** has grown so much that the latter's effectivity has been reduced considerably the valve **10** is moved to the left hand position shown with dotted lines. The valve **16** is opened whereas the valve **15** is closed. The pressure in the cabinet will then be kept down through the pipe **14**. Furthermore, the cooling surfaces **11** are supplied with cooling medium, whereas this is no longer the case for the cooling surfaces **12**. The valve **20** is closed and the valve **19** is opened. The water vapor from the goods will be precipitated as ice on the cooling surfaces **11**. The cooling surfaces **12**, which are separated from the rest of the cabinet, are connected to the container **21** by means of the pipe **17** and the valve **19**.

Water vapor is developed from the water in the container **21**, the vapor flowing up through the pipe **17** and condensing on the ice on the cooling surfaces **12**. Hereby heat is transmitted to the ice and the pressure in the chamber in question increases so that the ice melts successively. The water from the melting process flows through the pipe **17** down into the container **21**.

When so much ice has accumulated on the cooling surfaces **11** that the latter's effectivity has been considerably reduced, and when the cooling surfaces are completely free of ice, a changeover of the valves **15, 16, 19** and **20** and the hinged valve **10** is carried out so that cooling surfaces **11** and **12** exchange functions. In this manner a changeover between two groups of cooling surfaces is carried out until the drying is terminated. At this point the water will be removed to a desired degree from the goods in the trays **6** and will be collected in the container **21** that must have a correspondingly large volume. When the drying is terminated the pressure in the cabinet and the container is increased to atmospheric pressure. The dried goods are taken out and packed and the water is let out of the container until level **22** is attained. To control automatically the operation of respective closing members, the apparatus may have a control means which is adapted to control operatively the valves **10, 15, 16, 18, 19** and **27** according to a predetermined time program, or in dependency of a sensing member (not shown), such as for instance a member for sensing ice thickness on the cooling surfaces **11** and **12**, or a pressure sensing member in the drying cabinet.

Preventing Uptake of Atmospheric Oxygen in Dried Product

M.L. Brewster; U.S. Patent 3,401,466; September 17, 1968 provides a method and apparatus for preventing the uptake by or return to the vacuum freeze dried product of the previously removed atmospheric oxygen and moisture. The vacuum freeze dried material is saturated with purified dry nitrogen and then passed or transferred directly from a vacuum drying chamber means into a dry nonoxidizing nitrogen environment within which final handling, processing and packaging of the dried product is performed.

Figure 7.16a diagrammatically illustrates a plant layout for a vacuum freeze drying system having main walls **10** defining a main room **11**. Within the main walls are located a plurality of interior enclosures including a blast freezer **12** capable of attaining and maintaining a temperature of at least –40°F for freezing the wet material, a vacuum drying chamber apparatus **14** and an inert atmosphere room **16** which constitutes the dry material unloading area and final handling, treating and packaging room within which all final steps are performed so that the final product or material is never exposed to the hazards of moisture or oxygen once the material has been freeze dried and does not leave the final area **16** until it is protectively packaged and ready for storage, marketing or any other subsequent activities. Additional included enclosures comprise an entry or dressing room area **18**, and an airtrap passage **20**.

Disposed within the main area **11** defined by the enclosure walls **10** are a plurality of tracks and rails communicating between the various areas and apparatus for assisting in the handling of the material and the movable equipment.

The inert atmosphere for the dry room **16** is provided with a constant supply of fresh purified dry nitrogen, through at least one inlet port **56**, from an adjacent reservoir of tanks containing nitrogen under high pressure, comprising a fully automatic system. The purified dry nitrogen atmosphere pressure within the dry room is maintained slightly above constantly changing outside environmental atmosphere by means of a pressure modulator. Thus, any migration of atmosphere will be from the dry room outwardly, such as the tumbling of nitrogen atmosphere outwardly through the entry and exit passageway **20**. The environmental nitrogen atmosphere of the dry room is circulated and recirculated through typical air conditioning ducts constituting a completely closed system through wall floor outlet vent **58** and duct **60**, through the vapor and particle filter **62**, ultraviolet section **64** and cooling and/or heating section **66**, returning to dry room **16** through ceiling duct **68** and inlet port **56**.

The air conditioner and purifier remove the food powder dust created during the unloading, milling, blending and pack-out of the product, and removal of other contaminants. The dry room also serves as a sterile clean room. In order to provide for the comfort and safety of the personnel operating within the inert, non-life-supporting nitrogen gas environment of the packaging room **16** (see Figure 7.16b) the ceiling **70** of the room is provided with an outside atmosphere supply system comprising a plurality of housings such as the one indicated at **72**, each having an air-supply hose **74** coupled thereto and communicating through the ceiling with a plurality of personnel supply hoses **76** via a subjacent housing **78** of preferably rotatable construction, each personnel air hose being coupled to a helmet **80** adapted to be worn by the individual operator for supply of normal atmospheric air thereto.

FIGURE 7.16: PREVENTING UPTAKE OF ATMOSPHERIC OXYGEN IN DRIED PRODUCT

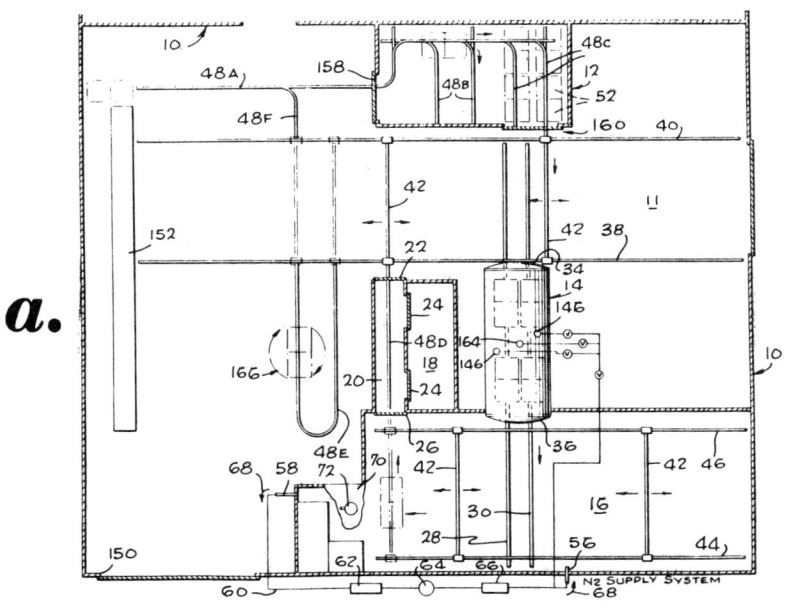

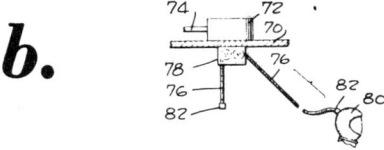

(a) Plant Layout of System
(b) Air Supply System for Operators of Packaging Room

Source: U.S. Patent 3,401,466

Preferably, the operator wears a total suit of airtight construction in communication with the helmet and the air hoses **74** and **76** include exhaust hose means whereby the air is circulated throughout the helmet and suit and exhausted exteriorly of the dry room 16 in order to avoid the possibility of contamination of the inert atmosphere within the dry room by moisture or other vapors emanating from the operator. The air hoses 76 are provided with a quick-detach coupling attachment 82 for coupling the hose to the helmet whereby the operator can dress within the entry room 18 and provide himself with a suitable auxiliary air supply until sequential entry through doors 24 and 26 into the dry room and coupling of his air hose to his face mask or helmet. Obviously, the reverse procedure is employed when the operator leaves the dry room.

ECONOMIES IN EQUIPMENT COST

Use of Multicylindrical Vacuum Chambers with Removable Product Cars

An object of the process developed by *J.A. Abbott and E. Thuse; U.S. Patent 3,132,930; May 12, 1964; assigned to FMC Corporation* is to reduce the equipment cost chargeable against the product, while maintaining a high production rate. This is accomplished by using a relatively long cylindrical vacuum chamber, a shape that is cheaply fabricated and can be designed to resist crushing at close to the minimum theoretical material cost. The product material is loaded on trays and the trays are placed in a long car that has liquid heated hollow shelves for supporting the trays, and heating the product. The loaded car is wheeled into the vacuum chamber. Vapor condenser plates occupy side sectors of the chamber at each side of the car, thereby providing maximum utilization of otherwise waste chamber space.

Another object of the process is to require a minimum vapor pressure drop as sublimed vapor flows from the product being dried to the condenser plates in the vacuum chamber. This is accomplished by having the condenser plates arranged at each side of the chamber with their inner edges lying in vertical planes disposed close to the corresponding edges of the food trays on the car.

To make possible optimum use of the available space in a cylindrical vacuum chamber for the condensing plates, coupled with efficient circulation of refrigerant through the plates, the sets of plates are mounted at each side of the chamber in longitudinally extending vertically spaced units, the plates in each set being divergent to form a fan-shaped assembly, with refrigerant inlet connections being made to the lower edges of all the plates.

To facilitate the defrosting of the condenser plates in the vacuum chamber, a removable car is used which provides operator access to the vacuum chamber. Means are provided to internally heat the condenser plates by hot gaseous refrigerant so that when the car is removed, the vacuum chamber can be entered, and the plates flushed free of defrosted ice crystals with a hose.

The car can be alternately supplied with heated and cooled liquid at a minimum heating cost by providing a warm and a cold liquid tank or reservoir. The procedure is to transfer the hot liquid from the car shelves to the warm liquid reservoir at the end of the drying cycle, when the car has been wheeled to a product unloading or drying room. In addition, when a car is in the cold or loading room, its shelves can be filled with circulating cold liquid supplied from a cold tank or reservoir. This cools down the car for loading with the frozen product.

To provide maximum utilization of the refrigeration and vacuum pump installations, a number of freeze drying chambers and their internal vapor condensers are connected in tandem, there being a single main refrigeration unit and both a heavy duty and a lighter duty vacuum pump unit, connected to the chambers by suitable valves. The heavy duty or pump-down vacuum pump unit is connected to each chamber at the start of the cycle, to remove the air trapped in the chamber for rapidly pulling down the vacuum, whereupon the lighter duty or holding unit is connected to the chamber to maintain the previously established vacuum.

Satisfactory temperature control is attained by mounting thermostat switches directly on some of the product trays, with the thermostats being connected in series and controlling the heat source for the heating liquid circulated in the hollow shelves that support the trays. As soon as one tray reaches the temperature at which the thermostats are set, the heat is turned off automatically and it is not turned on again until there are no trays at a temperature above the control point temperature. This precludes scorching of the product and makes possible excellent control of product drying time.

To facilitate defrosting of the condenser plates in the vacuum chambers while consuming minimum power in the form of heat loss, hot compressed gaseous refrigerant is diverted from the line between the refrigerator compressor and the refrigerant condenser. The hot refrigerant gas is passed through the vapor condenser plates of the unit to be defrosted, by the use of suitable valving. This defrosting has no effect on the refrigeration of the other units, and makes use of heat that would otherwise be lost.

The general arrangement of a multichamber freeze drying system is shown in Figure 7.17. The installation includes a main room M, in which four cylindrical vacuum chambers A, B, C and D are mounted. In a four vacuum chamber system, there will normally be provided five product-carrying cars **10**, all of these being identical. In the figure there is shown a pair of turntables **12** and **13** disposed in front of chambers B and C, and A and D, respectively. Tracks **14** lead from a cold, or product-loading room **16** to the turntable **12**. Tracks **15** lead from the turntable **12** to the chambers B and C, and a similar set of tracks **17** lead from turntable **13** to chambers A and D. Tracks **18** interconnect the turntables **12** and **13**, and tracks **19** lead from turntable **13** to a dry, or product unloading, room **20**.

In the figure, one of the cars **10** is shown just after removal from chamber A. At this time, there will be cars in chambers B, C and D, and a fifth car appears in the dry room **20**. Chamber doors **21–24** are provided for vacuum chambers A–D respectively, and the doors are mounted on overhead tracks **26** and **27**, for ready manipulation. The cylindrical vacuum chambers are each reinforced by a pair of spaced annular ribs **28**.

In the main room M is a heavy duty or pump-down vacuum pump P1 which connects to each vacuum chamber by a main vacuum line **31**, and downcomers **31a** leading to each of the vacuum chambers A–D. A valve **32** is mounted in each downcomer **31a**. Also in the main room M is a lighter duty or holding vacuum pump P2. This pump connects by main lines **33** and downcomers **33a** to the vacuum chambers, through individual valves **34** at each chamber. This arrangement makes it possible to connect the heavy duty vacuum pump P1 to any chamber for an initial pump-down operation, and thereafter the heavy duty pump can be replaced by the holding vacuum pump P2 by manipulation of the valves. In the broader aspects of the process, the chambers may be evacuated by steam ejectors, and the term vacuum pump is intended to refer to either ejectors or mechanical pumps unless otherwise specified.

A central refrigeration unit R1 cools hollow refrigerant conducting elements in the form of condenser plates mounted in the vacuum chambers A–D. The compressed, gaseous refrigerant from the refrigeration unit R1 is conducted to each vacuum chamber by an inlet header 40, having branches at each chamber leading to a valve 41. A branch 40a leads from each valve 41 to the associated vacuum chamber. Also connected to refrigeration unit R1 is a suction or return header 42 that has a branch at each vacuum chamber that is connected to a two-way valve 43. A branch line 42a connects each chamber to its two-way valve 43.

FIGURE 7.17: ECONOMIES IN EQUIPMENT COST

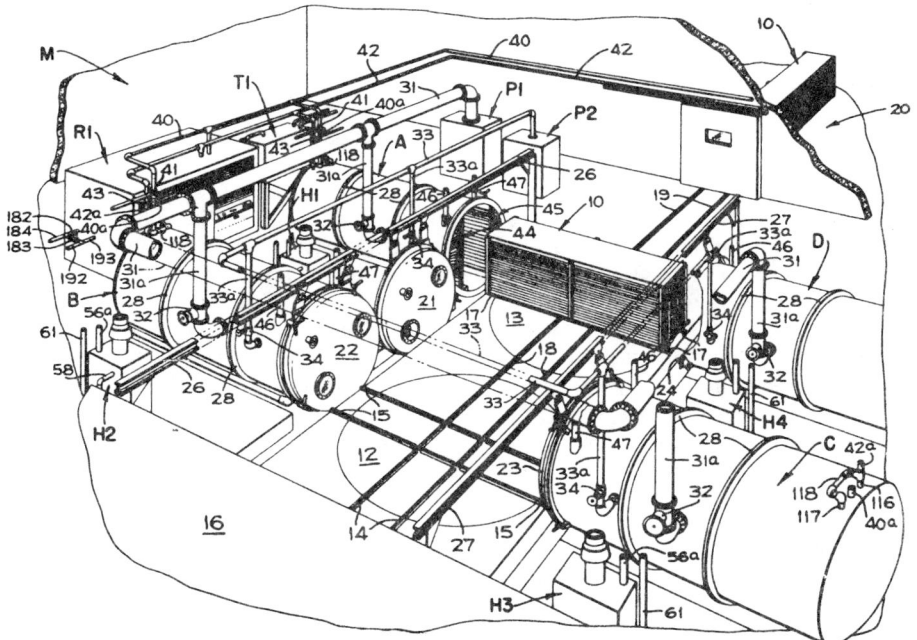

Source: U.S. Patent 3,132,930

As indicated at vacuum chamber A, each vacuum chamber is provided with left and right sets 44 and 45 of vapor condensing plates. These are hollow plates that are refrigerated to condense and freeze vapor sublimed from the product loaded in the cars in the vacuum chamber, during the drying process. With some products the product need not be frozen before introduction into the vacuum chamber. In this case a certain percentage of product moisture will be evaporated before the vacuum is pulled down sufficiently to cause the remainder of the product moisture to freeze. A hot liquid such as a glycol solution is circulated through the shelves of the cars during the sublimation process. In the main room M a storage or makeup tank T1 is provided for the hot glycol solution. By means of suitable valves and piping, the tank T1 serves as both a supply and expansion tank for each of four heaters, H1, H2, H3, and H4, provided for vacuum chambers A—D, respectively. Each of the vacuum chambers A—D has a hot glycol inlet line 46 and a glycol return line 47 connected thereto. Each of the cars 10 has a tier of hollow shelves for supporting trays of the frozen product to be dried, through which shelves the hot glycol solution circulates.

Example of a Cycle: The product, whose processing will be described, is a batch of $3/16$ x $5/8$ inch diameter sliced green asparagus. The product will have been frozen with the slices in a free-flowing condition, so that they are readily handled. A car will be in the cold room with its shelves cooled by the cold glycol system, ready for loading. The frozen product is loaded on the trays at a unit loading of 2 pounds per square foot, and with a total tray area of 500 square feet. The loaded car will contain 1,000 pounds of frozen asparagus.

The trays are supported on the shelves by insulating feet, as described, for radiant heating. The temperature of the frozen asparagus will be below freezing, at 20°F, and before the loaded car is placed in the vacuum chamber, the refrigeration unit will have brought the temperature of the vapor condenser plates within the chamber to −40° to −50°F, and except during the peak load period at the start of the cycle, the plates are maintained substantially at this temperature throughout the drying portion of the cycle.

The cold glycol is purged from the car shelves, the loaded car is wheeled into the cold vacuum chamber, and the hot glycol solution hose connections are made. If desired, thermocouples will have been embedded in the product at selected spots, and these will be plugged in when the car is in the chamber. The plug-in connections to the tray thermostat switches are made and the door is closed. The large, or pump-down, vacuum pump is connected into the vacuum chamber and the vacuum is brought down to abut 1 mm of Hg in about 15 minutes. At this time, most of the air has been evacuated from the chamber. The car shelves are filled with heated glycol solution, and the heated liquid is circulated as previously described, with which relatively rapid sublimation begins.

The large vacuum pump continues to evacuate the chamber, but most of the air has been withdrawn and the pump removes the last of the air and any small quantities of noncondensed water vapor. After about 20 to 25 minutes total elapsed time, the pressure in the chamber will have been brought down to 0.1 to 0.2 inch of Hg, at which time the vacuum valves are operated to disconnect the main vacuum pump and connect the holding vacuum pump to the chamber.

The thermostat switches on the product trays will have been set to turn the glycol heater off when the switches sense a temperature of 130°F. The temperature of the glycol solution will vary somewhat with the load, but will be maintained at a temperature in the order of 180°F during an initial period of the cycle. As the ice core in the product is reduced, the temperature of the glycol solution will progressively drop under control of the tray thermostats. If thermocouples have been installed on the shelves, in the product, and on the trays, and if the corresponding temperatures are measured and recorded, it will be found that as the sublimation process continues, the shelf temperature slowly rises from about room temperature to the temperature of the glycol solution, the latter temperature being reached by the shelves after elapse of about ½ of the total drying time, depending on the capacity of the glycol heater. The tray temperature drops at first, because of cooling of the product by sublimation as the vacuum is brought down, whereupon the tray temperature slowly rises and reaches 130°F, the control temperature. At about this time the shelves and the glycol solution temperature will be at a maximum. From this point on, the tray thermostats turn the heater on and off at intervals of increasing frequency, as previously described.

The product ice core temperature soon drops to the temperature corresponding to the vapor pressure in the chamber (-40° to -50°F) and stays there during a large part of the drying cycle. Near the end of the drying cycle, the ice in the product has been almost completely sublimed, and the product temperature rises relatively rapidly until it almost reaches the tray temperature. Because of the difficulty in locating thermocouples in the product so that they will be precisely in the center of the last remaining ice core, and because the thermocouple wires conduct heat to the ice core in which they were initially embedded, it is difficult to ascertain the true, or effective product temperature, as previously mentioned. This is the reason that control of the tray temperature, to insure that the product will be dried as rapidly as possible without scorching, is important. In the present example, drying is completed in about 18 hours.

After experience and instrumentation shows that the product is dry, the vacuum in the chamber is broken by the introduction of dry air or nitrogen in the conventional manner, and the door is opened. Suitable valves and piping (not shown) are provided for this purpose, in accordance with usual practice. The car with its dried product is wheeled into the dry or product unloading room, wherein it is purged of the heated glycol and unloaded. The dried product is placed in air tight containers, such as polyethylene bags, or nitrogen atmosphere cans.

As soon as the car with its dried product has been removed from the vacuum chamber, the refrigeration unit is disconnected and the chamber is defrosted and flushed out. After the shelves have been defrosted, the refrigeration system is reconnected to the shelves of the chamber, and the shelves cooled, ready for the next car. A newly loaded car is then introduced into the vacuum chamber, and the previously unloaded car is wheeled into the cold room and its shelves are filled with refrigerated glycol solution. The trays are loaded and introduced into the car, the cold glycol solution is blown out of the car shelves, and the car is wheeled into another vacuum chamber of the system for another cycle of product drying.

Use of Baffles in Internal Condenser to Prevent Stagnation of Air

In a process described by *J.H. Blake, J.P. Pelmulder and E. Thuse; U.S. Patent 3,382,585; May 14, 1968; assigned to FMC Corporation* stagnation of air within the vapor condenser of an internal condenser freeze drying unit is prevented by providing baffles at the ends as well as above and below the condenser, thereby directing all gases through the condenser. A vacuum port for exhausting air is behind the baffled condenser unit. The baffle units could be used in the system described above by Abbott (U.S. Patent 3,132,930). The process is explained with reference to Figure 7.18. Referring to Figure 7.18a, the vacuum drying chamber **10a** has a door **30** which can be considered to provide the leakage path for the air entering during the drying cycle. Figure 7.18b shows tracks **32** for the entry of a shelf cart (not shown) which has heated shelves **S** that are arranged in vertical tiers and support trays **T** for the frozen food **F** (Figure 7.18a). The outer edges of the heated shelves and the trays thereon are in close proximity to the inner edges of the condenser plates **C** of the condenser.

FIGURE 7.18: USE OF BAFFLES IN INTERNAL CONDENSER TO PREVENT STAGNATION OF AIR

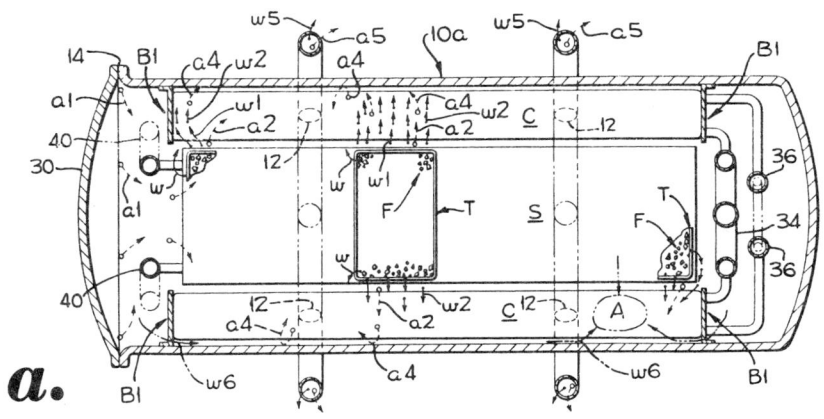

a.

Horizontal Section Through Drying Chamber

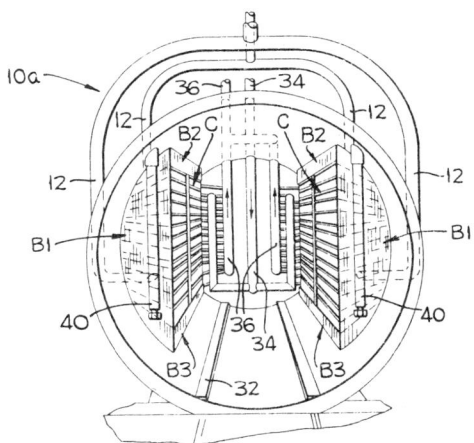

b.

Perspective End View of Chamber with Shelf Cart Removed

(continued)

FIGURE 7.18: (continued)

c.

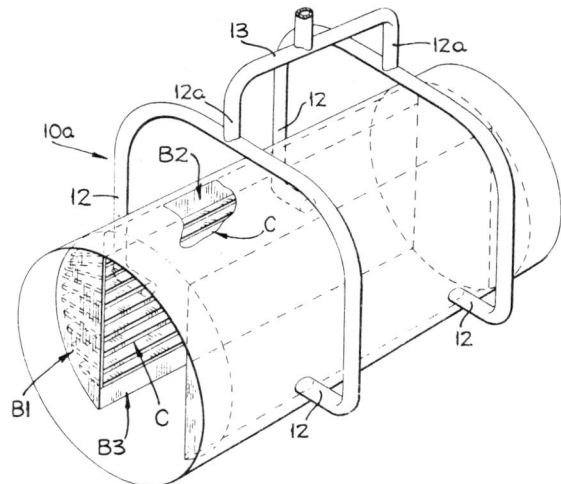

Diagrammatic Perspective of the Chamber

Source: U.S. Patent 3,382,585

In the preferred construction, the chamber **10a** is long, and there are four vacuum ports **12**, located in the lower half of the chamber, particularly where the chamber is cylindrical as shown. The ports are manifold to a common pipe **13** leading to the pump. This location of the vacuum ports assists in removing pockets of air, such as that indicated generally at **A** in Figure 7.18a. These pockets were previously found to have accumulated between the condenser plates **C**, particularly away from the vacuum ports between the lower plates. The relatively low mounting of the vacuum ports aleviates this condition. It has been found desirable to distribute two or more vacuum ports along the length of the chamber. The efficiency of the operation will be improved if ports **12** are placed at each side of the chamber.

It has been overlooked that there can be an objectionable amount of stagnant air trapped or remaining between the condenser plates, and that the function of the holding pump, which should theoretically remove noncondensable gases, is not efficiently affected. The assumption being that noncondensable gases will adequately find their way to the pump, simply by connecting it at any convenient zone in the drying chamber is fallacious.

As seen in Figures 7.18a and 7.18b, inlet lines **34** and outlet lines **36** are provided for circulating refrigerant to the condenser plates **C**. Also, flexible hoses **40** have connections (not shown) to a source of heating fluid for the shelves **S**, and quick attachable connections for connecting up to the shelves themselves, when the food cart is rolled into place in the drying chamber. One of the connections

40 will be an inlet connection and the other an outlet connection. Considerable improvement in the operation of the chamber just shown is provided by insuring that the only flow path available for water vapor evolving from the food on the heated shelves S is between or over condenser plates. It may be true that water vapor flow substantially ends at the outer (far) edges of the condenser plates, but this is only because all the water vapor in excess of that corresponding to the equilibrium pressure of the condenser plates will have been condensed up to these edges.

However, air which evolves during the process, or leaks into the chamber, and which can only be withdrawn by the vacuum pump, must also take the same path, and it is induced to take this path by the aforesaid flow of water vapor, even though the latter terminates as described. In other words, there is an air flow in the chamber sufficient to hold the total pressure down in the chamber to a point that sublimation is rapid and that no melting of the ice cores in the product occurs.

The optimum baffling system includes principally baffles B1 closing off the fore and aft ends of the condenser plates C. These baffles seal off the ends of the condenser plates such as those at the top or bottom designated B2 and B3 in Figure 7.18b. As seen in the upper left of Figure 7.18a, water vapor w evolving from the food, if it is to flow at all as directional flow (which is the type of flow that will normally occur under the low driving forces developed in the chamber) must flow between the condenser plates C. These are the very locations wherein pockets of air A, as shown at the upper right of Figure 7.18a, have previously been permitted to accumulate.

Any water vapor that finds its way into the end pockets of the chamber will merely stagnate there and will serve as a diffusion barrier to water vapor w leaving the food. The latter will therefore take the path of least resistance, which is to flow into and across the condenser plates.

The mere presence of vacuum ports 12 is not effective; it has also been found that the pairing of vacuum ports 12 on opposite sides of the chamber significantly reduces the total pressure in the chamber. For example, in a given installation where the baffles are installed as in Figure 7.18a, the chamber pressure dropped from 180 to 100 microns total by simply pairing the vacuum ports 12 as shown in the drawings. This difference apparently results because the process insures that the vacuum ports can serve their intended function and efficiently remove air, whereas previously they would not, and it made no difference where they were located, or even if more than one were provided.

Compact Plant Using Horizontally Elongated Conveyor in Parallel, Spaced, Superimposed Relationship

V.W. Lind; U.S. Patent 3,218,727; November 23, 1965 describes a freeze drying system with economies in time and floor space. The process is described with reference to Figure 7.19. The apparatus includes (see Figure 7.19a) a freezing room 1 and a sublimation room 2 each fully enclosed by walls. Within room 1 is a pair of identical conventional water defrosted, direct expansion air handling units 4 and 5. These may be positioned in space opposed relation spaced at opposite sides of a central plenum chamber 18. A conventional power drive in the upper portion of cooling unit 4 discharges the cold air through a side

Apparatus Modifications

opening in the unit through a horizontally extending duct **19** into the plenum chamber **18**, while air from the cooling unit **5** is adapted to be delivered through a similar duct **20** into the plenum chamber.

FIGURE 7.19: CONVEYOR IN PARALLEL, SPACED, SUPERIMPOSED RELATIONSHIP

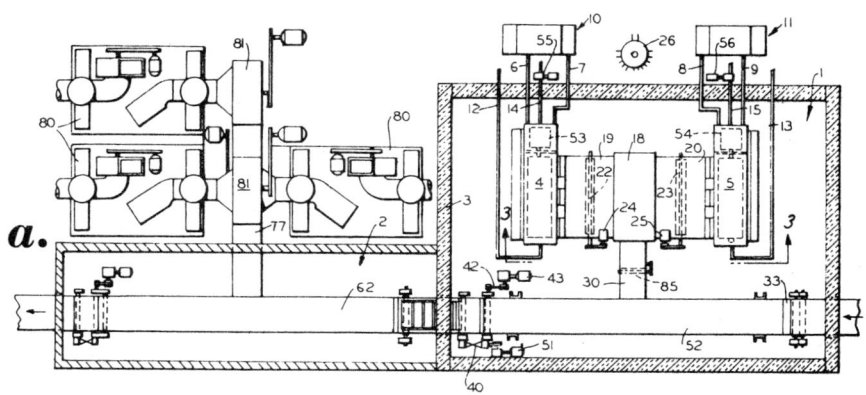

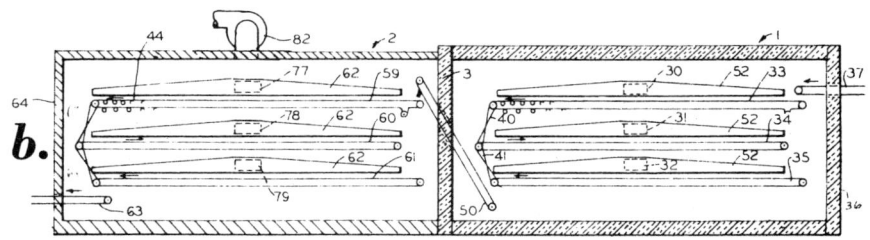

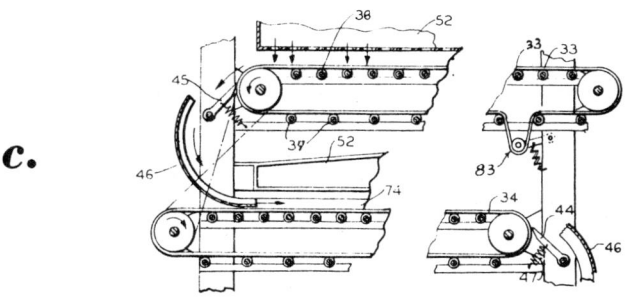

(a) Semischematic Plan View of Apparatus
(b) Semischematic Side Elevational View of Apparatus
(c) Sectional View, Partly Broken Away Showing Scraper Means

Source: U.S. Patent 3,218,727

Connected with the plenum chamber are three corresponding cold air ducts **30**, **31**, and **32** in a vertical row (Figure 7.19b) which ducts extend horizontally from the same side of the plenum chamber, the duct **30** being uppermost and duct **32** lowermost. The air that is discharged from the ducts is at a temperature of approximately -15° to preferably -30°F. Within the freezing room **1** are a plurality of vertically spaced, horizontally elongated endless conveyor belts **33**, **34** and **35** in superposed relation, with the conveyor **33** being indicated as the uppermost conveyor, while conveyor **35** is the lowermost.

The rooms **1**, **2** are horizontally elongated. In an actual installation, rooms 1, 2 may each be approximately 36 feet in length and approximately 16 feet in height, with each conveyor 7–9 being approximately 30 feet in length, making a total of approximately 90 feet of travel of material within room **1**, or within a room length of approximately 36 feet.

Conveyors **33**, **34** and **35** each extend at their ends over conventional rollers on pulleys the shafts of which are journaled for rotation in conventional bearings that may be supported on suitable frame members (not shown) within room **1**. The right hand end wall **36** of room **1**, as seen in Figure 7.19b or the end wall that is opposite to the partition wall **3**, is formed with an opening adjacent to the upper end of the wall through which extends the discharge end of a conventional, horizontally extending, endless feed conveyor **37**. The roller or pulley supporting the discharge end of the feed conveyor within the upper portion of room **1** adjacent to end wall **36** has its supporting shaft suitably supported for rotation within the upper end of room **1**.

The conveyors **33**, **34** and **35** are made preferably of antisticking material, such as Teflon. The thickness of such a belt may be approximately 0.006 inch. Preferably, the upper runs or flights of the conveyors are supported on relatively closely spaced rollers **38** that may also be carried by the frame supporting the belts, and the middle belt **34** is offset to the left, as viewed in Figure 7.19b, a sufficient distance so that material deposited on the upper flight of the upper conveyor **33** at the right-hand end of the latter will be discharged onto the left-hand end of the middle conveyor **34**, upon movement of the upper flight of the upper conveyor to the left, or in a direction away from the discharge end of the feed conveyor **37**. Rollers **39** may support the lower flight of each conveyor against sagging.

Also, the lower conveyor **35** is offset to the right, relative to the middle conveyor **34** so that material discharged from the right-hand end of the middle conveyor will be carried on the latter to the left. The movement of the upper flight of conveyor **33** is away from the discharge end of the feed conveyor or from right to left as seen in Figure 7.19a. The movement of the upper flight of the middle conveyor **34** is from left to right, as seen in Figure 7.19b, while the movement of the upper flight of the lower conveyor is from the right to the left.

The pulley shafts at the left-hand ends of the conveyors **33**, **34** and **34**, **35**, carry pulleys over which a pair of crossed belts **40**, **41** extend for simultaneous movement of belts **33**, **34** to the left, as seen in Figure 7.19b, and for movement of the belt **32** to the right, all at the same rate of speed since the pulleys mounting belts **33–35** are the same diameter. A pulley on the conveyor shaft at the left-hand end of the lower conveyor **35** is connected by a belt **42** with a motor **43** for so simultaneously moving the conveyor belts **33–35**. In the above structure alternate ends of the conveyors **33—35** are the discharge ends, with the discharge ends of the upper conveyor and the lower conveyors being at their left-ends, as seen in Figure 7.19b, and with the right-hand end of the middle conveyor being its discharge end. The ends of the conveyor **30—32** opposite to their discharge ends may be called their receiving ends.

A plastic scraper **44** (Figure 7.19c) similar to a doctor, extends across and substantially in sliding engagement with the outer surface of each of the conveyors **33—35** at the discharge end of the latter, each scraper being formed with an edge positioned adjacent to the belt and an upper surface **45** over which the material on the belt is adapted to be carried away from the belt for dropping against the concave surface of a downwardly extending guide plate **46** that, in turn, will deliver the material onto the receiving end of the conveyor therebelow in an inverted position. A spring **47** may connect each scraper with the conveyor supporting frame to yieldably urge each scraper into engagement with the outer or load supporting surface of each conveyor where the latter extends around the pulley, so that the pulley will constitute a rigid backing for the conveyor belt, and each guide plate **46** may be also supported on the conveyor frame.

The discharge end of the lower conveyor **35** is adjacent to partition **3**, and is adapted to discharge the material thereon, onto the lower end of an endless elevator conveyor **50** that slantingly extends upwardly from below the discharge end of conveyor **35** through partition **3** and into the upper end portion of the subliming room. Slats or equivalent conventional load carrying supports on the elevator conveyor **50** will thus receive the material from the discharge end of the lower conveyor and will carry it upwardly through the partition **3** for discharge into the subliming room **2** at the upper end of the latter adjacent to the upper end of wall **3**.

Also carried by any suitable frame means in the freezing room **1** in a position over each conveyor **33—35** is a horizontally elongated air discharge nozzle **52** preferably of the width of each conveyor and extending from end to end thereof. The lower side of each nozzle is provided with downwardly directed relatively closely spaced discharge openings for discharging air uniformly against any product supported on the upper flight of each conveyor. In fact the lower side of each nozzle is a perforated plate. The body of each nozzle above the discharge opening is preferably of progressively decreasing cross sectional area transversely of the length thereof in directions away from the central portion of the nozzle body so as to provide for the uniform discharge of air over and against substantially the entire upper surface of each of the conveyors **33—35**.

Conduits **30, 31** and **32** (Figure 7.19a) extend respectively from the plenum chamber **18** to the enlarged central portion of each of the nozzles **51** for supplying air to each of the nozzles under pressure from the blowers of the air conditioning units **4, 5**. The provision of the two air cooling or air refrigeration units **4, 5** enables a continuous discharge of cold air from nozzles **51** onto the products on conveyor belts **33—35**. The fan motors for the fans in units **4, 5**, are schematically indicated at **53, 54** (Figure 7.19a), and the solenoid valves in the spray lines **14, 15** leading to units **4, 5** are indicated at **55, 56**, all being in the circuit in which timer **26** is positioned. The high torque damper motors **24, 25** are, as already explained, also in this circuit. As an example, the timer control automatically alternately actuates the fan motors for periods of twelve minutes. When the fan motor 53 in unit 4 is actuated the damper motor 24 will automatically move dampers 22 to open position and motor 25 will automatically close the damper 23 and the solenoid valve 56 will be actuated to open the spray line 15 for defrosting the coils in unit 5. Solenoid valve 55 will be closed. At the end of the 12 minute period the timer 26 will break the circuit to fan motor 53 and close it to fan motor 54 and the damper motor 24 will close the damper 22 and motor 25 will be actuated to open damper 23, while the solenoid valve 56 will be closed and the valve 55 will be opened to defrost the coils in unit 4.

It should be noted that the timer is preferably adjusted to permit adequate time for draining the water from the drain pan in each unit 4, 5 before the fan motor in each cooling unit is started. Substantially the same conveyor and nozzle arrangement is provided within the sublimation room 2 as in the freezing room 1, there being a plurality of endless, horizontally extending, vertically spaced, parallel superposed conveyors 59, 60, 61 within room 2, and over each conveyor is a downwardly directed air discharge nozzle 62, that corresponds in structure to each nozzle 52. Each nozzle 62 is disposed over the upper flight of each of the conveyors 59–61 in the same manner as nozzles 52 are disposed over conveyors 33–35.

The conveyors 59–61 may be offset relative to each other in the same manner as the conveyors 33–35 so that the discharge end of the elevator conveyor will discharge the material therein onto the receiving end of the upper conveyor 59, and the discharge end of the lower conveyor 61, which is at the left-hand end as seen in Figure 7.19b, will discharge the material therein onto the receiving end portion of a discharge conveyor 63. The discharge conveyor is horizontally elongated, and extends through an opening in the end wall 64 of the housing, for carrying the material discharged thereon out of the sublimation room 2.

Any suitable frame may support the conveyors 59–61 and nozzles 62, and any other parts associated with the conveyors, within the sublimation room. A motor may be connected with the conveyors 59–61 for moving the latter in directions corresponding to the directions of movement of the correspondingly positioned conveyors 33–35 in the freezing chamber 1. A crossed belt 67 may connect the pulleys on the shafts at the left-hand ends of conveyors 59, 60 and a crossed belt 68 may connect the pulleys at the left-hand ends of conveyors 60, 61 for simultaneously moving the conveyors 59–61 for carrying products back and forth within the chamber 2 and for finally discharging the products onto the receiving end of discharge conveyor 63.

Where each of the three conveyors in the freezing room and in the sublimation room, respectively, is of substantially the same length, or approximately thirty feet, the three conveyors in each room have substantially the same capacity as a single conveyor that is substantially 90 feet long. The rate of travel of the load carrying flight of each conveyor 34–35 and 59–61 is preferably approximately 0.66 feet per minute, or a total of one hour in the freezing room and one hour in the sublimation room. During the time the product is carried through the freezing room, it is subjected to air from nozzles 52 at a temperature of approximately -15° to -30°F.

APPARATUS FOR CONTINUOUS FREEZE DRYING PROCESSES

DRYING TUBES

Rotating Drying Tube of Polygonal Cross Section

An important feature of the drying tubes described by *W.C. Rockwell, V.F. Kaufman and E. Lowe; U.S. Patent 3,303,578; February 14, 1967; assigned to The U.S. Secretary of Agriculture* is that they are polygonal in cross section. With such a construction the material is subjected to positive tumbling action as the material rolls from one face of the polygon to another. This means that all surfaces of the pieces of material are uniformly exposed to the heated walls of the drying tubes.

On the other hand, were one to use a tube of circular cross section, the material would tend to slide without any relative motion of the individual pieces with the result that nonuniform dehydration would take place. It is also to be observed that the polygonal configuration not only causes the desired tumbling action but also presents a smooth surface to the material being treated. Thereby the tumbling is achieved with a minimum of mechanical damage to the size and shape of the pieces of material being treated. It is preferred that the length of the drying tubes be at least 10, preferably 15 to 25, times the diameter thereby to achieve the desired efficiency.

In operation (see Figure 8.1), the material to be dried is fed through a conventional vacuum lock valve **23** into chute **24**. This chute extends into shell **1** through vacuum-tight static seal **24a** and into feed cylinder **20** through circular opening **25** in the end wall of the cylinder. Usually, the material is introduced at such a rate as to continuously maintain a bed of material in the cylinder, this bed extending almost up to the lower lip of the opening. Because of the circular wall of the cylinder, this bed of material remains at the base of the cylinder. Then as each feed tube **21** moves through the lower point of its orbit, the feed screw **22** associated therewith moves through the bed of material and scoops up a quantity of material which is then advanced by the screw into the feed tube and eventually into drying tube **17**.

FIGURE 8.1: ROTATING DRYING TUBE OF POLYGONAL CROSS SECTION

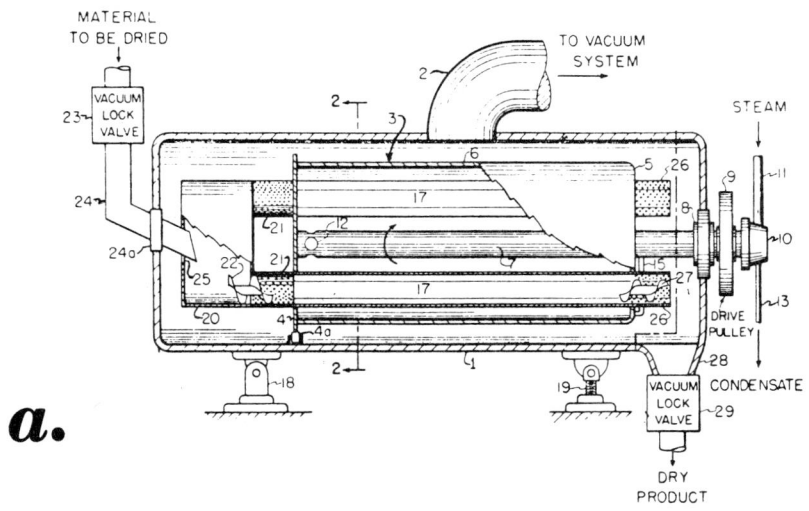

a.

Side View

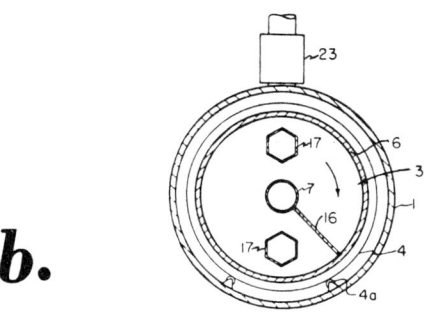

b.

Cross Section Along 2–2 of Figure 8.1a

Source: U.S. Patent 3,303,578

Although there is no relative motion between each screw **22** and each feed tube **21**, there is obtained a rotation of each screw as it revolves (or orbits) together with the rest of the assembly around the axis presented by shaft **7**. It is this rotation effect which makes the feed screws operative so that the material is continuously fed into each drying tube. Such items as the pitch of the feed screws, the speed of rotation of the shaft, the rate of introducing material into the drier, the slope of the entire drier, etc. are so correlated that the drying tubes are continuously maintained about $1/3$ to $2/3$ full of material.

To maintain tubes **17** filled to the desired degree there is provided a weir arrangement at the discharge (right-hand) end of the system. Thus each of the drying tubes is provided with a discharge tube **26** of polygonal cross section, fabricated from perforated sheet material to provide free passage of vapors from the drying tubes. Rigidly fastened in each of these discharge tubes is a discharge screw **27**. It is to be observed that the direction of twist of these screws is the opposite of that of feed screws **22**. Thus with a standard selected direction of rotation of drum **3** and associated parts, feed screws **22** are selected to impel material toward the right, whereas discharge screws **27** are selected to impel material to the left and thus act as weirs to restrain free discharge of material from the drying tubes and thus cooperate to keep the drying tubes filled to the desired level.

Example: The drier used in this run was essentially as described above except that there was only one drying tube **17**. It was 8 feet long, hexagonal in cross-section, internal diameter 7½" (or about 6½" measured across the flat). Steam was introduced under pressure into the steam drum to maintain the drying tube walls at 250°F. Speed of rotation of the drum and tube assembly was 1.7 rpm, the slope was 3/16 in/ft.

The starting material was ⅜" diced, frozen chicken meat, having a moisture content of 65%. It was fed into the system at the rate of 10 lb/hr. Retention time in the drying tube was calculated to be 2.5 hours. The product, having a mositure content of 2%, was produced at the rate of 3.6 lb/hr. The diced form of the chicken meat was retained in the product and taste tests indicated there was no damage to flavor. Also, because of the porous texture of the product, it absorbed water very rapidly on being reconstituted. It was calculated that the drying efficiency was 0.45 lb of water evaporated/hr/ft^2 of heat transfer surface. This is four times the efficiency obtained in the conventional shelf-type vacuum driers.

Rotating Drying Tube with Fins and Baffles

Here *V.F. Kaufman and W.C. Rockwell; U.S. Patent 3,308,552; March 14, 1967; assigned to The U.S. Secretary of Agriculture* again provide a rotating dryer in which a heated surface is provided and all surfaces of the material are contacted therewith by applying a tumbling action to the pieces of material. Heat-transmitting fins and baffles are provided, which are utilized efficiently by contacting material to be dried with both sides of these heat-transferring members.

The apparatus is described with reference to Figures 8.2a and 8.2b. Drying chamber **8** is rotated at 0.1 to 1.0 rpm by shaft **1** and is in communication through the hollow center of this shaft with vacuum line **7**. Within the drying chamber are mounted tubes **11, 12, 13** and **14**, each provided with four fins **11a, 11b, 11c, 11d, 12a**, etc., made of copper or other metal of high thermal conductivity. As shown in Figure 8.2a, these fins extend the full length of the drying chamber. Viewed transversely, in Figure 8.2b, each set of fins forms a cross and the corresponding fins of one tube are parallel to those of every other tube. Projecting radially from the interior wall of the drying chamber are a series of baffles **15** made of copper or other metal of high thermal conductivity. Each of these baffles extends the full length of the drying chamber.

To provide heat, steam from supply line **25** flows through the joint **24** and pipe **23** into plenum chamber **21**, from where it is distributed to the annular

198 Freeze Drying Processes for the Food Industry

space **22** and to tubes **11, 12, 13** and **14**. To prevent air-lock, each of these tubes contains a conduit **26**, open at its left-hand end and connected to a petcock **27** at the right-hand end.

FIGURE 8.2: ROTATING DRYING TUBE WITH FINS AND BAFFLES

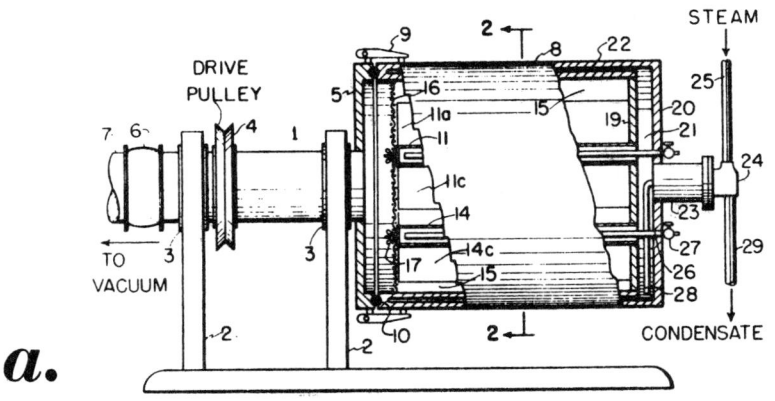

a.

Side View

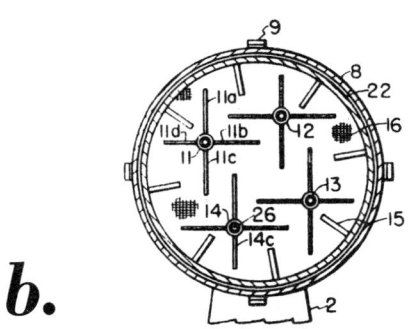

b.

Cross Section on Plane 2—2 of Figure 8.2a

Source: U.S. Patent 3,308,552

In operating the device, steam and condensate lines **25** and **29** (preferably of flexible braided metal hose) are disconnected. Latches **9** are released and drying chamber **8** is deposited in a suitable cradle with its open end up. Wing-nuts **17** are unscrewed and screen **16** lifted out. The material to be dried, for example, frozen fresh peas; is then placed in the chamber, filling it about $\frac{1}{3}$ to $\frac{2}{3}$ full. The screen is then replaced, secured, and the drying chamber again latched to head **5**. Rotation of shaft **1** is initiated which, of course, rotates the chamber. The vacuum system is activated and when the drying chamber is exhausted, the steam lines and condensate lines are connected. Heat from the steam is transferred to the walls of the drying chamber, to tubes **11, 12, 13** and **14**, to the fins extending from these tubes, and to baffles **15**. Thus, as the drying chamber rotates, the material rolls from one heated surface to another with uniform exposure thereto. This exposure to uniform heat, plus concomitant exposure to the vacuum existing within the chamber causes a rapid and uniform sublimation of moisture from each particle of the material. The evolved vapors pass through screen **16** and through shaft **1** to the vacuum system connected to conduit **7**. On completion of the drying, the chamber is detached from head **5**, the screen is removed, and the dried product poured out.

The fins and baffles function not only as supports, but also as guides to direct the particles in closed circuits within the small pockets formed by adjacent fins and/or baffles. The net result of this arrangement is that the drying chamber is effective at high loadings; the material is distributed quite uniformly over the network of fins and baffles.

Example: The drier used in the run was as shown in Figures 8.2a and 8.2b. The drying chamber had internal dimensions: 15½" diameter, 12" long. Fins **11, 12, 13**, and **14** were 12" long and 2¾" or 3½" wide. Baffles **15** were 12" long and 2" or 3" wide.

The starting material was frozen fresh peas, having a moisture content of 80%. 25 lb of this material was placed in the drying chamber, filling it about half full. During the dehydration, the drying chamber was rotated at 0.4 rpm and steam was introduced at 260°F except during the last half-hour of the run when the steam temperature was tapered off to 212°F. The vacuum was maintained at 0.5 mm Hg. In 4 hours the product was removed and found to have a moisture content of 3%. The product consisted of whole peas; there was no breakage of individual seeds or even removal of skins. Also, the peas were uniformly dehydrated; there were no moist centers. Taste tests indicated that there was no damage to flavor.

In a control run a batch of the same frozen fresh peas were dehydrated in a conventional vacuum shelf drier. It was found that it required 8 hours to dehydrate the product to a moisture content of 3% and yet 2% of the peas in the product still contained ice in their centers.

DRYING CHAMBERS

Upright Drying Chamber with Vacuum Locks at Entrance and Exit

W.P. Ullrich and W.E. Christison; U.S. Patent 3,243,892; April 5, 1966; assigned to Beverly Refrigeration, Incorporated describe an apparatus, which is capable of carrying out the freeze drying process on a continuous basis. This apparatus includes improved vacuum lock means for receiving the product into the drying chamber, and for providing a continuous flow of the product through the apparatus.

The vacuum lock construction is shown with reference to Figure 8.3. The vacuum chamber **54** includes an external casing **150** which defines a hopper for the product fed into the chamber. The upper portion of the casing includes an inlet **152** which may be coupled to an appropriate feed line for the introduction

200   Freeze Drying Processes for the Food Industry

of the frozen product into the compartment 54.

FIGURE 8.3: UPRIGHT DRYING CHAMBER WITH VACUUM LOCKS AT ENTRANCE AND EXIT

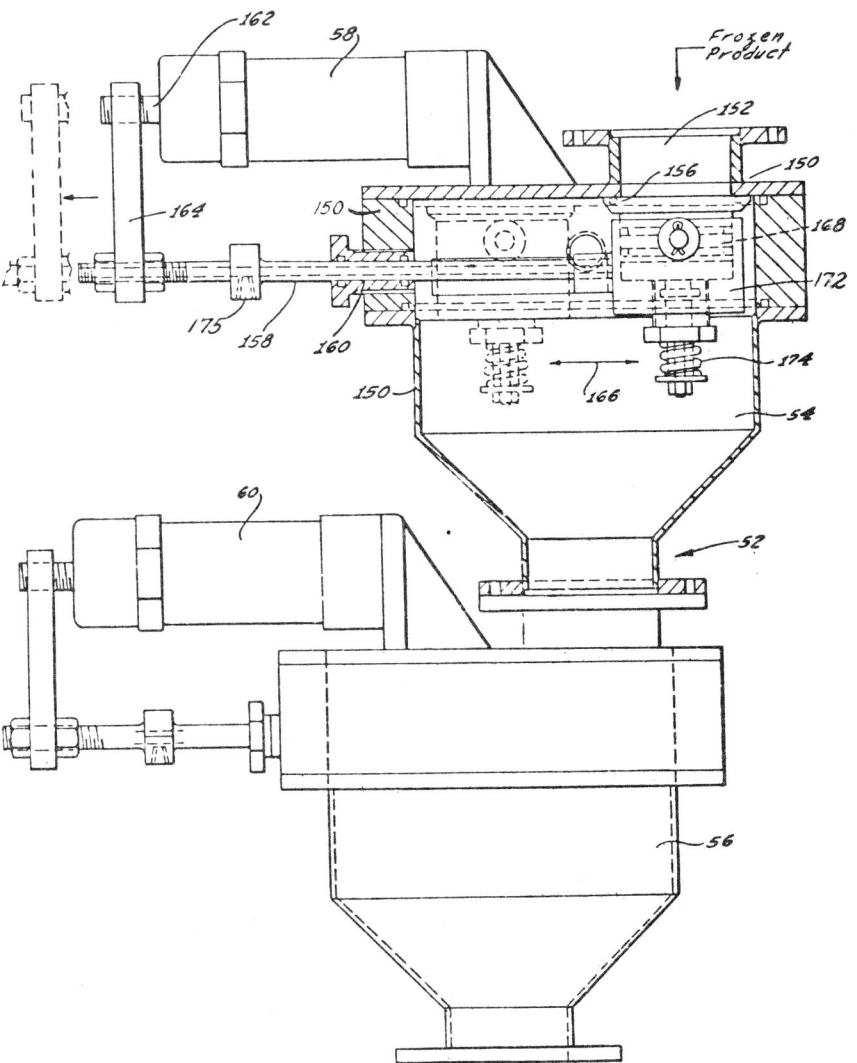

Source: U.S. Patent 3,243,892

The casing defines a valve seat at the lower end of the inlet and an O-ring **156** may be provided, for example, for sealing purposes. A tubular shaft member

158 extends into the casing 150 through a vacuum seal 160. The tubular shaft 158 is movable to the right and to the left under the control of the air cylinder 58. At this end, the shaft is coupled to the plunger 162 to the air cylinder by means of an appropriate linkage 164.

The air cylinder under the usual pneumatic control, may be actuated to move the shaft to the right or to the left as designated by the arrow 166. A valve member 168 is mounted on the end of the tubular shaft in the vacuum chamber 54. The valve member is movable between a position in which it is axially aligned with the valve seat defined by the O-ring 156, in a first position; and in which it is displaced to the left from the valve seat (as shown by the dotted representation) in a second position. When the valve member is in its second position, the product can be introduced into the compartment 54 through inlet 152.

The valve member is mounted in a cylinder 172, the latter being mounted on the tubular shaft. The valve member is movable vertically within the cylinder and it is normally biased to a lower position by means of a coil spring 174. When the valve member is in its lower position, it is displaced downwardly from the valve seat defined by the O-ring, so that the chamber 54 is in an unseated condition.

However, when pressurized fluid (such as compressed air) is introduced through a fitting 175 into the interior of the hollow shaft 154, this pressurized fluid is carried into the lower part of the cylinder 172, to overcome the force of the spring 174 and force the valve member up against the valve seat, so as to seal the vacuum chamber. By means of the valve mechanism the valve 168 may be first released from the valve seat, merely by interrupting the introduction of the pressurized fluid through the fitting 175. When the valve member is so relieved from the valve seat, it drops away from the O-ring and it may be moved to the left under the control of the air cylinder.

Conversely, to close the valve assembly, the air cylinder may be first actuated to bring the valve member into position directly under the valve seat, and then the pressurized fluid may be introduced into the fitting so as to close the valve, and seal the compartment 54. In the manner described above, the vacuum chamber may be conveniently opened for the reception of the frozen product, but with a minimum of wear to the valve member, or to the O-ring which normally seals the compartment, when the valve member is in its closed condition.

Elongated Drying Chamber Straddled by Inlet and Outlet Lock Chambers

U. Hackenberg; U.S. Patent 3,273,259; September 20, 1966 is concerned with a method and apparatus for continuous freeze drying applications. The process is described with reference to Figure 8.4. There is shown a freeze drying installation having a central drying chamber 11 formed by the elongated cylindrical tank 12 of circular cross section and by the housing 13 which is of larger cross section than the elongated tank. Along the entire length of and adjacent to the inner walls of the elongated tank and enlarged housing are a plurality of separate heating plates 14 which are adapted to be independently controlled by suitable heat sources (not shown).

An inlet lock chamber 15 having a hinged door 16 is connected to the elongated tank by the gate valve assembly 17. Connected for gas communication with the interior of the inlet lock chamber 15 through the valve 18 is the cold condenser 19 and the mechanical vacuum pump 21. A compressor 22 is connected to supply a refrigerant to the cold condenser. The vent valve 23 is positioned between the outer atmosphere and the interior of the inlet lock chamber.

The end of the drying chamber 11 adjacent the enlarged housing 13 is bifurcated into a pair of condenser housings 25. The condenser housings are connected to the drying chamber by isolating gate valve assemblies 26 and enclose the cold condenser elements 27. Communicating with each of the condenser housings through gas valves 28 is the mechanical vacuum pump 29. A pair of compressors 31 is connected to supply refrigerant to each of the cold condenser elements.

Extending laterally from the enlarged housing is the outlet lock chamber 32 which is connected thereto by the isolating gate valve assembly 33. The outlet lock chamber is provided with a hinged door 34 and an atmospheric vent valve 35. Connected for gas communication with the outlet chamber through the gas valve 36 is the cold condenser 37 and the mechanical vacuum pump 38. A compressor 39 is connected to supply refrigerant to the cold condenser 37.

In operation, a suitable product carrier (not shown) is first loaded with a frozen material to be dried and placed inside inlet lock chamber 15 through the open door 16 with the isolating gate valve 17 in a closed position. The hinged door is closed and the inlet lock chamber evacuated by the mechanical vacuum pump through open gas valve 18. The isolating gate valve is then opened and the loaded product carrier is automatically transported into the already evacuated drying chamber 11 on the conveyor rail 41 by a suitable motive means (not shown).

The isolating gate valve is again closed and the vent valve 23 opened to produce atmospheric conditions in the inlet lock chamber. Upon opening of the hinged door, another loaded product carrier is moved into the inlet lock chamber. After evacuation of the inlet lock chamber, the second loaded product carrier is also transported into the drying chamber through the opened isolating valve 17. This procedure is continued until the entire drying chamber is filled with product carriers filled with the material to be dried.

Inside the drying chamber the heating elements 14 apply heat to the frozen material causing sublimation of the moisture therein. This sublimating vapor is pumped through the length of the drying chamber and open isolating gate valves 26 into the condenser housings. The condensable portion of this vapor is condensed on the cold surfaces of the cold condensers 27 while the noncondensable portion is pumped out of the housings through open valves 28 by the mechanical vacuum pump 29. The use of the two independent condenser housings is advantageous as it allows one of the units to be isolated from the drying chamber for cleaning and deicing while the drying process continues with the other unit.

As the material is gradually transported through the drying chamber, a dry outer layer is formed on the material due to the removal of sublimating water vapor. This dry layer is very susceptible to thermal damage and becomes increasingly thicker as the drying process continues. It is therefore desirable to reduce the level of applied heat as the thickness of the dry layer becomes greater.

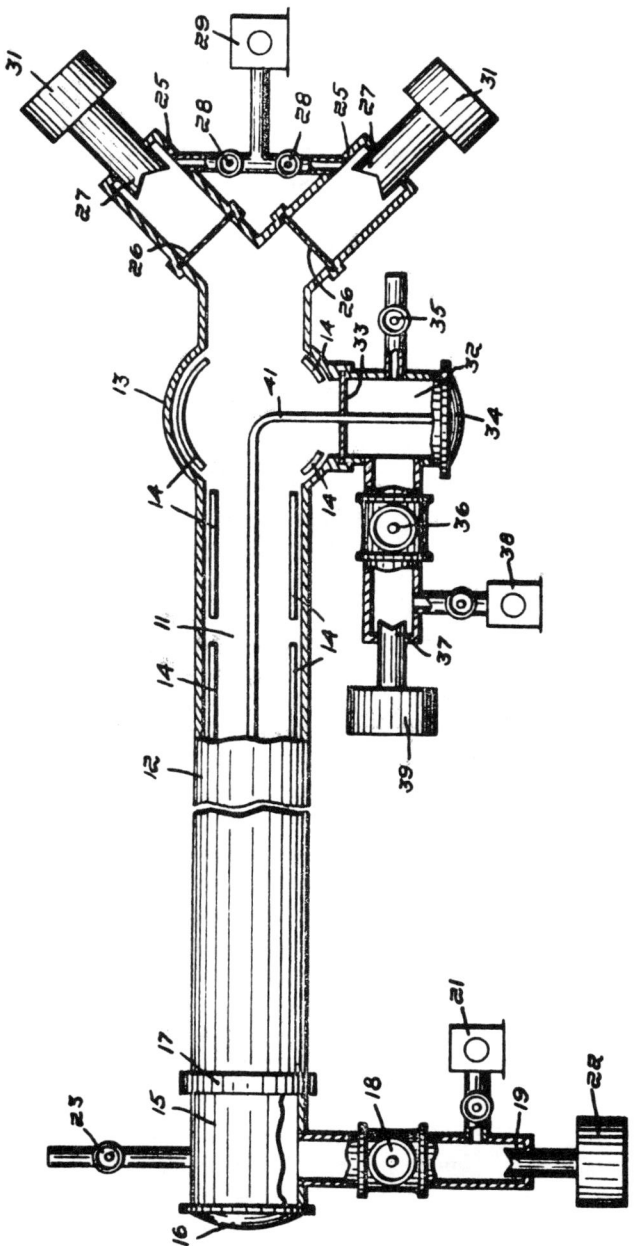

FIGURE 8.4: ELONGATED DRYING CHAMBER STRADDLED BY INLET AND OUTLET LOCK CHAMBERS

Source: U.S. Patent 3,273,259

The reduced heat level can be correlated to the material drying speed (in mm of water removed/min) which drying speed decreases as the thickness of the dried material layer increases. For this reason the heating units **14** are controlled so that each successive heating unit, going in direction from the inlet lock chamber **15** toward outlet lock chamber **32**, operates at a reduced temperature. For example only, with seven independent heating units the preferred operating temperature for each successive unit would be 280°C, 220°C, 170°C, 130°C, 100°C, 80°C and 70°C.

The advantage of the process is that vapor produced in the initial portions of the drying chamber which vapor is overheated by the higher temperature heating units is forced to travel the entire length of the drying chamber before reaching the cold condensers **27** and the vacuum pump **29**. While passing through the chamber a portion of the heat in this vapor will be absorbed by the freeze dry components within the advanced parts of the tank **12** and housing **13** and especially by the frozen material contained in these parts.

The heat absorption by the frozen material is particularly beneficial since it assists in providing the desired sublimation of the frozen moisture. The heat lost in this way greatly reduces the average temperature of the vapor reaching the cold condensers **27** and the mechanical vacuum pump **29**. This reduced temperature in turn reduces both the energy requirements for maintaining the cold condensers at their desired operating temperature of, for example, –40°C and the pumping capacity requirements for the mechanical vacuum pump.

As each product carrier reaches the enlarged housing, the material carried has been completely dried and the carrier is transported into the outlet lock chamber through open isolating gate valve **33**. The isolating valve is then closed and the vent valve **35** opened to produce atmospheric conditions within the outlet chamber. After removing the dried material, the hinged door **34** and the vent valve are closed and the outlet lock chamber is again evacuated by the mechanical vacuum pump **38** through the open gas valve **36**. The isolating valve is then opened preparing the outlet chamber to receive another product carrier containing dried material.

HEATING ARRANGEMENTS

Parallel Heating Plates Defining Tortuous Path

V.J. Janovtchik and C. Catelli; U.S. Patent 3,469,327; September 30, 1969; assigned to H.J. Heinz Company Limited describe an apparatus for the continuous drying of frozen products in containers which comprises a vacuum chamber through which the containers are advanced from an entrance airlock to an exit airlock.

The heating plates are preferably fixed parallel to one another in the vacuum chamber and the entrance airlock is located opposite one end of the topmost or lowest plate while the exit airlock is located opposite the other end of the lowermost or topmost plate. This arrangement ensures the longest possible path for the containers of products over the heating plates in the vacuum chamber. Transfer devices are mounted relative to the ends of the plates and are operable to transfer a container from one plate to the next in the path after a container

Apparatus for Continuous Freeze Drying Processes

has been introduced into the chamber. The process is described with reference to Figure 8.5.

As shown in Figure 8.5a, at the opposite side of the cylindrical outer wall of the vacuum chamber **1** there are two outlets **6** closed by valves in the form of sealing doors **14**, for connection to vacuum and vapor removing equipment. Inside the vacuum chamber there are a series of heating plates **7** to **13** fixed above one another and in parallel relationship.

The entrance airlock **4** has a shaped outer wall **15** and in one side of the wall there is a rectangular inlet **16** which is closed by a sealing flap door **17**. The top of the flap door is fixed to a pivot rod **18** connected by a lever **19** which is connected to the piston rod of a hydraulic motor in the form of a two-way piston and cylinder indicated generally at **20**. The pressure fluid supply to the motor **20** normally maintains the door pressed into sealing engagement with the inlet, and when the motor is reversed the flap door pivots upwardly so that a container **21** of frozen products to be dried can be placed in the entrance airlock.

The entrance airlock communicates with the main vacuum chamber through an inlet **22** also of rectangular form to permit the passage of the container, and normally sealed by a flap door **23** which is mounted on a pivot rod **24** which passes through a sealing bearing **25** in the wall of the entrance airlock. The pivot rod is connected through a lever **26** to the piston rod of a two-way piston and cylinder hydraulic motor **27**. This door pivots downwardly. Normally the door is held by this motor in sealing engagement with the inlet and when a container is to be moved into the vacuum chamber from the airlock the door is pivoted downwardly to open the inlet and the container can run over rollers **29** mounted on the plate **23** and an apron **30** mounted just inside the inlet (see Figure 8.5b) during its passage on to the end of the first of the heating plates.

The outer end of the wall of the entrance airlock is elongated to form a vacuum-tight casing indicated at **31** to house a push rod **32** which carries a flat pusher **33** at its inner end for engaging the back face of the container. The rod has rack teeth indicated at **34** (see Figure 8.5b) on its lower surface and these teeth engage a pinion **35** which is fixed to a shaft **36** which passes through a sealing bearing **37** in the wall of the entrance airlock and is driven through gearing **36a** by a hydraulic motor **38**. There are also two outlets in the wall of the entrance airlock. The outlet **40** is connected to vacuum equipment through a sealing valve **41** and the outlet **39** can be opened to the ambient atmosphere by opening its sealing door **42**.

In operation when a container is to enter the plant the entrance airlock is first isolated from the vacuum chamber by operating the motor **28** to maintain the flap door **23** closed. The outlet **40** from the airlock to the vacuum equipment is then sealed by closing the valve **41** and the vacuum inside the airlock is broken by opening the valve **42** which communicates with the outside atmosphere. The flap door **17** is then opened, by operation of the motor **20**, the container is introduced into the entrance airlock and the door **17** and the valve **42** are closed. The valve **41** is then opened so that a vacuum is applied to the airlock to put the airlock under the same vacuum as that prevailing inside the vacuum chamber.

FIGURE 8.5: PARALLEL HEATING PLATES DEFINING TORTUOUS PATH

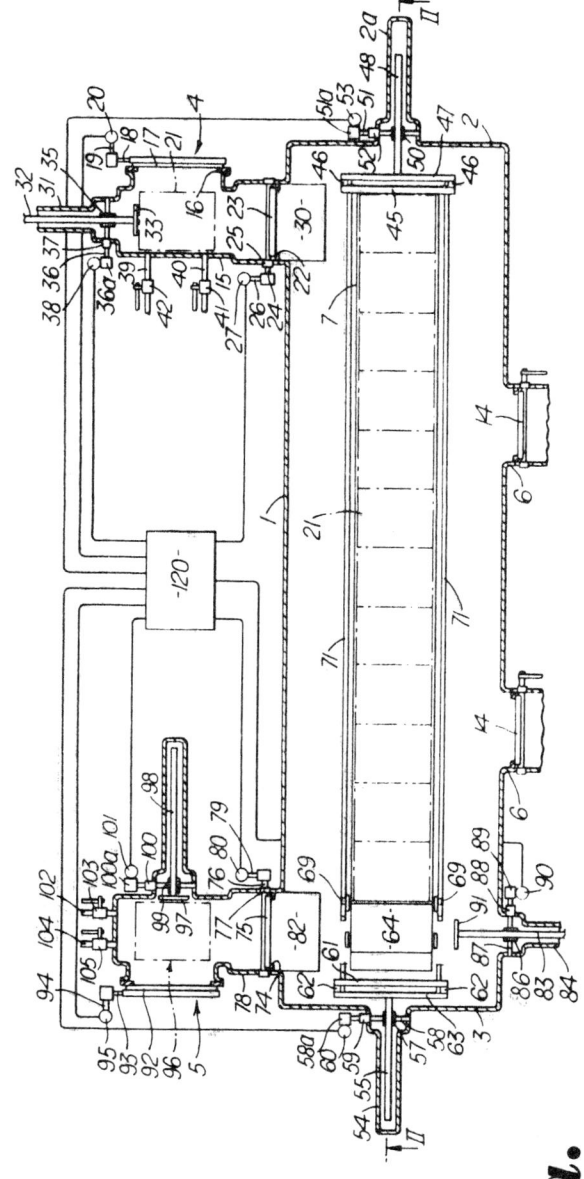

a.

Horizontal Section Through Freeze Drying Apparatus

(continued)

FIGURE 8.5: (continued)

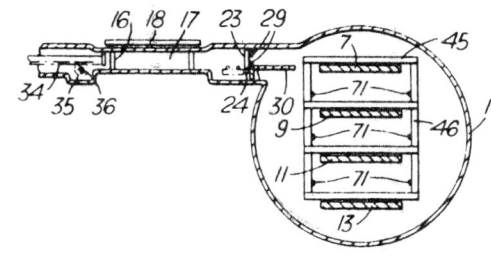

b.

Arrangement of Entrance Airlock with Push Bar Mechanism for Moving Containers on to the Heating Plates from Airlock

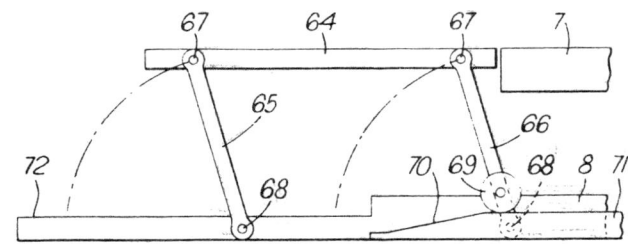

c.

Type of Transfer Device Mounted on One End of Heating Plates

Source: U.S. Patent 3,469,327

When this is achieved the flap door **23** is opened, the motor **38** of the push bar mechanism is operated and the pusher **33** pushes the container **21** over the rollers **29** on the lowered flap door **23** and the apron **30** on to the first heating plate **7**. This flap door is then closed, the valve **41** is closed and the valve **42** is opened to release the vacuum, and the entrance airlock is then ready for use to introduce another container of products to be dried into the apparatus.

The vacuum chamber is maintained at a very low pressure (1 mm of mercury) by means of the vacuum equipment which is connected to the vacuum chamber through the same ducts **6** as the vapor condensing equipment. The means for moving the containers comprises push bars mounted inside the chamber at opposite ends of the chamber. At the right hand end of the chamber as illustrated in Figure 8.5a near to the entrance airlock the end wall **2** of the vacuum chamber is formed with an outwardly extending and central hollow tube **2a** to accommodate push bar equipment.

The set of push bars mounted adjacent the wall are for engaging containers on the first of the heating plates **7** and on subsequent odd-numbered heating plates **9, 11** and **13**. The push bars **45** are mounted as cross members between two side struts **46** which are connected centrally by a strut **47** which is fixed to one end of a push rod **48** which is housed in the casting **2a** and whose undersurface is formed with rack teeth which mesh with a pinion **50** mounted on a shaft **51** which extends through a sealing bearing **52** in the wall of the casing **2a**. The shaft is connected through gears **51a** to a hydraulic motor **53** in the form of a two-way piston and cylinder which can be driven forwardly or in reverse in order to move the push rod by a fixed distance appropriate for the movement of the containers along the heating plates by the length of one container.

When the push rod moves into the vacuum chamber one of the push bars **45** engages the side of the container to push that container along the plate **7** by a distance sufficient to allow another container to be moved into the vacuum chamber from the entrance airlock and on to the plate. At the same time, the apparatus being full of containers, the push bars opposite the odd-numbered heating plates **9, 11** and **13**, also slide the end containers along those plates from right to left by the same distance.

At the other end of the vacuum chamber mounted in a tube **54** formed in the end wall **3** of the vacuum chamber there is a further push rod **55** which is formed as a rack with teeth meshing with a pinion **57** mounted on the wall of the casing **54** and is connected through gears **58a** to the output of another reversible hydraulic motor **60**. A second set of push rods **61** are mounted between side struts **62** which are held together by a cross strut **63** which is fixed to the end of the push rod **55**. In the particular embodiment described there are only three push rods **61** at the left hand side of the apparatus and these push bars cooperate with the even-numbered plates such as **8**, to push from left to right the end containers on those plates by the distance equal to the length of one of the containers.

The heating plates which may be made of several units are all of equal length and are mounted in staggered relation so that the ends of the even-numbered plates all protrude to the left beyond the ends of the odd-numbered plates that are at the end of the vacuum chamber opposite to the entrance airlock. The ends of the odd-numbered plates all protrude beyond the even-numbered plates at the right hand end of the vacuum chamber adjacent the entrance airlock.

Means are provided for transferring containers from one heating plate to another. Mounted on the protruding end of each of the heating plates, except the first plate there is a transfer device which is operable to transfer a container from one plate to the next below, so as to maintain the continuous path of the containers along the heating plates.

One of the transfer devices is illustrated in greater detail in Figure 8.5c and includes a container support frame **64** which is connected at both its sides to the protruding end of a heating plate shown as plate **8** by pivoted links **65** and **66**. These links are parallel to each other and are fixed by pivot pins **67** and **68** to the support frame and to the heating plate respectively. As shown in Figure 8.5a the support frame is of rectangular shape open at the side facing the oncoming containers and its dimensions are such that it receives snugly a container pushed along the heating plate **7** by the operation of the set of push bars **45**.

The pair of pivoted links **66** each carry a cam roller **69** near its lower end, and these cam rollers are engaged by wedge shaped cam surfaces **70** at the ends of actuating arms **71** extending along both sides of the heating plate **8**. When the actuating arms are moved towards the end of the heating plate, from right to left, the rollers ride up the surfaces and thereby cause the links **65** and **66** to pivot so that the support frame **64** is raised from its normal flat position in which the frame lies in a rabbet **72** cut in the protruding end of this plate to receive the frame. When the cam rollers have moved right up the surfaces the support frame is at the level of the heating plate **7** so that the end container on the heating plate is pushed along that heating plate by operation of the push bars **45** into the support frames.

In order to ensure the raising of the support frame prior to the pushing of the containers along the heating plate the actuating arms are preferably operated by the same mechanism as the push bars. This has the added advantage that no additional driving connection is necessary through the wall of the vacuum chamber, it being desirable to keep the sealed connections through the walls of the vacuum chamber to a minimum.

The actuating arms which extend along the sides of each of the even-numbered heating plates are connected to the side struts **46** which hold the set of push bars. The arrangement of these arms is such that when the push rod **48** is moved inwardly the cam rollers immediately begin to ride up the cam surfaces so that the support frames which are pivotally mounted on the ends of the even-numbered heating plates are already raised to the level of the ends of the heating plates **7, 9** and **11** by the time the push bars begin to move the containers along the odd-numbered heating plates.

When the inward movement of the push rod is completed a container is freely carried on the support frame and when the push rod is then retracted the support frame is lowered as the cam rollers **69** run down the cam surfaces. The frame then sinks into its rabbet and the container on the support frame is on a level with the upper surface of the heating plate **8**. The push rod **55** is then moved inwardly from left to right and the push bars **61** engage the containers on the even-numbered plates. The container on the support frame is then pushed off the frame and along the heating frame **8** so that the frame can be raised to receive another container from the plate **7** in the next cycle of operation.

There are actuating arms **73** similar to the arms **71** extending along the sides of the odd-numbered heating plates and these actuating arms are all connected to the support struts **62** for the push bars. Just before the push bars engage the containers at the left hand ends of the even-numbered heating plates the actuating arms **73** raise the transfer devices from the right hand ends of the odd-numbered heating plates so that they are ready to receive containers from the right hand ends of the even-numbered heating plates and transfer them down to the level of the odd-numbered heating plates.

In this way, a container of frozen products to be dried begins its journey by being pushed from the entrance airlock on to the right hand end of the first heating plate, then gradually progresses along the path defined by all the heating plates in series until it reaches the left hand end of the last of the odd-numbered heating plates. The exit airlock is constructed in a similar manner to the entrance airlock and communicates with the inside of the vacuum chamber through an outlet which is normally closed by a sealing flap door **75**, of similar construction to the door 23. Another push bar arrangement is provided to push a container off the heating plate 13 and into the exit airlock.

The operation of the exit airlock 5 is synchronized with the operation of the entrance airlock 4 in such a manner as to ensure the removal of containers of dried products at the same time interval as the introduction of containers of frozen products into the vacuum chamber. Also the operation of the push bar mechanisms are synchronized with the operation of the airlocks so that there is a continual movement of containers along the heating plates at regular time intervals corresponding to the period of operation of the airlocks. In order to effect this required sequence of operations the hydraulic motors 20, 27, 38, 53, 60, 80, 90, 95 and 101 are all connected through appropriate spool valves in known manner by pressure fluid supply lines to a central control device, indicated at 120 in Figure 8.5a, which automatically controls the operation of these motors at the correct times to control the sequence of operations described above.

In the modification described there are seven heating plates on each of which twelve containers can be accommodated. That is, a total of eighty-four containers are supported inside the vacuum chamber at any time. The time required to dry a frozen product is for example, 9 hours so that the time interval at which containers can be introduced into the entrance airlock is every 6½ minutes. That is, the output of the apparatus is approximately 9 containers per hour and as each container will carry a load of, for example, 20 lb of frozen products about 180 lb/hr of products can be processed by the plant.

Efficient Access to Heat via Use of Screw Conveyor

The object of the process of *H. Gottfried; U.S. Patent 3,731,392; May 8, 1973* is to provide means for exposing, continuously, each and every particle of triturated frozen material to a source of controlled heat, in a vacuum, so that all moisture in the material can be removed, and to perform this function in a completely automatic manner. Material to be processed is first frozen and ground. It is then introduced under vacuum to a treating chamber. Within this treating chamber are agitating means, e.g., screw conveyor means for moving the material longitudinally across the chamber to one end thereof, during this process exposing all particles of the material to the blades of the conveyor and walls of the container.

A return path for the material is provided to guide the material back from one end to the other end of the chamber. Reversible refrigeration means are provided to cool or heat the chamber walls so as to maintain the temperature within a predetermined range. A vacuum pump and cold trap arrangement constantly evacuates gas formed by moisture in the chamber and vacuum discharge means are provided to discharge the treated material from the chambers into packing containers. The process is described with reference to Figure 8.6.

The Input Zone and Freezing and Crushing Zone: The input zone 12 is fed by a spigot 42 into a liquid container 44 maintained at a constant level. Liquid slurry in the container drops along pipe 50 to a work unit. The pipe is surrounded by cooling coils 52 designed to keep the slurry cool, the coils being at about 0° to 5°C. This work unit is the freezing zone 16 and consists of an elongated cylindrical chamber 54, disposed at about a 30° inclination from the

Apparatus for Continuous Freeze Drying Processes

vertical, also surrounded by cooling coils **56**. The temperature of these cooling coils is about −40°C. This causes the material to freeze to a solid state. In the center of the cylindrical chamber is a screw conveyor **58** disposed at an angle to the vertical. Thus, pipe **50** conveys the raw material to the top of the chamber. In the screw conveyor, the last few upper convolutions **60** have a special cutting configuration to crush the icy material.

FIGURE 8.6: EFFICIENT ACCESS TO HEAT VIA USE OF SCREW CONVEYOR

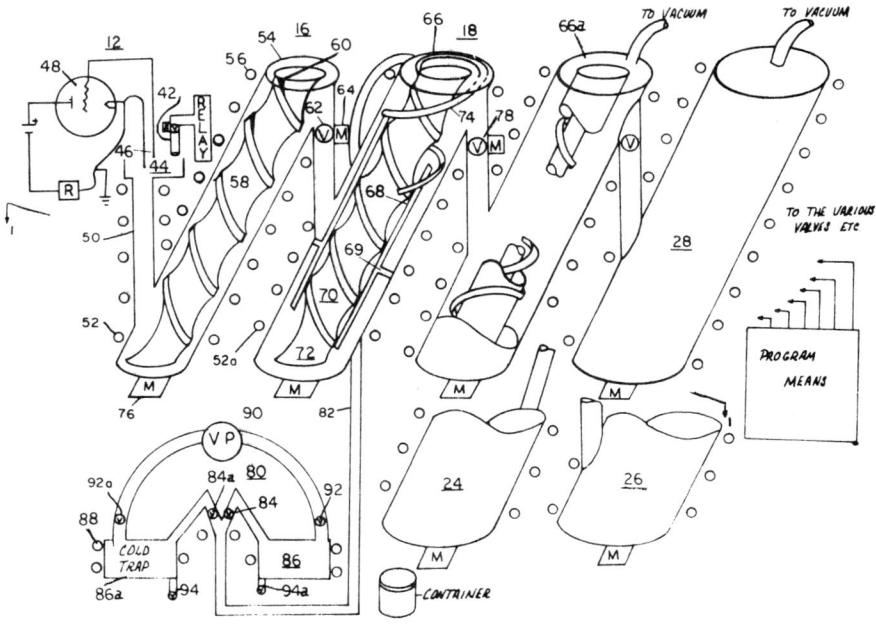

Source: U.S. Patent 3,731,392

The material moving along the screw conveyor is to be transferred to the lyophilization zone **18**. However, after the transfer, while the material is being processed in the lyophilization zone, the screw conveyor of the freezing zone **16** does not stop turning; the screw conveyor turns slowly at an idle speed of about 1 rpm to keep the material from freezing solid and jamming the conveyor. In this state, the contents merely slide around and do not advance.

First Lyophilization Zone: The material now enters the first chamber **66** of the lyophilizing zone through a ball valve **62** with a 4" opening which is opened and closed by a motor **64**. The crushed ice passes through the ball valve and enters the first lyophilizing chamber having cooling coils **52a**. After the lifting screw conveyor has turned a preset number of times, the first lyophilizing chamber contains an amount of frozen material, approximately equivalent to one-half its full capacity. At this point the motor is activated to close the ball valve.

continues until the gas pressure in the chamber reaches 200 microns or less at 50°C. This usually takes approximately two hours.

Thus, in the lyophilization chamber **66**, the material is circulated forward, up, back down, and forward again. In addition, the material is being lifted and circulated against the walls of the chamber in a spiral fashion. This results in intimate contact of the particles of material with the screw conveyor and container walls permitting efficient access of the heat to the material to be lyophilized. It is to be noted that the problem of getting heat to each particle is solved by bringing the heat to each particle by the mechanical motion of a mixing and circulating device. This produces a readily controlled temperature at the surface of each particle.

A central rod supports a first screw conveyor and, attached to the first screw conveyor is a second screw conveyor so that both move together. Rigidly attached to the wall of the container is a stainless steel cylindrical partition containing the first screw conveyor, while the second screw conveyor, which does not completely extend across the entire chamber, and indeed, may have only one or two convolutions, goes around the cylindrical partition.

The inner screw conveyor drives the material forward and, when it reaches the end of its travel, it piles up and is soon picked up by the outer screw conveyor which moves it in the opposite direction. In this way, the material circulates from front to back. Since the outer and inner screw conveyors also lift material on each rotation, the material describes a spiral motion forward and the material comes repeatedly in contact with the screw blades and the walls of the container.

While the ball valve **62** is closed, operation of the screw conveyor in the freezing zone **16** is in an idling position, rotating only slowly at 1 rpm, or less to prevent impaction of the ice, but not fast enough to lift it. At the same time no new liquid is coming in because the spigot **42** is closed. This permits the lyophilization chamber to cause circulation of the material until a vacuum of less than 200 microns is reached. This low pressure starts a timer switch. Lyophilization and circulation is continued for about an additional 1 hour. At this point valve **78** opens.

Communication then exists between the first and second lyophilization chambers **66** and **66a**. The second lyophilization chamber is at less than 100 microns, so that the vacuum is not broken. All the material in the first lyophilization chamber now empties into the second lyophilization chamber. This chamber is almost identical to the first chamber and need not be described in detail. In fact, for some applications this second chamber of the lyophilization may be omitted. After about 50% of the moisture has been removed in the first chamber, the material is transferred to the second chamber kept under a higher vacuum. Here, the process proceeds as in the first chamber, except that upwards of 90% of the moisture is then removed.

The Finishing Zone: When the material in the second lyophilization chamber of the lyophilization zone has reached a vapor pressure of less than 75 microns at 30° to 70°C it usually indicates that its moisture content has been lowered to less than 10%. More time is required to remove the last traces of moisture than is required to reduce the moisture content in the first two lyophilization

Apparatus for Continuous Freeze Drying Processes

Thus, lyophilization zone **18** comprises an elongated cylindrical chamber **66** also inclined from the vertical at about 30° and held in this chamber is a cylindrical partition **68** which is fastened to the chamber wall by spokes **69**. Within the cylindrical partition is a central screw conveyor **70** which has a central shaft **72**. The material treated travels up the central screw conveyor and over the cylindrical partition. Disposed around the cylindrical partition is an outer screw conveyor **74** which will push the material downwards along the partition wall. This outer screw conveyor does not extend fully along the entire length of the chamber. There are at most one or two convolutions.

The central partition does not extend the full length of the chamber so that the wall does not prevent the first and second outer convolutions of the outer screw spiral from passing from the central shaft over the wall. The outer screw conveyor extends only partly along the outer wall since it is only connected to the central shaft at one extremity. As the chamber is substantially vertical, material must be pushed upwards by the inner screw conveyor **70** until past the partition wall. The material will tumble in the space between the partition outer side and the chamber inner wall and will travel back downwards to the bottom of the chamber along the partition wall.

In addition to the continuous movement just described by the screw conveyor arrangement, the first and second chambers of the lyophilization zone have two other features namely: a reversible refrigeration arrangement, and a vacuum pump cold trap arrangement.

Thus, as material is transferred from the freezing and crushing zone **16** to the first lyophilization zone across valve **62a**, a vacuum pump arrangement **80** is started so as to form a vacuum in first chamber **66**. At this time the chamber is sealed. When the crushed ice reaches the first lyophilizing zone, it is only a few degrees below the freezing point. The application of the vacuum causes first a rapid evaporation process to proceed. This cools the ice further to temperatures below –10°C. The ice becomes harder as traces of liquid are frozen. During this process, the rate of evaporation of the ice is so rapid as to cool the ice rapidly. Contact with the rotating double screw conveyor supplies the heat required for the heat of vaporization.

As the screw conveyor rotates in the chamber, the mechanical energy is converted to heat. Thus, the rate at which heat is supplied is determined by the rate at which the rotor rotates. In addition cooling coils add or subtract heat and thus maintain a constant temperature at a constant speed of rotation. When a balance has been reached between the cooling process due to evaporation, and the heating process due to rotation, so that evaporation is taking place without melting of the ice, and with the temperature remaining constant, the lyophilizing conditions or process conditions have then been established. A motor **76** drives the screw conveyor in the vacuum chamber.

After permitting the lyophilization process to be established at ambient room temperature, the temperature is slowly raised in the lyophilization chamber by means of the coils from the reversible refrigeration means, so that the wall temperature reaches 30° to 70°C, depending upon the lyophilization rate and stability of the material. It must be pointed out that while the wall temperature is above room temperature, the temperature of the material remains well below freezing because of the rapid evaporation process. Treatment in the chamber chambers. The contents of the second lyophilization chamber are therefore transferred to a finishing primary chamber **24** under vacuum. This chamber is one of three finishing chambers. When the product in the second lyophilization chamber is again ready for transfer, this is then transferred to the finishing secondary chamber **26**. When a third charge is ready from the second lyophilization chamber **66a**, this is transferred to last finishing chamber **28**. The finishing chambers are maintained at temperatures ranging from 4° to 40°C.

The outer screw conveyors of the finishing chambers have a few turns which are wedge-shaped so as to pulverize the product, since some dried powders obtained from materials such as orange juice, tend to cake and form a compact mass as they approach dryness.

Eliminating Need to Supply Heat from External Heat Source

A.D. Passey; U.S. Patent 3,909,957; October 7, 1975 provides a freeze drying system substantially eliminating the need to supply heat from an external heat source for freeze drying a material. The system comprises in combination, a drying chamber, a means for passing material through the drying chamber, a heat pump system (including one or more of the first, second and third heat pump means), an evacuation means including the first heat pump means of the heat pump system, and a heat exchange means for supplying heat to the material such that the need to supply heat from an external heat source is substantially eliminated by recycling the heat between the judiciously selected heat sources and main heat sinks found within the freeze drying system, the recycling of heat being achieved by the heat pump system.

Further, the requirements for mechanical energy and cooling water are minimal thus resulting in tremendous savings in cooling water and mechanical energy, and in reduction in the amount of heat rejected to the environment. The process is described with reference to Figures 8.7a and 8.7b.

Referring firstly to Figure 8.7a, the drying chamber **51** comprises a cylindrical portion **2** and a conical portion **3** which are welded together to form a conical hopper. The chamber is supported vertically by legs **4** and has a nozzle at the bottom of the hopper for accepting a demountable vacuum lock assembly **24** from which the freeze dried product is withdrawn. The hopper is closed by a demountable head **1** to form the drying chamber which includes vacuum-proof windows **13** which may serve for inspection or manholes and includes one or more nozzles **15** with demountable flanges which are used to introduce vapors and/or gases as a means to influence the process occuring in the chamber.

The chamber is provided with one or more nozzles **25** through which water vapor and gases are removed from the chamber and their positioning in the chamber near the top allows an overall vertically upward flow of vapors surrounding the particulate material and in so doing aids in the suspension of the particles in the chamber. The angle alpha ($\alpha°$) of the conical portion depends upon the flow characteristics of the product and is small enough to provide a slope compatible with an easy flow of material.

The drying chamber is provided with a heat exchanger **5** and heating medium is brought to the heat exchanger by way of conduit **9**. Heating medium inlet coupling **8** provides a vacuum-proof connection between this conduit and conduit **7**, the latter serving to bring the heating medium to the distribution header **6**. After adequately traversing the heat exchanger **5** the heating medium leaves the heat exchanger by way of conduit **10**, heating medium exit coupling **11** providing a vacuum-proof connection between the conduit **10** and conduit **12**. Conduit **12** serves to carry the heating medium from the chamber for future use, e.g., return to the third heat pump means of evacuation means as will be explained hereinafter. Also the conical portion **3** of the chamber **51** may be similarly heated if desired.

The atomizable substance, e.g., a coffee extract, suitably at a concentration of 40 to 50%, after receiving any required pretreatment is fed to the chamber by way of feed line **23** which on reaching the spinning disc **20** of the centrifugal disc atomization system is atomized. The atomized material freezes almost instantly due to prevailing vacuum conditions compatible with freeze drying and the frozen particles that remain in the drying chamber dry to the desired moisture content while they are still in suspension and settling.

FIGURE 8.7: ELIMINATING NEED FOR EXTERNAL HEAT SOURCE

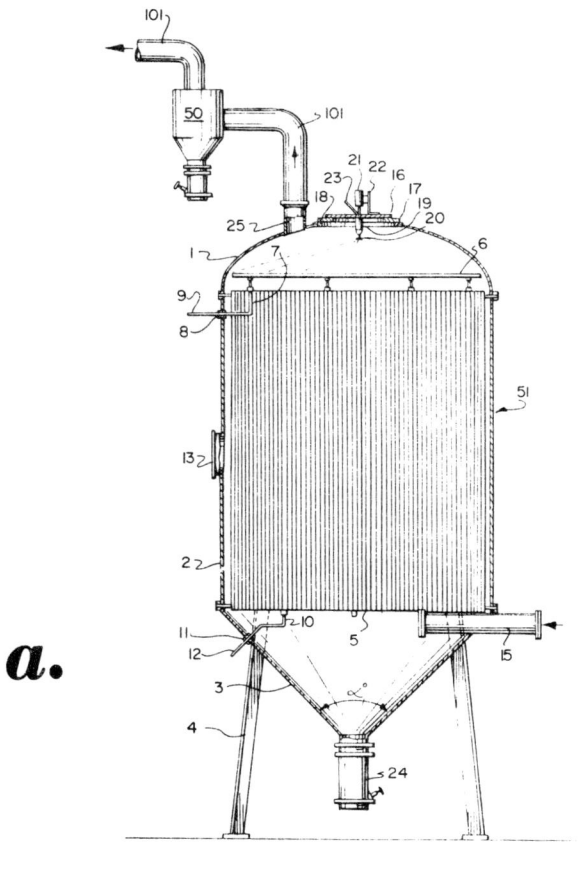

Drying Chamber of System

(continued)

FIGURE 8.7: (continued)

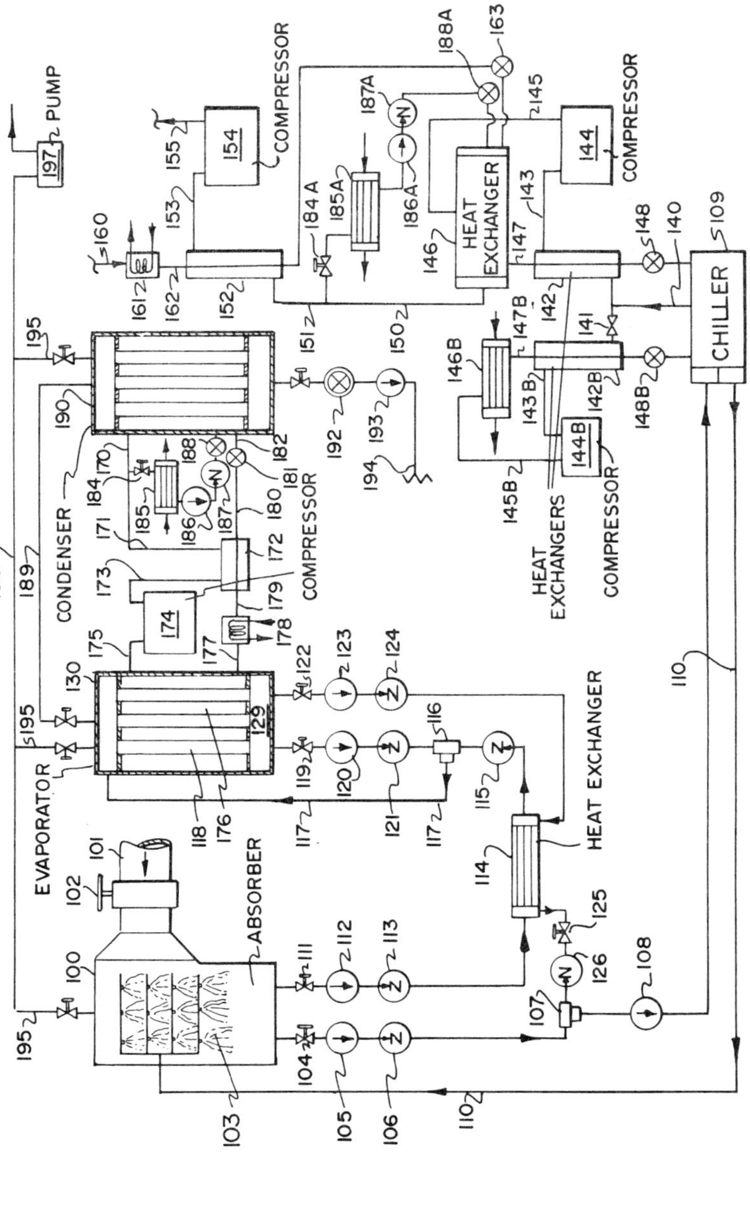

Flow Diagram of Evacuation Means

Source: U.S. Patent 3,909,957

Apparatus for Continuous Freeze Drying Processes

Particles that leave the drying chamber through outlet **25** with water vapor are led to the separation equipment **50**, such as a cyclone separator, wherein these smaller particles are separated from the vapors and are subsequently dried to the desired moisture content, the separator being also provided with the necessary heat for drying by means of the third refrigerant from the third heat pump means of the evacuation means.

The larger particles separated in the drying chamber **51** and the small particles separated in the separator settle down into respective shallow fluidized bed zones and these particles leave the respective chambers through suitable vacuum lock means into a product handling system. From the vapors that leave the separator a portion may be bled for recirculation to the chamber and/or the separator and the remaining vapors are led via vacuum line **101** as shown in Figure 8.7b through valve **102** to an absorber **100**.

Referring to Figure 8.7b the absorption of water vapor is accomplished in the absorber by a suitable absorbent medium such as a chilled aqueous solution of lithium bromide, maintained at the desired temperature by recirculating it through an absorbent chiller **109** and maintained at the desired concentration by regeneration of the absorbent medium in evaporator **130**. The control of the temperature and concentration of the absorbent medium serves the purpose of maintaining a sufficiently low vapor pressure in the absorber in order to overcome the resistance of the vacuum line to the flow of vapors between the drying chamber and the absorber. Thus the temperature and concentration of the absorbent medium serves as an external means to control the vapor pressure in the drying chamber.

A vapor pressure of 500 millitorrs in the absorber is consequently adequately low and can be expeditiously maintained by chilling the lithium bromide solution to about 0°C. The water vapor coming from the drying and separation chambers by way of vacuum line **101** passes through vacuum valve **102** and enters the absorber where it is subsequently uniformly distributed and absorbed by spray **103** of the chilled concentrated absorbent medium.

Pump **108** draws in a mixture of concentrated absorbent solution returning from the evaporator through valves **125** and **126** and a dilute absorbent medium from the absorber through valve **104**, pump **105** and valve **106**. Mixing has previously occurred in the mixing valve **107** in the desired proportions. The absorbent medium returning from the evaporator is circulated by pump **123** through flow control valve **122** and check valve **124** on its way to the mixing valve. The rate of flow of absorbent medium through the chiller is preferably kept stable at an optimum design value. The pump **108** forces the absorbent medium through the tubes of the absorbent chiller in which a refrigerant (second refrigerant) is evaporating on the shell side.

The chilled concentrated absorbent medium is then passed through line **110** and is suitably distributed and dispersed in the absorber in order to effect optimum rate of generation of absorption area per unit volume. The water vapor is thereby absorbed and as a result the absorbent medium becomes diluted and heated. Some heated and diluted absorbent medium is withdrawn at a suitable rate through the proportioning control valve **104** in order to be pumped by the pump **105** into the mixing valve **107** in which it mixes with the concentrated absorbent medium returning from the evaporator. The mixture is then returned to the absorber **100** after being chilled in the chiller **109** in order to repeat the cycle.

A large portion of the heated and diluted absorbent medium, corresponding to the difference between (a) the rate at which absorbent medium is supplied to the absorber plus the rate of absorption of water vapor therein, and (b) the rate at which a portion leaves the absorber through valve **104**, is withdrawn from the absorber through valve **111** by means of pump **112**. On its way to the evaporator **130** this dilute absorbent medium passes via a check valve **113** through a heat exchanger **114** in which it gains heat from the concentrated absorbent medium returning from the evaporator. As a result of this exchange, the concentrated absorbent medium is cooled while the dilute absorbent medium is heated.

This heat exchange therefore helps reduce the mechanical energy requirements during the subsequent reconcentration as well as chilling operations. From this heat exchanger, the partially heated dilute absorbent medium follows a line leading it through check valve **115** to a mixing valve **116** and in this mixing valve the absorbent medium coming from the absorber mixes with the absorbent medium recirculated by pump **120** and from the mixing valve **116** the absorbent medium passes through line **117** to the evaporator, which is uniformly distributed and descends in thin films on the inside of the tubes **118** exchanging heat with the refrigerant of a heat pump system (first heat pump means) condensing on the shell side **176** of the evaporator at a suitable temperature for example, about 104°C.

Some of the water contained in the concentrating absorbent medium is thereby evaporated at a vapor pressure of about 120 torrs and a concentrated absorbent medium collects in evaporator sump **129**. In order to maintain steady state operating conditions the concentrated absorbent medium is returned to the mixing valve **107** at the same rate as it is withdrawn from the absorber by pump **112** for supplying it to the evaporator less the rate of evaporation in the evaporator. This returning concentrated absorbent solution is cooled on the way in heat exchanger **114**.

The water vapor evolved from the concentrating absorbent medium leaves the evaporator by way of line **189** and is led to a cascade condenser **190** where the incoming water vapor condenses on the tube side at a temperature of about 55°C governed by thermodynamic considerations. The condensing water vapor loses heat to a refrigerant (first refrigerant) evaporating on the shell side at a suitable temperature in order to maintain the desired condensation temperature on the tube side.

A heat pump system (first heat pump means) uses the vapor of the evaporating first refrigerant for supplying the compressed first refrigerant vapor to the evaporator where it condenses on the shell side in order to provide the heat necessary for concentrating the absorbent medium. The condensed water is removed from the cascade condenser by means of pump **193** through trap **192** and is discharged along the line **194**. This condensate represents a potential source of available heat which can be used to supply process heat wherever needed.

The water vapor condensing, for example at 55°C on the tube side of the cascade condenser serves as a heat source for the first refrigerant evaporating on the shell side, and the absorbent medium concentrating on the tube side in the evaporator serves as a heat sink for the compressed first refrigerant vapor condensing, for example, at 104°C on the shell side of the evaporator. A heat pump system (first heat pump means) couples together the cascade condenser (which serves as the water vapor condenser and as the evaporator of first heat pump means) and the absorbent evaporator (condenser of the first heat pump means).

The liquid line 177 leads to subcooler 178 where it loses heat to a cooling medium which may be an external heat sink such as cooling water or some other medium which needs to be heated and is thereby subcooled. The liquid refrigerant then passes through liquid line 179 and is further subcooled in passing through liquid-vapor suction line heat exchanger 172 which ensures adequate superheating of the suction vapor. The sequence of subcooling in cooling medium subcooler 178 and in the liquid-vapor suction line heat exchanger may be interchanged if so required by the prevailing temperature conditions or some other thermodynamic considerations.

The subcooled liquid refrigerant passes along liquid line 180, is then expanded by an expansion valve 181 and enters the cascade condenser (evaporator of the first heat pump means) 190 on the shell side by way of line 182. The first refrigerant vapor leaves the cascade condenser by way of suction line 170. At this stage a portion of the suction vapor is bled by means of proportioning by-pass control valve 184. The unbled fraction of suction vapor follows suction line 171 and after sufficient superheating in passing through the liquid-vapor suction line heat exchanger 172 follows the suction line 173 to heat pump compressor 174.

After compression the refrigerant vapor is discharged along line 175 and the compressed refrigerant vapor condenses on the shell side of the evaporator 130. Size of the bled fraction of the suction vapor is controlled by the proportioning by-pass control valve to allow sufficient unbled suction vapor to be compressed by the heat pump compressor in order to meet the heat requirements in the evaporator at all times. The bled fraction is liquefied in passing through a condenser 185 by losing heat to a cooling medium which may be an external heat sink such as cooling water. The resultant liquid refrigerant is then returned by pump 186 through check valve 187 and expanded by expansion valve 188 into the shell side of cascade condenser 190 where it evaporates together with liquid refrigerant returning from the evaporator.

The absorbent medium to be chilled in chiller 109 serves as the heat source for a heat pump system that delivers heat to the freeze drying material or wherever required in the freeze drying process at a proper rate. The absorbent refrigeration system (second heat pump means) is designed to chill the absorbent medium to a proper temperature, for example, 0°C in order to maintain a water vapor pressure of about 500 millitorrs in the absorber 100. Indeed a temperature to which the absorbent medium needs to be chilled depends upon the thermodynamic considerations of the absorption-regeneration process and the water vapor partial pressure maintained in the drying chamber of the system among other things.

The liquid second refrigerant after having been subcooled in liquid-vapor suction line heat exchangers **142** and **142B** is expanded on the shell side of the shell and tube absorbent chiller **109**. This expanded refrigerant evaporates on the shell side by receiving heat from the absorbent medium being chilled and leaves the absorbent chiller as a vapor by way of suction line **140**. A controlled fraction of the suction vapor required to by-pass the main compressor **144** is bled by proportioning by-pass valve **141** while the remaining suction vapor goes to the main compressor by way of suction line **143**. On its way to the main compressor the vapor passes through liquid-vapor suction line heat exchanger **142** where it is superheated thereby causing a subcooling of the liquid refrigerant.

The control system is designed to ensure a desired degree of superheating of the suction vapor depending upon the refrigerant characteristics. The suction vapor compressed by the main compressor then follows the discharge line **145** and is let to the cascade heat exchanger **146** where it condenses on the shell side (serving as heat source for third heat pump means) by losing heat to the third refrigerant evaporating on the tube side. The evaporating third refrigerant is later compressed by a heat pump system (third heat pump means) for providing the heat required in freeze drying process or elsewhere. The second refrigerant condensed on the shell side of the cascade heat exchanger is passed through line **147** to reach expansion valve **148** and is subcooled in the liquid-vapor suction line heat exchanger.

On its way it may be preferably subcooled in a heat exchanger (not shown) by losing heat to an external heat sink such as cooling water. Depending upon the prevailing temperature conditions and some other thermodynamic considerations, the sequence of subcooling by means of cooling water and in the liquid-vapor suction line heat exchanger may be interchanged. After subcooling, the liquid refrigerant is expanded on the shell side of the chiller wherein it evaporates and leaves by way of suction line **140**.

The fraction of suction vapor beld by the proportioning by-pass control valve is likewise superheated in liquid-vapor suction line heat exchanger **142B** and passed along suction line **143B** for compression by means of a booster compressor **144B**. This compressed refrigerant vapor leaves the booster compressor by way of discharge line **145B** and is condensed in condenser **146B** by losing heat to an external heat sink such as cooling water. On its way to the expansion valve **148B** by way of liquid line **147B** the liquid refrigerant is subcooled in the liquid-vapor suction line heat exchanger **142B** which also ensures an adequate superheating of the suction vapor compressed by the booster compressor **144B**. The expansion valve **148B** expands this subcooled liquid refrigerant on the shell side of the chiller where it evaporates together with liquid refrigerant returning from the cascade heat exchanger.

The temperature and concentration of the chilled absorbent solution leaving the chiller through line **110** and its rate of recirculation are suitably controlled by properly interlinked control means in order to maintain the desired vacuum in the drying system while avoiding any problems of absorbent medium freeze up. The condensation and evaporation temperatures in the cascade heat exchanger should be maintained at levels permitting an optimum performance.

From refrigeration engineering and thermodynamic considerations it is obvious that a mean cascade temperature of about 50°C is compatible with the optimum performance which indeed is a much higher temperature than the temperature

Apparatus for Continuous Freeze Drying Processes

level of 32° to 38°C which can be maintained by the natural cooling medium (cooling water) at 27°C. In such cases the heat pump compressor **144** and the booster compressor **144B** are both needed to increase the efficiency and power economy.

Moreover, if one is willing to sacrifice the thermodynamic gains coming from the use of the heat pump compressor as well as the booster compressor, it is possible to compress in the heat pump compressor all the low pressure refrigerant vapor coming from the absorbent chiller **109** and then bleed the compressed refrigerant vapor rather than the low pressure vapor at a properly controlled rate for condensing it by losing heat to an external heat sink such as cooling water. This liquid refrigerant formed by condensing the compressed bled vapor is subsequently returned to the shell side of the absorbent chiller in the same manner as for the liquid refrigerant formed by condensing the bled suction vapor after compression in the booster compressor. The remaining compressed vapor as in the previous case condenses on the shell side of the cascade heat exchanger **146**.

The evaporating third refrigerant in the cascade heat exchanger gains heat from the compressed second refrigerant vapor of the second heat pump means condensing on the shell side of the cascade heat exchanger. The condensing second and the evaporating third refrigerants in the cascade heat exchanger are different when the thermodynamic or mechanical reasons so demand. The vapor of the evaporating third refrigerant leaves the cascade heat exchanger by way of suction line **150** and a properly controlled portion of this suction vapor may be bled by means of proportioning by-pass valve **184A** while the remaining vapor passes along suction line **151** through a liquid vapor suction line heat exchanger **152** where it gains heat from returning subcooled liquid third refrigerant.

The suction vapor is superheated to the desired level and led, via suction line **153**, to the compressor **154**, the compressed refrigerant vapor is led from the compressor by discharge line **155** to the heat exchanger means of the freeze drying system where it condenses at the desired temperature, for example, 100° to 150°C, to provide heat in the freeze drying chamber and/or the separation equipment, and the condensate still at high temperature may be further used to supply process heat whereby it is subcooled (first stage subcooling). This partially subcooled liquid refrigerant passes along liquid line **160** and is further subcooled (second stage subcooling) in a cooling medium subcooler **161** by losing heat to an external heat sink such as cooling water.

The refrigerant in liquid line **162** is subsequently subcooled again (third stage subcooling) in liquid-vapor suction line heat exchanger, and depending upon the prevailing temperature conditions and some other thermodynamic considerations the sequence of subcooling in the liquid-vapor suction line heat exchanger and in the cooling medium subcooler may be interchanged or one or more of the first and second stages of subcooling may be eliminated. The liquid third refrigerant is subsequently expanded by expansion valve **163** on the tube side of the cascade heat exchanger where it evaporates by gaining heat from second refrigerant vapor compressed by the compressor **144** of the second heat pump means condensing on the shell side of the cascade heat exchanger.

Under steady state conditions, the proportioning by-pass valve **141** would bleed just the right quantity of low pressure second refrigerant vapor coming by way of suction line **140** to leave a sufficient quantity of refrigerant vapor for compression by the compressor **144** of the second heat pump means in order to provide an adequate heat source to the third heat pump means.

When the saturation temperature of the low pressure suction vapor of third refrigerant leaving cascade heat exchanger **146** is greater than the temperature of the external heat sink, e.g., cooling water, it is sometimes simpler although less advantageous in the thermodynamic sense to adapt the following alternative scheme of balancing the thermal needs of the system. The portion of the low pressure suction vapor leaving the cascade heat exchanger is allowed to by-pass the compressor **154** of the third heat pump means.

The by-passed portion is controlled by a proportioning by-pass valve **184A** and is condensed by exchanging heat in a condenser **185A** with an external heat sink (cooling water). The resulting liquid refrigerant is then returned by a pump **186A** through check valve **187A** and expanded by expansion valve **188A** on the tube side of the cascade heat exchanger where it evaporates together with the liquid refrigerant returning from the heat exchanger means and expanded by expansion valve **163**.

Under steady state operating conditions the partial pressure of noncondensible gases in the freeze drying system is maintained at a desired level by means of a purge unit **197** which pumps the noncondensible gases from the absorber **100**, evaporator **130**, and cascade condenser **190** via lines **195** and **196**. In operation, the pressure in the drying chamber is lowered by purging the noncondensible gases by means of pump **197** through line **101** and recirculating the absorbent medium through the absorber and when the desired low pressure has been achieved the valve **102** in the line **101** is adjusted to provide the necessary pressure differential between the chamber **51** and the absorber.

Use of High Frequency Oscillators or Infrared Radiators

In a process described by *Y. Sahara; U.S. Patent 3,531,871; October 6, 1970* continuous freeze drying is accomplished by rotating a plurality of vacuum containers on a circular process line which is divided into four subprocesses of loading, preexhaustion, main exhaustion and refilling in cyclical sequence. The main exhaustion process has a heating field with atmospheric pressure where the object enclosed in such vacuum container is heated by the dielectric effect of high frequency flux or by the irradiation of infrared ray or other radiating heat to have the frozen moisture sublimated in the vacuum.

Vacuum containers of glass, clay, synthetic resin or other similar material having a substantially low dielectric loss factor (dielectric constant times dielectric power factor) for high frequency flux, pass through a heating tunnel which is open to atmospheric pressure and provided with sufficiently large heating capacity high frequency oscillators, thus heating the object at the inside of each vacuum container with high frequency flux applied from the outside of the vacuum container by making use of the low dielectric loss of such materials for high frequency flux.

Alternately vacuum containers of glass, synthetic resin or other material having a substantially high permeability for infrared ray, pass through a heating tunnel which is opened to atmospheric pressure and provided with sufficiently large heating capacity infrared ray radiators, thus heating the object at the inside of each vacuum container with infrared ray energy applied from the outside of vacuum container by making use of the high permeability of such materials for infrared ray energy.

The process is described with reference to Figures 8.8a through 8.8e. The apparatus has an annular turn-table **2** movable along a pair of annular rails **1** by means of wheels **31**. An upper main duct **3** is provided at the center of the turn-table and is rotatable by a driving mechanism A. The main duct is provided with a plurality of branch ducts **4** in a radial arrangement at the top end thereof. The branch ducts are supported on the turn-table at the open ends thereof, these ends opening upward. A short duct **6** is joined to the open end of each branch duct by the intermediary of an electric valve **5**.

FIGURE 8.8: USE OF HIGH FREQUENCY OSCILLATORS OR INFRARED RAY RADIATORS

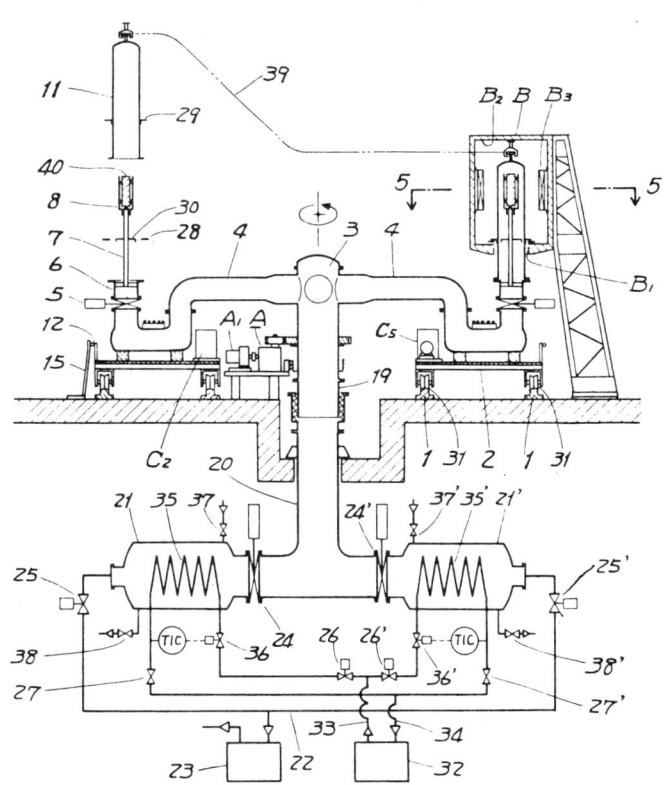

Vertical Elevation of Apparatus

(continued)

FIGURE 8.8: (continued)

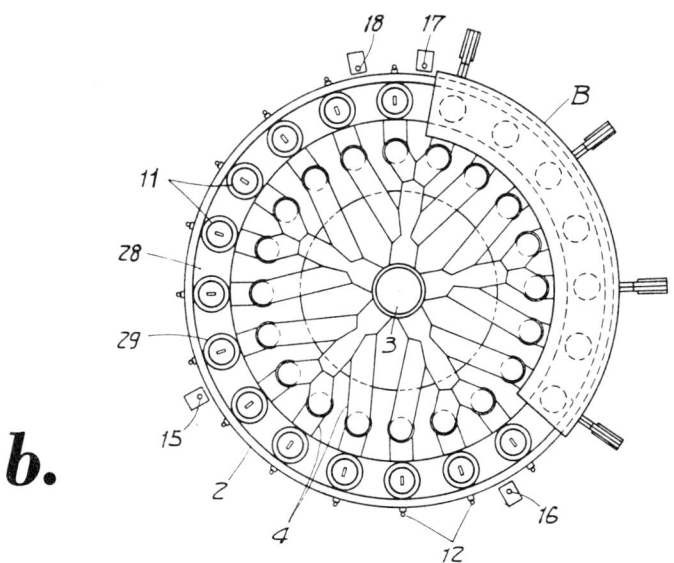

b.

Plan View of Apparatus

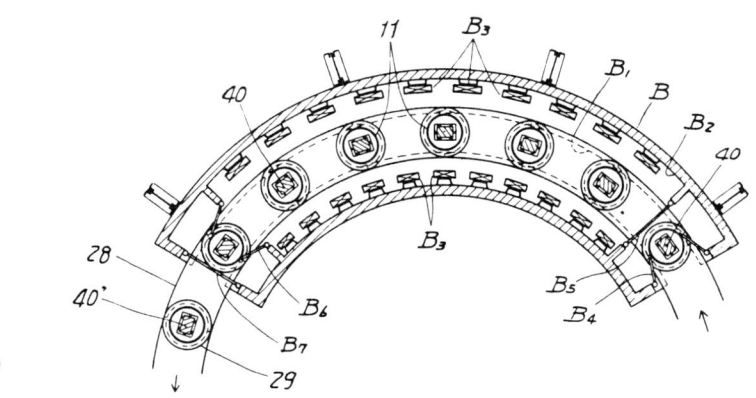

c.

Plan View on Line 5–5 of Figure 8.8a, Showing Heating Tunnel

(continued)

FIGURE 8.8: (continued)

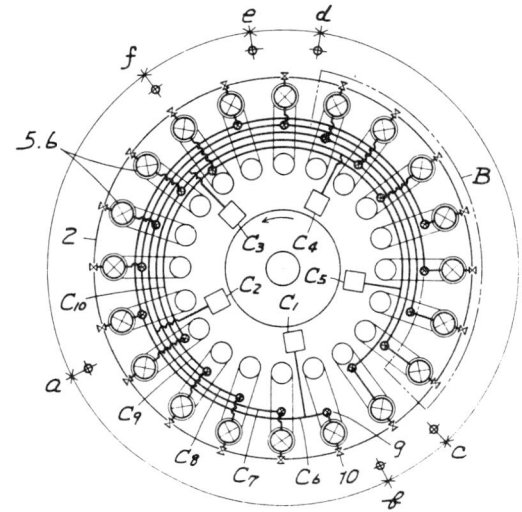

d.

Schematic View Showing Rotary Sequence of Process
and Piping for Preexhaustion of Air

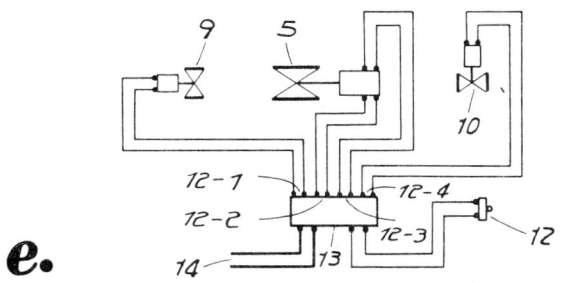

e.

Circuit Diagram for Automatic Changeover

Source: U.S. Patent 3,531,871

A rod **7** is secured to the short duct **6** by means of a rib at the lower end thereof, and is engageable with a holder **8** for the object **40** at the upper end thereof. The short duct is connected to an electromagnetic valve **9** for preexhaustion of air and also to another electromagnetic valve **10** for refilling of air. The upper end of duct **6** is provided with a flange to meet another flange of a bell jar **11** by the intermediary of packing in a freely detachable relation. The bell jar is a kind of vacuum container made of glass, clay or other similar material with substantially low dielectric loss factor.

A heating tunnel B which is under atmospheric pressure is installed over a part of the rotary locus of the bell jars carried on the turn table to permit the bell jars to pass through it as the turn table 2 is rotated. The tunnel B is lined with metallic reflectors B2 on the inner walls thereof to shield high frequency flux, and is provided with a slit B1 to the full length of the bottom thereof to guide the bell jars. A plurality of high frequency oscillators B3 are provided at the side walls of this tunnel. The turn-table is equipped with five vacuum pumps C1, C2, C3, C4, C5 for preexhaustion of air from the bell jars 11. The number of vacuum pumps are the same as the number of bell jars running in the preexhaustion process.

The rotatable upper main duct 3 is connected to a fixed lower main duct 20 by the intermediary of rotary joint 19. The lower main duct is then connected to a plurality of cold traps 21, 21' in parallel relation by the intermediary of valves 24, 24', and the cold traps are in turn connected to a vacuum pump 23 for main exhaustion of air by way of pipes 22 and stop valves 25, 25'. Each cold trap has its own cooling system which comprises the trap 21 or 21' itself, a cooling tube 35 or 35', an automatic flow regulating valve 36 or 36', a drain valve 38 or 38', a feed stop valve 26 or 26', a return stop valve 27 or 27', and a defrost valve 37 or 37'. The cooling systems are all connected to a refrigerator 32 by way of a feed pipe 33 and a return pipe 34. The cold traps are changed over by the stop valves 24, 24', 25, 25', 26, 26', 27, 27' to put one of them into service while the others are out of service.

In the operation, each relay 13 is initially set at a desired arrangement, for instance, to change over the circuit 14 to the contact 12—1 to actuate the valve 9 for preexhaustion of air the moment the microswitch 12 touches the microcontact 15 located at the point a of Figure 8.8d, while one of the cold traps, say 21, is brought into service by the control of relevant stop valves; and then the whole apparatus is put into operation. The vacuum pump 23 reduces pressure in the main exhaustion system which consists of electric valves 5, branch ducts 4, the upper main duct, the joint, the lower main duct, cold trap 21 and connection pipe 22.

The vacuum pumps C1, C2, C3, C4, C5 reduce pressure in the preexhaustion pipes C6, C7, C8, C9, C10, respectively. The upper main duct is rotated by the driving mechanism A while the lower main duct remains stationary. The turn table is rotated together with the upper main duct in the direction arrow-marked in Figure 8.8a, carrying the bell jars thereon. As the turn-table goes round, the bell jars pass the heating tunnel with atmospheric pressure on a given part of the guide rail 39 and are detached upward off the flanges of short ducts 6 on another given part of the guide rail.

While a bell jar is detached off the short duct, that is, in the loading process from f to a, a holder 8 with a frozen object 40 is put on the top of rod 7. Then the bell jar is lowered down to the flange of the short duct as it approaches to the end of the loading process, now enclosing the object therein.

At the point a the microswitch touches the microcontact to have the relay change over to the contact 12—1 to open the electromagnetic valve 9. Air is then exhausted out of the bell jar by way of the opened valve 9 and its relevant pipe C6 by the action of vacuum pump C1. In this manner in the preexhaustion process from a to b, the bell jar is exhausted of air nearly down to the negative pressure in the branch duct **4**, thereby balancing pressure at both sides of the electric valve **5** to facilitate the succeeding opening of valve **5** while unbalancing pressure at both sides of the wall of the bell jar to keep the jar firmly to the flange of short duct **6**.

At the point b the microswitch **12** touches the microcontact **16** to have the relay **13** change over to the contact **12–2** to open the electric valve while allowing the electromagnetic valve **9** to close, thus the bell jar is transferred from the pre-exhaustion process into the main exhaustion process without disturbing the negative pressure at both sides of the electric valve. In the main exhaustion process from b to e, the bell jar is always kept at a given substantial vacuum by the action of vacuum pump **23**.

The electric valve is kept fully opened from b to d by the action of a limit switch (not shown) provided therein. After passing the point c the bell jar opens the shielding doors **B4**, **B5** and enters the heating tunnel B along the slit **B1**. As the bell jar goes through the electric field provided by the high frequency oscillators **B3** in the tunnel, the object **40** which has a substantially high dielectric loss factor is heated internally by the effect of high frequency flux to sublimate the frozen moisture thereof under the negative pressure within the bell jar.

In the heating tunnel (see Figure 8.8c) the object which is kept in a vacuum of approximately 0.1 mm Hg within the bell jar **11** goes through an electric field defined by the shielding members **28, 29, 30, B2, B4, B5, B6, B7**, being exposed to high frequency flux flowing through the wall of the bell jar which is made of a material with a substantially low dielectric loss factor.

There then electric energy is converted into thermal energy in proportion to the strength of electric field, frequency of flux, and dielectric loss factor of the object, thereby the temperature of the object is raised approximately to −20° or −25°C which is higher than the saturation temperature of moisture at the standing negative pressure and therefore the frozen moisture is quickly sublimated out of the object. The more moisture is sublimated, the more the dielectric loss factor is decreased and accordingly the less the object is heated thereby overheating will be considerably avoided. Thus the object is dried sufficiently into a perforated solid product, before the bell jar opens the shielding doors **B6, B7** and leaves the tunnel.

At the point d immediately after the tunnel, the microswitch touches the microcontact **17** to have the relay change over to the contact **12–3** to begin to close the electric valve, which is completely closed before the microswitch touches the microcontact **18** at the point e thus isolating the bell jar completely from the branch duct where there is a given substantial negative pressure.

At the point e the microswitch touching the microcontact has the relay change over to the contact **12–4** to open the electromagnetic valve **10**. Clean dry air is then sucked into the bell jar with negative pressure by way of the opened valve **10** and a filter (not shown) provided on the valve. Thus atmospheric pressure is recovered within the bell jar and therefore pressure is balanced within and without the bell jar to facilitate the succeeding detaching of the bell jar from the flange of the short duct, ending the refilling process designated from e to f. After the refiling process, the bell jar is returned to the loading process designated from f to a.

OTHER EQUIPMENT

Shield Structure Between Freezing and Drying Regions

The process of *E.L. Rader; U.S. Patent 3,616,542; November 2, 1971* is concerned with a continuous type of freeze drying apparatus. The apparatus is so constructed as to enable removal of all evaporated moisture from the equipment without interruption of the main freeze drying process, and without any substantial buildup of condensed moisture within the freeze drying regions themselves. The evaporated moisture is condensed in a condensing chamber, which may be isolated from the main freezing and drying regions for removal of accumulated moisture from the condensing chamber while the overall drying process continues. Two such condensing chambers may be provided, so that one may be in use while moisture is being removed from the other, and vice versa.

The process is described with reference to Figures 8.9a, 8.9b and 8.9c. The main shell 11 of the freeze drying system defines a single closed vacuum chamber 17, which may be separated into a freezing region or compartment 18 and a sublimation region or compartment 19, with communication being maintained between these two regions or compartments past the lower edge 20 of a partition or shield structure 21, which acts to prevent the transmission of excessive heat between sublimation chamber and freezing chamber, and also serves to prevent the development of turbulence within the interior of the shell as a result of convective air or vapor currents which may tend to be produced by the different temperature conditions prevailing in the shell.

The refrigerant liquid, which may be any suitable noncorrosive heat transfer fluid, such as a diethylene glycol base coolant, is forced into the lower end of the spiral refrigerant passage 37 by a circulation pump 41 (Figure 8.9a), whose discharge line 42 first directs the coolant through a heat exchanger 43, from which the cooled fluid flows through a heat insulated line 44 which connects into the lower end of the spiral passage through the inlet opening 39. Within the heat exchanger, the refrigerant fluid is cooled by transfer of heat therefrom to a primary fluid which is cooled by a conventional refrigerating system 45.

The interior of the shell, including both of the regions or compartments, is maintained at a very low subatmospheric pressure, that is, at a very high vacuum, the pressure desirably being not greater than about 225 microns, and for best practical results as low as about 175 microns. When the highly atomized sprayed liquid is introduced into the upper portion of the conical freezing chamber, the evaporative cooling effect produced by evaporation of some of the moisture from the sprayed liquid tends to cool the sprayed droplets toward freezing temperature. The additional cooling effect attained by the refrigerant fluid within the spiral passage supplements the effect of the evaporative cooling, and causes the sprayed droplets to freeze very rapidly from liquid form to solid form, as they advance downwardly within freezing compartment 18.

The temperature within that compartment is maintained sufficiently below freezing (desirably at least as low as about –30°F, and for best results, between about –40° and –50°F) to assure a rapid enough freezing action to cause conversion of the minute droplets of liquid into a very low density snowflake form, in which there are large voids or pores extending into each snowflake, with the flake having projections extending out in different directions to provide a very large surface area on each flake. The sprayed material is all in solid particulate form before the frozen particles reach and fall onto the upper surface of a conveyor belt 52 with vibrating upper run 53 which extends across the underside of the freezing chamber.

To describe a complete cycle of operation of the apparatus, assume that the refrigerating and heating units have all been brought to their proper operating temperatures, and that the desired high vacuum condition has been developed by pump 67 within freezing region 18 and sublimation or drying region 19 of the compartment 17. Also, assume that butterfly valve 70 of Figure 8.9c is open, and the related valve 74 at the outer side of chamber 72 is also open, while the corresponding valves 78 and 80 associated with the second moisture removal chamber 79 are closed, and with the door 84 of chamber 72 closed. In this condition, vacuum pump 67 withdraws moist air from within chamber 17 through moisture removal chamber 72, to cause freezing of the moisture on the outside of the structure 82 within this chamber until a substantial block of ice is formed about that structure.

For this purpose, the three-way valves 101 and 106 are in a condition in which they pass cold refrigerant from pump 100 into and out of the structure within chamber 72, but close off the flow of any cold refrigerant into or out of the structure within the second moisture removal compartment. Also, it may be assumed in the starting condition of the apparatus that all ice which had accumulated in the second compartment on the previous cycle of operation has already been removed past the door and that the door has returned to closed position, with heated fluid pump 109 stopped and valve 111 closed, and with vacuum valve 78 closed and valve 80 open.

With the apparatus in this described condition, pump 16 of Figure 8.9b acts to inject a pulsating flow of atomized liquid into the upper portion of chamber 18, in the downwardly flaring spray pattern with rapid conversion of the sprayed tiny droplets into the form of frozen particles, and desirably snowflake type particles which then fall or drift downwardly within this chamber and onto the left end portion of the upper run of the conveyor belt. The conveyor belt then advances the falling particles continuously rightwardly beneath the infrared lamps 63, which supply enough heat to gradually sublime all moisture from the particles to vapor form, without converting that moisture to or through liquid form. The particles are thus gradually dried while in frozen form, to ultimately arrive at the right end of the conveyor in completely dried form, for discharge into receptacle 13.

The water vapor which enters the atmosphere within sublimation chamber 19 of Figure 8.9b is drawn by vacuum pump 67 through chamber 72, within which the moisture is frozen onto the outer surfaces of structure 82, so that the dried gases from the sublimation chamber may then pass through line 73 for discharge to the atmosphere from this pump.

After a sufficient predetermined timed interval has elapsed to allow for the accumulation of a predetermined quantity of ice on the outer surface of the structure 82 of compartment 72, timer 116, which turns continuously, acts through one of its switches 117 to shift the refrigerant controlling the three-way valves to reversed conditions, to direct the cold refrigerant from pump 100 into chamber

FIGURE 8.9: SHIELD STRUCTURE BETWEEN FREEZING AND DRYING REGIONS

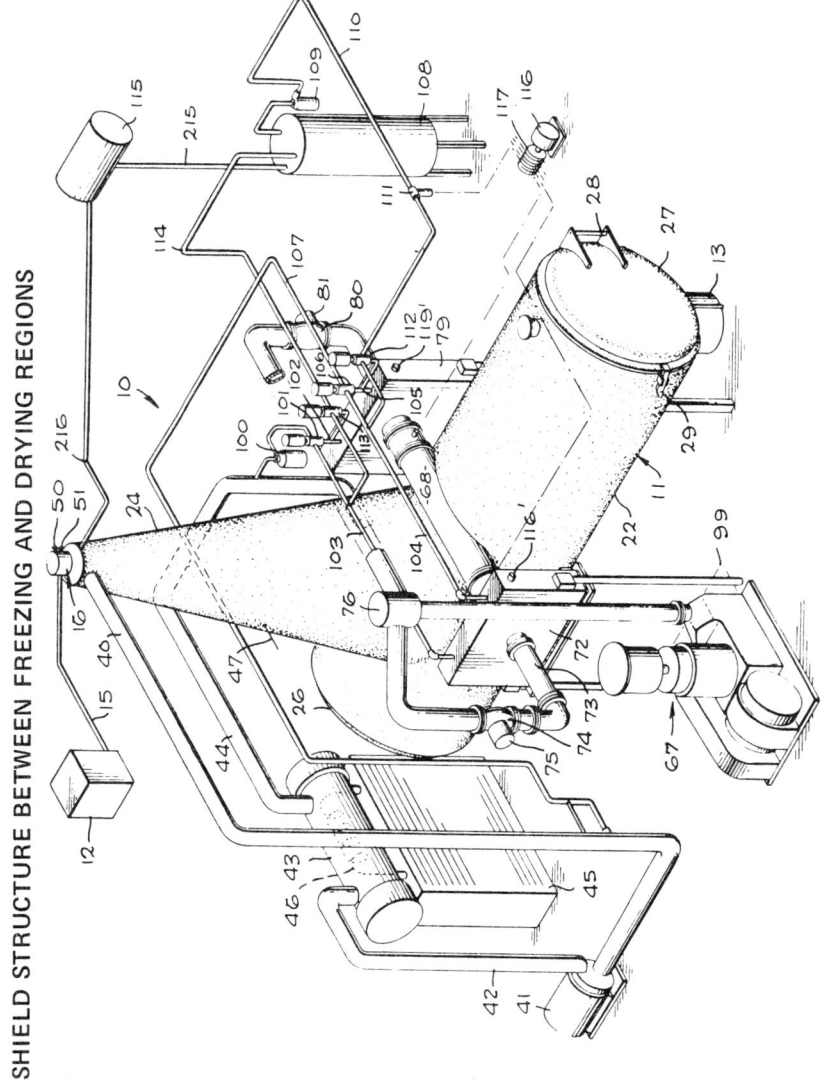

a.

Diagrammatic Perspective Representation of Freeze Drying System

(continued)

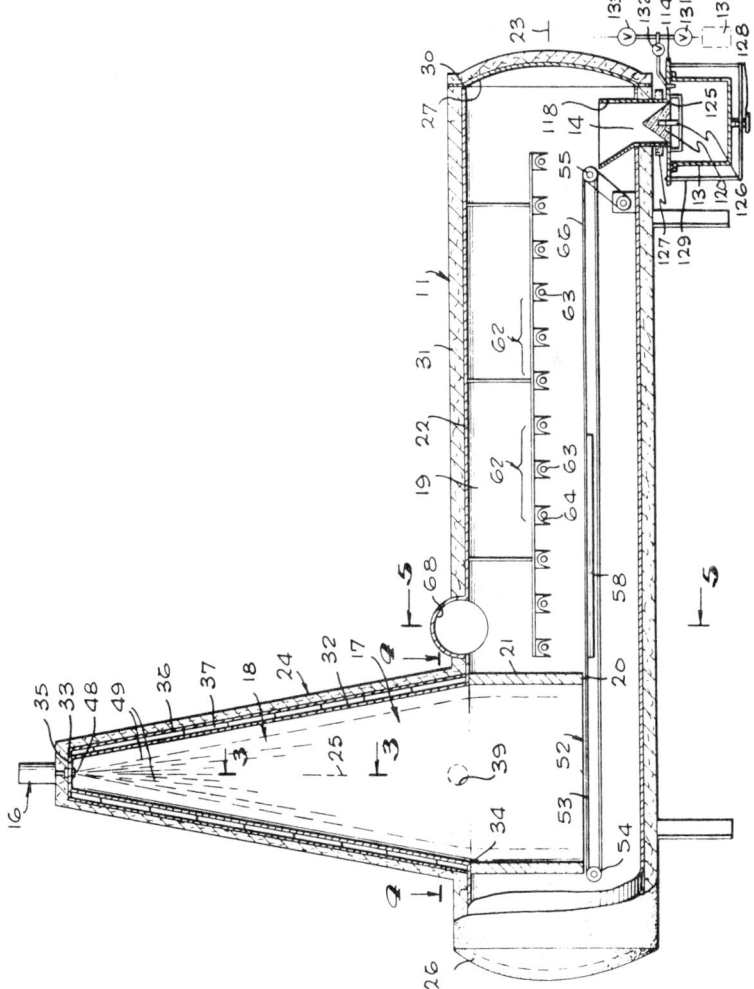

FIGURE 8.9: (continued)

b.

Central Vertical Section Through Main Vacuum Chamber

(continued)

FIGURE 8.9: (continued)

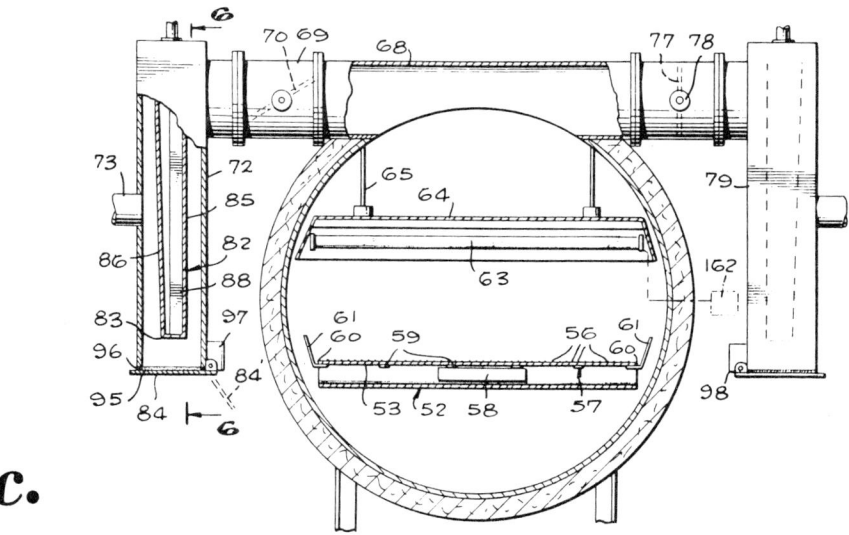

c.

Vertical Section Across Line 5—5 of Figure 8.9b Showing Moisture Removal Chambers 72 and 79

Source: U.S. Patent 3,616,542

79 instead of chamber 72, and return the refrigerant from chamber 79 rather than chamber 72, so that the structure 82 within chamber 79 is gradually refrigerated, while the ice on the structure within chamber 72 of course remains cool enough to continue to condense moisture from the moist gases even after its refrigerant is shut off.

After the three-way valves 101 and 106 have been reversed, and after the flow of cold refrigerant liquid to the structure within chamber 79 has continued for a long enough period to assure reduction of the temperature of that structure to a predetermined operating temperature, low enough to effectively freeze substantially all moisture which passes through chamber 79, timer 116 opers butterfly valve 77 of Figure 8.9c, so that moist gases from within the main sublimation chamber may commence to be withdrawn past that butterfly valve and through chamber 79. After such opening of the butterfly valve, both of the gas flow control valves 70 and 74 at the opposite side of shell 11 are closed simultaneously by the timer motor, so that thereafter the moist gases from within the shell can now only discharge through chamber 79, to commence the accumulation of ice in that chamber.

Next, the timer opens solenoid valve 111, and starts pump 109, to commence the flow of heated fluid from heater 108 through three-way valve 112 and line 104 into the interior of the structure within chamber 72, to partially melt the ice which has accumulated about the structure, with the heated fluid discharging from that structure through line **103** and the three-way valve **113** and line **114** back to the heater. Valves **112** and **113** may already be in proper positions to thus direct the heated fluid through chamber **72** and not chamber **79**.

After the heated fluid has passed through structure **82** within chamber **72** for a short period of time, the ice about the tapered structure will have melted slightly at the location adjacent the walls of the structure, and will fall downwardly from that structure. At approximately the time that this ice falls, timer **116** momentarily opens an electrical vent valve **116'** (Figure 8.9a) to release the vacuum within chamber **72**, and then actuates motor **97** (Figure 8.9c) to open door **84** at the bottom of this chamber, so that the ice block may fall downwardly past the door and to a position on tray **99**, in which position the upper portion of the block will be low enough to clear the door even when the latter is swung between its open and closed settings.

In connection with the release of the ice from about the structure, it should be noted that the upper ends of all of the walls of the structure **82** preferably connect directly to the top horizontal wall of this chamber, so that the structure has no upper surface on which ice can accumulate within this chamber, and therefore the accumulated block of ice can only be formed on the side walls and bottom wall of the structure, and will therefore fall downwardly very freely from about that structure when only slightly melted.

As soon as a predetermined time interval has expired long enough to assure sufficient melting for the ice to drop out of this chamber, the timer closes the door of this chamber, stops pump **109** and closes the solenoid valve, and then opens vacuum valve **74** to commence a reduction of the pressure in this chamber. The hot fluid three-way valves **112** and **113** may also be reversed at this time by the timer to be in a position for passing heated fluid to and from chamber **79**, though the fluid is not at this time actually delivered to the three-way valves.

After the apparatus has been in this condition for a period sufficient to enable accumulation of a predetermined quantity of ice on chamber **79**, and assuming that the pressure in this chamber has already been reduced to a vacuum condition corresponding to that in the sublimation chamber **19** itself, the refrigerant controlling three-way valves **101** and **106** are reversed back to their original condition by the timer, to again deliver cold refrigerant into chamber **72** rather than chamber **79**.

When the temperature of the structure within chamber **72** has then reached a predetermined moisture freezing normal operating temperature for that structure, butterfly valve **70** is opened by the timer to commence flow of the moist gases through moisture removal structure **72**, and the two gas flow controlling valves **77** and **80** at the opposite side of shell **11** are then closed to terminate the flow of moist gases through chamber **79**. Next the timer opens the solenoid valve, and commences operation of pump **109**, to force heated fluid into the structure **82** within chamber **79**, and subsequently opens an atmospheric vent valve **119'** on the chamber **79**, and the bottom door of that chamber, when approximately the right interval has expired for the ice to melt sufficiently to drop onto the tray.

After the door has remained in open condition long enough to assure dropping of the ice onto the tray, the door is closed by the timer, and valve **80** is opened to commence the development of a vacuum in chamber 79; and the apparatus has then returned to the condition which was assumed to be the initial starting condition in this description. Thus, a complete cycle of operation has been concluded, and the same cycle is then of course repeated after enough time has elapsed to accumulate a block of ice within chamber 72.

Elongated Vapor Passage Between Sublimation and Precipitation Chambers

The apparatus described by *R.C. Mace and F.J. Moore; U.S. Patent 3,088,222; May 7, 1963* comprises a substantially closed vessel defining an elongated sublimation chamber and also defining an elongated precipitation chamber extending in generally parallel adjacent relation to the sublimation chamber. A helical screw conveyor agitates an elongated mass of the material in the sublimation chamber. The precipitation chamber opens into the adjacent sublimation chamber along substantially the entire length of the elongated mass to form an elongated vapor passage between the chambers.

The vessel has a vapor-precipitating wall at least partly defining the precipitation chamber, means for chilling the wall, and a suction line leading from the interior of the vessel for maintaining in the chambers a partial vacuum sufficient to extract sublimation vapors from the frozen liquid. These vapors pass substantially directly and laterally into the precipitation chamber through the vapor passage from each point of the elongated mass and precipitate in part as ice on the wall. The process is described with reference to Figure 8.10.

FIGURE 8.10: ELONGATED VAPOR PASSAGE BETWEEN SUBLIMATION AND PRECIPITATION CHAMBERS

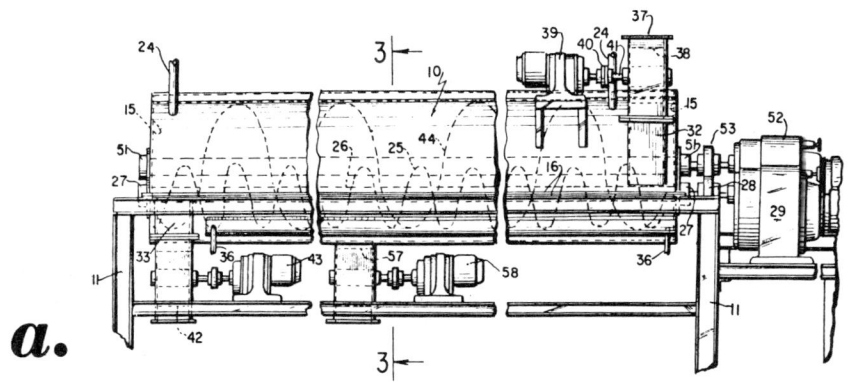

Side Elevational View

(continued)

Apparatus for Continuous Freeze Drying Processes 235

FIGURE 8.10: (continued)

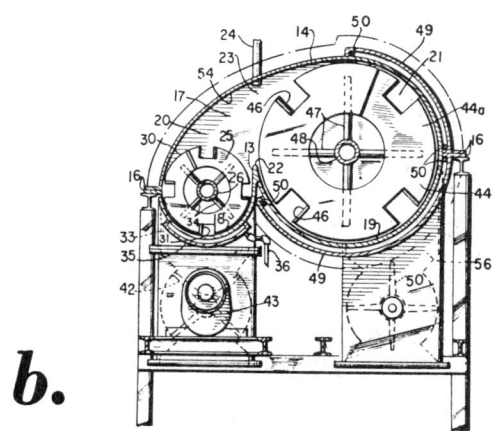

b.

Cross-Sectional View Along Line 3—3 of Figure 8.10a

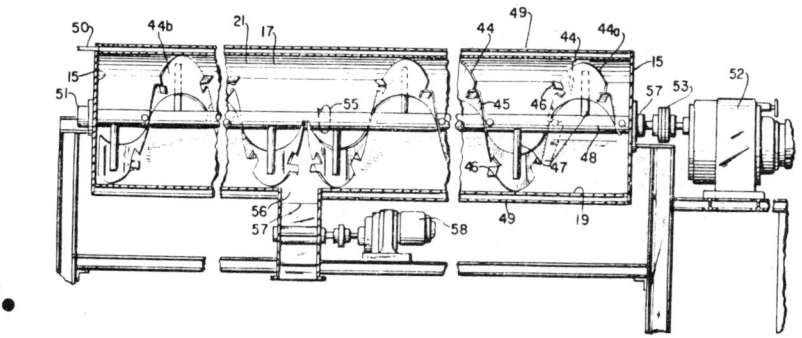

c.

Side Elevational View

Source: U.S. Patent 3,088,222

Vessel **10** comprises a lobate lower half **13** and an arched upper half **14** with vertical end walls **15** in each half. The halves are provided with flanges **16** which when brought together, utilizing a gasket or other sealing means, define a pressure-type enclosure surrounding an interior space **17**.

The lower half has a smaller trough **18** and a larger trough **19** respectively defining the lower portions of a sublimation zone **20** and a precipitation zone **21**. A virtually common wall **22** separates the two troughs at the bottom of an intermediate space **23** forming an unrestricted passage whereby the upper portions of the respective zones or chambers are in full and open communication with each other between the end walls. The entire space **17** is maintained at a substantially uniform pressure, absolute, by means of a vacuum pump connected to space **17** through suction line **24** shown as positioned adjacent the outlet end of the sublimation chamber where the sublimed vapor concentration would normally be lowest.

A helically bladed screw conveyor **25** is positioned in sublimation chamber **20** so as to fit trough **18**. The conveyor is provided with a longitudinally extending shaft **26** suitably supported in bearings **27** and one end of the shaft is connected by a flexible coupling **28** to a selectively variable speed power source **29** in the form of an electric motor speed reducer set by means of which the speed of rotation of the conveyor can be predetermined, e.g., in the light of the material being dried, the quantity of vapor to be sublimated therefrom and the capacity of the precipitation chamber **21** to handle such vapor.

As shown, the conveyor has its helical blades in the form of a screw ribbon **30** portions of the periphery of which are cut and folded to provide peripheral paddles **31** extending axially so that as they are rotated in a clockwise direction as viewed in Figure 8.10b, the particles of frozen material being desiccated will be tumbled as they are progressed along the length of chamber **20** from an inlet **32** to an outlet **33** therefrom. Spider arms **34** connect the blade **30** to the shaft.

As the blade sweeps the inner surface of trough **18** it will pick up frozen material and tumble and expose all surfaces thereof to the drying action and at the same time inhibit prolonged engagement of any portion of such material with that trough surface. A heating jacket **35** is positioned against the wall of trough **18** for the circulation therethrough of a heating fluid at selected temperature, circulation pipes **36** being provided at opposite ends of the jacket. In that way, the loss of heat to sublime frozen-state liquid in the particles is made up while such particles remain frozen during the drying operation.

In operation, the frozen edible particles themselves are fed into an inlet hopper **37** in which a rotary vane valve **38** is rotated at a selected speed by an independent variable speed motor-reducer set **39**, the output shaft of which is connected by a flexible coupling **40** to a shaft **41** which rotates the blades of valve **38**. After the frozen particles have been moved the length of chamber **20**, in the course of which they are constantly being tumbled by the conveyor and paddle action, they drop as dried frozen particles into the outlet, the lower end of which is connected to a casing for a rotary vane valve **42** operated by an independent variable speed motor-reducer set **43**. The discharge side of valve **42** is provided with a flanged conduit for connection to a receiving vehicle, vessel, conveyor or other equipment to remove the dried frozen particles for storage, use, further handling or other disposition.

The interior surface of trough **19** on the precipitation side is kept relatively clear by the rotation of a helically ribbon bladed screw conveyor **44** which also is provided with a cut and fold screw flight **45** having longitudinally extending scraper paddles **46** thereon, the screw flight being maintained in position by spider arms **47** fixed to a conveyor shaft **48**. This trough and a part of the upper portion of precipitation chamber **21** is surrounded by refrigeration jackets **49** having circulation pipe connections **50** to the respective ends thereof for the circulation through such refrigeration jackets of a refrigerating fluid at a selected temperature sufficient to precipitate sublimed vapor as frozen-state vapor or ice in precipitation chamber **21**.

Apparatus for Continuous Freeze Drying Processes

The rotation of conveyor **44**, the shaft of which is journaled in bearings **51** is provided for at correlative and selected speed by means of an independent variable speed motor reducer assembly **52** connected to the conveyor shaft by a flexible coupling **53**. As this conveyor rotates in a counterclockwise direction as viewed in Figure 8.10b, the blade and paddles will scrape the inner surface of the precipitation zone relatively clear and will toss ice fragments up into the space in zone **21** to serve as nuclei for augmented precipitation of sublimed vapor which moves through passage **23** substantially without hindrance from the sublimation chamber **20**.

Thus, the concave underside **54** of the arched wall **14** generally faces precipitation chamber **21** and appears to augment the movement of sublimed vapor from chamber **20** promptly and extensively, as formed, toward and into the precipitation chamber, as though a sublimed vapor particle were a ray which upon striking surface **54** would be reflected toward this chamber. Presumably, the surface is extremely smooth and may even be coated with a substance such as Teflon for that purpose to facilitate the virtually immediate movement of sublimed vapor out of chamber **20** into the precipitation chamber.

Such movement is promoted by the freeze precipitation going on in chamber **21** to aid in drawing vapor into contact with the ice precipitation means in that chamber. Significant advantages appear in that there is relatively fast removal of sublimed vapor from the sublimation chamber and substantially immediate, efficient precipitation of the vapor in the precipitator which promotes a shortening of time and expense and effectiveness of such freeze drying in vessels which, further, may be maintained at substantially the same negative pressure closer to the vapor pressure of the ice formed therein than heretofore considered practical.

In operation, the condenser portion of vessel **10** receives sublimed vapor for precipitation directly to the frozen state as ice substantially over the entire volume thereof as is apparent from Figures 8.10a and 8.10c. The screw conveyor **44** therein is provided with right-hand and left-hand flight portions **44a** and **44b**, respectively, which thereby act to sweep all ice therein in the course of the aforementioned counterclockwise rotation, shown by arrow **55** (see Figure 8.10c), to move all ice particles toward and out through ice outlet **56** intermediate the ends of chamber **21**.

It is in the course of such progression that ice particles are tossed into space in the precipitation zone to act as the aforementioned nuclei whereby the vapor precipitation surface and precipitation are augmented. Outlet **56** is provided with a rotary vane valve **57**, which acts as a pressure seal, rotated at a suitable predetermined speed by an electric motor speed reducer set **58**. The outlet on the discharge side thereof is provided with a flange for connection to a receiving vehicle, vessel, or conveyor or other equipment to remove the ice for disposal, heat exchange or other treatment or purpose.

Detachable Conveyor and Heater Means for Easy Cleaning

In an apparatus described by *E.L. Rader; U.S. Patent 3,601,901; August 31, 1971* the conveyor for advancing the particles through the drying zone is mounted for very easy bodily removal from the vacuum chamber, through the access opening at one end of the vacuum chamber, and to its exterior. The conveyor may be supported in the vacuum chamber in a manner enabling its horizontal sliding movement longitudinally of the chamber and from its end. For best results, the conveyor includes a frame movably mounted on spaced rollers on which an endless belt is carried, with the frame being supported for the desired horizontal sliding movement from the vacuum chamber and with the carried rollers and belt.

To further facilitate removal of the equipment for repair, and for cleaning of the removed parts and the interior of the vacuum chamber, the heater elements for drying the frozen particles are also preferably mounted to the conveyor unit for removal therewith from the vacuum chamber, and are desirably easily separable from the conveyor unit after such removal. The process is described with reference to Figure 8.11.

The freeze drying machine includes a vacuum chamber having a horizontally extending and horizontally elongated cylindrical portion defining a drying compartment, and an upwardly projecting portion defining and containing a freezing compartment. The walls of both of the horizontally extending and upwardly projection portions of the chamber are heat insulated and the sidewalls of the upwardly projecting portion are refrigerated, as by flow of a low temperature refrigerator coolant through the passages formed in the sidewall of the upstanding portion of the vacuum chamber.

A liquid to be freeze dried is delivered from a supply tank to a pump which forces the liquid under pressure through a spray nozzle at the top of the upwardly projecting portion of the vacuum chamber, to thereby spray the liquid downwardly into the interior of the freezing compartment, within which the liquid is rapidly frozen to the form of a large number of small frozen particles, which fall downwardly within the compartment and onto the left end of a conveyor assembly.

This conveyor assembly advances the particles progressively past a heater assembly, which delivers radiant heat energy to the particles at a rate sublimating the moisture from the particles while they remain in frozen or solid condition, so that by the time the particles reach the discharge end of the conveyor, they are in freeze dried form, desirably at substantially ambient temperature, to fall downwardly into a discharge hopper and through a sealed outlet valve structure into a collection receptacle which is maintained under the same vacuum as the main chamber.

After an extended period of use, it may be desirable to clean or repair the conveyor assembly **21** and/or heater assembly **22**. When this becomes desirable, the right-hand end door **54** of the apparatus is unlatched and swung to its open position of Figure 8.11a, locking pin **51** of Figure 8.11b is withdrawn upwardly from the registering apertures in flanges **48** and **49**, to release the parts for removal, and the entire conveyor assembly and carried heater assembly **22** are pulled rightwardly toward their Figure 8.11a removed positions. An operator may easily pull these parts rightwardly, as by grasping the two vertical arms **62** of one of the inverted U-shaped heater mounting elements **60**. As the conveyor and heater assemblies move rightwardly, they are guided for the desired movement by sliding engagement of flanges **48** of Figure 8.11b on the upper surfaces of flanges **49** provided within the vacuum chamber.

Apparatus for Continuous Freeze Drying Processes

When the conveyor and heater assemblies have reached a position at which they project outwardly from the vacuum chamber a short distance, the right end of frame **32** may be connected to a rigid support element or rod **74** (Figure 8.11c) attached at its upper end to a roller carriage **75** mounted on a track **76** extending parallel to and directly above the axis **34** of the cylindrical portion of the vacuum chamber.

Support element **74** may be shaped as shown in Figure 8.11c, to extend first laterally to a side of the heater and conveyor assemblies, then downwardly at that side, and then inwardly to form a horizontal rigid lower portion **77**. This portion is typically connectable into an opening **78** in flange **48**, and may project into one of the crosspieces in a relation supporting the conveyor assembly from only one side but in a horizontal position.

FIGURE 8.11: DETACHABLE CONVEYOR AND HEATER MEANS FOR EASY CLEANING

a.

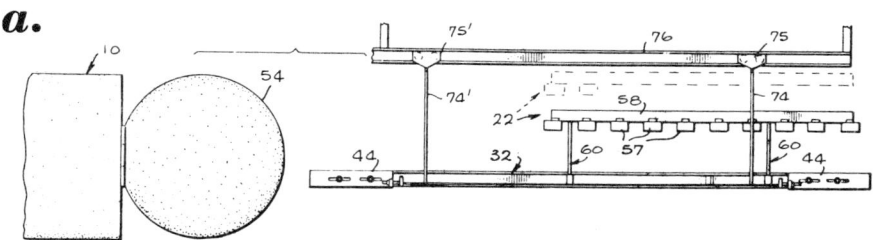

Conveyor and Heater Removed from Vacuum Chamber for Cleaning

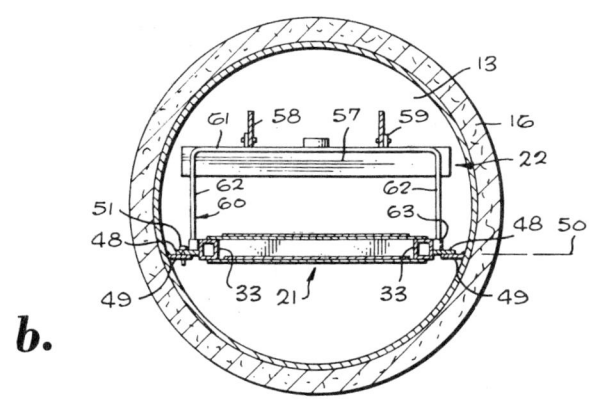

b.

Transverse Vertical Section Through Drying Chamber

(continued)

FIGURE 8.11: (continued)

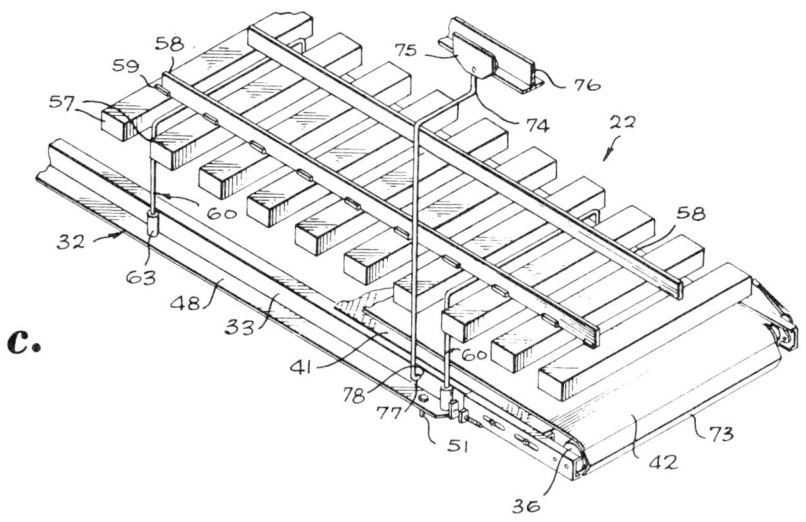

c.

Perspective View of Conveyor and Heater Partially Removed
from Vacuum Chamber

Source: U.S. Patent 3,601,901

Element **74** is of a length effectively supporting the withdrawn assemblies at their original level. When these assemblies have been withdrawn further to a position in which flanges **48** have almost reached the ends of flanges **49**, a second support element **74'**, identical with element **74** carried by a second roller carriage **75'** on track **76** is connected to frame **32** at another location near its inner end, so that the entire assembly may be moved outwardly to the Figure 8.11a completely removed and still horizontal position, in which the interior of the vacuum chamber is accessible for easy cleaning and repair, and the assemblies **21** and **22** are also completely accessible and exposed for cleaning.

After such removal, the heater assembly **22** may be separated upwardly from conveyor assembly **21**, as to the broken line position of Figure 8.11a, (and without interference by elements **74** and **74'**) to thus withdraw mounting arms **62** of the heater assembly from sockets **63**, and enable the heater assembly to be moved to any desired location for repair, cleaning, or the like. Subsequently, the heater assembly may be again positioned on and connected to the conveyor assembly, and the entire combination of these parts may be slid leftwardly to their original fully installed positions, with the door **54** then being closed and the entire apparatus thus being ready for further use.

Pair of Traps Alternately in Communication with Vacuum Manifold

V.A. Liobis and D. Freedman; U.S. Patent 3,516,170; June 23, 1970; assigned

Apparatus for Continuous Freeze Drying Processes

to New Brunswick Scientific Corporation describes a freeze drying apparatus capable of automatic and continuous operation.

The freeze drying apparatus includes a vacuum manifold means and at least two condensing trap means adapted to be placed alternately in communication with the manifold means. A refrigerating means and a vacuum-creating means both coact with each of the trap means. The automatic control means of the process provides, during a first period of time, coaction between one of the trap means and the refrigerating means and vacuum-creating means while simultaneously, during this first period of time, providing communication between the manifold means and only this one trap means, with the other trap means cut off from communication with the manifold means during the first period of time.

During a second period of time which is subsequent to the first period of time the automatic control means provides coaction between the other trap means and the refrigerating means and vacuum-creating means, and during this second period of time the automatic control means provides communication between the manifold means and only this other trap means while simultaneously preventing, during the second period of time, communication between the manifold means and the first trap means.

The process is described with reference to Figure 8.12a. The manifold drum **10** is in the form of a hollow cylinder closed at its top end in a fluid-tight manner by a cover **12**, and the cylindrical wall of the drum is provided with a plurality of short pipes **14** which communicate through the wall of the drum with the interior thereof and which have outer open ends. On the outer open ends of the pipes, which form ports communicating with the interior of the drum, there are respectively mounted valve assemblies **16** each of which is in the form of an elongated sleeve **18** fluid-tightly mounted on and surrounding the outer open end portion of a pipe.

The outer end of each sleeve is closed, and between its ends, situated outwardly beyond the open end of the pipe each sleeve carries a flexible resilient flat tubular member **20** which normally has its opposed side walls pressed tightly against each other due to the outer atmospheric pressure, the interior of the drum being under vacuum, so that in this way each valve assembly **16** is normally maintained closed. However, when a sample is to be freeze dried, this sample which is in a frozen state within a bottle or flask **22**, is placed in communication with the interior of the tube by placing the latter around the open outlet of the bottle with this flexible tubular valve member fluid-tightly surrounding the outlet neck of the flask or bottle which contains the sample which is to be freeze dried. The friction between the neck of the bottle and the flexible tubular valve is sufficient to support the bottle with the sample therein in a manner indicated in Figure 8.12a.

The bottom wall of the manifold drum rests on the top wall **24** of a cabinet **25** which contains the remainder of the apparatus and which is provided with suitable doors for access to the interior of the cabinet. Also the cabinet is provided at its upper front wall portion with a suitable control panel for supporting various controls by which the operations are regulated as well as for supporting gauges to indicate temperature, pressure, and the like.

The bottom wall of the manifold drum has a central opening from which a manifold outlet tube extends downwardly through the top wall 24 into the interior of the cabinet. This manifold outlet branches into a pair of conduits 26 and 28 which are fluid-tightly connected with the manifold outlet as by a suitable T-connection. The conduit 26 is provided with a valve V_1 while the conduit 28 is provided with a valve V_2. The valves are operated in a manner described below for the purpose of opening and closing the conduits. The conduit 26 communicates fluid-tightly with a trap 1 which forms a condensing trap means, while the conduit 28 fluid-tightly communicates with a trap 2 which forms also a condensing trap means.

The trap 1 communicates fluid-tightly through a conduit 30 with a vacuum pump 32 while the trap 2 communicates fluid-tightly through a conduit 34 with the same vacuum pump. The conduit 30 can be opened and closed by way of a valve V_3, while the conduit 34 can be opened and closed by way of a valve V_4.

FIGURE 8.12: PAIR OF TRAPS ALTERNATELY IN COMMUNICATION WITH VACUUM MANIFOLD

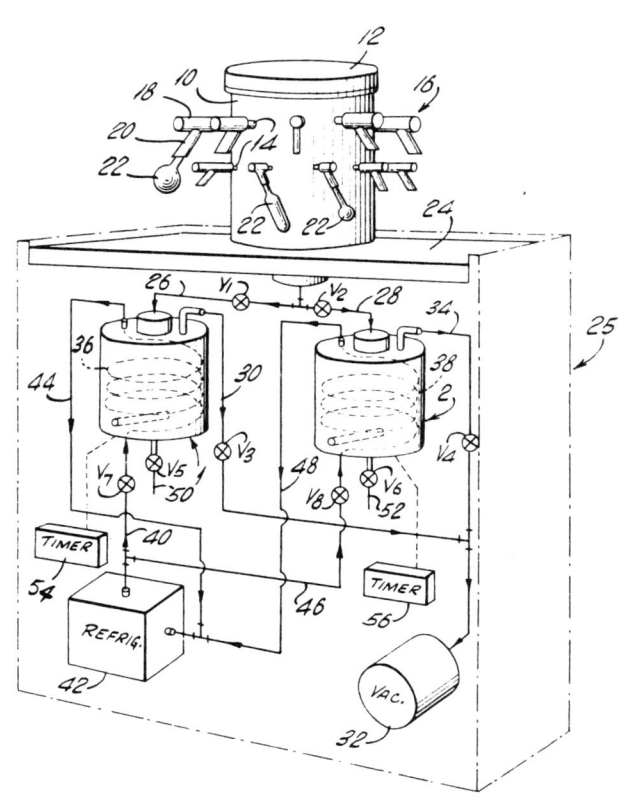

a.

Schematic Illustration of Apparatus

(continued)

Apparatus for Continuous Freeze Drying Processes

FIGURE 8.12: (continued)

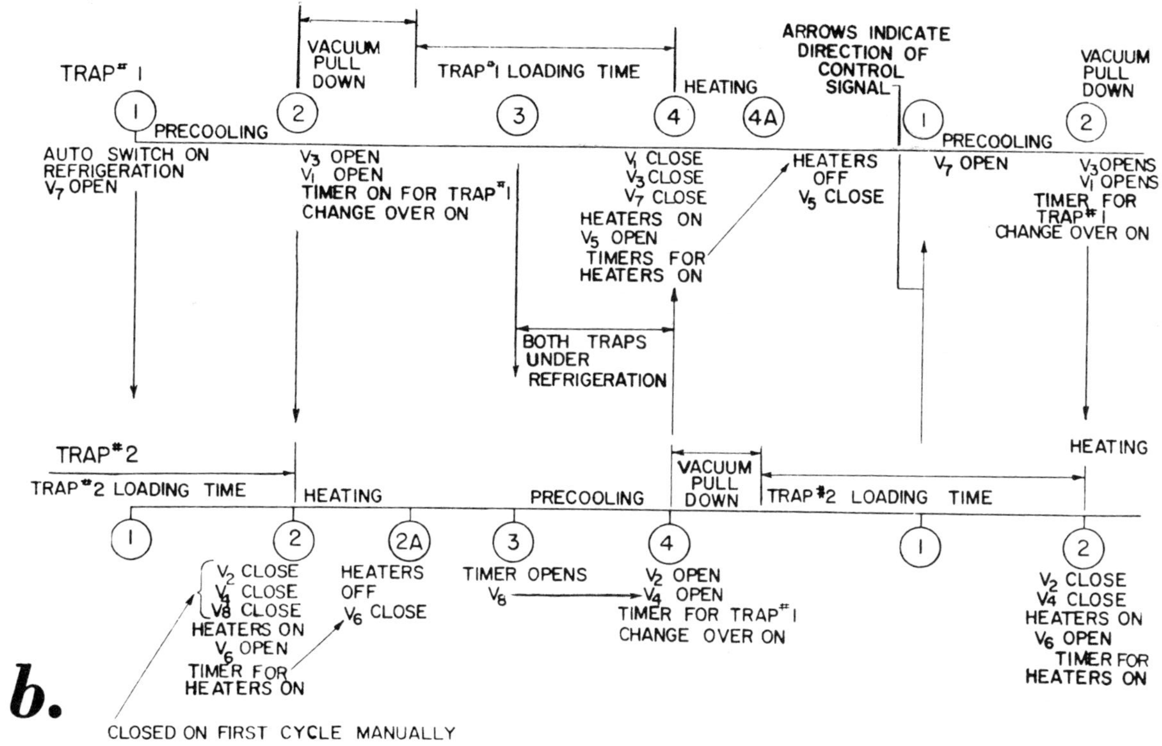

Diagrammatic Representation of Automatic Control System

Source: U.S. Patent 3,516,170

Each trap has a refrigerating coil (refrigerating coil **36** of the trap **1** and the refrigerating coil **38** of the trap **2**). The refrigerating coil **36** communicates through a conduit **40** with the outlet of a refrigerating means **42** so that a suitable refrigerant, such as Freon, will flow through the conduit **40** into the coil **36**. This coil communicates with a return conduit **44** which serves to return the refrigerant back to the refrigerating means **42**. The outlet of the refrigerating means also communicates through a conduit **46** with the coil **38** of trap **2** for supplying the latter coil with the refrigerant, and this coil communicates with a return conduit **48** through which the refrigerant is returned to the refrigerating means. The conduit **40** is provided with a valve V_7, while the conduit **46** is provided with a valve V_8, and these valves are operated for opening and closing the conduits.

Each trap is also provided with a valved drain outlet through which defrosted condensate can flow out of each trap. Thus, the trap **1** is provided with a drain outlet **50** provided with the valve V_5, while the trap **2** is provided with a drain outlet **52** provided with the valve V_6.

Each trap means includes an inner enclosure, an outer receptacle within which the inner enclosure is located, the outer receptacle being spaced from the inner enclosure for defining therewith a heat-exchanging chamber situated between the inner enclosure and outer receptacle. The refrigerating means include a refrigerant conduit within which a refrigerating medium flows, the conduit being located within the heat-exchanging chamber spaced from the inner enclosure so as to be out of direct contact therewith. A heat-exchanging medium is situated in the heat-exchanging chamber for providing heat transfer between the inner enclosure and the refrigerant conduit without direct engagement therebetween.

Part III. Freeze Drying of Specific Foodstuffs

FREEZE DRYING OF COFFEE

COFFEE EXTRACTION

E. Pitchon, M. Gottesman and R.W. Meier; U.S. Patent 3,655,398; April 11, 1972; assigned to General Foods Corporation found that by using only one or two intercolumn heaters in a percolator set and reversing the direction of flow of the extract through the columns, in the spent end of the percolator set, it is possible to attain a high solids concentration extract (solids content of 30-41%) with little if any loss of yield and without encountering operation difficulties necessitating frequent shutdowns. This extract is suitable for freeze drying.

The process may be best described as utilizing a standard percolator set, with one or two intercolumn heaters. In the spent end of the percolator set the coffee grounds are contacted with a hot extraction medium, typically hot water, as in any standard percolation operation. The extract produced in this section of the percolator set is very dilute. Generally the extraction medium flows through the extraction columns from the bottom of a column up through the column and out the top of the column.

After the extract has passed through at least one extraction column, and prior to introducing the extract into the next adjacent extraction column, it is passed through a heat exchanger (intercolumn heater) wherein the temperature of the extract is raised about 5° to 30°F. If a second intercolumn heater is being used, the flow of extract will be diverted through this heat exchanger and the temperature of the extract will again be raised about 5° to 30°F.

The extract, after passing through an intercolumn heater, will always flow through at least one extraction column prior to passing through the second intercolumn heater. As in a standard percolator operation, the extract will flow through a cooler prior to being introduced into the extraction column containing the freshly roasted and ground coffee (fresh column). The cooler is used to adjust the temperature of the extract and coffee in the fresh stage so that the temperature of the final concentrated extract drawn off from the fresh stage is maintained at a maximum temperature 210°F, and typically between 170° to 210°F.

The percolator set may be considered divided into three sections. The first section is the spent section, shown in Figure 9.1 as encompassing two extractor columns 5 and 6. The flow of extract and temperature drop across this section is similar to that encountered in a standard percolation operation. The second section of the percolator set, the intermediate section, encompasses all of the remaining extractor columns other than fresh stage. In the process, before the extract is allowed to enter an extractor column in the intermediate section, it is passed through a heater wherein the temperature of the dilute extract coming off column 5 is heated. The arrangement of intercolumn heaters 7 and 8 prior to columns 4 and 3 are a preferred aspect of the process.

As the extract passes through a heater the outlet temperature is controlled at a preselected temperature dependent upon the coffee being extracted, the DOF (drawn off factor) being used and the desired concentration and yield. In a typical run the average increase in temperature across a heater will vary from about 5° to 30°F. The actual increase in temperature will be greatest at the beginning of a cycle (e.g., when extract is being fed to a fresh stage in order to fill the fresh stage) and will be lowest at the end of a cycle (e.g., when the spent stage is being blown down), as the temperature of the extract coming out of the column preceding the heater will decrease from the start to the end of the cycle.

The heaters are utilized at all times while the percolator set is in operation. Therefore, the manifold arrangement is such that extract can be fed from any column through any heater, so that as the position of the columns change from fresh column to spent column, the flow of extract to a heater can be from the appropriate column. This manifold arrangement is desirable as there is a tendency for coffee solids (including waxes and tars) to settle out and carbonize in a heater, when the heater is not in use.

FIGURE 9.1: HIGH CONCENTRATION EXTRACT USING STANDARD PERCOLATOR SET WITH FLOW MODIFIED

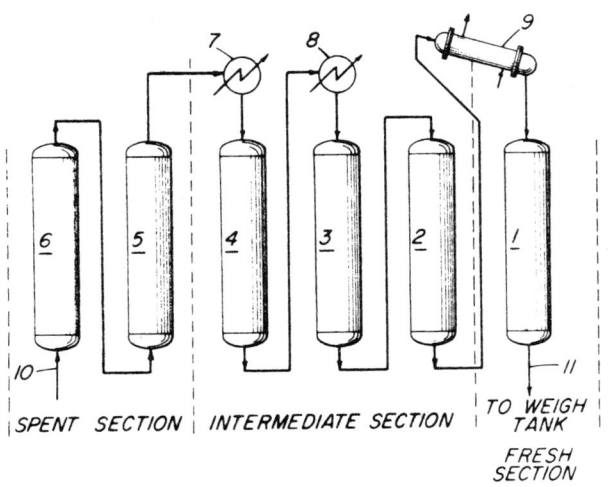

Source: U.S. Patent 3,655,398

Unexpectedly, and a key to the process was the discovery that if the flow of extract through the percolator columns is reversed at some point in the percolator set, at least before entering the intermediate section, the pressure problems otherwise encountered when using intercolumn heaters are alleviated. The precise reason for this is unknown, although there is speculation that it involves reversing the migration of fines within a percolator column, back-flushing the bayonets and loosening the bed of coffee within the extraction column. Thus, the flow of the extraction liquor in the spent section is into the bottom of column **6**, out the top of column **6**, into the bottom of column **5** and out the top of column **5**. The flow is then reversed after passing through heater **7** and set into the top of column **4** and all of the other columns in the set.

When practicing the process, it has been found that the concentration of the dilute extract coming off column **5** is generally about 5 to 15% solids by weight of the extract, that the concentration of the extract coming off column **4** is about 10 to 20% and that the extract concentration prior to the fresh column is about 15 to 35%.

REMOVAL OF UNDESIRABLE SOLIDS FROM COFFEE EXTRACT

Clarification in a Desludger Type Centrifuge

R.A. Chaplow and R.A. Hodgman; U.S. Patent 3,765,910; October 16, 1973; assigned to General Foods, Limited, Canada relates to a process for removing undesirable insolubles from coffee extract during a process for the production of freeze dried coffee which utilizes freeze concentration.

The original coffee extract is, for example, obtained from a percolator and has a solids content of generally 24 to 28%. A filter removes solid matter from the extract as it exits from the percolator at elevated temperatures usually above 150°F. The extract is then cooled in a cooler to room temperature or below to prevent flavor degradation, reduce the heat load on freeze concentration equipment and minimize precipitation of wax-like solids such as waxes, tars, sediments, etc., which might tend to clog processing equipment.

The extract is then supplied to a crystallizer which is of the scraped wall type having a minimum residence time and a relatively high ΔT (35° to 50°F). This type of crystallizer generates fine ice crystals and is not susceptible to clogging by the wax-like solids. The residence time in the crystallizer is from about 3 to 10 minutes, which is relatively low in comparison to the other types of crystallizers in which the extract is held for approximately two to four hours. Such low residence time minimizes precipitation of wax-like insolubles and avoids clogging. The effluent from the crystallizer consisting of a slush of ice crystals and concentrated coffee extract is delivered to a storage tank from which it is supplied to a separator.

The separator is, for example, a basket type batch centrifuge. This type of centrifuge traps wax-like solids in the ice layer which is plowed out of the basket. This prevents such wax-like insolubles from clogging the centrifuge screens.

Concentrated extract from the separator is fed to a storage tank. The plowed ice crystals are directed to a melting tank. The dilute extract in the melting tank is directed to a concentrator of the evaporative type. Water is removed and the concentrate derived from the ice cake melt is added back to the storage tank. The flavor degradation caused by evaporation does not materially affect the flavor of the combined extract in the storage tank because it is a low percentage of the combined extract. Flavor degradation can be minimized, if desired, by utilizing a vacuum type evaporator. The combined extract is maintained in the storage tank a relatively short time normally between 15 minutes and 1 hour.

The combined concentrated extract having a solids content of approximately 35 to 45% by weight is then heated to a temperature preferably about 50° to 80°F to facilitate clarification in the desludger type centrifuge. Undesirable insolubles are then easily removed in a clarifier. If these undesirable insolubles are not removed before freeze drying, they speckle the product with undesirable black spots which appear as unacceptable foreign matter in a reconstituted cup of the coffee. The clarified concentrate extract is frozen and ultimately dried in a freeze dryer to provide clean and flavorful freeze dried coffee.

Filtration Through Medium Duty Type Filter

R.A. Chaplow and R.A. Hodgman; U.S. Patent 3,843,823; October 22, 1974; assigned to General Foods, Limited, Canada describe the filter used to remove matter from the extract as it exits from the percolator (see U.S. Patent 3,765,910).

Coffee extract (24-30% solids) obtained from percolation is filtered while still hot (190° to 215°F). Filtration is accomplished in a medium duty type filter, such as is used in swimming pool water systems to remove all visible sediment and organic matter and greatly reduce bacteria and other organic material. Such a filter removes all visible solids, flocculant material, coffee bean chaff, etc. which act as nuclei for formation of insoluble wax and tar globules, which might clog freeze concentration or other apparatus used in making freeze dried coffee. A pair of such filters is provided to permit one to be cleaned while the other is being used.

Such a filter is operated at pressures ranging from 30 to 70 psig and flow rates of 6 to 11 gpm. It includes, for example, seven filter leaves which can be precoated with diatomaceous material but are usually not precoated to perform the coffee extract filtration. The leaves are approximately 2 to 3 feet in diameter and the extract is conducted from the outside to the inside. The leaves are covered by screen mesh, for example, of 80 mesh stainless steel. Insoluble solids are deposited on the outer surfaces of the mesh. When the pressure drop through the filter is prohibitively increased by solid deposits, the flow is switched to the standby filter and the blocked filter leaves are removed and cleaned. The filter is removed from the stream and cleaned when the pressure rises to 70 psig and operates at 30 psig after cleaning.

The temperature of the filtered extract is then reduced in a cooler to 40° to 70°F to reduce the heat load on the subsequent crystallizer and also to prevent flavor degradation which might result if the percolated extract is held too long at elevated temperatures. The extract is then supplied to the crystallizer; the remainder of the process is that described in U.S. Patent 3,765,910.

Removal of Tars and Waxes

When coffee extract is cooled to near its initial freezing point, a gummy or waxy solid material often precipitates from solution prior to ice formation. This wax or tar is carried along with the ice crystals and coffee extract into the centrifuge and collects on the centrifuge basket, eventually plugging the basket and preventing the complete separation of ice from concentrated extract.

R.G. Reimus and A. Saporito; U.S. Patent 3,381,302; April 30, 1968; assigned to Struthers Scientific and International Corporation provide a process in which coffee extract, free of insoluble elements at the freezing temperature of ice, is subjected to freeze concentration. This is accomplished by extracting coffee beans at elevated temperatures with water to produce a coffee extract, cooling the extract to precipitate materials insoluble at the temperature at which ice forms in the brew, removing insolubles from the resulting mixture and then subjecting the insoluble free mixture to concentration by partial freezing of water therefrom.

The coffee extract to be concentrated by freeze concentration can have a coffee solids content as high as 40 to 50%, preferably 10 to 30%. Because the initial freezing point of the coffee extract is dependent on the total coffee solids content of the extract, the temperature to which the extract should be chilled to precipitate the gums, tars and waxes varies with the exact nature of the extract. Ordinarily, however, minimum temperatures of 40°F and below may be employed in the process. If it is not desired to form ice in the extract during the precipitation process, the temperature to which the extract is chilled should not be below the initial freezing point. For a coffee extract having 30 weight percent coffee solids, the temperature is about 27°F.

In practice, it has been found that the tars, waxes and gums will begin to precipitate from coffee at temperatures as high as about 80°F and that the major portion of insolubles will precipitate at temperatures above 32°F. Preferred temperatures for the chilling operation are between about 45° and 32°F because this temperature range insures virtually complete removal of insolubles.

Depending upon the exact nature of the coffee blend employed in preparing the extract and the extract itself, from less than 0.5 to about 5% by weight of tars, gums and waxes will form during the precipitation step.

The filter apparatus may be a conventional batch filtration unit or, where the precipitated wax or tar can be handled on a centrifuge basket, may comprise a rotating basket centrifuge or ordinary laboratory or batch-type centrifuge.

Example 1: Control — A batch of coffee extract containing 24% coffee solids was precooled to below 40°F and admitted to a tubular heat exchanger having an internal agitator. The coffee extract was fed to the heat exchanger at the rate of 3 gpm and coffee extract was recirculated around the heat exchanger until ice crystals had formed due to the cooling internally. Recirculation of slurry and mother liquor was continued as coffee extract was fed to the crystallizer, the internals of which were maintained at a temperature of 27°F.

A slurry of ice and coffee liquor containing about 30% coffee solids was then continuously removed from the crystallizer and fed to a continuous type rotating basket centrifuge. The centrifuge basket was in the form of a perforated screen which allowed coffee liquor to pass through the ice cake, thus separating the ice from the coffee liquor. After 10 minutes of operation, the centrifuge screen plugged with tars and waxes which were precipitated from the coffee extract during the cooling and crystallization process. This caused the operation to stop, due to improper separation of ice and mother liquor.

This procedure is repeated except that the extract is fed to the crystallizer without first prechilling. Tars and gums form along with the ice and the centrifuge plugs in the same manner as if the extract had been prechilled.

Example 2: Advantage of Removing Tars and Waxes — The general procedure of Example 1 was repeated except that the chilled coffee extract was filtered through fine cheesecloth prior to introduction into the crystallizer. In this filtration process, approximately 1.9% of the original extract was removed as tars, gums or waxes. The resulting precipitate-free coffee liquor was then admitted to the crystallizer vessel and the slurry produced in the crystallizer was centrifuged continuously. The ice-mother liquor separation was not impeded by plugging of the centrifuge screen with precipitate.

Excellent results are also obtained when coffee liquor to be concentrated is precooled from 80° to 36°F, and when the extract has from 12 to 32% coffee solids.

In a related process, *R.G. Reimus and A. Saporito; U.S. Patent 3,449,129; June 10, 1969; assigned to Struthers Patent Corporation* utilize a coffee extract containing 10 to 50% solids (preferably 15 to 30%). The process includes the steps of:

(a) chilling the extract to a temperature range of from about 80° to about 45°F and maintaining the extract in the temperature range with agitation for a period of time sufficient to precipitate at least a portion of the insoluble matter;

(b) removing the precipitated insoluble matter formed within the range of about 80° to about 45°F from the extract by centrifugation.

FREEZE CONCENTRATION OF COFFEE EXTRACT

It is highly advantageous to concentrate coffee extract prior to freeze drying it. This minimizes the amount of water vapor that must be removed, generally facilitates freeze drying and heightens its efficiency. Freeze concentration is a method of concentrating the extract, which removes water in the form of ice crystals. It is important to minimize the amount of coffee solids removed from the extract with the crystallized ice and to facilitate such removal.

Growing Large and Uniform Ice Crystals

D.E. Dwyer, Jr.; U.S. Patent 3,845,230; October 29, 1974; assigned to General Foods Corporation describes a continuous process of freeze concentrating coffee extract in a single pass through a heat exchanger to 35 to 50% solids content.

Clarified coffee extract is freeze concentrated under conditions which provide a relatively uniform and large ice crystal ranging from about 10 to 80 mils in size. The ratio of crystallizer volume to refrigerated heat exchange surface ranges from 1:0.5 to 1:3.5 preferably 1:0.75 to 1:1.50 and is advantageously 1:1. The "U" factor (Btu's per hour, per square foot refrigerated heat exchange surface area, per °F) is between 25 to 100 and advantageously 50. The delta T (temperature differential) between extract and refrigerated heat exchange wall, for concentrations of from 30 to 50% solids and usually 35 to 45% solids by weight of extract varying in ice point from 29.4°F (−2°C) to 19.4°F (−7°C), ranges approximately between 25° and 60°F, preferably 30° to 40°F and is advantageous in the neighborhood of 35°F.

The extract is retained in the crystallizer, which is of the scrape wall type, from 1 to 6 hours and usually from 1.5 to 3.5 hours. The slurry of extract and crystals from the crystallizer is centrifuged in consecutive load and spin cycles to separate them and to minimize the coffee solids retained in the crystals. The slurry is first spun at a relatively lower speed in a loading spin cycle, ranging approximately from 400 to 800 rpm and developing forces ranging approximately from 100 to 400 g.

The ice cake, thus built up to a thickness of from 0.5 to 2 inches in approximately 3 to 5 minutes, has a solids content of approximately 25% by weight. The ice cake is washed by dilute extract or water during the terminal portion of the loading spin cycle to reduce the solids content to about 5 to 10%. Then the ice cake is final spun at speeds in excess of 800 rpm to create elevated g forces in excess of 500 g which deplete the ice cake solids content to less than 5 to 10% and usually in the neighborhood of 1 to 3%.

Multistage Freeze Concentration of Extract

The process of *N. Ganiaris; U.S. Patent 3,620,034; November 16, 1971; assigned to Struthers Patent Corporation* provides for the centrifuging of ice crystals from a multistage freeze concentration system for coffee which produces a coffee solution at a concentration of over 37% from a feed at a concentration of about 26%. This is achieved by centrifuging and removing the ice crystals from the system in a first centrifuge which separates crystals from a coffee solution at a concentration of less than 30% leaving an early or first stage crystallizer. The solution from the first centrifuge is passed through an additional crystallizer and a second centrifuge from which a product stream of a coffee solution of over 37% is withdrawn, ice from the second centrifuge being recycled by being added to the feed. The process is described with reference to Figure 9.2.

Referring to the drawing in detail, a feed solution of 26% by weight coffee dissolved solids enters feed tank **10**. Pump **12** draws this solution and passes it through pipe **13** to enter crystallizer **14**. The solution will be from 13 to 24% coffee dissolved solids. This lower coffee content results from the fact that the solution is diluted because ice enters tank **10** from centrifuge **15** through pipe **16** and because wash water from centrifuge **17** passes into tank **10** through pipe **18**. Thus wash water and the melted ice dilute the feed solution entering crystallizer **14**.

A pump **19** may partially recirculate through pipe **20** the slurry of ice crystals and coffee solution leaving crystallizer **14** until the coffee solution contains 26 to 30% coffee dissolved solids. This slurry then passes through pipe **21** to enter centrifuge **17**, where mother liquor is separated from the ice crystals and passes through pipe **22** to the mother liquor tank **23**. Wash water enters centrifuge **17** through pipe **24** to wash ice crystals therein and flow to tank **10** through pipe **18**.

FIGURE 9.2: MULTISTAGE FREEZE CONCENTRATION OF EXTRACT

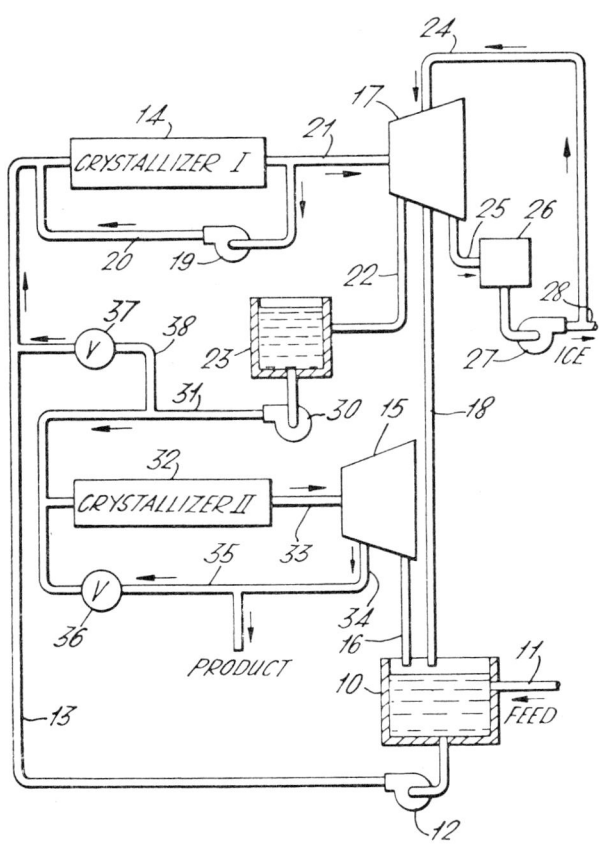

Source: U.S. Patent 3,620,034

Ice passes from centrifuge **17** through pipe **25** to enter the ice tank **26**. Some ice melts in tank **26** to be recirculated by pump **27** through pipe **24** as wash water. Ice passes from the system through pipe **28**.

The 26 to 30% coffee dissolved solids solution in mother liquor tank **23** is pumped by pump **30** through pipe **31** to the second stage crystallizer **32**. A slurry of ice crystals in at least a 37% coffee solution leaves crystallizer **32** through pipe **33** to enter centrifuge **15**. The 37% coffee dissolved solids solution leaves centrifuge **15** through pipe **34** as a product. By means of pipe **35** and valve **36**, this solution may be recirculated through crystallizer **32**.

By means of valve **37** and pipe **38** mother liquor may be recirculated through crystallizer **14**. This recirculation is required to raise the concentration of the solution in pipe **13** entering crystallizer **14** above the concentration of the feed while keeping it below 30% coffee dissolved solids. Excessive recirculation through pipe **20** could also serve to concentrate the feed from pipe **13**, but this would result in too high a percentage of ice crystals in the crystallizer **32** which would reduce the rate of crystal growth.

Recirculation through pipes **20** and **38** should be balanced and controlled so that the percentage of ice crystals by weight in solution in crystallizer **14** is between 20 and 30%. If the percentage of ice crystals is too low, below 15% by weight, as when pipe **20** is not used for recirculation, excessive nucleation and small and nonuniform crystals result which are difficult to centrifuge. When centrifuging ice crystals from the system from a 26 to 30% solution as described, less than 1% coffee solids by weight will be lost with the ice.

Minimizing Loss During Washing of Ice Crystals in Continuous Process

N. Ganiaris; U.S. Patent 3,283,522; November 8, 1966; assigned to Struthers Scientific and International Corporation developed an apparatus consisting of a number of stages each having a crystallizer to freeze ice crystals out of solution and means to remove the crystals from the resulting more concentrated solution, the ice crystals from at least one or more later stages being mixed with some incoming solution and then being returned as a slurry to be mixed with the feed of an earlier crystallizer. The growth of crystals is nucleated in a concentrated slurry and further grown in a less concentrated slurry from which the resulting crystals are washed free from solution.

For example, in other processes ice crystals removed from a 46% coffee solution in a centrifuge would have had to be washed, preferably with cold water. Crystals separated from a 46% solution may have up to 40% of the weight of the crystals of coffee solution adhering to them. Since this adhering solution must be washed from the crystals as the coffee extract or solute is the desired and valuable end product, the crystals must be very thoroughly washed if they are extracted from a relatively concentrated 46% solution.

However, in this process the ice crystals which are washed and removed in the centrifuge are washed from a 16% coffee solution. Since ice crystals in a 16% solution of coffee extract will only have about 4% by weight of the crystals of the less viscous coffee solution adhere to them, they are much more easily washed and the resulting loss of coffee in the entire process is less than 1%.

The crystals which are separated from more viscous solutions in the centrifuges are not necessarily washed but they are mixed with the incoming feed and recycled through the crystallizer. Thus, the product is removed at the end of a series of stages of concentration while ice separated out in later stages is mixed with incoming feed solution and passed through a first crystallizer from which ice is removed from solution.

Recovery of Coffee Adhering to Ice

N. Ganiaris; U.S. Patent 3,531,295; September 29, 1970; assigned to Struthers Patent Corporation describes a process which comprises extracting coffee beans at elevated temperature with water, removing part of the water by partial freezing, separating the resulting ice from the mother liquor, and melting the ice with subsequent recycling of the solution produced to the extraction step so as not to expend valuable coffee concentrate which adheres to or is occluded in the ice.

The freeze concentration of the extract can be accomplished in either a single stage or in a plurality of stages each of which comprises a crystallizer in which heat is removed from the coffee extract to form a slurry of ice crystals and concentrated mother liquor. After each crystallization stage, ice is removed from the slurry and the concentrated extract either goes to further processing in the preparation of soluble or powdered coffee or is further concentrated in a subsequent stage of the freeze concentration process.

The crystallization of ice from the extract is preferably carried out in a tubular heat exchanger, the outside surfaces of which are cooled by a circulating refrigerant. The internal section of the tube is ordinarily fitted with a shaft on which is located agitator paddles. Alternatively, a conventional scraped surface tubular heat exchanger may be employed. For separating the concentrated coffee extract from the resulting ice crystals, a rotating basket centrifuge is preferred.

Example 1: Ground coffee and water are charged to a countercurrent extractor and extracted to result in an extract containing 24% soluble coffee solids. This extract was continuously metered into a crystallizer wherein 10 pounds per minute of water formed individual ice crystals to result in a slurry of concentrated coffee, having 32% soluble coffee solids and ice. This slurry was conveyed to a centrifuge having a basket rotating at 2,200 rpm and a wash water rate of 4 pounds of fresh water per minute. The concentrated solution of coffee was discharged through the perforated rotating basket as product at the rate of 30 pounds per minute and contained 32% soluble coffee solids. The ice, upon melting, was found to contain 1.3% soluble coffee solids.

Example 2: The general procedure of Example 1 was repeated except that the dilute coffee solution resulting from melting the ice was recycled to the extractor. The resulting extract having 24.5% soluble coffee solids was metered into the crystallizer. The slurry produced was centrifuged and the concentrate leaving the centrifuge as product at the rate of 30 lb/min was found to contain 32.5% soluble coffee solids, thus providing 13 more pounds of coffee as product per 1,000 pounds extracted.

Processing Ice Stream to Recover Trapped Soluble Coffee

M. Gottesman and F.D. Pascal; U.S. Patent 3,684,532; August 15, 1972 found that a freeze concentration system can be successfully operated in such a manner that any waxlike material tending to precipitate out of a coffee extract can be removed with the ice stream and any soluble solids trapped by the waxlike precipitate can be recovered along with other soluble solids in the ice stream.

The process comprises cooling coffee extract obtained via standard commercial percolation technique to below 32°F, further removing heat from the extract in order to cause part of the water to freeze out as ice crystals, thus forming a slush of concentrated coffee extract and ice crystals, separating the ice crystals from the concentrated extract and then further processing the ice stream to recover trapped soluble coffee solids. Crystallization should be performed in a continuous type crystallizer wherein the holdup time of extract is minimized.

Coffee extract having an original concentration of 15 to 35% solids by weight of the extract is thus concentrated to a solids content of 30 to 65%.

The ice stream in addition to containing the ice crystals and soluble solids trapped by these crystals is also found to contain waxlike materials precipitated out of the extract during crystallization and additional soluble solids trapped by these waxlike materials. It is critical that centrifugation be performed in a manner which will cause the ice to bridge, thus trapping the waxes in the ice and preventing the screen from blinding. The ice stream after removal from the centrifuge is heated in order to melt all of the ice crystals and obtain a very dilute solution of coffee extract along with insoluble material.

This solution is then subjected to centrifugation in a solid bowl centrifuge, such as a "Westphalia" or other similar types of desludging centrifuges, in order to remove the waxlike precipitates and any other insoluble material contained therein. The clarified dilute extract is then concentrated and blended back with the main stream of concentrate obtained via freeze concentration.

Example: Coffee extract containing 25% soluble solids by weight of the extract was freeze concentrated and freeze dried in the following manner.

The coffee extract obtained from the percolators was maintained at a temperature of 55° to 65°F and pumped directly from the percolators to a crystallizer feed tank. Holdup time of extract in the tank was 25 minutes. The tank was jacketed and the extract temperature in the tank was controlled by circulation of chilled water through the jacket.

Extract from the feed tank was pumped through a crystallizer (a jacketed, scraped-surface heat-exchanger) wherein it was cooled by circulation of chilled brine through the jacket. In the crystallizer, water crystals were frozen out of the extract onto the refrigerated surface and these frozen crystals were scraped off the surface and mixed into the extract forming a slush. Holdup of extract from inlet to outlet of the crystallizer was about 5 minutes. Temperatures and flow rates were controlled such that the slush discharging from the crystallizer consisted of extract at a 35% concentration by weight of the unfrozen extract and the outlet temperature was about 26°F.

The slush from the crystallizer was fed into a centrifuge holdup tank and when a sufficient quantity of material for a centrifuge load was obtained, the slush was centrifuged in a vertical batch centrifuge, the concentrated extract (35% solids by weight of extract) was sent to a tank for further processing in the freeze drying system and the ice stream was treated separately.

The ice stream was heated in a jacketed tank in order to completely melt the ice crystals. The resultant melted ice stream contained about 11% solids. This melted ice stream was then subjected to centrifugation in a solid bowl desludging type centrifuge and all of the waxlike precipitate which was formed during crystallization, plus other sediment normally found in extract after percolation, were separated out from the melted ice stream.

The clarified melt was then concentrated in a vacuum evaporator to obtain a 35% concentration extract and the concentrated melt ice stream was added back to the main concentrated extract stream.

An average of 3,600 lb of extract per hour (or 900 lb of soluble solids per hour) was processed in the foregoing manner for about 18 hours. Unrecovered soluble solids, the waxlike precipitate and other insoluble material separated out of the melted ice stream, averaged about 2.0% of the solids in the ice stream or 0.5% of the starting material (e.g., 4.5 lb of soluble solids per hour).

This represented a reduction in losses of about 1.5% when compared to a system wherein the percolator extract was first chilled and tempered and then centrifuged to remove waxes prior to crystallization, the system having losses of about 18 lb of soluble solids per hour.

The concentrated extract was frozen on a continuous belt freezer and the frozen sheets of extract were ground into particulate form and freeze dried. The freeze dried instant coffee was found to have a flavor closely resembling that of the original extract and was free of undesirable insoluble matter.

Use of Coated Aluminum or Tin Heat Transfer Surface

A serious problem encountered in crystallizing the solvent in heat sensitive substances has been the ease with which the solvent crystallizes on the heat transfer surfaces thus reducing the ability of the equipment to transfer the heat of crystallization from the solution to the cooling medium.

R.H. Hedrick; U.S. Patent 3,335,575; August 15, 1967; assigned to Struthers Scientific and International Corporation has developed a process and means that prevent the deposition of the solvent as crystals on heat transfer surfaces. The discovery is based on the work (energy) required to form a deposit on the surface correlated with the surface energy of the heat transfer surfaces while in contact with a slurry of ice crystals uniformly dispersed throughout an aqueous solution. It has been found that if the surface energy of a heat transfer surface is less than the work required to form ice deposits, none will form on the surface providing the surface is also in contact with a slurry of crystals.

An aqueous solution at a temperature below its freezing point and in the presence of nuclei will form or grow crystals. The heat of crystallization in the solution will be released principally either in the creation of nuclei or as crystal growth. The nuclei may be added to or formed in the solution or may form on the heat exchange surfaces. In accord with the process it has been found that the energy required to form crystals can be controlled by regulating the temperature difference between the average temperature of the solution and the temperature of the heat transfer surface.

Studies were carried out using tin, anodized aluminum, aluminum, and aluminum-brass alloy, copper, and stainless steel as heat transfer surfaces. The permissible temperature differences between refrigerated crystal solutions and the various heat transfer surfaces were determined using orange juice, beer, and coffee as the experimental solutions. Tin and anodized aluminum were found to give the best results and are recommended for use on the surface of heat exchangers.

Freeze Concentration of Extract, Followed by Freeze Drying

J.G. Muller; U.S. Patent 3,404,007; October 1, 1968; assigned to Struthers Scientific and International Corporation has developed a process for the preparation of a dehydrated coffee beverage product which is readily soluble in cold water, which comprises:

(a) preparing an aqueous coffee extract containing about 10 to 30% by weight of dissolved solids;

(b) subjecting the extract to concentration by partial freezing to form ice crystals and a more concentrated extract containing about 30 to about 50% by weight of solids;

(c) separating the more concentrated extract from the ice crystals by centrifugation; and

(d) subjecting the more concentrated extract to relatively complete dehydration by freezing the extract to a solid mass and freeze drying to a moisture content of about 1 to 5% at temperatures between about $0°$ to $-45°C$.

Example: Ground coffee and water are charged to a countercurrent extractor and extracted to result in an extract containing 24% soluble coffee solids. This extract was continuously metered into a crystallizer wherein 10 lb/min of water formed individual ice crystals to result in a slurry of concentrated coffee, having 32% soluble coffee solids and ice. This slurry was conveyed to a centrifuge having a basket rotating at 2,200 rpm and a wash water rate of 4 pounds of fresh water per minute. The concentrated solution of coffee was discharged through the perforated rotating basket as product at the rate of 30 lb/min and contained 32% soluble coffee solids.

This solution is then poured onto trays and frozen at a temperature of $-40°F$. Then the trays are inserted into a freeze drying apparatus. There, the ice in the frozen mass is sublimed as the pressure is reduced. The final pressure vacuum is 250 mm. Heating is applied to the shells on a programmed basis, with temperature ranging from $250°F$ at the beginning of the cycle to $90°F$ at its conclusion. Duration of the cycle is approximately 8 hours. The resulting dry coffee product has about 3% moisture and is readily soluble in water. It has a flavor preferable to soluble coffee prepared by other methods.

FREEZING STEP MODIFICATIONS

Formation of Extrudable Slush

S. Katz; U.S. Patent 3,619,204; November 9, 1971; assigned to General Foods Corporation found that an extract containing 20 to 60% solids can be prepared in the form of a slush and extruded in the form of continuous ropes into a vacuum system wherein the slush may be finally frozen to below the eutectic point of the extract via evaporative cooling. By cooling the extruded slush in a vacuum the product is made to undergo some degree of puffing and the puffing results in some characteristics of appearance and also affords some degree of density control over the final freeze dried product.

The degree of puffing can be controlled by varying the amount of ice formed in the slush operation prior to extrusion and by varying the vacuum to which the ropes are exposed.

Generally the degree of puffing will vary inversely with the degree of hardness of the slush being extruded. To achieve a lower density by the expansion of the product, a soft slush should be used. After the puffing and hardening of the extruded rope has been achieved the completely frozen expanded ropes can be cut to any desired particle size distribution by standard slicing or grinding techniques. Also, the degree of puffing will vary with the degree of vacuum. Higher vacuums will cause more violent evaporation and result in a greater degree of puffing.

Alternatively, if it is desired to minimize puffing or to slice the ropes as they are discharged from the extruder prior to puffing then a hard slush should be used. When freeze drying coffee extract it has been found that an extract concentration of 20 to 45% and an ice content of 20 to 60% by weight of the slush are the parameters within which a hard slush can be achieved.

Example 1: Coffee extract containing 35% solids by weight of the extract was chilled in a scraped surface, jacketed heat exchanger in order to form a slush. The slush was equilibrated at 14°F on discharge from the heat exchanger and was fed to a chilled piston type extruder. The slush from the extruder was discharged through a die in the form of ropes into a vacuum chamber maintained at a pressure of 250 microns of mercury. Puffing was found to occur at the surface of the ropes and produced a rough scaly appearance.

The frozen and puffed ropes were coarse ground and freeze dried. The final dried particles retained the same rough scaly appearance as the frozen ropes.

Example 2: Coffee extract containing 25% solids was slushed and extruded as in Example 1. The extract was equilibrated at a temperature of about 18°F prior to extrusion. The ropes were extruded into a vacuum chamber maintained at 350 microns of mercury. It was found that there was some interior puffing but that the surface of the ropes was smooth. After freeze drying, the particulate pieces had a unique bitextural appearance.

Extruding Slush into Cold Gaseous Atmosphere

B.E. Elerath; U.S. Patent 3,637,398; January 25, 1972; assigned to General Foods Corporation found that the hard slush can be extruded in the form of ribbons which retain their shape and can be easily frozen before or after cutting without the necessity of using belt freezers. Typically the final freezing can be accomplished by extruding the ribbons into a cold gaseous atmosphere at a temperature below the eutectic point of the extract. The gas used can be air but also it may be an inert gas which may afford additional protection to volatile aromatic constituents.

It has been found that the ice content may be varied from about 20 to 60% by weight of the slush when the extract concentration varies from about 20 to 45% solids by weight of the initial extract. The desired degree of hardness will result from a combination of increasing viscosity due to the ice crystals being formed and the increasing viscosity of the concentrated extract which constitutes the liquid phase of the slush. Normally the higher the initial solids content in the extract being frozen, the smaller the amount of ice which must be crystallized in order to form a hard slush.

With this process the texture and color of the particle surfaces can be varied.

This can be done by allowing a surface thawing of the ribbons in the extruder die head. On such products as coffee extract this surface thawing can result in a darker color. The surface appearance can also be changed by plating the ribbons of slush as they are extruded. A water spray for example would tend to quickly freeze on the surface of the ribbons and prevent potentially tacky particles from sticking to each other.

An extract spray would tend to have the same advantages as the water spray and in addition can effect the color and appearances of the dry particles. Alternatively, the slightly tacky surface of the extruded ribbons could be coated with dry product in order to achieve unique surface effects.

Example 1: Coffee extract containing 25% solids by weight of the extract was chilled in a continuous scraped surface, heat exchanger. The resultant slush was chilled to a temperature of 20°F prior to extrusion. This slush was found to be a hard slush suitable for extrusion and contained 44% ice by weight of the slush. The liquid portion of the slush had a concentration of 44.5% solids by weight of the liquid.

The slush was extruded in the form of $1/16$-inch rods into an atmosphere of −30°F air. The rods were cut with an oscillating wire to produce $1/16$ to $1/8$-inch long pieces. These pieces were thoroughly frozen to −30°F and collected in a hopper. The frozen pieces were then loaded into trays and freeze dried at a vacuum of less than 500 microns of mercury to a moisture content of 2.5%. The resultant product was very regular in appearance, easily handled in packing operation, and essentially free of dust.

Example 2: Coffee extract containing 25% solids was freeze concentrated to produce an extract containing 35% solids by weight of the extract. The concentrated extract was then treated as in Example 1. The slush was chilled to a temperature of 15°F and the resultant hard slush contained 30% ice crystals by weight of the slush. The solids concentration of the liquid portion of the slush was 50%.

It should be noted that a hard slush in this example was attained with only 30% ice as compared to 44% ice in Example 1 because the solids concentration of the initial extract was greater. The resultant dry product was again found to be very uniform and easily handled in packing operations.

Slow Partial Initial Freezing, Quick Completion

In this process of *B.E. Elerath and E. Pitchon; U.S. Patent 3,373,042; March 12, 1968; assigned to General Foods Corporation* coffee extract is partially frozen with agitation into a slush in a first conductive cooling zone, the slush is then placed in a second conductive cooling zone for completion of the freeze and the completely frozen product is then freeze dried.

The extract in the first zone can be frozen to an extent wherein the extract retains its shape and form due to partial freezing by any dynamic freezing method, e.g., agitation of the body of coffee extract with a stirrer or agitator means or a swept-surface heat exchanger wherein a thin film of coffee extract is distributed and frozen onto a cylindrical heat exchange surface and continually scraped off this surface by a series of revolving scraper blades. The cooling temperature of the heat exchange surface in the first zone is usually in the range of 15° to 25°F, and cooling is continued generally until a product temperature of between 15° to 20°F is achieved. At this product temperature the ice content of the extract will be above 20%, more usually about 30 to 45%, and the extract may be molded to any shape and form having maximum surface area available for drying while being capable of easy handling.

During the partial freezing of the coffee extract in the first cooling zone, it is essential that the extract be continually agitated or stirred to assure a homogenous blending of the aromas in the coffee, soluble coffee solids and newly formed ice crystals. This agitation of the extract must be carried out in the absence of atmospheric conditions, and is preferably blanketed with an inert gas, such as nitrogen, argon, or carbon dioxide during the partial freeze.

During the partial freezing of the coffee extract in the first cooling zone, it is preferred to slowly freeze the extract in order to develop a larger ice crystal growth thereby assuring a darker colored freeze dried product. A suitable initial freezing time in the first cooling zone would be 15 to 20 minutes and at least 10 minutes to partially freeze at least 20% of the available water. Heat removal should be controlled within this stage to a range of 1 to 3 calories per g per cc per minute and preferably less than 1.5 calories per g per cc per minute. Completion of the freezing may then be accomplished rather quickly in a period of 5 to 10 minutes to achieve a suitable dark color which approaches the dark color of roasted and ground coffee.

Alternatively, the coffee extract may be frozen initially at a rapid rate of 1 to 3 minutes and then may be frozen more slowly in the second cooling zone over a period of about 20 to 30 minutes and at least 15 minutes, the heat removal being at the rate of less than 3 calories per g per cc per minute and preferably about 1.5 calories per g per cc per minute.

Example: Coffee extract containing about 1% by weight expressed coffee oil and about 1.5% of steam volatile aroma was passed through a scraped surface heat exchanger consisting of a 2 foot long stainless steel screw in a jacketed 2 foot by 2 inch diameter shell. Propylene glycol solution chilled to $-30°F$ was circulated through the jacket.

A batch of 4 gallons of extract was recycled through the heat exchanger for 35 minutes at which time the effluent had acquired the consistency similar to soft ice cream. The ice crystals appeared to be slightly larger than ordinary table salt or sugar. The slush was somewhat thixotropic (flowed through the exit pipeline, but held its shape when at rest). The batch was divided into two parts and further frozen until completely solid in about 20 additional minutes. One portion was formed as a ½" thick slab, the other as a ribbed slab having the same average weight per area as a ½" slab, but with ½" x ½" ribs extending above a ¼" thick base.

Each of the above slabs was dried in a freeze dryer at the maximum rate such that the product ice temperature did not exceed $-10°F$ and the dried portions did not exceed 90°F. The slab dried in 15 hours and the ribbed slab in 9 hours. Control samples of ½" slabs formed by freezing similar extract without the slushing step were dried in 17 to 19 hours using the same drying criteria.

Examination of the dried product showed that the usual surface film of an oily or waxy material was not present. The cross sectional color was uniformly dark. When ground through an 8 mesh screen the product appearance was that of porous, irregularly shaped but somewhat spherical chunks. The static frozen liquid extract when dried tended to fracture into platelets of thin cross section.

Separation of Ground Frozen Extract into Coarse and Fine Fractions

J.L. Anderson, J.A. Gaedtke and A.R. Mishkin; U.S. Patent 3,573,929; April 6, 1971; assigned to Societe d'Assistance Technique Pour Produits Nestles SA, Switzerland provide a process for preparing dry tea and coffee extracts, comprising solidifying an aqueous tea or coffee extract by freezing, grinding the frozen extract and separating two fractions of ground material, freeze drying one of the fractions, compressing the second fraction at low temperature to provide a homogeneous solid, subdividing the solid and freeze drying at least a part of the subdivided solid.

Preferably, the particle fraction which is compressed comprises the undesired fine particles, which in the case of coffee are those which pass through a sieve of 0.2 to 0.4 mm aperture. Preferably, the particles are compressed continuously into a solid such as a ribbon or tube which is subsequently subdivided. After screening the subdivided solid, the required fraction is added to the material retained by the initial screening and the mixture is freeze dried.

Example 1: An aqueous coffee extract containing 28% soluble solids is frozen to $-45°C$ in 60 minutes. The frozen product is then broken and ground, after which the ground extract is screened on two sieves of 0.3 and 1.8 mm aperture respectively. The fraction retained by the 1.8 mm sieve is returned to the grinder, the product passing this sieve and retained by that of 0.3 mm is freeze dried and the particles smaller than 0.3 mm are collected. This fraction represents about 25% by weight of the total subdivided product.

The fine particles are compressed in the frozen state, in a tablet press. The pressure in the press is 200 kg/cm^2 and the temperature of the product is held at $-30°C$. After compression, a homogeneous frozen solid is obtained having the form of a disc or tablet 20 cm in diameter and 15 mm thickness.

The discs obtained are broken up, ground and, after sieving the ground product, the fraction retained by the 0.3 mm sieve is freeze dried. The dry extract particles from the first and second sieves are thoroughly mixed and a dried coffee extract having a specific gravity of from 200 to 300 g/l is obtained.

Example 2: A coffee extract containing 45% soluble solids is frozen to about $-3°C$ and carbon dioxide is blown into the extract until homogeneous foam is obtained. This foam is frozen in about 10 minutes on a refrigerated metal belt to obtain a solid product having a temperature of $-45°C$. The frozen extract is broken up, ground and sieved as described in Example 1, except that the fraction retained by the 0.3 mm sieve is not dried immediately and that the fraction of fine particles represents about 28% of the total ground product.

An extrusion apparatus is fed with the fine particle fraction, which is compressed continuously. The temperature inside the compression chamber is maintained at about $-30°C$ and the extrusion head produces a ribbon of frozen extract of

10 mm thickness and a width of about 10 cm. The ribbon is a homogeneous solid, and has substantially the same appearance as the initial frozen extract. However, its density is slightly higher.

After breaking and grinding, the subdivided extract which results from the compression of the fine particles is sieved. The fraction retained by the 0.3 mm sieve is added to that retained during the first sieving and the mixture is dried by sublimation. A freeze dried coffee extract is obtained having the appearance of roast and ground coffee, practically free from fine particles and having a specific gravity between 200 and 240 g/l.

Composite Porous Material Consisting of Continuous Phase and Dispersed Phase

L.R. Rey, M. Dousset and F. Chauffard; U.S. Patent 3,579,360; May 18, 1971; assigned to Societe D'Assistance Technique pour Produits Nestle SA, Switzerland provide a composite material comprising a continuous phase consisting essentially of a porous structure of a dry substance, there being included within the pores of the structure at least one further substance, different from the first, which has been dried by lyophilization "in situ."

The process is especially suitable for the preparation of lyophilized coffee extracts, starting, for example, from an emulsion consisting of an aqueous solution of coffee solubles and a solution of coffee aromatics and coffee oil in an organic solvent such as a fluorinated hydrocarbon (Freon). The emulsion is frozen rapidly in relatively thin layers and lyophilized under vacuum in a suitable apparatus.

Alternatively, the starting material may be a homogeneous system consisting of an aqueous solution of coffee solubles and a solution of the aromatic and lipid constituents of coffee in dioxan. The two solutions are stirred together and the system frozen in thin layers at −40° to −50°C. The frozen system is then lyophilized under vacuum until all the frozen liquid phase has been sublimed.

In a further modification, the lipid and aromatic coffee constituents may be dissolved in liquid carbon dioxide instead of dioxan. This solution is mixed at a pressure of at least 5.5 kg/cm^2 with an aqueous solution of coffee solubles. By slowly lowering the pressure, at least partial freezing of the mixture may be induced, which may be completed by outside refrigeration. Lyophilization may be effected under vacuum or at atmospheric pressures.

The coffee extract is obtained as a porous product of finely heterogeneous structure aromatized in depth with aromatic and lipid constituents of coffee. When powdered extracts are being prepared, it is preferable to grind the frozen material before lyophilization.

Example 1: A concentrated aqueous solution of coffee solids (about 45%) is transformed into a foam by blowing in gaseous carbon dioxide. The foam is rapidly frozen at about −45°C in relatively thin plates which are freeze dried to a dry product having a porous structure. The plates of freeze dried coffee are impregnated with a concentrated solution of the lipidic and aromatic constituents of coffee in carbon tetrachloride or Freon 113 so as to obtain a concentration of 0.5 to 1% of these constituents in the final product.

The impregnated plates are frozen at a temperature of about −70°C and the frozen diluent is eliminated by vacuum sublimation in a conventional freeze drying apparatus. In a modification of the process described above, the porous dry coffee extract may be impregnated at a pressure not below about 5.5 kg/cm^2, with a solution of lipidic and aromatic constituents of coffee in liquid carbon dioxide.

By lowering the pressure a spontaneous freezing of the diluent is obtained and it is left to sublime under atmospheric pressure. Freezing may be accelerated by using a refrigerant.

A dry coffee extract is obtained in the form of a porous material having a specific gravity between 0.2 and 0.3 and which contains, in a continuous phase, the soluble constituents of coffee and, in the dispersed phase, the aromatic and lipidic constituents of coffee which are nevertheless intimately mixed within the pores of the continuous phase.

Example 2: 43.5 kg of roasted and ground coffee are moistened with 4.5 kg of water and stripped with steam until 4.35 kg of concentrated aroma are obtained. 370 ml of this solution are twice extracted with 23 ml portions of Freon 113 and 7 g of coffee oil are dissolved in the organic phase. 45 ml of the first solution are emulsified in 3.6 kg of aqueous coffee extract containing 45% coffee solubles and the emulsion is rapidly frozen at about −70°C. The plates of frozen extract are broken up and ground to a particle size between 0.25 and 2.0 mm.

The frozen particulate extract is then lyophilized in a suitable apparatus under reduced pressure (5×10^{-2} torr). Complete removal of the liquid phase takes about 6 hours, whereupon a powdered coffee extract consisting of porous particles is obtained.

Example 3: 800 ml of aqueous coffee extract containing 20% coffee solubles are mixed with a solution of 1.6 g of coffee oil and 20 ml of aromatic coffee distillate in dioxan. 100 cc of the mixed solution are rapidly frozen at a temperature of −40° to −50°C in the form of a thin film which is then lyophilized at reduced pressure (7×10^{-2} torr). The lyophilization takes about 1½ hours whereupon a lyophilized coffee extract is obtained, having a porous structure and containing the aromatic and lipid constituents of coffee in intimate admixture with the water-soluble constituents.

Shock Freezing of Sprayed Drops with Cryogenic Gas

The method developed by *J.W. Casten and S.H. Shimabuku; U.S. Patent 3,573,060; March 30, 1971; assigned to Hills Bros. Coffee, Inc.,* consists of preparing a concentrated coffee extract (20 to 60% solids) by any suitable method. The coffee concentrate is then instantly or shock frozen by being discharged from a spray nozzle into drops which are passed into a low temperature zone created by a surrounding spray of liquid cryogen gas (preferably liquid nitrogen).

The last step is preferably carried out in a free falling system at atmospheric pressure. The particles of frozen drops fall under gravity and are collected after which they are freeze dried to remove the moisture content by sublimation.

The concentrate sprayer can be of any suitable type which will provide a hollow cone spray pattern. The concentration of the coffee extract concentrate determines the size of nozzle which should be used for the instant spray freezing step prior to freeze drying so that bulk density of from about 0.20 to 0.25 per cc is achieved.

Varying extract line pressure influences the color of the freeze dried coffee by spraying a varied quantity of coffee concentrate per unit time. The product color differs in that at the low line pressure the Agtron reading is 10 and at the higher line pressure the Agtron reading is 29 (Agtron reading increases as product becomes darker). The Agtron color test is based on the light reflection principle wherein the percentage of 640 mμ light reflected by the sample is measured on the Agtron Model F Color Meter. The color range of 72 to 75 is approximately that for commercial, ground, roast coffee.

The freeze drying process conditions are as follows: Chamber pressures are maintained below 4.0×10^{-1} torr; condenser temperature at $-50°F$ and plate coil temperature maintained at $200°$ to $210°F$ or about 12.0 to 14.5 psia steam pressure. The drying time varies with the amount of water to be removed. At concentration of 30% solids the drying time is 6 hours, at 50% concentration the drying time is 4½ hours, etc.

The product temperature is controlled to prevent melting during the sublimation period. After sublimation has ceased, the product temperature is allowed to rise to 100°F for an hour before the drying cycle is complete. The properties of the resultant products under preferred conditions are set forth in the following table.

Sample	Nozzle Size, inches diameter	Solids, %	Sprayer Nozzle, psig	Drying Time, hours	Density of Dried Product, g/cc	Color of Dried Product, Agtron	Ratio Extract to Liquid Nitrogen
A	0.136	20	<5	6	0.20	12	1:1
B	0.128	30	<5	5.75	0.18	15	1:1
B	0.136	30	<5	5.5	0.22	25	1:1
C	0.081	40	15	5	0.18	80	2:1
D	0.081	50	25	4.25	0.21	47	1.7:1
E	0.0595	55	100-150	4.25	0.20	61	–
E	0.0595	55	100-150	4.25	0.23	65	–
E	0.0595	55	125	4.5	0.22	–	–
E	0.0595	60	175	4	0.20	48	1:2.1
E	0.0595	60	150	4	0.21	19	1:4.1
E	0.0595	60	150	4	0.21	82	–

Example: Preparation of a Commercial Instant Coffee — A high quality coffee which is freeze concentrated is used to form one component of the extract concentrate. The other portion of the extract concentrate is made from a lower quality coffee which may be evaporatively concentrated to a higher solids content than is commonly achieved with freeze concentration.

Normally, the volatiles lost in evaporative concentration represent loss of desirable constituents but in the case of somewhat lower quality coffees, the loss of such volatiles may well be an improvement. Roast coffee of high quality, labeled A, is processed through typical milling, extraction and freeze concentration steps. The volatiles that are evolved in these steps are captured. Other flavor and aroma volatiles may be supplied as desired. These constituents are held out of the process until the dried product has been developed.

In a separate operation, roast coffee of somewhat lower quality labeled B, is subject to milling extraction and evaporative concentration operations to a solids content higher than about 25%. The coffee concentrates from A coffee and B coffee are then mixed to the desired solids concentration (20 to 60%) so as to be suitable for the spray feeezing operation previously described.

The solids concentrate is then sprayed into drops through a low temperature chill zone to shock freeze the drops into discrete frozen particles in the manner described. A suitable liquid cryogen is sprayed to create the low temperature zone. The frozen particles are collected and freeze dried to sublimate their ice content and the resulting product may be directly used as a dried coffee extract product or ground to separate the discrete particles that have agglomerated to assure that the dried product has a size less than that which will pass through #10 mesh screen.

The structure of the freeze dried product of the process, because of its porous structure, has an extraordinarily high affinity for sorbing volatile flavor and aroma constituents. Accordingly, the dried product is contacted with the volatlle constituents which were previously collected and aromatized by sorbing such volatiles.

Spray Freezing Using Two-Fluid Nozzle

A.R. Mishkin and W.S. Symbolik; U.S. Patent 3,620,776; November 16, 1971; assigned to Societe d'Assistance Technique Pour Produits Nestle SA, Switzerland provide a method of spray freezing and drying of liquids containing up to 60% by weight of solids, which comprises spraying the liquid into a zone of subatmospheric pressure in a current of gaseous fluid thereby to form frozen particles of the liquid and subsequently freeze drying the frozen particles.

The liquid is preferably sprayed through a two-fluid nozzle of the suction type together with the current of gaseous fluid and its path intersects that of the fluid at a short distance from the nozzle. In this manner the fluid breaks up the liquid jet into fine particles and at the same time prevents its lateral dispersion and inhibits the freezing of liquid around and on the nozzle.

The gaseous fluid may be any gas which is inert with respect to the liquid being dried, such as air, nitrogen or an inert gas. A condensable gas, especially steam, is preferred since it may be easily evacuated from the system by condensation on a cold surface thereby maintaining a suitably low pressure within the spraying zone.

As shown in Figure 9.3 the apparatus comprises a vertical freezing chamber **1** of metal construction, with cylindrical walls and conical bottom. The dimensions of the chamber may, for example, be 3 meters in diameter and 10 meters in height.

At the upper end of the freezing chamber is mounted a spray system comprising a two-fluid nozzle **2** and feed lines **3** and **3a** for supplying, respectively, the solution to be dried and the gaseous fluid which is preferably steam. At the bottom of the chamber is a hopper **4** if the chamber is directly connected to a freeze drying chamber or an air-lock if the freeze drying chamber is not directly connected.

FIGURE 9.3: SPRAY FREEZING USING TWO-FLUID NOZZLE

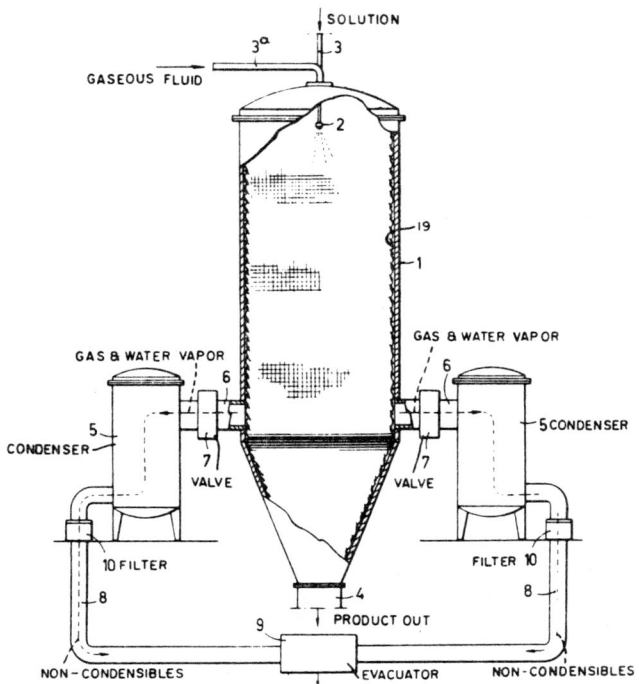

Source: U.S. Patent 3,620,776

Ducts **6** connect the lower part of the chamber to condensers **5**, disposed symmetrically. These may be two, four or six in number, depending on their dimensions and those of the chamber. A valve **7** is mounted on each duct so that each condenser may be individually isolated for cleaning or defrosting without necessarily breaking the vacuum within the chamber.

The condensers **5** are also connected by ducts **8** to a vacuum system **9** comprising one or more pumps of adequate capacity to maintain an absolute pressure within the freezing chamber of 50 to 500 microns. When steam is used as the gaseous fluid, the pumping system need not be excessively large since the water vapor formed can be condensed out of the system and need not be pumped.
A filter **10** is mounted on each duct **8** between the condenser and vacuum system to trap any fine particles which may have been sucked out of the chamber.

Freezing of the particles of liquid formed at the upper end of the chamber **1** should be rapid. When the liquid to be dried is a coffee extract containing 30 to 50% by weight of dissolved coffee solids, the particles should be frozen solid after a free fall lasting about 0.5 to 1.5 seconds. Particles of a tea extract of similar concentration require about 0.5 to 1.0 seconds.

Example 1: Tea — A concentrated extract containing 65% by weight of soluble tea solids is diluted to a concentration of 43% by addition of an aromatic tea distillate. This solution is preheated to about 55°C and sprayed through a two-fluid nozzle (Spraying Systems, No. 60.100) into a freezing chamber as described in Example 1. The tea solution is sprayed at a pressure of 0.21 kg/cm^2, simultaneously with steam at a pressure of 0.7 kg/cm^2. The absolute pressure within the chamber is maintained at 180 microns of mercury.

A small quantity of gaseous carbon dioxide is incorporated in the liquid extract to adjust the density of the finished product. The particles produced by spraying are frozen solid when they reach the bottom of the chamber after a fall of some 10 meters. They are collected in a hopper and distributed automatically on the transporter of a continuously functioning freeze drying unit which is directly connected to the freezing chamber.

The diluent is sublimed and the product reaches the end of the transporter in the form of a freeze dried tea extract containing around 3% by weight of moisture. It is removed through an air-lock and packed in suitable containers. The product is slightly flaky in appearance and has an attractive color; on reconstitution with water, the product has a flavor and aroma which are superior to conventional spray dried tea extracts.

Example 2: Coffee — An aqueous solution containing 45% by weight of soluble roasted coffee solids is preheated to a temperature of about 60°C and sprayed through a two-fluid nozzle into a freezing chamber in which an absolute pressure of 100 microns of mercury is maintained. The nozzle is placed at 0.76 meters below the top of the chamber, which has a diameter of about 3 meters. The coffee solution is sprayed at a pressure of 0.35 kg/cm^2 simultaneously with steam at a pressure of about 2.1 kg/cm^2.

Under these spraying conditions liquid particles or globules between about 500 and 300 microns in diameter are produced. These particles are frozen to a solid state by the time they arrive at the bottom of the freezing chamber from which they are removed through a suitable air-lock. The frozen product is then distributed on trays and dried in a batch freeze drying unit. The resulting dry coffee extract has attractive dark brown color and a density of 190 g/l.

Particles of Concentrate Frozen into Prills by Refrigerated Gas

In a process developed by *K.M. Grover and N. Ganiaris; U.S. Patent 3,431,655; March 11, 1969; assigned to Struthers Scientific and International Corporation* a concentrated solution of the coffee to be freeze dried is formed into relatively uniform droplets or particles in the top of a column. A circulating current of a refrigerated gas is passed through the column to freeze the falling liquid particles to form prills. These prills are collected and freeze dried by the sublimation of their frozen liquid contents. The process is described with reference to Figure 9.4.

The coffee solution to be freeze dried is introduced through feed pipe **10** into a conventional crystallizer **11**. The refrigerant **13** enters jacket **14** through pipe **15** and is withdrawn through pipe **16**. The coffee solution and crystal slurry thus formed passes to centrifuge **18** which removes ice crystals through pipe **19**. Concentrated and cooled coffee solution enters the liquid storage tank **20**.

FIGURE 9.4: PARTICLES OF CONCENTRATE FROZEN INTO PRILLS BY REFRIGERATED GAS

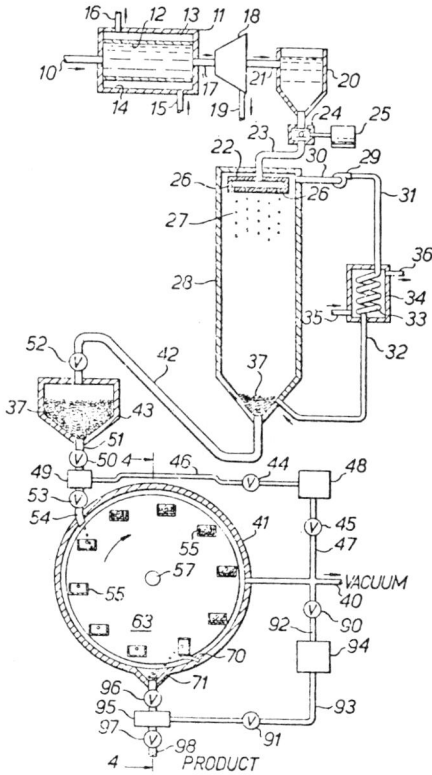

Source: U.S. Patent 3,431,655

Tank **20** feeds coffee solution to the prilling head **22** through pipe **23**. A rotary valve **24** driven by motor **25** periodically interrupts the flow of liquid through pipe **23** to provide periodic pulsations of pressure in the prilling head **22**. The pulsations of pressure may vary from two to thirty per second. Apertures **26** in the bottom of prilling head **22** allow uniform and fairly large droplets **27** of coffee solution to form which fall downward within prilling column **28**.

A conduit or duct **30** leads from the top of column **28** to blower **29** which forces a gas through ducts **31** and **32** and coil **33**. Coil **33** is disposed in a shell **34** into which refrigerant is introduced through pipe **35** and from which it is withdrawn through pipe **36**. Thus, it may be seen that a gas is flowed upward in column **28** to be withdrawn, refrigerated, and recycled. This gas, preferably nitrogen which is inert, freezes the falling droplets **27** to form small spherical beads or prills **37** of coffee solution which collect at the bottom of column **28**. The gas used to freeze the prills **37** is preferably at atmospheric pressure.

Line **40** leads from sublimation cabinet **41** to a vacuum source. Line **42** extends from the bottom of column **28** to a cold insulated hopper **43** for frozen prills **37**. If the valves **45** and **44** in the vacuum lines **47** and **46** are successively opened and closed, a vacuum will be drawn in the vacuum reservoir **48** which, in turn, will draw a vacuum in the prill charge metering leg **49**. When valve **50** is opened, at least a partial vacuum will be drawn in hopper **43** and frozen prills will slide downward to fill leg **49**. Valve **50** is then closed. The partial vacuum in hopper **43** may be used to draw prills from the bottom of column **28** into hopper **43** as in conventional air slide delivery devices.

Valve **53** in line **54** is opened to fill each tray **55** with prills as it moves under a laterally flared end of line **54**. As each tray **55** of prills revolves, a suitable time elapses to vacuum dry or otherwise allow sublimation of the prills. A spring arm **70** catches passing trays **55** to dump the dried prills into a product hopper **71** formed in the bottom of sublimation cabinet **41**.

To assist in the sublimation of the prills in the trays, a low capacity heating element may be laid in the form of coils in the bottom of each tray. To remove the product, sublimated or freeze dried prills of coffee, valves **90** and **91** in lines **92** and **93** are successively opened and closed to draw a vacuum in vacuum reservoir **94** and the product delivery leg **95**. Valve **96** is opened to allow the product to fall into product leg **95** and then closed. Valve **97** may then be opened to allow the product to fall from line **98**.

FACILITATING SEPARATION OF FROZEN EXTRACT FROM RETAINING SURFACE

Freezing Precoat of Aqueous Film

Before freeze drying coffee it is necessary to cool the roasted extract to below the eutectic temperature of the coffee in preparation for the freeze drying step. In freezing, the extract comes in contact with a belt or tray which retains the extract in the desired shape and form during the freezing operation. The surface of the belt or tray which comes in contact with the frozen extract many times sticks or adheres to the extract and presents a separation problem in removing the frozen extract from this surface.

R. De George; U.S. Patent 3,253,420; May 31, 1966; assigned to General Foods Corporation found that adhesion of frozen coffee extract to a retaining surface is avoided by applying a film of an aqueous liquid on the surface to thereby coat the surface, freezing the film, applying a body of liquid coffee extract having a temperature sufficiently low to avoid melting of the frozen film to the coated surface, and maintaining the film in a frozen state while freezing the coffee extract. Figure 9.5 shows the freezing belt with the side-skirts for retaining the liquid extract on the freezing belt until it is frozen.

Usually a belt having a length of between 20 to 120 feet and a width of 1 to 5 feet is sufficient to freeze up to a one inch layer of coffee extract slowly over a period of at least 15 minutes so as to give a dark colored final product.

Example 1: Coffee extract having a soluble solids concentration at 27% and product temperature of 37°F was applied to a moving stainless steel belt about

FIGURE 9.5: FREEZING PRECOAT OF AQUEOUS FILM

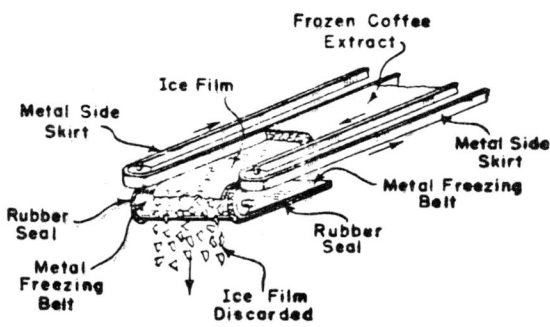

Source: U.S. Patent 3,253,420

3 feet from the starting end of the belt. The stainless steel conveyor belt was 2 feet wide, was operated by 2 pulleys located about 50 feet apart and provided a 60 minute contact time for the coffee extract to be completely frozen. The belt was cooled by contact with 2 cold brine tanks located on the underside of the belt, one tank having a temperature of 20°F and the other tank a temperature of −30°F. The first half of the belt using a 20°F brine solution had a contact time of 30 minutes while the second half of the belt employing a −30°F brine solution also had a contact time of 30 minutes.

Additional freezing was provided by refrigerated fins having a temperature of −35° to −25°F located 3 inches above the coffee layer. The coffee layer was applied at a thickness of about 0.5" and the liquid extract was retained by rubber side skirts operated by terminal pulleys at opposite ends of the belt. The skirts moved at the same rate as the belts.

The thin water film was applied to the initial portion of the belt before ntroducing the coffee extract. This was accomplished by spraying water onto the belt in a thin film having a thickness of about 0.01 inch (preferred range 0.01 to 0.04 in). The thin water film was frozen instantly by passing the steel belt over copper coils which contained −35°F brine. Under these conditions the steel belt had a frozen film of water thereon when the belt contacted the body of liquid coffee extract. The frozen film was retained in a solid state while the body of coffee extract was cooled from its ice point of about 27°F to below its eutectic point of −13.5°F in a period of 60 minutes.

At the terminal portion of the belt it was observed that the 0.5" slab of frozen coffee extract parted easily and cleanly from the steel belt as it curved around the terminal pulley. The thin layer of frozen water fractured and fell cleanly off the belt. The slowly frozen extract was conveyed directly as a slab into a commercial freeze drying unit wherein the frozen extract was freeze dried by sublimation at a vacuum of below 500 microns to a stable moisture content. The dried product was uniformly dark brown and very coffee-like in color.

Example 2: The procedure of Example 1 was followed with the exception that stainless steel sheet metal side-skirts were used in place of the rubber side-skirts. The steel side-skirts were 0.6 inch in height, had a thickness of 0.1 inch and included an 0.2 inch rubber lip adapted to contact the freezing belt to provide a watertight seal. The two steel side-skirts were each operated by a pair of terminal pulleys adapted to move the side-skirts at the same rate as the freezing belt.

The steel side-skirts were cooled by brine tanks having a temperature of $-40°F$ at the initial portion of the belt and a thin water film was applied to the side skirts at the same time the film was sprayed onto the freezing belt to give a frozen ice film about 0.01" thick. Under these conditions the frozen films on the side-skirt and freezing belt did not melt when $37°F$ extract was applied to the belt and slowly frozen. The frozen slab of coffee extract was found to separate cleanly and easily at the terminal portion of the belt and provided a distinct improvement over the rubber side-skirts of Example 1 which had no ice coating.

Cooling Below $-50°F$ and Flexing Retaining Surface

The process of *H. Guggenheim; U.S. Patent 3,408,919; November 5, 1968; assigned to General Foods Corporation* is founded on the discovery that separation of coffee extract which has been frozen to a retaining surface at mild freezing temperatures may be facilitated by a process which comprises applying a body of liquid coffee extract to the retaining surface, cooling the extract from its ice point to below its eutectic point at freezing temperatures of above $-20°F$ to thereby completely freeze all the water in the extract, further cooling the extract to a temperature below $-30°F$ (preferably $-50°F$) and then flexing the retaining surface to obtain release of the frozen coffee extract.

Preferably, the extract is cooled from its ice point to below its eutectic point over a period of at least 15 minutes, say 30 to 45 minutes in order to assure a darker, more coffee-like color in the final dried product. If the extract is cooled sufficiently low, $-60°F$ or below, little or no flexing of the belt or retaining means is necessary to obtain release.

Example 1: Coffee extract having a soluble solids concentration of about 27% and a product temperature of about $35°F$ was applied to the upper surface of a freezing belt about 1 to 3 feet from the starting end of the belt. The freezing belt was made of stainless steel 2 feet wide and was operated by 2 pulleys located about 60 feet apart. The belt was adapted to move through a freezing zone over a total period of 60 minutes. The extract was frozen by contacting the steel belt with 3 cooling tanks containing cold brine. These tanks were located on the underside of the freezing belt. Three tanks were used, each 20 feet long, the first tank having a temperature of $20°F$, the second tank having a temperature of $-20°F$ and the third tank a temperature of $-40°F$.

The initial one-third of the belt using a $20°F$ brine solution for a coolant had a contact time of about 20 minutes. The second one-third of the belt, employing a $-20°F$ brine solution, had a contact time of 20 minutes. The final one-third of the freezing belt also had a contact time of 20 minutes, but the coolant had a temperature of –40°F. The coffee layer was applied at a thickness of about ½" and the liquid extract was retained on the freezing belt by rubber side members operated by terminal pulleys at opposite ends of the belt. The side members moved at the same rate as the freezing belt.

A second freezing belt was used as a control. The control belt employed three brine tanks having a 20°F temperature for the first brine tank and –20°F temperature for the second and third brine tanks. Total freezing time was 60 minutes and contact time with each brine tank was 20 minutes.

The frozen extract, having a product temperature of between 35° and 40°F, was found to release quite easily in the form of slabs from the –40°F cooled belt surface as the belt curved around its terminal pulley with only slight sticking of the frozen extract to the freezing surface. The control sample, frozen to the –20°F cooled belt surface, experienced serious stickage problems and severe breakage of the frozen extract as it was separated from the freezing surface. The final product when ground and freeze dried according to this procedure, had a dark, coffee-like color.

Example 2: The procedure of Example 1 was followed with the exception that the third brine tank employed brine solution having a temperature of –60°F. The control sample was kept the same.

The frozen extract having a product temperature of between 55° and 60°F at the terminal portion of the freezing belt was found to release easily and cleanly from the freezing belt. Release of the frozen extract was easily accomplished without flexing the belt by merely breaking the frozen extract in the form of 3' x 2' x ½" slabs before the belt curved around its terminal pulley.

The control sample was very difficult to separate from the freezing surface before it curved around the terminal pulley. Separation of frozen extract was achieved only at the expense of severe breakage of the slabs due to sticking of extract to the freezing surface of the belt.

DRYING MODIFICATIONS

Increasing Drying Rate by Disrupting Surface of Frozen Slab

B.E. Elerath; U.S. Patent 3,443,962; May 13, 1969; assigned to General Foods Corporation found that the rate of vacuum freeze drying frozen slabs of coffee, or other solids-containing liquids, is appreciably increased by scraping or disrupting the surface of the frozen slabs prior to freeze drying the slabs.

Disrupting, as used here, means removing, abrading, cutting, scratching, perforating, breaking, cracking or otherwise disturbing part of the surface film of a body of frozen coffee extract in order to provide a more direct and open path for the removal of water vapors by vacuum sublimation from the interior portions of the frozen coffee.

Example: A small laboratory size freeze dryer having a drying chamber of 15" x 24" x 24" was used with an external condenser maintained at –40° to –60°F. Slabs of roasted coffee extract having 26% soluble solids were cast in a mold using

−30°F brine. The slab dimensions were ½" x 6" x 6". Slabs of extract were placed on an open mesh screen made of expanded aluminum. Heat was supplied from oil heated platens above and below the product at a distance of ¾". The screen was arranged on an internal scale so that weight of product could be measured as drying proceeded. Thermocouples were placed in the center of slab so that temperature could be recorded. Pressure was reduced to less than 300 microns of mercury and heat applied at such a degree that the ice layer approached but did not exceed −10°F.

In the first sample, an unscraped coffee slab was dried and it was found that 120°F platen temperature initially resulted in a −11°F product ice temperature. Small modifications in the platen temperature were made as drying progressed to keep the frozen coffee at about −10° to −12°F. When the ice layer had disappeared, the platen temperature was reduced to 110°F and maintained at this temperature until no further weight loss was noted.

The weight of water removed was plotted against time of drying so that percent of total water times original thickness in sixteenths of an inch was called the thickness dried. It was found that until the ice layer disappeared that drying time was controlled by the formula $T = 1.3L$ where T was the drying time in hours and L the thickness in sixteenth inches. The overall drying time was 17.5 hours and this corresponds to $T = 2.2L$.

In the second sample, the same technique was used except that the top surface of the slab was completely scratched to a depth of about $1/32$ inch with a cooled hacksaw blade. The loosened material remained on the surface and became frozen to it. In this case, platen temperatures of about 350°F were inadequate to provide enough heat and the ice temperature remained at −12° to −14°F. It was found that the rate of drying was controlled by the formula $T = 0.9L^2$ until disappearance of the ice layer occurred. Overall drying time for this slab was 9.6 hours and this corresponded to $T = 0.15L^2$.

Increasing Drying Rate by Compressing Frozen, Granulated Material into a Porous Slab

H. Schwartzberg; U.S. Patent 3,468,672; September 23, 1969; assigned to General Foods Corporation describes a process for increasing the rate for removing water vapor from a frozen extract during vacuum freeze drying without substantially decreasing the heat transfer rate of the frozen extract.

This comprises grinding the frozen extract, concentrate, puree, juice or other material to be dried into small granules, compacting the particles to form a porous slab and subjecting the porous slab to vacuum freeze drying to sublime the frozen liquid in the extract. The term compacting is meant to include forcibly compressing fritted material into a porous slab. The compacting can be carried out by the use of a ram and a mold, the compacting pressure being applied by either a hydraulic press or clamping device.

The size of the frozen granules should be large enough so that the interparticle pores in the subsequently generated porous slab are substantially (e.g., 2 to 3 times) larger than the pores produced within the particles by the sublimation of their ice content. Stated in other terms, the particle size should be large enough so that in conjunction with the degree of compaction used in generating the fritted slab there is little or no resistance offered to the efflux of sublimed water vapor.

Example 1: Porous slabs of a frozen coffee extract were made by compacting ground particles of the frozen extract. The frozen particles had a 28% soluble solids concentration. The compacting was carried out by the use of wooden rams and a wooden mold, the pressure being applied by a hydraulic press. The resultant porous or fritted slabs were one-half inch deep by three inches wide and three inches long. The slabs were made at porosities ranging from 10 to 30%.

The fritted slabs were then placed in a vacuum freeze dryer and dried at the maximum platen temperatures compatible with maintaining the slab surface temperature at or below 110°F. After eight to nine hours approximately 90% of the water had been removed and it is estimated drying would have been complete in 10 hours. In no instance was there evidence of meltback during the drying process. The temperature of the undried portion of the slabs remained at about −36° to −40°F with a box pressure of 40 to 50 microns, indicating very little resistance to vapor release.

Example 2: A porous slab was prepared from 45% extract by the process described in Example 1. The slab was five-sixteenth inches deep, three inches wide and three inches long. The slab had a porosity of about 20% and thus contained about as much soluble matter as in the previous example.

The slab was placed in a vacuum freeze dryer similar to that of Example 1 and dried at a slab surface temperature at or below 110°F. After two and one-half hours, about 90% of the water in the slab was removed and, according to the drying rate, the drying would have been completed in a total of about three to three-and-one-quarter hours. The drying was accomplished using a heating platen temperature of 260°F, which temperature is greatly in excess of the temperature which would normally cause melting in solid slabs of this concentration.

Rapid, High Temperature Freeze Drying

E.R. Hair and D.A. Strang; U.S. Patent 3,486,907; December 30, 1969; assigned to The Procter & Gamble Company found that the practice of maintaining relatively low temperature in the dried portion of the extract is not essential to the attainment of satisfactory flavor retention during the freeze drying process.

The process for rapidly freeze drying a frozen aqueous coffee extract comprises maintaining the frozen extract at a temperature of less than about −10°F and at an absolute pressure of less than about 500 microns of mercury, transferring heat to the frozen extract at a rate sufficient to promote sublimation of the frozen water in the extract, and varying the rate of transferring heat whereby:

(a) the maximum temperature of the dried portion of the extract exceeds about 120°F during some portion of the freeze drying process;

(b) the maximum temperature of the dried portion of the extract never exceeds about 200°F; and

(c) the length of time for which any part of the dried portion of the extract is held at and above any temperature below about 200°F does not exceed the time shown in Figure 9.6, curve A.

Freeze Drying of Coffee

FIGURE 9.6: RAPID, HIGH TEMPERATURE FREEZE DRYING

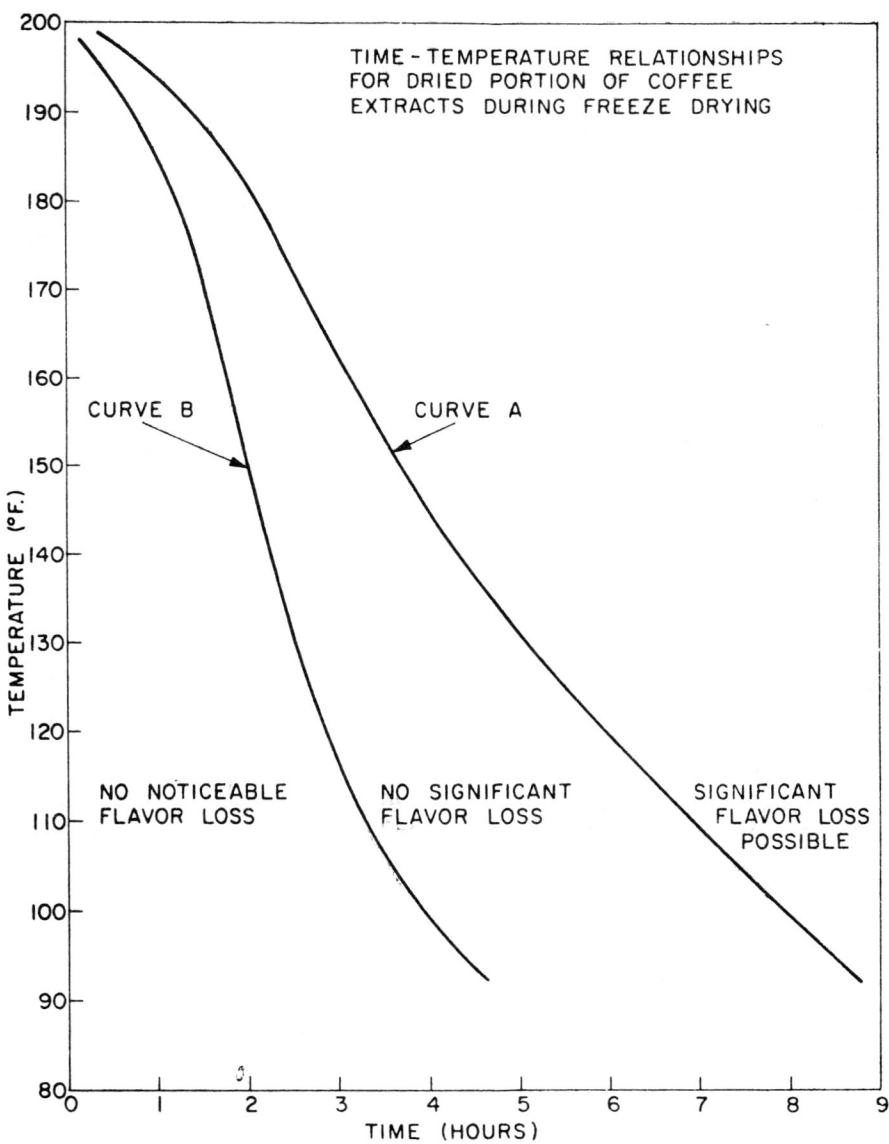

Source: U.S. Patent 3,486,907

Preferably, the length of time for which any part of the dried portion of the extract is held at and above any temperature below about 200°F does not exceed the time shown in Figure 9.6, curve B. In this manner freeze dried coffee is produced which has a flavor that is substantially identical to freeze dried coffee of the prior art even though a time savings in the freeze drying step of as much as about 35% is achieved.

Example: A 600 pound batch of roast and ground coffee (through #8 mesh; 95% by weight on a #20 mesh Tyler screens) was used to prepare an aqueous coffee extract in a conventional pilot plant size countercurrent extraction train. The extraction system consisted of 6 stainless steel columns, each 6 feet high and 6 inches in diameter, connected in series for continuous countercurrent operation. Each column held approximately 26 pounds of the roast and ground coffee. The inlet temperature of the water was 365°F; the extract cooled by natural heat losses as it passed through the system and was withdrawn at 210°F. The extract contained 25% by weight coffee solubles.

Twenty pounds of the extract obtained above were separated for freeze drying. The extract was poured into 8 inch by 12 inch aluminum trays to a depth of ¼ inch and the trays were suspended in a −35°F Dry Ice-acetone bath for 15 minutes to freeze the extract. The frozen slabs of extract were chopped by hand into small pieces (approximately ⅛ inch cubes) and the pieces were placed, at a loading factor of 0.75 pound per square foot, in the trays of a Sublimator 40 laboratory scale freeze dryer.

Thermocouples were placed on the top and bottom of the layer of frozen extract pieces. The absolute pressure in the freeze drying chamber was reduced to 100 microns of mercury absolute.

Heat was then transferred to the frozen particles via radiant heaters above the trays and conduction heaters below the shelf on which the trays were placed. The transfer of heat was controlled by the thermocouples so as to produce a maximum temperature in the dried portion of the frozen particles of 150°F. This temperature was maintained for one and one-half hours. The total drying time was two and one-half hours. The length of time for which the temperature of any part of the dried portion of the extract was held at and above any temperature below 150°F, did not exceed the time shown in Figure 9.6, curve B. The final moisture content of the freeze dried product was 3.5% by weight.

The freeze dried particles were removed from the dryer and tested to determine the degree of change in flavor character, if any, which occurred during the freeze drying operation. Sufficient water was added to the freeze dried particles to prepare an aqueous coffee solution with a concentration of 25% by weight coffee solubles, the same concentration as the original extract from which the freeze dried particles were prepared.

The flavor of the two coffee solutions was compared by an expert panel. The solution prepared from the freeze dried coffee was found to have an excellent brew-like coffee flavor which was substantially identical to the flavor of the original extract.

Two-Phase Freeze Drying at Different Temperatures and Pressures

D.E. Dwyer; U.S. Patent 3,653,929; April 4, 1972; assigned to General Foods Corporation provides an economical process for freeze drying coffee which makes the ultimate product flavorful and appealing in color. Percolated coffee extract (20 to 30%) is freeze concentrated (30 to 50% by weight), frozen, ground, and separated so that about 90% of the particles are sized from about 200 to 4,000 microns (most preferably 600 to 1,500 microns) prior to freeze drying. Insoluble solids are removed before or after freeze concentration. The ice separated in freeze concentration may be melted, concentrated and returned to the processing stream to recover solids contained therein.

The concentrated extract may be bulked before freezing to facilitate control of its flow and to optimize its granular structure for grinding. Bulking may be attained by inserting gas or fines or a combination thereof into the extract. The ground particles may be screened prior to freeze drying to obtain the aforementioned fines and to obtain a uniform particle size for freeze drying. Freeze drying is accomplished in two phases, under the following conditions:

(a) initially, employing high heat energy input at a platen temperature of about 200° to 250°F and an absolute pressure below about 500 microns until a particle moisture content of about 40 to 50% by weight is attained, generally in two to four hours, and

(b) finally, lowering the heat energy input at a platen temperature between about 100° to 120°F (preferably 110°F) and an absolute pressure of below about 200 microns until a stable particle moisture content of about 1.0 to 2.5% by weight is attained.

The freeze drying is accomplished in from about 10 to 16 hours. The condenser temperature in step (a) is maintained at approximately −50°F, in step (b) it is lowered to −60°F.

Use of Auxiliary Heaters to Avoid Formation of Liquid

The process of *F. Manaresi; U.S. Patent 3,242,575; March 29, 1966* relates to the lyophilization of products having a high moisture content, and in particular, but not exclusively to the lyophilization of coffee infusions. There is provided apparatus for lyophilizing a mass of material within a chamber connected to a vacuum source, wherein lyophilization is effected initially in zones more distant from the vacuum source intake and is subsequently effected in zones which are closer to the intake. This arrangement is designed to avoid any formation of liquid in the material submitted to lyophilization which would be harmful to the material, whose correct lyophilization must exclude any liquid stage.

Figure 9.7 shows a refrigeration chamber 1, which is capable of cooling the coffee solution to be lyophilized, for instance down to a temperature of −50°C. The lyophilization chamber 4 is provided with an access door 4a opposite the access door 1a of the refrigeration chamber. In a side wall 4b of the chamber 4, a suction duct 6 is provided which leads to a vacuum pump (not shown), the pump being capable of producing a high vacuum in the chamber and extracting the vapors resulting from the lyophilization process. On a wall 4c opposite the wall 4b are arranged heating means 7 in the form of serpentine elements or other heat radiating means. The chamber is also provided with cooling means **8** by which the chamber when it is at ambient pressure can be cooled to a temperature of as low as -20°C.

FIGURE 9.7: USE OF AUXILIARY HEATERS TO AVOID FORMATION OF LIQUID

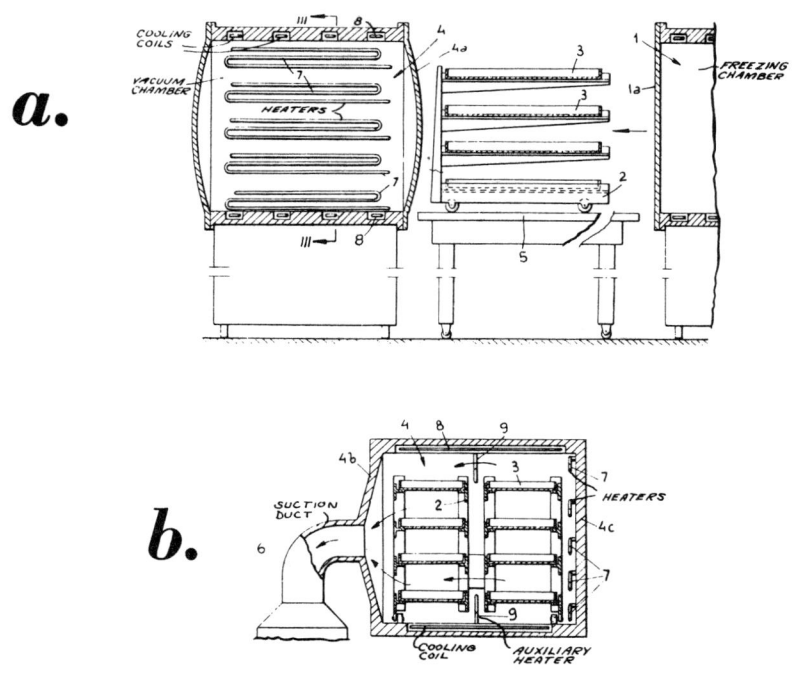

(a) Side Elevational View of Apparatus, Partly in Section
(b) Section Along Line III-III of Figure 9.7a

Source: U.S. Patent 3,242,575

Additional auxiliary heating means **9** (see Figure 9.7b) are provided intermediate the walls **4b** and **4c** lying in the upper and lower parts of the chamber in the region where the gap is provided in the central zone of the structure **2** referred to above. In use, the structure is inserted in the refrigeration chamber and brought to a temperature, for instance down to -50°C, but at least such that the trays **3** and the material therein contained assume a temperature of approximately -40°C. Simultaneously, the chamber is cooled by the cooling means **8**, to a temperature of -10° or -20°C, the pressure being maintained at ambient pressure. When the desired temperature is obtained in chamber **4**, the structure **2** with the tanks **3** is transferred from the refrigeration chamber **1** to chamber **4** for instance by transport on the trolley **5**.

With chamber **4** closed the low temperature is maintained and the vacuum pump is started to reduce the internal pressure substantially to zero. When this operation is completed, heating is started using only the side heating means **7** to effect lyophilization of that part of the material in the tanks **3**, which is near to the wall **4c**, where the heating means **7** is located, while the material in the tanks **3** nearer to the passage **6** remains practically solid but not heated.

Consequently, the vapors resulting from the lyophilization process adjacent the heating means are not condensed during movement through the chamber towards the passage **6**, owing to the practically complete absence of vapors in the region traversed. As the lyophilization process in the zone next to the heaters is completed, the lyophilization of the adjacent zones is started and subsequently completed.

In order to accelerate the lyophilization process, at a certain time after the start, for instance after two-thirds of the overall time required for the lyophilization and when the lyophilization of the material in the trays **3** next to the heaters is completed, heating is effected by the auxiliary intermediate heaters **9**; at the same time some of the heaters may be switched off and eventually all the heaters switched off. Thus, the lyophilization of the material in the trays **3** more distant from the heaters is accelerated, and the lyophilization process completed for the material in these trays.

After-Dryer for Drying Dust Recovered in Process

J. Schimpfle; U.S. Patent 3,936,952; February 10, 1976; assigned to Krauss-Maffei AG, Germany found that it is possible, without loss of a valuable component of the product to be dried, to carry out freeze drying with recovery of the dust, by collecting the dust in a separator between the dryer and the usual condenser, and subjecting the dust to an after-drying independently of the main dryer.

In a system having a main sublimation type dryer, the dust separator communicates with the dryer and the dust separator is provided with an after-drying device with its own discharge arrangement for the dry dust.

The after-dryer arrangement is a heated worm conveyor having a housing of a cross section widening progressively in the direction of displacement of the dust by the worm, with the vapor outlet opening into this housing at a location of minimal vapor turbulence. This approach ensures especially effective drying of the dust, minimum residual moisture and an effective separation of the dry dust from the vapor.

The process is described with reference to Figure 9.8. Lateral passage or ducts **10** communicate with the side wall of the dryer and open into a dust separator **11** of the cyclone type. The dust separator has its free space above the conical discharge hopper of the separator connected with one or more condensers **12** which, in turn, are chilled by a refrigerating cycle not shown in detail. The condenser communicates with a suction source which maintains the reduced pressure in the entire system which is necessary for freeze drying. At the bottom of the dust separator **11** there is a discharge opening or outlet **14** which communicates with an after-drying device **13**.

FIGURE 9.8: AFTER-DRYER FOR DRYING DUST RECOVERY

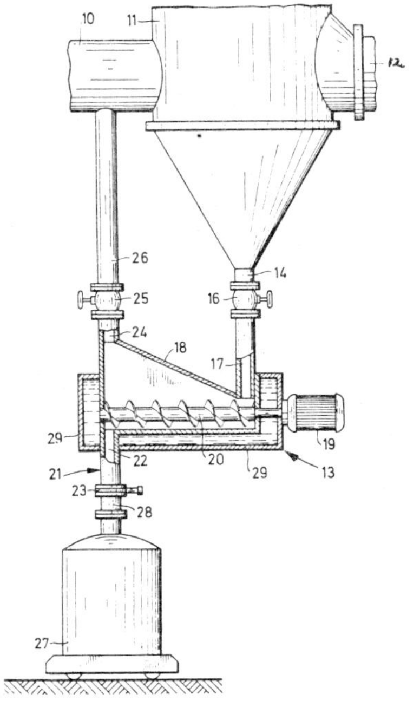

Source: U.S. Patent 3,936,952

The discharge opening **14** of the dust separator **11** is connected via a ball valve **16** with a dust-conveying tube **17** which opens downwardly into the small cross section end of a housing **18** which widens horizontally away from the point at which the dust supply tube **17** communicates with the housing.

A worm **20**, driven by a motor **19** extends horizontally in the housing and is rotated so as to displace the dust from the narrow end of the housing to the wide end thereof. At the wide end of the housing there is provided a further discharge arrangement **21** which may be in the form of a simple vertical downwardly opening discharge pipe **22** connected through a vacuum gate **23**, such as a tube-blocking duct, to a canister **27** in which the dust can be collected. A detachable fitting **28** couples the canister **27** with the duct **22**.

As noted, the housing **18** widens in the direction of advance of the product by the worm **20** toward a discharge opening **24** and the upper part of the housing. This opening **24** is connected by a further ball valve **25** to a feedback line **26** which opens directly into the dryer or into one of the ducts **10** communicating between the dryer and the dust separator **11**. The housing **18** is jacketed to provide a heating chamber **29**.

The product dust (coffee) containing residual moisture and collected in the dust separator **11** is led to the right-hand side of the worm or screw **20** and is thereby displaced across the heated zone formed by the housing **18**. The vapor generated by the heating of the dust travels with reduced speed and turbulence through the housing and out through the duct **24** to the passage **10**.

Freeze Drying of Coffee

Any dust entrained with this vapor is thus recycled to the dust separator. The after-dried product cascades through tube **21** and the discharge gate **23, 28** into the replaceable canister **27**. The ball valves **16** and **25** and the vacuum slide **23** enable the required pressure differentials vis-a-vis the main freeze drying apparatus to be established for operation in the manner described.

Spray Freeze Drying System

In a process developed by *E. Thuse, L.F. Ginnette and R.R. Derby; U.S. Patent 3,362,835; January 9, 1968; assigned to FMC Corporation* coffee extract is continuously dried into instant coffee by spraying the extract into a directly condensing vacuum freezing chamber so that the spray freezes in transit, conducting the frozen particles through a directly condensing drying chamber which also contains the heat source, and removing the dried product from the drying chamber.

The spray freeze drying system is essentially a one step process, in that the solids-bearing liquid is spray frozen in one unit of the system, whereupon the frozen particles pass directly to a freeze drying unit of the system, and hence out of the system through an air lock. The freeze drying unit includes a specially constructed vibratory conveyor which minimizes drying time, and prevents adhering of the frozen particles to the conveying surfaces.

The operation of the spray freeze drying system is described with reference to Figure 9.9. Assume that the apparatus has been shut down. An empty receptacle **16** will be fitted to the air lock **14**, and the vacuum pump **32** started up with butterfly valves **80** and **97** open, the air inlet valve **96** to the air lock closed, and the air lock vacuum valve **92** closed. The refrigeration unit is started up. The heaters are turned on and a water spray introduced through the nozzle to prevent overheating the pans.

In a short time the noncondensible gases are exhausted from the freezing chamber, the drying chamber, the air lock and the receptacle, and the pressure within the freezing chamber **10** and drying chamber **12**, as indicated by gauges **35** and **72** respectively, will be equalized at a very low value, considerably below the triple point pressure. The pan vibrating motor **138** is now turned on, the product liquid pump **42** started, and the liquid supply valve **45** adjusted to produce the desired spray pressure for the nozzle **38**, as indicated by the gauge **45a**.

The spray freeze drying process now begins. The particles of atomized liquid emerge in a cone-like envelope of finely atomized particles. These particles are directed upwardly in the freezing chamber **10**, and due to the low pressure now existing within the freezing chamber, most of the particles freeze soon after leaving the nozzle **38**, but they do not freeze at the nozzle itself. Most of the particles of the product will have been frozen before they strike the wall or jacket of the condenser **22**, with which they bounce off the wall and start falling down towards the outlet neck **48** of the freezing chamber.

Particles that are not fully frozen when they reach the wall of the condenser **22** will soon freeze thereafter, and all particles will have been completely frozen by the time they fall to the hopper **20** at the bottom of the freezing chamber. These completely frozen particles fall directly into the input end of the vibrating pan **54**. The action of the pan drive assembly **60**, and the angular relation of the links **56**, are such as to cause the particles **P** to advance along the pan **54** toward the delivery end of the drying chamber **12** in a progressive motion.

FIGURE 9.9: SPRAY FREEZE DRYING SYSTEM

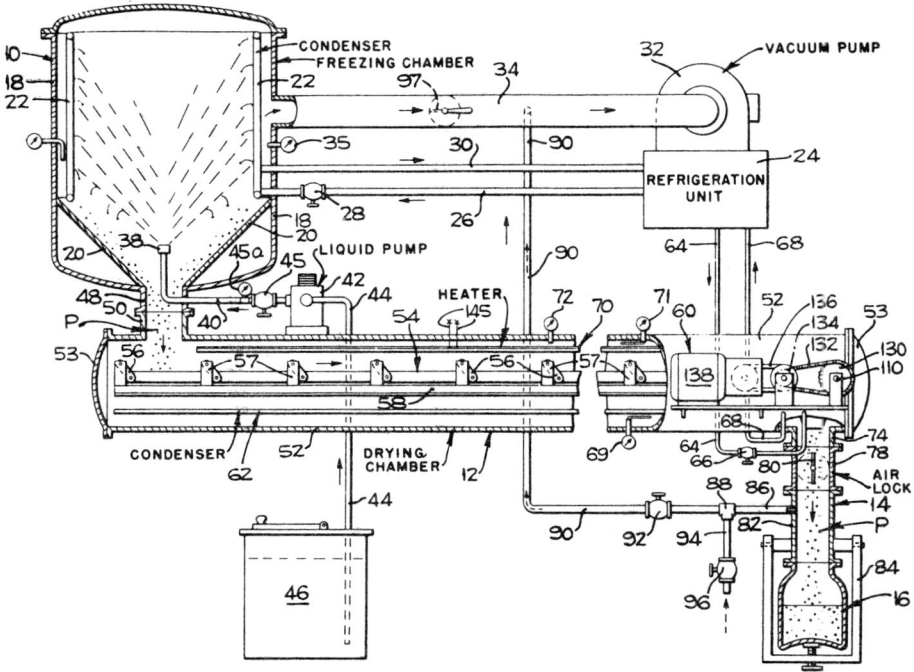

Source: U.S. Patent 3,362,835

As the particles advance along the pan, the heat of sublimation is supplied by the heater **70**, and the sublimed vapor is condensed directly in the chamber, by the internal condenser **62** in the drying chamber. The refrigeration loads for the condenser **62** in the drying chamber and **22** in the freezing chamber are adjusted so that the pressures within these chambers are substantially equal. It is particularly important that the pressure within the drying chamber **12** not substantially exceed that which exists in the freezing chamber.

If such were the case, sublimed water vapor could flow upwardly through the inlet neck **50** of the drying chamber, and the outlet neck **48** of the freezing chamber, thereby entraining frozen particles **P** attempting to fall into the drying chamber from the freezing chamber. Since approximately 20% of the water in the product is removed during the freezing process, in the freezing chamber **10**, the drying chamber requires correspondingly less refrigeration, per pound of liquid starting product, than would be required if the product were introduced in the state of prefrozen particles.

When the particles reach the end of the pan **54**, they are vibrated out of the end of the pan and into the funnel hopper (not shown), whereupon they fall through the air lock **14** into the receptacle **16**. When the receptacle has been filled to the desired extent, the air lock and valving system permits changing receptacles without breaking the vacuum in the freezing and drying chambers.

Experience with the freeze drying of coffee extract has shown that the geometry of the link mounting for the pan **54** is somewhat critical. It has been found that the angle which the links **56** make with the horizontal plane should be substantially 65°. As a result, the average direction of pan motion is at the same angle of 65° with the vertical plane.

Example: Optimum Operating Conditions for the Spray Freeze Drying of Instant Coffee Extract — The freezing chamber **10** is approximately five feet in diameter and six feet high. The pan **54** is approximately 16 feet long and one foot wide and the drying chamber **12** is two feet in diameter. Best results are obtained when the solids content is about 20% by weight. The refrigerant temperature should be adjusted to maintain a pressure in the freezing and drying chambers in the order of 250 microns. The pressure range of 250 to 270 microns gives a residence time in the order of 40 seconds and is found to produce the best overall results. The 40 second residence time is short enough to prevent heat damage to the product, and long enough to insure 100% drying during the process.

The spray nozzle **38** and the pressure range in which it operates are selected to give an evaporation rate in the order of 0.35 pound per hour per square foot of pan surface in the drying chamber **12**. This nozzle is operated at a pressure of 200 pounds per square inch gauge from the pump **42**, as adjusted by the liquid inlet valve **45**, and indicated on the gauge **45a**.

The temperature of the refrigerant in the condensers **22** and **62** is maintained at -50°F or thereabouts. If the temperature of the refrigerant is allowed to become as high as -30°F, difficulties with sticking and long residence time are encountered. The temperature of the heater **70** is maintained at approximately 300°F, and a lowered temperature unnecessarily increases the residence time of the product, but if the temperature is permitted to reach 325°F, the product sticks to the conveyor.

The product gradually becomes warmer as it reaches the delivery end of the pan **54**. A delivery temperature of 150°F at the pan is selected. This temperature is measured by a thermocouple (not shown) on the delivery end of the pan. The pan **54** is vibrated at 500 cycles per minute (500 rpm of shaft **110**) over a horizontal stroke of ¼", and with the links **56** forming an angle of 65° with the horizontal plane.

COLOR OF COFFEE PRODUCT

Darker Coffee by Controlled Melt-Back

J.W. Johnson, G.B. Ponzoni and W.P. Clinton; U.S. Patent 3,244,529; April 5, 1966 describe a process for producing a dark colored freeze dried coffee which resembles the color of roasted and ground coffee. This comprises cooling a body of coffee extract containing coffee solids and aromatics from its ice point to below the eutectic point of the extract, subliming the water in the extract to initiate development of a porous structure amenable to the removal of water vapor, liquefying a minor portion of the frozen extract at the surface portions of the porous structure, while leaving the balance of the extract in a frozen state, refreezing the liquefied portion of extract and continuing the sublimation to thereby dry the coffee extract to a porous state and a dark coffee-like color.

More specifically, this is accomplished by subliming the water ice from coffee extract at a temperature which is below the eutectic point (-15°F) of the coffee extract, about -16°F, and under a vacuum of less than 500 microns, and periodically fluctuating the temperature and pressure during freeze drying to achieve partial melting of eutectic solution thereby allowing less than 3%, preferably less than 1%, by weight of water ice to be liquefied and distributed throughout the extract.

By this method of controlled melt-back the porous cellular structure in the extract is not seriously disturbed to an extent which will effect efficient drying operations. The noneutectic water may stil be sublimed at a rapid rate and the eutectic water may also be evaporated at a rapid rate without serious aroma loss.

Alternately, coffee extract may be frozen, a substantial proportion of the water ice dehydrated by sublimation to achieve a total terminal moisture of 12 to 20% and a water ice content of 0.1 to 8%, and more preferably 0.1 to 3%, and then subjecting the partially dried extract to controlled melt-back prior to refreezing and drying thereby achieving a darker colored coffee product.

The water ice may be distributed in a liquefied state throughout the surface portions of dried material by either elevating the temperature or the pressure within the freeze-drying chamber or preferably by elevating both temperature and pressure to a point wherein the surface of the frozen eutectic solution (composed of eutectic water, coffee solids and coffee aromatics) is caused to melt and wet the dried coffee solids. After the material is darkened, preferably by distributing 0.1 to 3% of noneutectic water throughout the material, the melted portion is refrozen and dried by sublimation in a vacuum of less than 500 microns.

After the noneutectic water is removed, the remaining water which is present as eutectic solution may be removed at drying temperatures above the eutectic point and as high as 80° to 120°F at the termination of the drying cycle, the product temperature being raised at a uniform rate of 8 to 12 degrees, say 10 degrees, for each 1% of eutectic water removed. This eutectic water may be redistributed into noneutectic water by melting the eutectic and in this case the eutectic point itself will be increased dependent upon the amount of eutectic water which is removed.

Example: About 28,000 cc of commercial coffee extract containing 8,850 g of soluble coffee solids (27% concentration) was poured into seven polyethylene pouches having dimensions of 20" x 24". The pouches were placed in freezing molds having a thickness of $7/16$", the molds being immersed in dry ice having a temperature of -109°F. About three minutes was allowed for the extract to freeze. The pouches were removed from the freezing molds and the frozen extract which had a light tan color was removed from the pouches and transferred to meshed stainless steel drying trays. The drying trays were then placed in a commercial freeze dryer and the coffee was dried under a vacuum of less than 500 microns, a condenser temperature of between −40° and −70°F and a platen temperature which was controlled to provide sublimation of noneutectic water in the extract at a product temperature of below −15°F. The noneutectic water was sublimed for a period of about three hours in which time a terminal moisture content of 12 to 14% was achieved, the terminal moisture being measured by weight differential in the extract as it dried.

The freeze drying chamber was then opened to atmosphere. The frozen extract at this point still retained its light tan color and was superficially dry to the touch. The extract was exposed to atmospheric conditions for about 12 minutes when the surface of the extract appeared to become moistened and visually darkened. The extract at this point still retained its shape and form and its porous cellular structure did not appear to be disturbed. No excessive wetting or dripping of water from the extract was noted.

The chamber was then immediately closed to atmosphere and the darkened extract was again subjected to a vacuum of less than 500 microns and a platen temperature which assured sublimation of water ice at a temperature of below −15°F for two additional hours until the terminal moisture was reduced to 10%. The vacuum was then increased to between 50 and 150 microns while the platen temperature in the freeze drying chamber was elevated gradually at the rate of 10° for each 1% of eutectic water removed until the extract had a terminal moisture content of 2% and a product temperature of 80°F. About 10 hours further drying time was required to reduce the terminal moisture content to this level. The final freeze dried product was uniformly dark in appearance and resembled the color of roasted and ground coffee.

Darker Color by Slow Freezing

The process of *G.J. Lutz; U.S. Patent 3,399,061; August 27, 1968; assigned to General Foods Corporation* is founded on the discovery that the color of freeze dried coffee is determined by the rate of freezing the liquid coffee extract. Darker, more coffee-like color in freeze dried coffee can be produced by a process which comprises slowly cooling the coffee extract from its ice point to below its eutectic point over a period of at least 15 minutes to thereby form a crystalline structure of substantially pure water ice distributed in a matrix of a eutectic mixture of water, coffee solids and aromatics, and freeze drying the frozen extract.

The ice point of coffee extract will vary with its solids concentration; dilute extracts (20 to 28% soluble solids) will begin to form crystals of water ice at temperatures slightly below the freezing point of pure water, while more concentrated extracts will begin to form water ice crystals at temperatures further below the freezing point of water. However, regardless of the initial coffee solids concentration in the extract, the eutectic temperature will always be constant at about −13.5°F.

When slowly cooling the extract, it is preferably that heat removal be uniform throughout the freezing step since this assures a uniform growth of ice crystals throughout the body of extract. In the case of a 0.5" thick layer of coffee extract and a 15 minute cooling step, the heat removal rate should be uniform at 2 calories per gram per cubic centimeter of frozen extract per minute.

Since the cooling step provides a gradual precipitation (crystallization) and separation of water present in the extract, the aromatic substances which are present in the liquid portion of the extract are progressively concentrated to a liquid eutectic composition wherein the aromatics can be preserved during subsequent drying. This eutectic composition assumes the form of a liquid matrix of water, aromatics and coffee solids (soluble and insoluble) having relatively pure crystals of water ice distributed in the interstices of the matrix. When the extract is finally cooled to a temperature below the eutectic point for the coffee extract, the frozen extract will be ready to be freeze dried. Figure 9.10 shows the preferred nonordered dendritic ice crystal structure formed by freezing coffee extract slowly in accordance with this process.

FIGURE 9.10: NONORDERED DENDRITIC ICE CRYSTAL STRUCTURE

Source: U.S. Patent 3,399,061

In cooling the liquid material, it may be advantageous to seed the material with previously formed ice crystals in order to initiate controlled ice crystal growth, prevent supercooling of the liquid material and suppress the spontaneous nucleation of ice crystals, thereby promoting the growth of a smaller number of large ice crystals resulting in a darker colored product. Before seeding the extract, the extract may be agitated as it is cooled. Ice crystals are added as the extract is depressed to below its ice point and are uniformly dispersed throughout the extract. Agitation is preferably continued until at least partial ice crystal formation has begun. With this process the color of freeze dried coffee extract approaches the color of ground, freshly roasted coffee.

Example: Coffee extract having a soluble solids concentration of 27% coffee solids was separated into two portions and placed in two aluminum freezing molds having dimensions of 40" x 20" x ½". The coffee extract at this point had a product temperature of 55°F. One freezing mold was immersed in liquid nitrogen having a temperature of –329°F and was completely frozen within a period of one to three minutes. The second mold was placed in a freezing room in contact with a freezing platen having a temperature of –40°F. Ambient temperature of the freezing room was about –10°F.

About 15 to 20 minutes were required for the coffee extract to reach its ice point of 26° to 28°F and approximately 60 to 90 minutes were required to completely cool the extract to below its eutectic point of –13.5°F. Both samples of extract were then removed from their molds and freeze dried to a terminal moisture of less than 5% in the presence of vacuum of less than 500 microns. Drying time was about 18 hours and the product temperature of the frozen product was kept below −13.5°F during sublimation in order to avoid any melting of the product.

The first extract, frozen with liquid nitrogen, was found to have a light tan, uncoffee-like color upon drying, while the second extract which was slowly frozen in the freezing room was found to have a uniformly dark brown color which was identical in appearance to regular roasted and ground coffee. The second sample had a Munsell color rating of between 15 and 17.5 in hue, between three and five in value and four and seven in chroma or 15, 3/5 to 17.5, 4/7.

Cross sections of each extract were viewed under a microscope having a magnification of 200 times the original size and the ice crystal structure was determined from the open spaces left in the dried structure by the sublimated ice crystals. The fast-frozen extract was found to have a highly ordered arrangement of substantially parallel needle-like pores while the slow-frozen extract had a highly random and nonordered distribution of openings which were of substantially larger size than the needle-like pores of the fast-frozen product. The openings in the slow-frozen product resembled the large dendritic ice crystal structure of Figure 9.10.

The color differences in the product persisted upon grinding the freeze dried slabs of coffee extract in a Fitzpatrick mill to a mesh size distribution wherein 90% of the particles were retained on a 40 mesh U.S. Standard Sieve Screen. The slow-frozen product retained a dark brown color identical to that of regular roasted and ground coffee.

Slow Freezing Followed by Subdivision to Coarse Particle Size

W.P. Clinton, J. Mahlmann and G.B. Ponzoni; U.S. Patent 3,438,784; April 15, 1969; assigned to General Foods Corporation found that a stable, dark-colored freeze dried coffee can be produced having the appearance of roasted and ground coffee by freezing coffee extract having a concentration of below 50% soluble solids slowly from its ice point to below its eutectic point over a period of at least 15 minutes to develop a nonordered distribution of dendritic ice crystals, the crystals being characterized by nonparallel main stems, smaller extending branches from the main stems, and an absence of discrete ice crystals of nondendritic form in the eutectic mixture located between the dendritic ice crystals, subdividing the frozen coffee to obtain a granulated product having at least 97% by weight of the particles greater than 80 mesh (177 microns) and then vacuum freeze drying the slowly frozen granulated coffee.

In the case of above 50% solids extract, the coffee may be frozen in a period of 3 to 15 minutes prior to being ground and freeze dried while still obtaining a dark-colored coffee. A suitable ambient freezing temperature for slowly freezing a ½" thick slab of coffee extract having a dimension of 40" x 20" and a product temperature of 35°F is −30° to −40°F. Complete freezing of the extract is accomplished over a period of about 120 to 150 minutes, the ice crystal formation taking 100 minutes. After freezing, the extract is granulated or subdivided to a coarse particle size. In subdividing the frozen extract, it is preferred to use pressure equipment which compresses and abrades the frozen coffee into a granular product as distinguished from equipment which slices or cuts the frozen extract into the desired particle size. The use of pressure equipment is believed to give a case-hardened surface to the coarser particles (those above 40 and 80 mesh) which is preserved on freeze drying, thus giving better stability to the product.

The avoidance of particles having a mesh size of less than 80 mesh, preferably less than 40 mesh, is most important since the presence of particles of this size presents disadvantages relative to controlling the stability, bulk density, color and freeze drying conditions to be employed.

Example 1: Aromatized coffee extract containing a mixture of expressed coffee oil and volatile steam distilled aromas, and having a soluble solids concentration of 27% and a product temperature of 60°F was poured into a stainless steel freezing tray having dimensions of 40" x 20" x ½". The freezing tray was then placed in a freezing room having an ambient temperature of -30°F and chilled for about 20 minutes until it reached its ice point (28°F).

The extract, which was still in a substantially liquid state, was then seeded with about 1.5 lb of frozen extract having a particle size of less than 40 mesh (U.S. Standard Sieve Screen). About 90 minutes was then required to depress the temperature of the extract from its ice point (28°F) to below its eutectic point of -10°F. The freezing tray was removed from the freezing room, and the frozen slab separated from the freezing tray in preparation for the grinding operation.

The frozen slab was then ground in a Fitzpatrick mill to a final particle size distribution of between 12 to 80 mesh. The grinding operation was conducted in a freezing room wherein the ambient temperature (as well as the temperature of the Fitzpatrick mill) was below -30°F. The under 40 mesh fraction was then separated from the granular extract and used to initiate controlled freezing of the next charge of liquid extract.

The granular frozen extract was placed in a ½" stainless steel drying tray having dimensions of 40" x 20". The bed of frozen granular material had a height of ½". The drying tray was placed on a heating platen in a freeze drying chamber having a vacuum of below 500 microns. The platen temperature was raised uniformly over a period of two hours to 160°F, held at this temperature for four hours and reduced to a temperature of 110°F for the final four hours. The product temperature of the dried portions of the product during the freeze drying was kept below 105°F while the product temperature of the frozen product was kept below -10°F. The product reached a terminal moisture of less than 2% in about 15 hours.

The final product was uniformly dark in appearance and resembled a natural blend of roasted and ground coffee in its shape, color and appearance. The freeze dried product had an average bulk density of about 0.20 g/cc and was found to approach the average bulk density of conventional spray dried soluble coffee. A teaspoon of the freeze dried product when added to a cup of boiling water reconstituted to an aromatic and flavorful coffee identical in all respects to a freshly prepared cup of brewed roasted and ground coffee.

Example 2: The procedure of Example 1 was followed except that conventional unaromatized coffee extract was used. The extract was slowly frozen with seeding, subdivided in a Fitzpatrick mill, had its under 40 mesh fraction removed and recycled into the extract to initiate controlled freezing, and was then dried. The final product produced was similar in color, shape and appearance to the product of Example 1. The dried product, while not reconstituting to the aroma and flavor level of the Example 1 sample, still provided a unique and improved flavor and aroma when compared to a cup of reconstituted spray dried coffee.

Temperature and Freezing Rate Dependent on Concentration of Extract

H.R. Simon and S. Barnett; U.S. Patent 3,443,963; May 13, 1969; assigned to General Foods Corporation found that freeze dried coffee having a dark color is produced by controlling the freezing rate of the extract within a critical temperature range depending on the concentration of the extract. The process comprises cooling a water extract of coffee solids to from 25° to 10°F over a period of at least 10 minutes, preferably with seeding and agitation, to thereby develop large crystals of water ice, further cooling the extract to below its eutectic point, and then freeze drying the frozen extract.

In the case of extract having a concentration of 15 to 30% soluble solids a critical portion of the cooling curve will be found in the range of 25° to 20°F. If extract is kept within this temperature range for at least eight minutes, a dark-colored final product will be obtained. In the case of 30 to 50% soluble solids extract, the critical portion of the cooling curve will be found between 20° and 15°F. Extract kept within this temperature range for at least eight minutes during freezing will have a dark color on drying.

Example 1: Coffee extract having a soluble solids level of 27% coffee solids was separated into two 75 ml portions and placed in separate three inch aluminum foil molds having a thickness of about ½". The extract at this point had a temperature of 65°F. One mold was immersed in liquid nitrogen (-327°F) and found to freeze within a period of less than one minute.

The second mold was placed in a constant temperature bath maintained at 10°F and allowed to reach a temperature of 27°F in a period of two minutes. The extract was agitated with a small propeller during the initial ice formation to accomplish uniform dispersion of ice crystals throughout the extract and prevent supercooling of the extract. Agitation was ceased when the temperature of the extract reached 25°F. The partially frozen extract was held for 20 minutes at between 20° and 25°F and then placed in liquid nitrogen. The two samples were then freeze dried at sublimation temperatures of below -13.5°F.

The sample which was held at between 20° and 25°F for 20 minutes had a dark-brown color (similar to freshly roasted and ground coffee) while the sample frozen in liquid nitrogen had a light tan color. The freeze dried slabs were then ground to a particle size distribution approaching that of regular roasted and ground coffee. The color differences in the ground samples remained the same, the fast frozen particles having a light tan appearance while the slowly frozen product had an appearance similar to roasted and ground coffee.

Example 2: The procedure of Example 1 was followed with the exception that the slowly frozen extract was seeded while being cooled to below its ice point. The extract was seeded with 0.2 g of ice frost prepared by allowing moisture from the air to form on a chilled metal plate. Seeding was accomplished while the extract was being cooled to below its ice point with agitation and while the extract was still in a substantially liquid state, the temperature of the extract being about 25° to 28°F.

The small ice crystals dispersed uniformly throughout the extract due to the agitation and were found to suppress spontaneous nucleation and supercooling of the coffee extract. The partially frozen extract was held for about 10 minutes at between 23° and 20°F while agitation was continued and then frozen in liquid nitrogen. The two samples were then freeze dried at sublimation temperatures of below -13.5°F. The slowly frozen product was found to be uniformly dark brown in color and identical in all respects to the slow frozen product of Example 1 while the fast frozen product was a light tan color.

Example 3: The procedure of Example 2 was followed with the exception that no agitation was employed as the extract was seeded. The slowly frozen sample was seeded upon being cooled to a temperature just below 28°F (the ice point). Seeding was discontinued at 25°F and the extract was then held at between 25° and 20°F for 15 minutes. The slushed extract was then frozen in liquid nitrogen and dried. The slowly frozen product and the fast frozen product had the same color differences which were found in the products of Examples 1 and 2. The color differences in the two products persisted upon grinding the extract to a particle size distribution similar to that of regular roasted and ground coffee.

Freezing in a Plurality of Layers on a Continuous Belt

S.N. Katz and D.E. Dwyer, Jr.; U.S. Patent 3,966,979; June 29, 1976; assigned to General Foods Corporation are concerned with a process for producing a dark colored, freeze dried coffee at a high freezing rate. A coffee extract is obtained, and held at 32° to 85°F. The freezing surface to be used is a continuous stainless steel belt. Underneath the freezing surface of the belt is a brine system. The brine is continually sprayed against the belt. It is preferred that the temperature of the brine, and thus, the entire freezing surface of the belt be maintained at the lowest possible temperature permitted by the brine system (-30°F).

Located above the belt is a means for applying the extract evenly to the belt, preferably a series of spray nozzles along almost the entire length of the belt. Extract is continually sprayed onto the moving belt at a rate such that a slab of frozen extract between about ¼ to 1 inch thick is obtained at the end of the belt. No extract is sprayed onto the belt along approximately the last 20% of its freezing surface. This is to insure that the product exiting the belt is completely frozen. It will be apparent to anyone with an ordinary skill in the art that various combinations of extract temperature, spray rates, and belt speed are contemplated.

As the first extract contacts the beginning of the moving belt, it forms a fast frozen layer. However, as the belt progresses, additional warm extract is continuously applied. Thus, as the warm extract contacts each previous layer of extract, the upper portion of that layer is constantly remelted. This remelting permits the growth of large dendritic ice crystals and a darker final product. The most critical period for ice crystal growth is during the initial freezing. What must be avoided is supercooling and its resultant spontaneous nucleation. The ideal environment for the desired crystal growth occurs in a narrow temperature between the ice point and the eutectic point of the extract (about 22° to 28°F). Using the warm extract itself to constantly remelt the product permits the product to remain in the critical temperature range for a relatively long period of time.

Capacity is increased on the order of about 20 to 50% over the conventional method of applying the extract to only one end of the belt and, thus, initially forming a liquid bed.

Example: The freezing surface used is a continuous stainless steel belt. The belt is about two feet wide and is operated by two pulleys located about 50 feet apart. The belt is cooled by contact with the plurality of cold brine tanks located on the underside of the freezing surface of the belt. All tanks are maintained at -30°F. Directly above the belt are located 22 spray nozzles so positioned that when coffee extract is passed through them, they provide a continuous, even spray along about 40 feet of the freezing surface of the belt.

Coffee extract at 26% soluble solids by weight and the temperature of 52°F is passed through the spray nozzles onto the moving belt at a rate that results in a final slab thickness of 0.5 inch. The completely frozen coffee slab is then ground to below 8 mesh (U.S. Standard Screen) and dried in a commercial freeze drying unit in a vacuum of below 500 microns. The resulting coffee product appears dark brown in color.

Coating Frozen Particles of Extracts with Soluble Coffee Powder

J.P. Mahlmann; U.S. Patent 3,565,635; February 23, 1971; assigned to General Foods Corporation found that an improved soluble coffee in regard to appearance and color is obtained by coating frozen particles of coffee extract with a soluble coffee powder and then drying the coated extract. The amount of soluble coffee used in relation to the frozen coffee particles may vary over a wide range (from 5 to 85%) depending on the ultimate properties desired. When the aim is to modify a freeze dried coffee, the level of spray dried coating should be kept below 50%, preferably 10 to 35%. However, if the aim is a spray dried coffee improvement, the level of spray dried coffee can be increased to above 50%, say 60 to 75% by weight of the frozen particles.

Example: Aqueous coffee extract is quick frozen with liquid nitrogen to a temperature below -13°F. The frozen extract is then ground to a particle size such that it passes through a 12 mesh screen but is retained on a 40 mesh screen. The ground, frozen extract is then placed in a tumbler and 10% by weight of soluble coffee powder having a particle size below 200 microns and a temperature of -13°F are added.

The temperature is raised to 0° to 5°F during tumbling to partially thaw the ground, frozen extract thereby enabling the soluble coffee powder to coat the surface thereof. When all of the powder has been absorbed on the surface of the ground, frozen extract, the coated extract particles are quickly frozen to reduce their temperature to -13°F. The coated extract particles are then freeze dried. The freeze dried particles are darker than conventional soluble coffee and approach the color of fresh roasted and ground coffee.

Extract Dispersed into Immiscible Liquid Refrigerant

B.E. Elerath; U.S. Patent 3,961,424; June 8, 1976; assigned to General Foods Corporation describes a method for freezing coffee extract which produces a finished product which is dark and coffee-like in color, and which eliminates the grinding and fines removal processing steps of conventional freeze dried coffee processing.

A coffee extract is first obtained, held at a temperature of about 32°F. The extract is then introduced to an immiscible liquid refrigerant. The temperature of the refrigerant is maintained such that the product temperature is held at about 24° to 28°F for at least four minutes and preferably more than eight minutes. This is to permit the growth of large ice crystals. By suitable adjustment of pumping parameters small droplets of extract can be formed in the refrigerant.

Another method is mixing the extract in a venturi device with the refrigerant in the high velocity zone and the extract entering the low pressure zone. The output of the venturi is an extract/refrigerant mix in which the extract is highly dispersed in the refrigerant in droplet form.

Another method of freezing the product is to pump an extract/refrigerant mix through a tubular heat exchanger. Another method of controlling the temperature is by passing the extract/refrigerant through progressively lower pressure zones causing the refrigerant to evaporate and the temperature to drop to the desired level. The evaporated refrigerant would be recovered.

In any of the freezing methods, once the product has remained at the critical temperature for ice crystal growth for a sufficient amount of time, it may then be brought down to a temperature below the eutectic point of the coffee extract which is about $-13.5°F$. The product is next separated from the refrigerant by suitable means and freeze dried. The liquid refrigerant may be recycled.

The ratio of extract to refrigerant can be between about 1:5 and 1:50 and preferably about 1:10. Since the extract exists within the liquid refrigerant as droplets, the relationship of the surface area of the extract droplet to the refrigerant permits the product temperature to equilibrate with the refrigerant temperature almost instantaneously. Thus, substantially all of the coffee extract may be controlled at a desired temperature.

Example: 50 ml of 35% concentration coffee extract is placed in a 500 ml beaker and 200 ml of R21 ($CHCl_2F$, BP 47°F) is added. A stirrer is used to disperse the extract into small globs. Dry ice is used to reduce and maintain the temperature of the extract/refrigerant mix to between 24° and 28°F for four minutes. Then the temperature of the extract/refrigerant mix is reduced to below $-20°F$. The extract, now frozen and granular in form, is removed by pouring the mixture through a screen after first removing the remaining dry ice. The frozen extract is then freeze dried in a small Stokes laboratory freeze dryer, keeping the pressure less than 200 μ of Hg. The resultant product is dark brown and coffee-like in color.

Frozen Ground Particles Screened and Fines Dried as Bottom Layer

In the preparation of freeze dried coffee from comminuted frozen extract, it has been found that fines which will pass through an 80 mesh U.S. Standard Screen have undesirable properties and characteristics when subjected to freeze drying. It is desirable that the final freeze dried product contain no more than 3% by weight of fines that will pass through an 80 mesh U.S. Standard Screen, and which are light in color and aroma-deficient.

H.A. Oldenkamp and C.D. Watson; U.S. Patent 3,556,818; January 19, 1971; assigned to FMC Corporation found that this standard is met by processing the frozen material before drying so that a bed of frozen particles is formed wherein the fines are at the bottom of the bed. This processing is preferably accomplished by screening the frozen and ground particles into at least two fractions, one fraction constituting a layer of fines which will pass through an 80 mesh screen. These 80 mesh fines are spread as a bottom layer on precooled trays, whereupon the coarser frozen particles are spread over the bottom layer of frozen fines.

The trays filled with this multilayered bed of particles are placed in a freeze drying chamber. The upper strata of the bed (the coarser particles) receive the heat of sublimation by direct radiation. The lowermost stratum of the bed (the fines) receive the heat of sublimation via the trays. As a result the surrounding pressure in the layer of fines is greater than that in the voids between the larger particles. This retards drying of the fines relative to drying of the larger particles, so that ice cores remain in substantially all of the particles, fine and coarse, during a large part of the sublimation-drying cycle. When the disparate sized particles are dried under these conditions, undesirable properties and characteristics are either lacking or significantly improved.

Example 1: Single Screening — Liquid coffee extract containing about 25% solids was frozen into slabs about 1" thick. The slabs were broken into chunks with a hammer and fed to a Fitzpatrick mill. The mill was set so that about 4% by weight of the fines produced would pass through a standard U.S. 80 mesh screen. The fines were evenly distributed on the bottom of trays precooled to a temperature below the eutectic point of the extract, namely to about −20°F. The coarser particles, retained on the screen were evenly deposited over the layer of fines, filling the trays to a depth of about ¾".

Before product melting could begin the trays were loaded in a freeze drying chamber and condensable gases exhausted. Radiant heaters at about 300°F supplied heat directly to the coarser granules of the top layer in the trays, as well as to the tray bottoms. The vapor pressure in the chamber initially reached 500 microns but soon settled down to about 300 microns throughout the majority of the drying cycle, dropping still further near the end. The sublimation drying was continued for about six hours, after which the trays were removed from the chamber.

The product had the natural brown color and aroma of freshly ground coffee. Substantially all of the fines had the same color and aroma characteristics as the larger particles. Substantially less than 3% by weight of the fines was lighter in color than the other particles and there had been no appreciable loss in fines, thus meeting commercial goals.

Example 2: Double Screening — The process given in Example 1 was followed except that a standard U.S. 40 mesh screen was interposed between the Fitzpatrick mill and the 80 mesh screen. The grind was such that over 10% by weight of the particles passed through the 40 mesh screen and over 3% passed through the 80 mesh screen. The appearance of the 80 mesh fines was as before, and the 40 mesh fines likewise had the desired brown color and full aroma.

PRODUCT OF CONTROLLED DENSITY

Freezing Slushed and Foamed Extract

According to *S. Barnett and T.T. Mak; U.S. Patent 3,804,960; April 16, 1974; assigned to General Foods Corporation* an extract is prepared for freeze drying by concentrating the extract to a solids content of at least 35% by weight of the extract, partially slush freezing the extract, foaming the partially slushed extract and freezing the slushed and foamed extract prior to freeze drying.

The method suitable for producing a freeze dried coffee product having a controlled density comprises the steps of:

(a) chilling, in a jacketed, scraped-surface heat exchanger, the coffee extract having a soluble solids content of at least 35% by weight to below the ice point of the extract,
(b) forming ice crystals by freezing water onto the surface of a jacketed scraped-surface heat exchanger,
(c) scraping the ice crystals off the wall of the heat exchanger,
(d) agitating the extract to completely disperse the ice crystals in the extract,
(e) continuing the freezing, scraping and agitation until a soft slush containing at least 7% but less than 15% by weight (preferably 13 to 15%) of ice is obtained,
(f) feeding the soft slush into a second scraped-surface heat exchanger along with a controlled amount of gas (preferably nitrogen) to cause the slushed extract to foam to at least a 40% volume increase as it is agitated,
(g) freezing, scraping and agitating in the second heat exchanger until the hard slush contains at least 20% by weight of ice,
(h) extruding the hard slush into a desired shape,
(i) freezing the shaped slush to below its eutectic point, and
(j) freeze drying the frozen shaped slush.

The density of the final product is inversely proportional to the degree of expansion. This is apparent by the results obtained when 35% concentration extract was formed into a soft slush in a Votator scraped surface heat exchanger using a refrigerant temperature of about 22°F in order to obtain a slush containing about 13% ice. The slush was then foamed by whipping nitrogen gas into the slush until a desired degree of expansion was achieved. The foamed samples were spread in a pan, frozen, freeze dried and ground and the densities of a –20 mesh +30 mesh fraction were compared. The following results were obtained:

Sample	Density, g/cc
No gasification	0.26
10% volume increase	0.24
20% volume increase	0.21
40% volume increase	0.18

Example: Coffee extract which had been concentrated to a solids content of 35% was chilled in a scraped, jacketed mixing bowl for approximately 20 minutes, using a coolant temperature of 10°F, forming a viscous slush containing about 13% ice by weight at a temperature of 24°F. The bowl was then placed in a nitrogen atmosphere and agitated with a scraper-stirrer whipping gas into the slush. The volume increase due to entrapped air was proportional to the stirring time.

The foamed slush was transferred to ½" deep molds maintained at 0°F for two hours without loss of volume and then transferred to a –30°F room for a final freeze. The ½" slabs were then freeze dried in six hours as compared to 12 hours for a comparable load of unconcentrated, frozen extract. The bulk density of a –20 mesh +30 mesh fraction of dried product which had been subjected to a 40% volume increase was 0.18 g/cc compared with 0.26 g/cc for the unfoamed 35% concentration extract.

Alternatively, the initial slushing operation could be performed in a scraped surface, double pipe heat exchanger such as a Votator using a refrigerant temperature of about 22°F to produce a slush having an ice content of about 13% by weight and the soft slush could then be fed into a second scraped surface heat exchanger along with a controlled amount of gas to cause the extract to foam as it is agitated. The extract may be further slushed in this second heat exchanger by using a refrigerant temperature of about 10°F and a slush containing about 40% ice by weight could be extruded, cut into desired lengths and freeze dried.

Slurry Containing Ice of Controlled Crystal Size and Mother Liquor of 40 to 50% Soluble Solids

H.T. Easton and D.E. Dwyer; U.S. Patent 3,682,650; August 8, 1972; assigned to General Foods Corporation found that a dark colored freeze dried coffee of controllable density can be produced by a process which comprises preparing a coffee extract slurry of small sized ice crystals and mother liquor, maintaining the slurry under gentle agitation at a temperature between 14° to 24°F for a period of time (generally at least two hours) sufficient to produce a slurry of extract containing water particles having an average diameter of at least 0.5 mm, a mother liquor concentration of 40 to 50% soluble solids, and subsequently freezing this mixture by conventional techniques to produce a frozen coffee extract suitable for rapid conventional freeze drying.

Where high extract concentrations (35 to 40% solids) are processed and gas is added to the slurry, the density of the slurry should be 0.6 to 0.85 g/cc. The exact extract slush density required for a given extract concentration is easily determined by adding sufficient gas in stages to produce a final freeze dried product of 0.18 to 0.22 g/cc bulk density, the product having 90% of its particles between 400 and 4,000 microns.

Example 1: Ice Crystals — Coffee extract (2,000 g) having a soluble solids content of 27% is rapidly cooled to a temperature of about 27°F to form a slurry of fine ice crystals. The slurry is transferred to a precooled Hobart ice cream machine maintained at about 21°F using a brine bath. The slurry is maintained under gentle agitation by manually turning the agitator while visually observing the ice crystal formation. Agitation is adjusted to maintain a relatively uniform slurry without retarding the development of large ice crystals. After 3½ hours of crystallization in the ice cream machine, the slurry is removed and transferred to a tray chilled to about 20°F. The removed slurry has a mother liquor concentration of about 42% solids and contains a majority of ice crystals having a diameter of greater than 0.5 mm.

A layer of slurry about ½ inch thick is formed in the tray which is then immersed in a –35°F brine bath. The slurry is frozen solid to below –20°F, the slab is removed and is freeze dried as a single slab using conventional vacuum freeze drying techniques. After drying and grinding, the product has the appearance of roasted and ground coffee exhibiting a pleasing dark color and density approaching that of conventional soluble coffee.

Example 2: Ice Crystals plus Inert Gas — Coffee extract from a conventional ice crystal concentration of coffee percolate and containing 39% soluble solids is rapidly cooled in a Hoyer scraped film heat exchanger. Nitrogen gas is introduced into the Hoyer and contacted with the extract to form a foamed extract which is discharged continuously into an agitated insulated crystallization tank. Fine particles of frozen extract are incorporated with the foamed extract to maintain a foamed slurry of frozen extract particles in mother liquor at a slurry temperature of about 19°F.

The slurry is maintained at constant temperature by adjusting the quantity of frozen particles added to the tank, and the foamed extract temperature. The slurry is maintained in the tank and contains about 22% of the available water in the form of ice crystals having an average diameter of at least 0.5 mm and a mother liquor concentration of about 45% solids. The gas introduced with the foamed extract is coalesced in the slush to develop a smaller number of larger gas bubbles contained within the slurry. The small ice particles added continuously to the tank are melted by the gentle agitation.

The extract slurry density is about 0.72 g/cc. Slurry is continuously pumped from the tank to a Sandvick freezing belt while maintaining a constant extract slurry level in the crystallization tank. The slurry is rapidly frozen into slabs. The slabs are ground and screened. Fine frozen extract from the screening is returned to the crystallizer tank while large particles are reground. The desired particle size, simulating roasted and ground coffee (90% have a particle size from 400 to 4,000 microns) are freeze dried in trays using conventional techniques.

The final freeze dried product is dark in color and has a bulk density of about 0.19 to 0.21 g/cc. The small ice crystals in the initial slush can be formed by adding fines alone as in the previous example or by using the Hoyer heat exchanger to form an extract slurry of fine ice crystals, or by any combination of art recognized methods.

AROMATIZED COFFEE

Extract Aromatized Before Freezing

W.P. Clinton, J.W. Johnson, F.W. Meyer, R.A. Pfluger and G.E. Jacobs; U.S. Patent 3,903,312; September 2, 1975; assigned to General Foods Corporation found that a stable, aromatized, freeze dried coffee may be produced by a process which involves removing coffee oil and volatile aromas from freshly roasted coffee, extracting the dearomatized coffee to obtain an aqueous extract of coffee solids, mixing 0.2 to 2% of coffee oil by weight of soluble solids in the extract with the extract, homogenizing the mixture of oil and extract, adding 0.5 to 5% of volatile aromas to the extract, freezing the aromatized extract and then freeze drying the extract under conditions which remove at least a portion of the unstable aromatic compounds added to the coffee.

Example 1: About 600 lb of blended green coffee beans were subjected to a conventional roast at between 400° to 440°F for about 15 to 20 minutes to yield 530 lb of roasted coffee. The roasted beans were then separated into a major portion (480 lb) and a minor portion (50 lb). The minor portion (50 lb) of whole roasted coffee was introduced into a commercial oil expeller and expressed at 10,000 psig in a screw press wherein the screw had flights traveling within a complementary perforated cage or screen concurrent to the feed of coffee.

About 4 lb of crude oil was obtained, which was then clarified by passing it through a commercial pressure filter to give 3.3 lb of clarified oil. The oil expression and clarification was carried out in a carbon dioxide atmosphere to prevent oil degradation. The clarified oil was then stored at 10°F under carbon dioxide until ready for use. The expeller cake resulting from the expression of the coffee was then pelletized by extruding it through 3/8" die holes and cutting it into pellet lengths in the order of 3/8 to 1/2".

The major portion of roasted coffee (480 lb) was then ground to a particle size distribution wherein 90% by weight of the particles were retained on a 20 mesh U.S. Standard Screen. The pellets were added to the ground coffee and the mixture was introduced into a conventional commercial coffee extractor approximately 20" in diameter and 20' high to give a total coffee charge of about 500 pounds. Steam at a pressure of 5 to 10 psig was introduced into the bottom of the percolator and permitted to pass through the column of coffee to wet the coffee and distill volatile aromatics.

The steam was allowed to pass through the column for a period of at least 25 minutes. Then, vapors from the top of the column were allowed to pass through a fractionator column (packed with 1/2" glass rings) 8' tall and 6" in diameter. The vapors coming off at above 180°F, about 200° to 230°F, were condensed in a brine condenser at 35°F under a nitrogen atmosphere. About 1,100 ml of the steam distillate was collected as a liquid condensate. The steam distillate collected in this manner was then held at 35°F until it was used.

The steamed coffee was extracted with water under conventional coffee percolation techniques used in soluble coffee production. About 200 lb of soluble solids were extracted and collected as a liquid extract which weighed about 500 pounds. This extract had a soluble solids concentration of about 27%. About 50 lb of extract was removed from the main extract stream. About one-fourth (0.8 lb) of the expressed oil was warmed to 65°F and dispersed in the 50 lb portion of extract by homogenization at 2,000 psig. The homogenized oil was then added to the extract stream and the 1,100 ml of steam distillate were mixed into the extract.

The aromatized extract now containing 0.4% of homogenized oil and 0.8% of steam distilled aroma was then frozen into a 1/2" thick layer of coffee extract by means of a freezing belt made of stainless steel. The belt was about 75' long,

30" wide, and was operated by two terminal pulleys which allowed the liquid extract about 60 minutes to be transported from one end of the belt to the other. Side skirts or retaining walls were used to keep the liquid extract on the belt as it was frozen. The belt was cooled by contact with three brine tanks, the first having a temperature of about 20°F for the initial one-third portion of the belt, the second having a temperature of 5°F, and the third about -10°F for the final one-third portion of the belt. In order to facilitate release of the frozen extract at the terminal portions of the belt, a water film (about ½" thick) was sprayed onto the initial portion of the belt and frozen to the belt.

Then, liquid extract having a temperature just above the ice point (31°F) was applied to the freezing belt and cooled to below its eutectic point of -13.5°F in about 40 minutes by passing the extract over the brine tanks. The frozen extract issued from the terminal portion of the freezing belt at a product temperature of about -20°F. The frozen extract was removed from the belt by breaking off frozen slabs about 20" long. This gave individual slabs having a dimension of about ½" x 20" x 30".

The slabs of frozen coffee extract were then placed in a freeze dryer equipped with horizontal shelves and an internal condenser. The frozen extract was heated by platens spaced about 1/16" from the frozen extract. A vacuum of 300 m crons of mercury was drawn on the chamber, a condenser temperature of -40°F was applied, and the platen temperature was raised to 120°F.

The pressure was not allowed to rise above 500 microns. The condenser temperature of -40°F was maintained for about 12 hours until the moisture of the coffee was reduced to about 10%. The platen temperature was then lowered to 95°F, the condenser temperature was lowered to -60°C and the pressure reduced to below 150 microns. These conditions were maintained for about five to six hours until the coffee was dried to a moisture level of about 1.7%.

The freeze drying chamber was released to atmosphere by injecting nitrogen into the system. The nitrogen thus penetrated into the pores or channels left by the ice sublimed from the extract. The dry slabs of coffee were then ground to a size wherein at least 90% of the particles were retained on a 40 mesh (U.S. Standard Sieve screen), and plated with the remaining portion (2.5 lb) of the expressed coffee oil which was concentrated to a fivefold fraction weighing 0.5 lb.

Fivefold oil is obtained by distilling the aromas from expressed oil and then adding the aroma back to one-fifth of the original oil content. The freeze dried coffee was not exposed to ambient conditions for more than five minutes prior to being packed in glass jars under a carbon dioxide head space, having less than 1% oxygen. The freeze dried coffee, having a final in-jar moisture content of about 2.2%, was stored at 95°F for three months and, at the end of this time, exhibited no appreciable change in regard to flavor or flowability (caking).

Example 2: The procedure of Example 1 was repeated with the exception that the coffee extract was concentrated by freeze concentration to 45% solids. The concentrated extract was then combined with the volatile aroma and oil according to the procedure of Example 1 and frozen slowly on the freezing belt in about 20 minutes. The frozen product was then ground as in Example 1 and freeze dried. The final product was similar in all respects to that of Example 1.

Part of Extract Stream Aromatized and Freeze Dried

W.P. Clinton, T. Kraut and E. Pitchon; U.S. Patent 3,244,533; April 5, 1966; assigned to General Foods Corporation found that a superior soluble coffee of the aromatic variety can be produced by removing coffee oil and volatile aromas from freshly roasted coffee by extracting the coffee to obtain a dearomatized extract of coffee solids, dispersing the coffee oil in the extract at a homogenization pressure of above 1,000 (preferably 1,000 to 3,000) psig to thereby avoid separation of the oil, mixing the volatile aromas with the homogenized extract while avoiding degradation of this aroma source, and then drying the aromatized extract to obtain an instant coffee of improved flavor and taste.

The level of homogenized coffee oil employed in the soluble coffee will be between 0.2 to 2%, preferably 0.5 to 1%, by weight of the soluble solids in the final coffee product. It is preferred to employ steam distilled aromas at a level of 3 to 5% with the homogenized oil. Most preferably, the steam distilled aroma is that described in U.S. Patent 3,132,947.

After the homogenized oil and volatile aromas are added to the extract, the aromatized extract may be dried, either by spray drying, freeze drying, or any other known technique. Also, the extract stream may be divided into two portions, one portion aromatized, frozen and then freeze dried while the remaining dearomatized portion is spray dried and then combined with the freeze dried product. In this manner the vapor load on the freeze drying equipment can be relieved at a small sacrifice in quality. Preferably, between 5 to 50%, say 20%, the extract stream is aromatized and freeze dried.

In freeze drying the aromatized extract, the extract is first frozen to below its eutectic point (-10° to -13.5°F) and then freeze dried under a vacuum of less than 500 microns and a product temperature of below 110°F. The frozen product may be freeze dried in the granulated form or in slab form and the freezing rate can be controlled to extend over 15 minutes if a dark colored freeze dried product is desired.

Example: Two hundred pounds of whole roasted coffee was introduced into a commercial oil expeller and expressed at 10,000 psig in a screw press wherein the screw had flights traveling within a complementary perforated cage or screen concurrent to the feed of coffee. Sixteen pounds of oil was obtained, which was then clarified by passing it through a commercial pressure filter. Twelve pounds of clarified oil and four pounds of fines were obtained. The oil expression and clarification was carried out in a carbon dioxide atmosphere to prevent oil degradation. The clarified oil was then stored at 50°F under carbon dioxide until ready for use.

The expeller cake resulting from the expression of the coffee was then pelletized by extruding it through 3/8" die holes and cutting it into pellet lengths in the order of 3/8 to 1/2". The pellets were added to 800 lb of roasted and ground coffee. The mixture was introduced into a conventional commercial coffee extractor approximately 20" in diameter and 20' high. Steam at a pressure of 2 to 3 psig was introduced into the bottom of the percolator and permitted to pass through the column of coffee to wet the coffee and distill volatile aromatics. The steam was allowed to pass through the column for a period of 25 minutes. The vapors from the top of the column were collected during the last 10 minutes steaming and condensed in a brine condenser at 35°F under a carbon dioxide atmosphere. Five thousand cubic centimeters of the steam distillate was collected. The steam distillate collected in this manner was then held at 35°F until it was used.

The steamed coffee was extracted with 3,000 lb of water under conventional coffee percolation techniques used in soluble coffee production. Three hundred pounds of soluble solids was extracted and collected as a liquid extract which weighed 900 lb. The extract was cooled to 60°F and 720 lb of the extract was then spray dried by conventional coffee spray drying wherein the extract was sprayed at 200 psig through an atomizing nozzle while air was introduced concurrently with the extract spray at a temperature of 500°F to obtain a soluble coffee powder having 3% moisture.

The remaining 180 lb of the extract was divided into two 90 lb portions. Six pounds of the expressed oil was warmed to 65°F and dispersed in 90 lb of extract by homogenization at 2,000 psig. The steam distillate was added to the other 90 lb of extract and mixed. The two 90 lb batches were then combined and frozen at a thickness of 1/8" in trays in a period of about one hour to a temperature of –30°F. The mixture of extract and aromas was placed in a commercial freeze drying unit. A vacuum of 100 microns absolute pressure was drawn on a freeze dryer and heat gradually applied to permit the temperature within the freeze dryer and of the product to rise to no more than –10°F during the first six hours.

The temperature was then permitted to rise gradually to 80°F during the next six hours, while the absolute pressure was permitted to decrease from 150 to 15 microns. The freeze dried product removed had 3% moisture. The freeze dried product was combined with the spray dried powder obtained from the 720 lb of extract which was obtained by spraying the extract. The cup flavor of the rehydrated product had desirable strength and body with the flavor notes of regular brewed coffee.

High Aroma Extract via Ultrasonic Cold Extraction

E.N. Bilenker; U.S. Patent 2,949,364; August 16, 1960; assigned to Curtiss-Wright Corporation describes a process for producing a dry coffee extract having distinctive flavor and aroma characteristics which comprises forming a slurry of freshly roasted finely divided coffee in the presence of cold liquid medium to yield a cold suspension of fine coffee particles, passing the suspension through an extraction zone in which the suspension is subjected to insonation to produce a suspension of coffee particles in coffee extract, separating the extract from the partially exhausted coffee solids, and drying it.

The partially exhausted coffee solids are subjected to further extraction with a relatively hot extraction medium in order to produce a comparatively low aroma extract. The resulting extract is dried to produce a low aroma dry extract. The low aroma coffee dry extract thus obtained is blended with the high aroma coffee dry extract obtained from the cold extraction step.

Alternately, the coffee extract resulting from the relatively hot extraction step is combined with the coffee extract resulting from the initial extraction step carried out under conditions of relatively low temperature, and the resulting combined extract dried. The process is described with reference to Figure 9.11 which

FIGURE 9.11: HIGH AROMA EXTRACT VIA ULTRASONIC COLD EXTRACTION

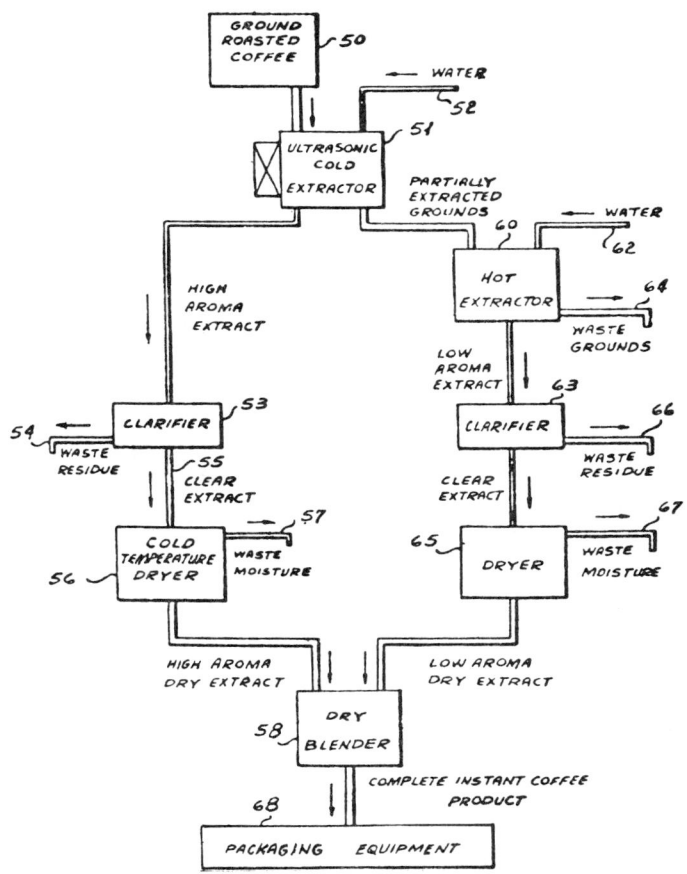

Source: U.S. Patent 2,949,364

Figure 9.11 shows a ground roast coffee source **50** from which the ground roast coffee is delivered into an ultrasonic cold extractor **51**, wherein the ground roasted coffee is subjected to ultrasonic insonation in the presence of water supplied from the inlet **52**. The resulting high aroma liquid extract is separated from the partially extracted ground coffee, and then is fed into a clarifier **53**, wherein the waste residue is removed from the clarifier via the waste residue outlet **54**, and the clear extract is fed via conduit **55** into the cold temperature dryer **56**.

The waste moisture is removed from the cold temperature dryer via the waste moisture outlet **57**. The high aroma dry extract is transferred from the cold temperature dryer **56** into the dry blender **58**. The ultrasonic cold extractor **51** is equipped with suitably oriented transducers in batch, continuous, cocurrent, countercurrent, or other appropriate type of extractor. The partially extracted grounds removed from the ultrasonic cold extractor **51** are fed into the hot extractor **60**, and in the presence of hot water fed from the water source inlet **62** result in a low aroma extract which is fed into the clarifier **63**, while the waste grounds from the hot extractor **60** are removed via the outlet **64**.

The clarifier **63** provides clear extract which is fed into the dryer **65**, while the waste residue of the clarifier is removed through the outlet **66**. The dryer **65** which removes the unwanted moisture from the clear extract via the waste moisture outlet **67** produces a low aroma dry extract which is fed into the dry blender **58** along with the high aroma dry extract. The output of the dry blender **58** is a complete instant coffee product, which thereafter is packaged as desired by suitable packaging equipment represented by block **68**.

Any suitable frequency either in the range of audible or above audible frequencies may be used for insonation of the process material. Particularly useful frequencies for the insonation of some process materials have been utilized in the range of 16,000 to 800,000 cycles per second. Preferred temperatures for some ultrasonic cold temperature extractions have been between $-10°$ and $80°F$.

In a modification of the process, hot extractor **60** can be replaced by an ultrasonic hot extractor. The cold temperature dryer can be an ultrasonic cold temperature dryer. In another variation, the clarified high aroma extract is cooled until a slush of ice crystals in concentrated coffee extract syrup is formed, the ice crystals separated from the residual concentrated syrup and the concentrated coffee extract dried in a cold temperature dryer.

Roll Milling of Instant Coffee-Aromatizing Oil Blend into Flakes

J.R. Andre, F.M. Joffe and D.A. Strang; U.S. Patent 3,625,704; December 7, 1971; assigned to The Procter & Gamble Company found that an oil-aromatized, free-flowing, easily measurable and spoonable instant coffee composition can be prepared by a process which comprises roll-milling a mixture of instant coffee particles and coffee oil to form oil-containing flakes.

The aromatized, free-flowing instant coffee composition containing instant coffee flakes has a thickness within the range of from about 0.002 to 0.01 inch and a density within the range of from about 1.00 to 1.5 g/cc, the flakes having incorporated therein from about 0.1 to 20% of an aromatizing coffee oil. A preferred amount of oil in the flakes for an aromatized, free-flowing composition is 0.2 to 15% with 0.2 to 1% being especially preferred.

The oil-containing flakes can be used per se as an instant coffee composition or combined with up to about 95% (preferably 10 to 90% and most preferably 30 to 70%) by weight of conventional instant coffee particles. The process is described with reference to Figure 9.12. Conventional instant coffee, such as spray dried particles **1** or freeze dried particles **2** serve as preferred starting materials **3** in preparing the flake-containing instant coffee product.

Irrespective of source, the coffee starting material is subjected to a densification step **4**. The densified product is size classified **5**, wherein the smaller or fine particles **6** are separated for blending **7** and the larger particles **8** are separated for final mixing **9** with flakes **20**, optionally passing through a conventional aromatization step **10** with oil **12** to form aromatized particles **8-A**.

Freeze Drying of Coffee

FIGURE 9.12: ROLL MILLING OF INSTANT COFFEE-AROMATIZING OIL BLEND INTO FLAKES

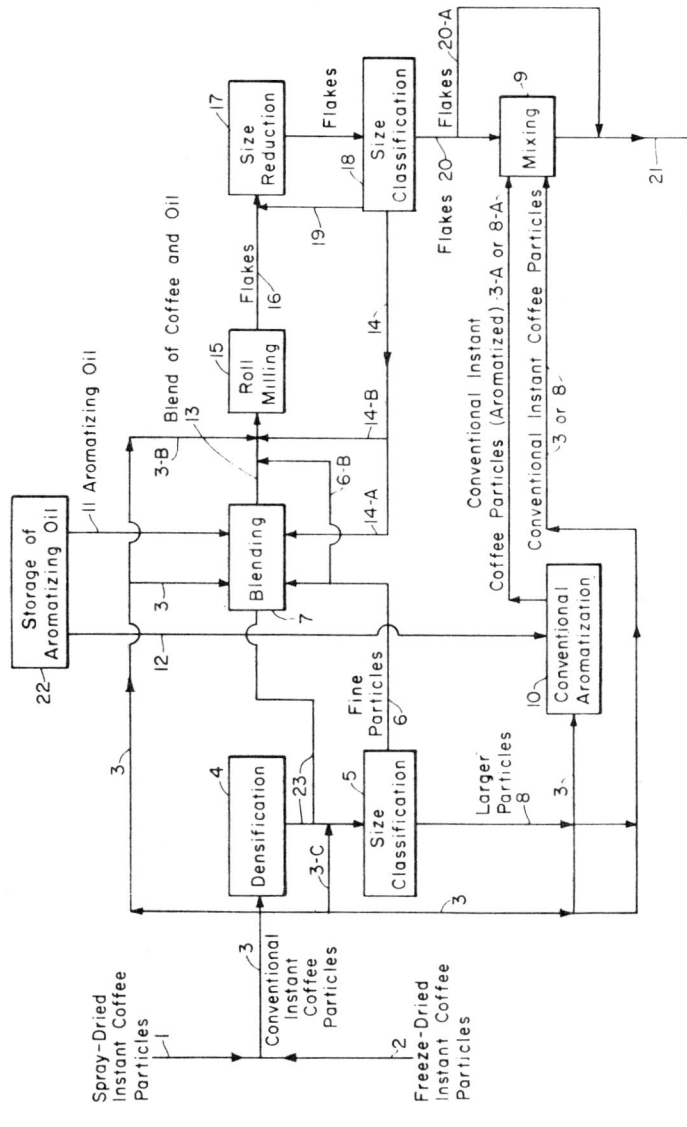

Source: U.S. Patent 3,625,704

Just prior to the time when the aromatizing oil **11** is to be used **7** in accordance with this process, it is heated to above the congeal temperature range of the oil (within the range of 60° to 70°F). The oil **11** is blended **7** in desired amounts with the coffee particles **6** and/or **3** and/or **23** to form a blend of coffee and oil **13**, which is preferably homogeneous.

In the key step of this process, the instant coffee particles **6-B** and/or **3-B** and/or the blended mixture of coffee and oil **13** are subjected to roll milling **15**. The material is fed into the nip between two rolls of a roll mill which are rotating so that the coffee material is pulled into the nip and compresses into flakes **16** which can then be removed from the roll.

Especially preferred conditions for the milling step, when it is desired to prepare flakes having a thickness within the range of from about 0.002 to 0.01 inch and a density within the range of from about 1.0 to 1.5 g/cc, are as follows when the coffee or coffee-oil blend to be milled has the specified oil content:

(A) For coffee containing less than 0.1% oil:
roll surface–very highly polished
roll diameter–4 to 22 inches
roll speeds–10 to 80 fpm, 1:1 ratio
nip pressure–1,000 to 3,000 lb/in
roll temperature–80° to 200°F

(B) For blends containing 0.1 to 1.0% oil:
roll surface–highly polished
roll diameter–4 to 22 inches
roll speeds–20 to 400 fpm
nip pressure–100 to 3,000 lb/in
roll temperature–70° to 150°F

(C) For blends containing greater than 1% oil:
roll surface–moderately polished
roll diameter–4 to 22 inches
roll speeds–20 to 200 fpm
nip pressure–100 to 1,600 lb/in
roll temperature–65° to 100°F

The milled flakes **16** are preferably size reduced **17**, and are then size-classified **18** and those deemed too small **14** can be recycled **14-A** into the blending step **7** or directly **14-B** into the milling step **15**. Those flakes deemed excessively large for the desired purpose can be recycled **19** for further size reduction **17**. As a final step in the process, the properly-sized instant coffee flakes **20** can be used per se **20-A** or can be mixed **9** with instant coffee particles **3**, **3-A**, **8**, **8-A**, or mixtures thereof, to form the desired flake-containing instant coffee composition **21**.

Freeze Drying Mixture of Spray Dried Coffee and Steam Distillate of Flavor and Aroma Constituents

R.J. Carbonell; U.S. Patent 3,554,761; January 12, 1971; assigned to Standard Brands Incorporated provides a process for producing an instant coffee which when reconstituted with hot water has substantially all of the flavor and aroma characteristics of freshly brewed, ground, roasted coffee. This is attained by subjecting ground, roasted coffee to a flavor and aroma extraction process, collecting the extracted flavor and aroma, and mixing the extracted flavor and aroma with coffee solids to provide an aqueous mixture having a solids concentration of above about 20% by weight, freeze drying the aqueous mixture, mixing sufficient amounts of the freeze dried aqueous mixture with an instant coffee to impart thereto, when reconstituted in hot water, substantially the flavor and aroma of freshly brewed, ground, roasted coffee.

Example 1: Approximately 1,500 lb of roasted and ground coffee having a particle size of about 8 mesh was introduced to fill a stainless steel, high-pressure extraction column. The column was about 200 inches high and 32 inches in diameter. Thirty gallons of water was added to the top of the column to prewet the coffee and provide an excess of unabsorbed free water at the bottom of the column. Saturated steam was introduced at the bottom of the column. The steam was at a pressure of about 5 psig measured in the steam feed line at the column and had a flow rate of about 190 lb/hr. The coffee was steamed for about 60 minutes. The volatiles started to exit from the column in 20 to 30 minutes, and were condensed in two condensers connected in series.

The first condenser was of a shell-and-tube type, having 47 square feet of surface, cooled by circulating water at about 201°F. The second condenser was of a similar type and size as the first condenser, but was cooled by circulating glycol at about 40°F which was sufficient to condense substantially all of the remaining vapors that were produced during the steaming operation. The distillate condensed in the first condenser was further cooled to a temperature of about 50°F using a subcooler. The distillate obtained with the second condenser had a temperature of about 50°F when it was collected.

After steaming, the extractor containing the steamed coffee was connected in series as a fresh coffee stage of a standard countercurrent high-pressure extraction battery, wherein it was extracted in the usual fashion. About 3,200 pounds of extract was collected, having a solids concentration of about 18% and an extraction yield of about 38% based on the weight of roasted and ground coffee charged. The pH of this extract was adjusted to about 4.8 by the addition of about 1,500 ml of a 45% solution of potassium hydroxide. This extract was spray dried in a conventional manner.

The first condenser yielded about 100 lb of hot condensate and the second condenser yielded about 40 lb of cold condensate. All the cold condensate was combined with sufficient hot condensate to provide a total condensate weight of 80 lb. The condensates and sufficient spray dried coffee to provide a 30% solids concentration therein were mixed in a ribbon mixer. The aqueous mixture was frozen, particulated and freeze dried in the conventional manner. The freeze dried material was then incorporated into the spray dried coffee at a level of 2.78% by weight (range 2.5 to 8%, preferably 2.5 to 3.5%).

The resultant product appeared to have captured substantially all of the delicate flavor and aroma of freshly percolated coffee. The cup strength of the product when reconstituted in hot water was greater than that obtained when an equal weight of commercially available spray dried instant coffee was reconstituted.

Example 2: The process set forth in Example 1 may also be carried out by mixing 100 lb of freeze concentrated coffee extract containing 40% coffee solids with 80 lb of the total condensate mixture obtained in Example 1, adding 20 lb of spray dried coffee solids in order to provide a freeze dryer feed having 30%

solids concentration. The mixture is then frozen, particulated and freeze dried as in Example 1 and incorporated into spray dried instant coffee at a level of about 6% by weight. The finished product was superior in flavor quality to that obtained in Example 1.

Extraction of Aromatic Fatty Constituents with Gaseous or Liquid CO_2

According to *L.-R. Rey, G. Pictet and A. Morand; U.S. Patent 3,532,506; Oct. 6, 1970; assigned to Societe d'Assistance Technique pour Produits Nestle SA, Switzerland* an extract of aromatic and fatty constituents from coffee is prepared by flowing gaseous or liquid CO_2 through the vegetable matter and collecting the extract in solution in liquid CO_2. CO_2-soluble extracts are used to aromatize aqueous extractives subsequently obtained from the CO_2-extracted vegetable matter by the addition of the CO_2 extract dissolved in CO_2 before, during or after the concentration of the aqueous extractives. The CO_2 is then removed.

Example 1: 1,000 g of roasted coffee, ground to a particle size of about 1 mm are moistened with 100 g of water and then placed in a double-jacketed tubular extractor. The apparatus operates as a closed circuit, and the pressure is raised to 20 atm by the admission of gaseous nitrogen. The temperature of the extractor is raised to 95°C by circulation of boiling water in its jacket, and a stream of carbon dioxide gas preheated to 45°C is passed through the batch of coffee for a period of about 40 minutes. The gas enriched with the volatile constituents of the coffee is condensed at 0°C. Approximately 3 liters of solution are collected and stored in a reservoir cooled to 0°C.

The same batch of coffee is subjected to a second extraction cycle as follows: The double jacket surrounding the extractor is completely emptied and the boiling water is replaced by a chilled brine solution so as to lower the temperature of the batch to slightly below 0°C. Gaseous carbon dioxide is rapidly condensed at around 0°C in a quantity sufficient to give approximately 5 liters of liquid carbon dioxide which is fed into the extraction column. Extraction is completed in 20 minutes, and the liberated volatile and fatty constituents in solution in liquid carbon dioxide are fed to the reservoir where they are mixed with the volatile constituents obtained by the first extraction.

900 g of the carbon dioxide solution prepared as described above are added under pressure and at 0°C to 1,000 g of a coffee extract obtained by aqueous extraction and concentrated to a solids content of 45% by weight. The mixture is stirred thoroughly to form a stable emulsion and then the pressure is gradually lowered to atmospheric pressure. The product is in the form of a cooled foam which is frozen rapidly into a relatively thin layer. The solidified extract is then broken up and ground into a powder with a particle size from 0.25 to 2 mm. Finally, this powder is spread on a tray and dried in a conventional freeze drier.

The aromatized extract, in the form of a dry powder, has a specific gravity of 0.23 to 0.25 g/cm^3. It has the appearance of a powder of roasted and ground coffee beans. The product is instant and, after reconstitution with a suitable quantity of boiling water, provides a beverage having both the aroma and favor of one prepared from freshly roasted and ground coffee beans.

Example 2: A solution of the volatile and fatty constituents of roasted coffee in carbon dioxide is prepared as described in Example 1. 1,500 g of this liquid solution are sprayed onto 1,000 g of a freeze dried powdered coffee extract spread out in an extremely thin layer over a conveyor belt passing through a chamber maintained under a pressure of approximately 8 atm. The powder contacted with the particles of solution is then exposed to atmospheric pressure, and the carbon dioxide is left to sublime gradually until it has almost completely disappeared.

The aromatized product is in the form of a dry powder having a specific gravity from 0.2 to 0.3 g/cm^3 and characteristics similar to those of the material obtained according to Example 1.

Example 3: 1 liter of a solution of volatile aromatic constituents of roasted coffee, obtained by steam stripping either of moistened roast and ground coffee or of an aqueous suspension of roasted and ground coffee, is charged into a tank. This tank is maintained under an inert gas pressure of 50 atm and about 4 liters of condensed liquid carbon dioxide are added to the solution. The two immiscible phases are maintained emulsified together for 15 minutes. To facilitate liquid-liquid extraction, it is advantageous to saturate the solution with crystalline sodium chloride.

After separation of the layers the spent, aqueous solution, which is heavier than liquid carbon dioxide, is drawn off from the bottom and discarded. The carbon dioxide solution, enriched with the volatile aromatic constituents of roasted coffee is maintained under a pressure of about 40 atm. It may be used for the aromatization of coffee extracts by any of the processes described in the preceding examples.

AROMA AND FLAVOR STABILITY

Coating Freeze Dried Beads with Fat

C.M. Buchzik; U.S. Patent 3,672,917; June 27, 1972; assigned to FMC Corporation provides a process for preparing a freeze dried product, such as coffee, which will not have a tendency to become stale quickly. In accordance with this process, a coating is applied to a freeze dried product to cover the voids in the product and to seal the voids from oxygen. An aqueous solution of soluble food concentrate such as coffee is sprayed into chilled hexane or water-immiscible liquids to freeze the concentrate into beads. The beads are freeze dried to remove ice therefrom by sublimation in a subatmospheric pressure chamber.

The freeze dried beads are placed on a screen in a chamber from which air is evacuated. A stream of warm nitrogen is introduced into the chamber and directed up through the screen on which the beads are placed. The stream of warm nitrogen agitates the particles, causing them to tumble about on the screen. A spray of a molten organic coating material, such as coconut-oil fat with or without addition of antioxidants, is introduced into the stream of nitrogen and carried thereby to the freeze dried particles.

The coating may also be a hydrogenated coconut-oil fat, or a hydrogenated coffee oil. Alternatively, the coating material may be a mixture of coffee oil and coffee bean wax. The process is described with reference to Figure 9.13.

FIGURE 9.13: COATING FREEZE DRIED BEADS WITH FAT

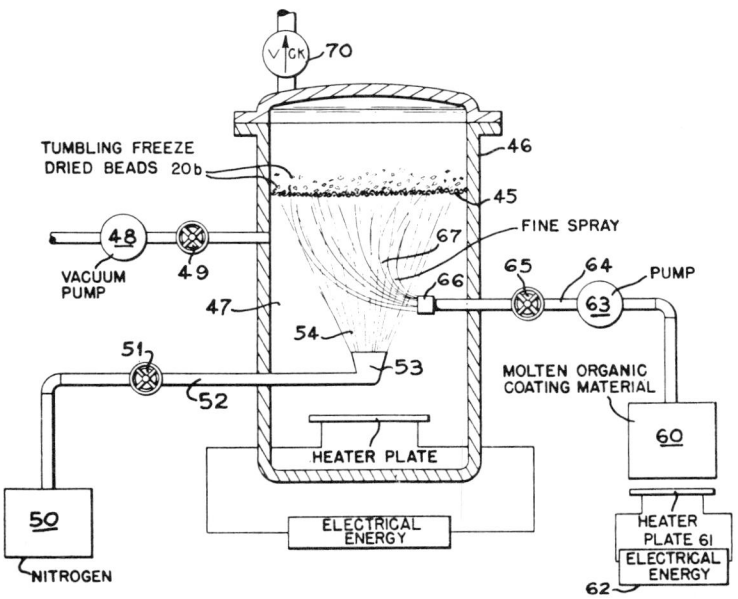

Source: U.S. Patent 3,672,917

The freeze dried beads (indicated at **20b**) are placed on the screen **45** in vessel **46**. The vessel is then sealed tight and valve **65** is closed. With valve **49** open, vacuum pump **48** is started to draw air from chamber **47**. After the air is evacuated from the chamber, valve **49** is closed and valve **51** is opened. This introduces the stream of nitrogen which is directed up through the freeze dried beads **20b**. The valve **65** is then opened to pump a spray of molten organic coating material into the nitrogen stream for elevation to the freeze dried beads which are tumbled on screen **45** by the nitrogen stream. The nitrogen (and the spray not deposited on the beads) pass through check valve **70** to escape from the vessel.

The freeze dried beads have voids, or capillary passages. These voids fill with nitrogen from stream **54** and then are sealed, but not filled, by the coconut-oil fat spray **67** which strikes the tumbling freeze dried beads to form a coating thereon. The coated bead takes on a shiny, deep-colored appearance which greatly enhances the commercial acceptability of the freeze dried coffee particles. The particles have large interior capillary area for quick rehydration, but with the surface sealed with a limited amount of fatty coating. When the coated particles are added to hot water, the coating melts quickly from the surface to permit water penetration. The fatty coating contributes to the superior mouthfeel of the product.

Pressure and Temperature Lowered for Four Hours During Vacuum Drying

W.W. Kaleda and J.W. Johnson; U.S. Patent 3,443,961; May 13, 1969; assigned to General Foods Corporation found that a stable, aromatic, freeze dried coffee may be produced by a process which comprises cooling an aqueous extract of coffee from its ice point to below its eutectic point to form a frozen eutectic mixture of water, coffee solids and coffee volatiles, this frozen eutectic serving as a matrix for crystals of water ice distributed throughout the extract; subliming the crystals of water ice at a pressure of below 500 microns of mercury; and a condensing temperature of below −30°F; and vacuum-drying the matrix to a moisture content of between 1 to 1.7% without melting the mixture. The pressure and temperature are reduced to below 200 microns and a condensing temperature of below −60°F for at least four hours during the vacuum-drying step to thereby remove at least some of the unusable volatiles from the coffee.

During the sublimation stage and the vacuum-drying (desorption) stage of drying, it is essential that the dried portions of the frozen coffee extract be maintained at a product temperature of between 60° and 120°F, preferably 60° to 108°F, in order to preserve the desired flavor of the coffee.

Example: About 1,400 lb of green coffee were introduced into a roaster of the type conventionally employed in the coffee industry. The coffee was roasted for about 18 minutes to a terminal roast temperature of 415°F to yield approximately 1,250 lb of roasted coffee beans which were then ground to a particle size range where 95% remained on a No. 20 U.S. Standard Sieve Mesh Screen and 5% remained on a No. 8 U.S. Standard Sieve Mesh Screen. Approximately 200 lb of this ground coffee was then introduced to fill a stainless steel extraction column, 15 feet high, 10 inches inner diameter.

Steam at between 1 to 10 psig was introduced at the bottom of the column and the steam pressure maintained at input within this range throughout the steam flavor volatilization cycle, which lasted approximately 30 minutes. During this cycle, approximately 40 lb of steam was supplied to the column. The volatile materials passed out of the top of the column and into a multitube vertical condenser. The condenser was cooled with brine at a temperature of 35° to 50°F and at substantially normal atmospheric pressure.

After removing the volatiles, the steamed coffee was subjected to aqueous extraction by the introduction of 5,400 g of an aqueous coffee extract produced by a plurality of previously separated extractions and having a solids content of about 26%. The extract was then divided into two portions, one portion (about 20%) was combined with expressed coffee oil obtained by pressing roasted coffee beans. The oil-extract mixture was then homogenized at about 1,500 psig and then added back to the extract stream. The volatile steam aroma was then added to the extract at a level of about 3.6 ml per pound of soluble solids. The expressed coffee oil was added at a level of about 0.4 g per pound of soluble solids.

The aromatized extract was then frozen into a ½" thick slab of coffee by means of a stainless steel freezing belt which was cooled by contact with cold brine having a temperature of about −40°F. The belt was about 50 feet long and 20 inches wide. The extract was frozen to below its eutectic point of 13.5°F in about 30 minutes and issued from the terminal portion of the freezing belt at a product temperature below −20°F. The frozen extract was removed in slab form at a dimension of about ½" x 40" x 20".

The slabs of frozen coffee extract were then placed in a freeze dryer equipped with horizontal shelves and an external condenser. The frozen extract was heated by platens spaced about 1/16" from the frozen extract. A vacuum of 300 microns of mercury was drawn on the chamber, a condenser temperature of –40°F was applied, and the platen temperature was raised to 120°F. The pressure was not allowed to rise above 500 microns. The condenser temperature of –40°F was maintained for about 12 hours until the moisture of the coffee was reduced to about 10%. The platen temperature was then lowered to 95°F, the condenser temperature was lowered to –60°C and the pressure reduced to below 150 microns.

These conditions were maintained for about five to six hours until the coffee was dried to a moisture level of about 1.7%. The freeze drying chamber was released to atmosphere by injecting nitrogen into the system. The dry slabs of coffee were then granulated, further processed and packaged under a minimum of oxygen and moisture exposure. The dried product was not exposed to ambient conditions for more than five minutes prior to being packed in jars. The coffee was granulated to a size approaching that of roasted and ground coffee (90% on a 40 mesh U.S. Standard Sieve Screen) and sealed in glass jars containing less than 1% oxygen in the head space. In jar moisture of the product was 2.2%.

The freeze dried coffee was stored at 95°F for three months and, at the end of this time, exhibited no appreciable change in regard to flavor or caking. A similar sample prepared under the same conditions but dried at a condenser temperature of –40°F throughout the freeze drying cycle, dried to a moisture content of 3% and packaged at 3.5% moisture, was found to have deteriorated both in regard to flavor and caking after only two weeks of storage.

Addition of Ungelatinized Modified Corn Starch to Extract Before Freezing

The work of *W.W. Kaleda; U.S. Patent 3,482,988; December 9, 1969; assigned to General Foods Corporation* relates to improving the stability of an aromatized freeze dried coffee extract by a process which comprises incorporating a sufficient quantity of starch or a starch derivative in contact with a liquid extract having aromatic carbonyl containing compounds to thereby react the carbonyl compounds in the aromas with the hydroxyl groups in the starch to form relatively stable acetal groups. The stabilized extract may then be frozen and freeze dried to yield a stable soluble coffee of improved flavor. The starch is present at a level of 0.5 to 15% (preferably 0.5 to 8%) by weight of the solids.

Example: About 4,000 ml of aqueous coffee extract obtained by conventional commercial extraction and having a soluble solids concentration of 27% was mixed in a homogenizer with 80 ml of expressed coffee oil and about 11 g of cornstarch (96% amylopectin and 4% amylose) at a pressure of 1,500 psig while protected with a nitrogen gas atmosphere. Steam distilled aromas obtained by steaming an elongated bed of roasted and ground coffee to obtain a reflux and rectification of the bed according to the process of U.S. Patent 3,132,947 were then blended into the homogenized extract and starch mixture at a level of about 8 ml.

The aromatized extract was then frozen to well below its eutectic point of –10°F and freeze dried in a commercial freeze dryer under a vacuum of less than 300 microns and a sublimation temperature below –10°F while keeping the temperature of the dried portion of the extract below 110° to 120°F.

The freeze dried coffee was removed from the dryer under ambient conditions of 85°F and 40% relative humidity and ground to a particle size which resembled roasted and ground coffee. The coffee was then packaged under inert conditions and stored in a 110°F storage cabinet. The total exposure time was 45 minutes.

After seven weeks of storage the above product developed no significant off-flavors or rancid notes and reconstituted to a flavorful and aromatic cup of coffee having no off-flavors. A control sample processed in the same manner without the starch as a stabilizer developed significant off-flavors after several days and the reconstituted sample had significant off-tastes.

Retaining Volatile Aromatics During Foam Freezing Operation by Avoiding Evaporative Cooling

R.A. Pfluger, M. Schulman and M.S. Hertzendorf; U.S. Patent 3,482,990; Dec. 9, 1969; assigned to General Foods Corporation found that an improved freeze dried product can be produced by retaining the volatile aromatics during the foam-freezing operation. This is done in a simple manner by avoiding evaporative cooling of the extract during the foaming and freezing step and then freeze drying the frozen foam.

Specifically, the process involves foaming the aromatic aqueous liquid to a substantial overrun while avoiding evaporative cooling of the aqueous liquid, freezing the foam to below its eutectic point while avoiding evaporation of the aqueous liquid, subliming the aqueous liquid from the frozen foam to reduce the moisture of the foam to at least 10 to 20%, and further drying the foam to a stable moisture content.

Evaporative cooling can be avoided by incorporating solid CO_2 (dry ice) in the aromatic liquid causing a one step foaming and freezing. Alternatively, a gas can be whipped into the liquid to create a foam and the foam then frozen while keeping the liquid under substantially atmospheric conditions. The gas may be carbon dioxide, nitrogen, nitrous oxide, air or any other gas, preferably a gas which does not react with or is inert to the material being dried.

Similarly, a refrigerated gas such as air or nitrogen, can be introduced into the aromatic liquid to cause freezing thereof incident to foaming the material. The frozen foam preferably has a high overrun whereby the density of the solution or suspension is changed from above 1.0 to between 0.1 and 0.7 g/cc. Any method of foam-freezing can be used as long as evaporative cooling is avoided and substantially all the aromas are retained in the frozen foam.

Example 1: Chilled roasted coffee extract (40°F) having a 30% solids content and containing 1% expressed coffee oil and 0.5% of steam distilled coffee aroma (recovered in the manner taught in U.S. Patent 2,562,206) was placed in an autoclave under a carbon dioxide gas headspace pressure of about 700 psig. The pressurized coffee extract was then discharged from the autoclave by means of a valve communicating with a 1/8" nozzle through which the pressurized extract issued continuously in the form of a foam. The nozzle was submerged in a liquid nitrogen bath which instantly froze the foam into ribbons. The frozen ribbons floated to the surface of the bath and were then atmospherically freeze dried. This was done by loading the frozen foam to a 4" depth in the form of a bed supported on a wire mesh screen. Dehydration was carried out by recirculating dry air at a temperature of -10°F through the bed for a period of 15 hours, followed by two hours of dehydration at an air temperature of 0°F and concluded by two hours of dehydration at an air temperature about 12°F to produce a cellular porous product having 7.5% moisture.

Velocity of the drying air throughout the operation was 500 feet per minute measured downstream from the bed, the air being circulated downwardly through the bed and the stream being recirculated after conditioning so as to assure that the air had a dew point upon entering the dehydration chamber in the neighborhood of -65°F. The product reconstituted to a cup of coffee with an improved flavor and aroma.

Example 2: The chilled coffee extract of Example 1 was fed to an elongated auger screw chamber adapted to retain the extract therein while it foamed under the influence of chopped dry ice. About 1 part by volume of chopped dry ice was fed through a suitable hopper to the advancing flights of the rotating screw for each 2 parts by volume of extract fed to the screw. The mixture of dry ice and extract was agitated to avoid localized freezing and foamed due to sublimation of the dry ice in the extract.

The foamed extract then issued through a nozzle in the form of continuous strips of foamed ribbon which were frozen by depositing the ribbons upon a chilled heat exchange belt having a surface temperature of -30°F. The frozen ribbons were placed on spaced heating shelves and vacuum freeze dried by using individual trays. The trays were loaded to a 1" bed depth. A vacuum of approximately 150 microns (Hg) was then drawn and the heat controlled in the initial stage of the operation to assure a product temperature of below -10°F.

After seven hours in this vacuum chamber the product was dehydrated to a moisture content of about 15% whereafter the temperature of the product was slowly raised to 80°F for an additional five hours to produce a dried product having a moisture content of 2%. These dried extrudates had a porous texture, could be granulated to any desired particle size, and upon reconstitution provided a most flavorful cup flavor and aroma.

FREEZE DRYING OF OTHER FOODSTUFFS

MEATS

Heat-Conductive Pin System for Drying Sliced Material

A.I. Nelson and E.V. Kwiat; U.S. Patent 3,446,635; May 27, 1969; assigned to University of Illinois Foundation provide a method and apparatus for freeze dehydration of materials in slice form, particularly foodstuffs such as meat, whereby the drying time is cut by 50% or more compared to conventional methods, without adverse effect on the quality of the material. The process is particularly suitable for use in drying relatively thick slices of about three-fourth inch, or more.

A plurality of heat-conducting metal pins or needles are introduced into the product to be dried and are inserted gradually more deeply into the slice as dehydration continues so as to keep contact with the gradually receding ice front within the material. These pins provide high conduction pathways for the heat energy required for sublimation of the ice, so that a high rate of sublimation can be maintained without the heat damage which would otherwise occur. Suitable apparatus is shown in Figure 10.1.

The slice of material to be dried, **42**, in frozen condition, is inserted between the pins in assemblies **22** and **23**. The pressure within the chamber **11** is evacuated through conduit **12** to a suitably low value below about 4 mm and plates **22** and **23** (made of heat-conducting metal such as steel-dried stock) are heated to a temperature suitably within the range of about 125° to 250°F or more.

By manipulation of valve **37**, atmospheric air is permitted to enter toroidal bladder **31** which expands and causes an upward force to be exerted on platform **27**, thus causing the pins to enter the frozen slice **42** to an extent determined by the total upward force which is applied.

As an alternative to the use of bladder **31**, any means such as springs, pneumatic or hydraulic cylinders and the like, can be used to create the necessary force for this purpose. Plates **22** and **23** are heated by conduction from heating plates **17** and **18**, the heat passing into the interior of slice **42** by conduction through the pins.

FIGURE 10.1: HEAT-CONDUCTIVE PIN SYSTEM FOR DRYING SLICED MATERIAL

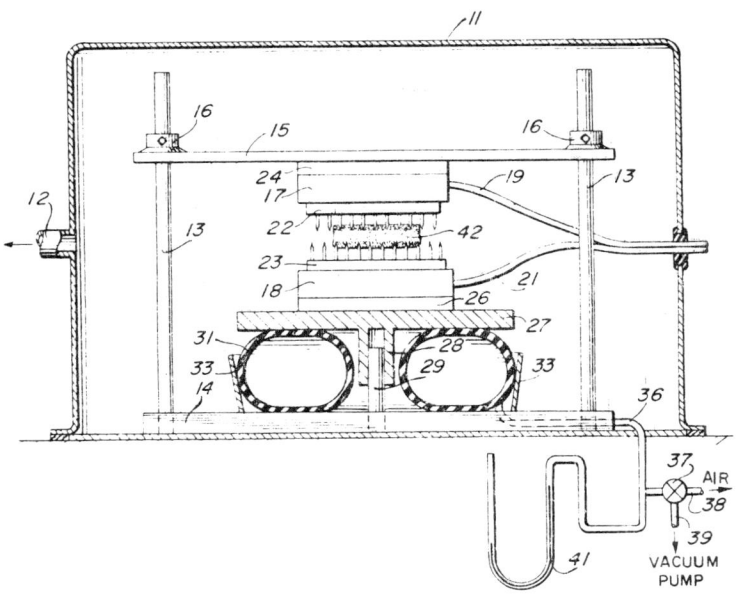

Source: U.S. Patent 3,446,635

Heater Rack

F.P. Mehrlich and R.R. Haugh; U.S. Patents 3,199,221; August 10, 1965 and 3,169,070; February 9, 1965; both assigned to U.S. Secretary of the Army provide a rack or holder for shaping, holding and heating material to be freeze dehydrated comprising superposed tiers each including a plurality of thermoelectric heating elements of angle shape in transverse section disposed to define a material holding space therebetween of substantially sinusoidal or undulatory shape in transverse section. The process is described with reference to Figures 10.2a through 10.2c.

Each heater rack **16** comprises opposed upper and lower tiers **22** and **24**, respectively, each including a series of substantially duplicate thermoelectric heating elements **26** of angle shape in transverse cross section, the angle between the sides of each heating element preferably approximating 120° and the sides being joined along an apex **28**. The heating elements **26** in each tier are rigidly fixed at opposite ends thereof to front and back crosspieces, respectively, by means of bolts **34**, the heating elements **26** in each tier being disposed in side-by-side or parallel relation uniformly spaced from each other laterally so as to define slots **36** between the contiguous or opposed edges of adjacent heating elements.

FIGURE 10.2: HEATER RACK

a.

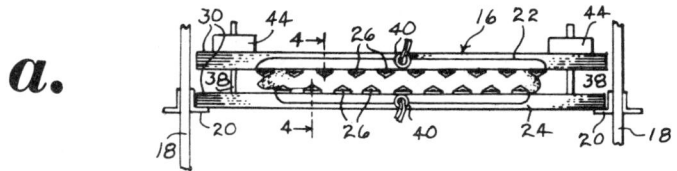

Front Elevational View of Heater Rack

b.

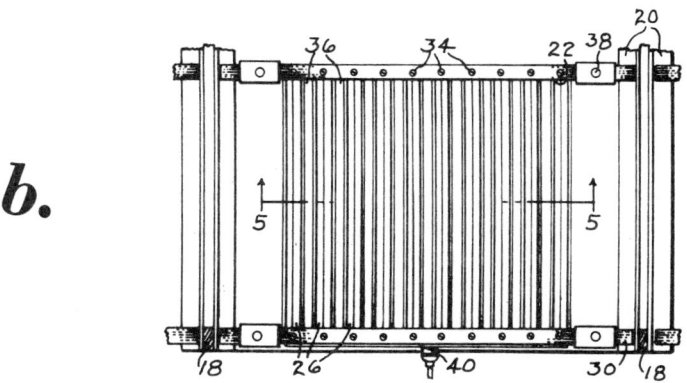

Top Plan View of Heater Rack

c.

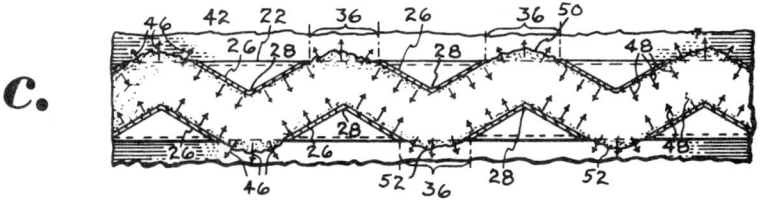

Longitudinal Section Along Line 5–5 of Figure 10.2b

Source: U.S. Patent 3,169,070

The separate heating elements in each tier may be connected to a socket **40** on the front end of the tiers. Each of these sockets in turn is connected to a suitable source of current through a lead disconnectable from the socket **40** to permit the individual heater racks **16** to be loaded upon or unloaded from the supporting rack.

Example: Using the Heater Rack — The heater rack included an upper tier **22** consisting of ten duplicate stainless steel heating elements **26** of angle shape in cross section each 18" long, ⅞" wide at the base, 0.0179" thick and a lower tier including nine similar heating elements.

The separate heating elements in each tier were disposed in parallel relation with the contiguous or opposed edges of adjacent heating elements laterally spaced to define slots 36 one-half inch wide between these opposed edges; heating elements in the upper tier were laterally offset with respect to the lower so as to center the apexes 28 of the heating elements in one tier longitudinally relative to the slots 36 in the other (when the tiers were disposed in superposed relation).

To dehydrate meat, (steak), the steak should be in the form of slices approximately one-half inch thick. These slices are placed on the lower tier 24 in unfrozen condition, and the upper tier 22 then slidably engaged upon the aligning rods 38 and pressed downwardly by means of weights 44 (See Figure 10.2a) capable of imposing a pressure of approximately 2 psi on the steak.

The resultant compressing force on the steak causes it to conform to the shape of the space between the upper and lower tiers as shown in Figure 10.2c and to be reduced in thickness from one-half to approximately three-eighths inch at least between the opposed sides of the superposed upper and lower heating elements 26.

The unfrozen steak is forced into an undulatory shape by the upper and lower tiers 22 and 24 thereby to form or define a first series of upwardly directed apical portions or peaks 50 as viewed in Figure 10.2c and a second series of downwardly directed apical portions or peaks 52. The capacity of the heater rack 16 constructed and loaded as above described ranges between 2 and 3 pounds of steak.

After loading of the heater racks 16 the steaks are frozen by placing the latter in a conventional blast freezer where the steak is subjected to about $-30°F$ for about 4 hours. The frozen steaks are then dehydrated in a vacuum chamber under an initial 3 mm pressure and later a 500 micron pressure at which pressures the frozen moisture will sublime directly to vapor. During this step, the thermoelectric heating elements are energized by a current of 20 amp at 10 volts. This causes these elements and the dehydrated portion of the steak to become heated to a temperature of approximately $200°F$, automatic control equipment of conventional construction being provided automatically to reduce the current as this temperature is approached so as not to cause overheating.

The heat thus furnished amounts approximately to the latent heat of vaporization of the frozen moisture in the steak. The slides 20 in the supporting rack 14 should be vertically and laterally spaced so that vaporized moisture can readily escape from around the heater racks 16.

Among the advantages of the process are the fact that the escape of moisture from steaks loaded in heater racks 16 is facilitated due to the fact that the steak is exposed along the slots 36, and it is flexed or stretched in these exposed apical or peak areas 50 and 52 due to the fact that there is an apex 28 of a heating element 16 longitudinally centered with respect to each slot 36. This causes the steak to be opened up along these areas 50 and 52 for the easier escape of vaporized moisture, and the slots 36 being relatively wide and closely spaced provide a wide area for escape of moisture from the heater rack 16 and shorten the path over which the moisture must travel to escape as indicated by the arrows 46 in Figure 10.2c.

This is particularly true in view of the staggered relation of the slots **36** in the upper and lower tiers **22** and **24**, respectively, since it leaves the major portion of the area of the steak exposed to the direct escape of moisture from either the top or bottom thereof.

The direct contact between the heating elements **26** and the meat shortens the path along which transfer of heat must occur as indicated by arrows **48**. By virtue of the shortened path of travel for both the escape of vaporized moisture from the meat in the heater rack and the transfer of heat thereto, more efficient operation is obtained which cuts the time of freeze dehydration by at least one-half the time required by known apparatus thus greatly reducing the cost of the process.

Infrared Sublimation

F. Oppenheimer; U.S. Patents 3,271,874; September 13, 1966 and 3,233,333; February 8, 1966 is concerned with a method and apparatus for freeze drying meats where the water of composition of the frozen product is removed by sublimation through the use of infrared energy radiated in a spectral region selected for optimum results.

The first step in the process is called precooling. The food should be refrigerated to a point close to freezing evenly throughout, i.e., the food before insertion in the vacuum chamber should be in a temperature range of +1° to +4°C, and preferably at about 2°C.

In the second step, called degasification, the cold and humid food is placed in a vacuum chamber, and the pressure therein is reduced in a succession of steps which serves to cause removal of gases from the internal structure of the food including oxygen from the cells and some moisture, but without substantially lowering the temperature of the food.

In the third step, called evaporation freezing, a full vacuum is drawn in the chamber, and the degassed food is frozen solid by evaporative cooling, an ice pack being formed which extends throughout the body of the food and is contiguous with the faces or surfaces thereof.

In the fourth and final step, called sublimation, radiant heat is applied to one surface of the food, the food having been initially placed on and pressed against a platen which is effectively transparent to infrared radiation in such a manner as to seal the contacting pores thereof. Vapors can then issue only from the free surfaces of the food. In this way the infrared rays impinge on the surface of the ice pack, engaging the platen and are conducted by the pack throughout the food product.

With sublimation then occurring only from the free surface of the food, the ice phase recedes not from the platen-contacting surface exposed to the rays, but from the free surface. Water vapor passes out through a porous layer of the material in order to escape into the vacuum chamber.

The ice boundary therefore moves inwardly and unidirectionally from the free surface to the contacting surface until the food is entirely desiccated. Thus until the ice at the very bottom boundary is sublimated, the food is not completely dried and the vapors passing through the fibers prevent cooking thereof. Figure 10.3 shows an apparatus for carrying out the process. The apparatus comprises a vacuum chamber **10** having a door or cover for admitting food.

FIGURE 10.3: INFRARED SUBLIMATION

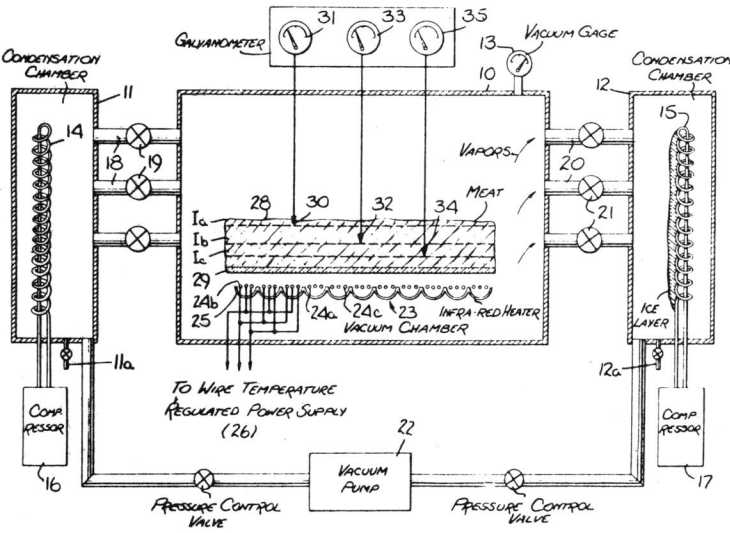

Source: U.S. Patent 3,271,874

The condensation chambers **11** and **12** each contain a freezing coil **14** and **15** for removing sublimated vapors from the chamber, the coils being connected to conventional compressors **16** and **17**, respectively, adapted to pump a boiling refrigeration fluid therethrough, e.g., propane. The condensation chamber **11** is coupled to the vacuum chamber through large ducts **18**, **20** provided with valves **19**, **21** respectively.

The condensation chambers are arranged so that all vapor must flow past it in order to reach the vacuum pump **22**. As drying proceeds, a layer of ice is built up on the condenser coils. The condensation area of the condenser should therefore be large enough so that the ice thickness is not excessive.

Horizontally mounted within the vacuum chamber is an infrared heater **23** constituted by an array of wire-like heating elements **24** disposed in parallel relation, each lying at the focal point of an individual parabolic reflector **25**. The heater elements are preferably of the high-voltage type, and are supplied from a power source **26** through a switch and a suitable voltage-adjusting element, such as a saturable core reactor. The elements **24** are black bodies which are caused to glow between 250° and 1500°C to emit at least 60% of their total radiation in the spectral range of 1 to 10 microns.

Freeze Drying of Other Foodstuffs

The food 28 to be dried is supported directly above the heating unit 23 on the flat, nonporous platen 29 which is effectively permeable to infrared energy and is not heated thereby. Preferably the material is a solid plastic sheet which is effectively transparent to infrared radiation in the selected region, such as one constituted by polymerized propylene material. The food supporting members should have minimum heat capacity so that there is no greater evaporation of water in the liquid state than is necessary to freeze the material.

The temperature of the food within the chamber is sensed by three thermocouple probes. Preferably the probes are constituted by very thin gauge wires (i.e., 25 to 50 microns in diameter) enclosed in hypodermic needle tubing, in order to minimize heat conduction and thereby obtain true readings. Probe 30 tests the temperature on the top surface of the food. Probe 32 penetrates the food at the half depth point, while probe 34 lies halfway between probe 32 and the bottom surface. All are coupled to recording means.

Example: Freeze Drying of Steak — The steak is kept in a humid condition in a refrigerator, and before being placed in the chamber, it is at a temperature in the range of +1° to +4°C, preferably at 2°C. After taking the steak out of the precooler, it is pressed firmly down on platen 29 to seal off all of its under surface pores. Hence the vapors can emanate only from the free surfaces of the steaks.

With the precooled steaks in the vacuum chamber, a vacuum is pulled for a period of 5 to 10 minutes, at which the pressure is about 10 mm. The temperature of the steak, as indicated by the three thermocouples inserted in one of them, will not change significantly at this point. Then the pressure is further reduced to 5 millimeters, and held at this level for about 10 to 15 minutes, during which gases and some liquid in the steaks are withdrawn. The temperature, as indicated by the thermocouples will still read about 2°C.

Now that the steaks have been degassed, the pressure is further reduced to 3 mm, and the temperature drops to about -2°C (thermocouple 31), -2°C (thermocouple 33), and -2°C (thermocouple 35), and then proceeds to move downward. At this point the meat is frozen and full vacuum is slowly applied (less than 200 microns) and in about 10 minutes the temperatures are now down to about -28°C, depending on the final vacuum and the water vapor pressure controlled by the temperature of the cooling coils.

The heater unit 23 is then activated to supply the latent heat of sublimation. Vaporization from the under surface is blocked by the platen 29, hence the vapors are emitted from the free surfaces of the food, and the ice boundary represented by dash lines Ia, Ib and Ic, recedes progressively from the top surface, the vapors passing through the dried pores of the food and into the vacuum chamber and from there to the condenser 11 where they form ice on the coils. When condenser 11 reaches its full capacity, the valves thereof are closed, and the valves of condenser 12 opened to put this condenser into operation.

During sublimation, the temperature of the thermocouples will remain at about -20°C for 3 to 4 hours and the surface temperature will then rise to about -18°C, then -15°C, and at the end of 6 hours it will reach 0°C. When the bottom thermocouple reaches about +15°C, the wattage of the heater is cut down stepwise until no further rise in temperature occurs. In the drying cycle, the temperature, as indicated by the surface probe, should not be permitted to rise above 15°C.

Example 2: Pork Chops — Starting with pork chops at an initial temperature of +6°C, +6°C and +6°C, the chops are pressed down on the platen to close off the under-surface pores, and then placed in the vacuum chamber, very much as in the case of the steak in Example 1. The pressure in the chamber is reduced in about 2 minutes to 10 millimeters and kept at this level for about 10 minutes, the temperature remaining at +6°, +6°, +6°C. Then the pressure is brought down to 5 millimeters in about 2 minutes, and there kept for 10 minutes, the temperatures falling to about +2°, +2°, +2°C, during which degassing occurs.

The pressure again is dropped, this time down to 3 millimeters in 1 minute, the temperature dropping to -2°C, -2°C, -2°C for about 9 minutes and then to -5°C, -5°C, -3°C. Then at full vacuum (300 microns), within 1 minute the temperatures are -13°C, -13°C, -6°C. Four minutes later (250 microns) it is -23°, -28°, -13°C. Eleven minutes later (150 microns) it is -27°, -33°, -29°C. The heater switch is then turned on and heating proceeds, for several hours, being careful never to allow any of the temperatures to rise above +15°C, until full desiccation is accomplished.

Use of Cryogenic Gas System

A basic feature of the processes of *M.R. Jeppson; U.S. Patents 3,304,617; February 21, 1967 and 3,222,796; December 14, 1965 both assigned to Cryodry Corporation* is the use of a cryogenic gas system for initially freezing the product, for imparting heat thereto in the course of drying, and for withdrawing water vapor which is released as ice within the product sublimes.

The product is placed within a drying cabinet on a gas manifold which has a large number of perforated hollow needles that penetrate into the product. The manifold, and thus the injection needles, are connected with a supply adapted to deliver gas at an adjustable temperature. In a preferred form the supply is a Dewar of liquid cryogenic gas, liquefied nitrogen being an advantageous example inasmuch as it is a readily available by-product of steel making and of liquid oxygen rocket fuel manufacture and can therefore be obtained at a relatively low cost.

The cabinet gas manifolding is connected with the supply through a first conduit which provides for the initial injection of liquid gas, or very cold vapor, directly into the product to effect rapid freezing thereof. Following freezing, the constituent water in the product is present in the form of minute ice crystals which, under appropriate temperature and pressure conditions, will convert directly to water vapor without passing through an intermediate liquid phase.

In contrast to the prior practice, evacuation of the drying cabinet is unnecessary for establishing pressure conditions under which sublimation will occur. What is required is that the partial pressure of water vapor in the cabinet be reduced to a negligible value, the presence of dry gases such as completely dehumidified air being unobjectionable. Accordingly, the use of a cabinet pumping technique which primarily withdraws only water vapor allows the process to be performed at atmospheric pressure or at any other desired pressure.

Cryogenic pumping is ideally suited for this purpose and is an advantageous technique in view of the availability of liquid gas. Thus the pumping of water vapor from the product is performed by communicating the cabinet with a pumping chamber into which the liquid gas is continually sprayed, collected and recirculated. Water vapor from the cabinet is thereby condensed and deposited on the wall of the pumping chamber in the form of frost.

To counteract sublimation cooling and accelerate the drying process, heat is delivered directly to the interior of the product by injecting relatively warm gas either continuously or in periodic bursts. This is most conveniently accomplished by connecting the cabinet manifold with the gas supply through a second conduit which includes a heat exchanger.

The injection of warm dry gas directly into the product largely avoids reliance on heat conduction across dry porous surface regions thereof and thus provides much more efficient heat transfer. In addition, the gas injection promotes drying by still another effect. The injected gas diffuses through the product to the surface thereof which gas flow promotes the removal of water vapor and does not itself have any appreciable effect on the product inasmuch as it is dry and inert.

The partial pressure of the water vapor in the product is not increased but is decreased owing to the purging action of the injected gas. With this technique the steps of refrigeration, heating and pumping may all be effected from a source of low cost liquefied gas, and drying may be done at atmospheric pressure if desired so that much of the costly equipment heretofore required in a freeze drying plant may be dispensed with. Figure 10.4 is a sectional view within the freeze drying cabinet showing details of the product support and gas injection means.

FIGURE 10.4: USE OF CRYOGENIC GAS SYSTEM

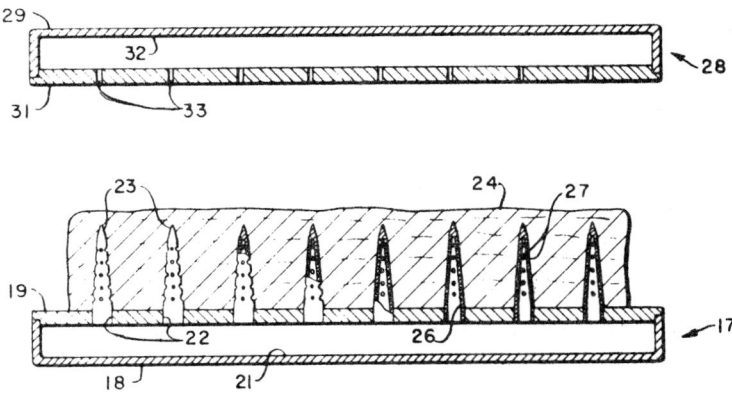

Source: U.S. Patent 3,304,617

A flat hollow rectangular gas manifold **17** is disposed within the cabinet, in a horizontal position, for receiving the products which are to be freeze dried.

The manifold is provided with a dished base member **18** which is closed by a flat top plate **19** forming a chamber **21** for connection with a temperature-regulated gas supply. The top plate is formed with a large number of apertures **22** which are closely spaced, a spacing of five-eighths inch between apertures being typical.

To provide for the injection of gas into the product, a plurality of thin hollow needles **23** are utilized, one being entered in each of the apertures and projecting directly upward therefrom for a distance sufficient to extend into the overlying product **24** to a level a small distance beneath the top surface thereof. The product which may be a cut of meat for example, is thus impaled on the needles **23** during processing.

Each of the needles is formed with an axial passage **26** which communicates with the chamber **21** in gas manifold **17**. Distributed along the length of each needle, and around the circumference thereof, are a plurality of minute gas emission passages **27** which connect with the axial passage **26**. To avoid clogging of passages **27** when the product is forced downwardly onto the needles the passages may be angled downwardly. To provide heat conduction into the product additional to that supplied by injected gas, the needles may be formed of a suitable metal such as stainless steel.

Although the product may be frozen solely by injecting cryogenic gas through the needles freezing may be accelerated by spraying the surface of the product with additional gas. Accordingly, a hollow flat rectangular spray manifold **28** is disposed within the freeze drying cabinet above manifold **17** and in parallel relationship therewith.

The manifold **28** is spaced above the tips of the needles a distance sufficient to allow the product to be easily emplaced and removed from the needles. Manifold **28** includes an inverted dished upper member **29** closed by a flat bottom plate **31** and forming a chamber **32** which is connected with the cryogenic gas supply. The bottom plate is transpierced by a plurality of narrow passages **33** which are distributed throughout the portion of the plate overlying the product and which serve to direct sprays of fluid against the upper surface of the product.

Frozen Particles on Porous Carrier Contacted with Gaseous Drying Agents

N. Tangsrud and O.G. Devik; U.S. Patent Application B 363,337; January 28, 1975; assigned to Sintef, Norway have developed a process for preparing a dehydrated powder of a meat or fish product by carrying out in sequence the steps of freezing the product to be dehydrated, subdividing it into a powder of fibrous particles, distributing this powder on a carrier means and drying it by contacting it at atmospheric pressure with a gaseous drying agent having a temperature above the thawing point of the frozen powder and moving relative to the fibrous particles of the powder.

The powder is distributed as a layer on a porous carrier means through which the drying agent is passed with an even distribution and with a velocity of about 0.5 meter per second which is sufficiently high to lift and maintain the powder layer in porous condition but sufficiently low so that the powder is not carried away by the flow of drying agent. The drying agent causes a thawing and drying of the fibrous particles which result in adhesion between the individual particles where they touch each other to form a firm porous mat having a thickness of about 15 mm. After the formation of the porous mat, the velocity of the drying agent is gradually increased up to about 5 meters per second and the drying is continued at the increased velocity to bring the dry content thereof up to about 95%, after which the mat is torn up to a dehydrated powder.

However, it is known that after a critical limit for different products has been reached, about 80%, the further drying may take place with a more free choice of drying conditions. It is therefore advantageous to complete the drying in a different manner, for instance after tearing up the mat.

Example 1: Fish — 200 grams of cod fillet, frozen to –20°C, was subdivided in a nail mill to a defibrated mass which was immediately transferred to a porous bottom, whose porosity and resistance against passage of the drying air was sufficient to secure an even velocity of the drying air forced through the bottom.

For building up the layer of the product the velocity of the drying air was adjusted to about 0.5 meter per second which was the highest velocity that could be used without carrying away the food product. After the layer had been built up to a thickness of 15 mm the temperature of the drying air was increased to 30°C. As the thawing proceeded it appeared that the fish fibers adhered to each other so that the velocity of the air could be substantially increased, up to 5 meters per second. After 15 minutes the food product was taken out for assay. The dry content had then increased from 20 to 85%.

The taste and the smell of the dehydrated fish powder were essentially better than of fish powder prepared by previously known drying methods of similar type, and the storage stability was the same or somewhat better. Upon an after-drying at 52°C, the dry content was brought up to 95%.

Example 2: Meat — Clean cut beef was cut into cubes and steamed for 2 minutes at 105°C, whereafter the wet meat was frozen down to –28°C. After subdivision on a nail mill the resulting powder was spread on to a porous bottom in an even layer. For building up the layer of the meat the velocity of the drying air was adjusted to 0.3 to 0.5 meter per second. The building up of the layer was carried out as in Example 1 with the exception that the increase of the air velocity and temperature was somewhat slower, although the final values were the same. The resulting total dry content was 78%. Upon an after-drying at 52°C, the dry content was brought up to about 92%.

Drying Gas Maintained at Subatmospheric Pressure and Controlled Temperature and Velocity

The method of freeze drying developed by *G. Tooby; U.S. Patent 3,487,554; January 6, 1970* includes the provision of a cylindrical drying chamber that is adapted for both freezing a product undergoing drying and freeze drying the product. The drying chamber includes a portion adapted to store a product to be dried wherein the product is subjected to a dry, inert gas at a velocity, pressure and temperature to supply the heat of sublimation of the fluid in the product and simultaneously carry away the vaporized fluid from the surface of the product. This is effected without altering the characteristics of the product due to heat received from the supporting structure.

To this end, the product stored within the drying chamber is subjected to a controlled cold stream of an inert gas, such as nitrogen, to freeze the product including the liquid or water within the product. The freezing is controlled to produce ice crystals throughout for defining capillaries upon sublimation of the ice crystals.

The nitrogen gas is then exhausted from the drying chamber and the exhausted chamber is filled with an inert dry gas, such as hydrogen, introduced therein at a subatmospheric pressure within the range of 20 to 80 millimeters of mercury, absolute pressure, and at a preselected velocity. The temperature of the gas is proportioned to provide the heat of sublimation to the product and passed thereover at a velocity to cause the sublimation vapors to be simultaneously carried away.

Specifically the temperature of the hydrogen gas (20° to 50°C) and the velocity (1,000 to 6,000 ft/min) with which it is passed over and in contact with the product undergoing drying are proportioned to maintain the rate of heat transfer and the rate of diffusion of the liquid in balance to cause the product to dry uniformly, completely and linearly with time. The internal temperature of the product during the drying period is by this means controlled to not only preserve the basic product characteristics but also the capillaries defined during freezing. The process is described with reference to Figure 10.5.

The drying chamber **10** is a generally, cylindrical metallic shell **11** capable of withstanding an external pressure of one atmosphere. The drying chamber can be considered to be divided into a cooling or refrigeration section R, a heating or heat exchange section H and a product storage section S for storing the product to be dried.

The drying chamber **10** is defined with the aforementioned three sections and a central or axial passageway through the drying chamber to allow the gas introduced therein to be conveyed horizontally through each of the sections. The drying cylinder is further defined to allow the entrapped gas to circulate outside of the axial passage to the outer walls of the chamber through the storage section S.

The gas then passes backward through the sections S, H, and R, in the reverse direction, in the passageway defined between the outer walls of the shell **11** and the walls defining the central passageway. In this fashion, the freezing or drying gas introduced into the chamber can be successively passed through the refrigeration section R, the heat exchange section H and the product storage section S whereby its temperature is successively modified and then recirculated back in the reverse order to again be successively modified.

Although the process is applicable to various products, it will now be described in terms of dehydrating a meat patty, which consists of raw, ground beef of a conventional grade and has been formed into a patty approximately 4½ inches in diameter and ⅜ of an inch thick. These meat patties are then placed on storage trays which are then introduced into the drying chamber **10** through the open door **38** and stored in the storage sections S_L and S_R. The thermometer is then inserted into the center of the meat patty, after which the door is closed and clamped shut to provide a gas-tight chamber.

Freeze Drying of Other Foodstuffs 325

FIGURE 10.5: DRYING GAS MAINTAINED AT SUBATMOSPHERIC PRESSURE AND CONTROLLED TEMPERATURE AND VELOCITY

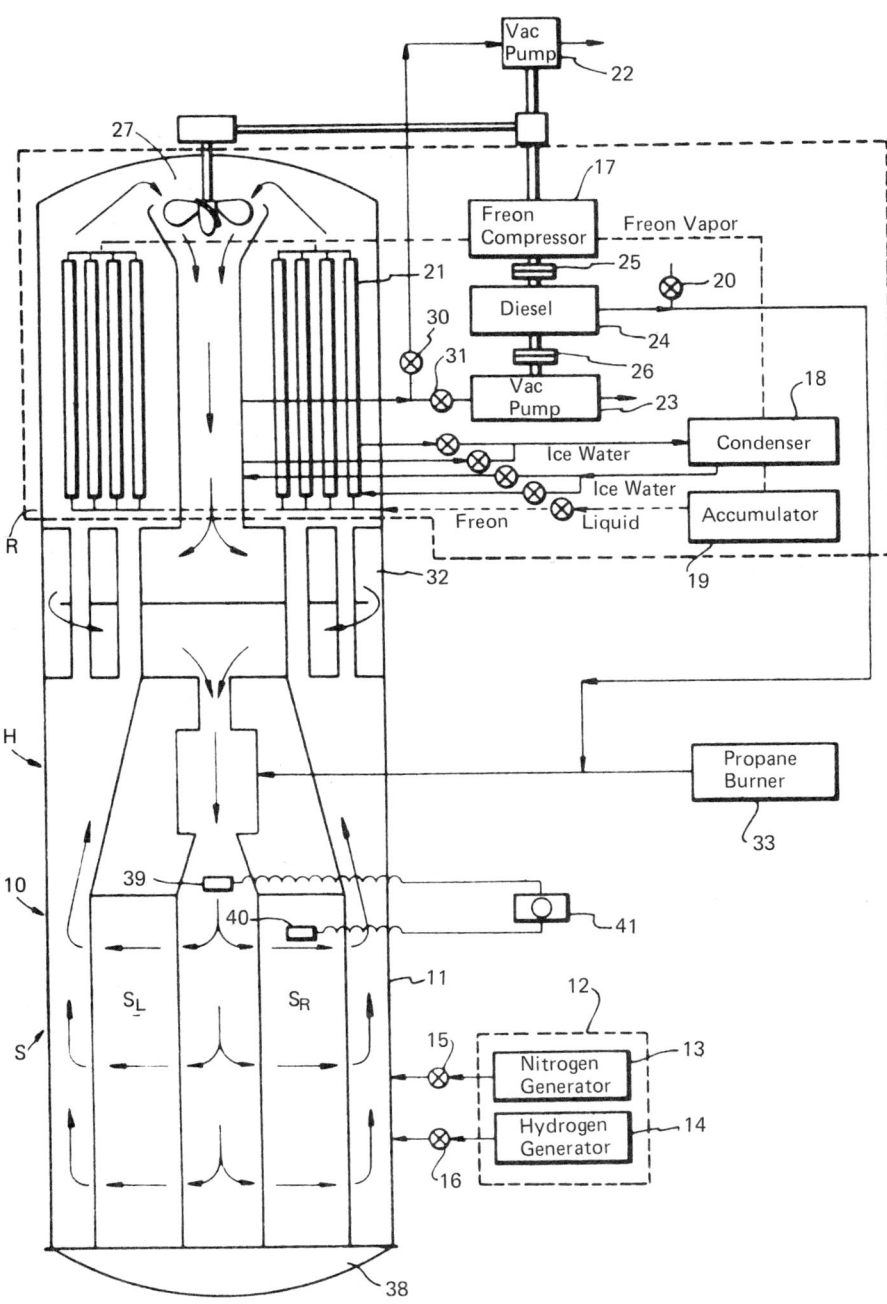

Source: U.S. Patent 3,487,554

To initiate the dehydration procedure, the diesel engine 24 is started and its exhaust valve 20 is operated to cause its exhaust to be vented directly to the atmosphere. With the operation of the diesel engine, the clutch 26 is engaged to cause the large capacity vacuum pump 23 to be placed into operation. The vacuum pump will exhaust the drying chamber to an absolute pressure of approximately 5 millimeters of mercury in about 10 minutes.

The drying chamber 10 is exhausted in this fashion to remove the air and, more particularly the oxygen, from the interior of the drying chamber to minimize any oxidation of the product undergoing drying, i.e., the meat patty. After the drying chamber is exhausted to the desired pressure, the clutch is disengaged to stop the vacuum pump. At this time, the control valve 15 for the nitrogen generator 13 is opened to introduce the nitrogen gas into the drying chamber 10 and to fill the chamber with a dry, oxygen-free nitrogen to substantially atmospheric pressure.

After the drying chamber 11 is filled with the inert nitrogen gas, the clutch 25 is engaged and thereby simultaneously places in operation the Freon compressor 17, the low capacity vacuum pump 22 and the fan 27. The Freon compressor coacts with the condenser 18 and the accumulator 19 and the ice accumulator and water cooler 21 to chill the inert dry nitrogen in the chamber. The nitrogen gas is preferably chilled to a temperature on the order of 20°F.

The pump 22 and the valve 30 will maintain the gas at the preselected pressure. With the operation of the fan 27, the nitrogen gas is circulated at a high velocity on the order of 500 feet per minute, throughout the drying chamber. This gas velocity may be provided by any conventional means including controlling the speed of the diesel engine. This stream of cold nitrogen is blown or conveyed over the meat patty for approximately 30 minutes to cause any fluid or moisture in the patty to be frozen into ice crystals. The temperature and velocity for the nitrogen gas have been selected to produce the controlled freezing to cause the ice crystals to form completely throughout the meat patty, for defining the capillary channels upon sublimation of the crystals.

After the meat patty is subjected to the cold stream of nitrogen gas for 30 minutes, the clutch 25 is disengaged and the clutch 26 is engaged to operate the vacuum pump 23 for evacuating the drying chamber to a pressure on the order of 5 millimeters of mercury.

Once the drying chamber 10 has been evacuated, it is filled with hydrogen from the source 14 by the operation of the hydrogen control valve 16 to introduce the hydrogen into the chamber. Once the hydrogen is introduced into the chamber, the clutch 25 is engaged to start the refrigeration system, the fan 27 and the small vacuum pump 22. At this time, it is desired to maintain the hydrogen gas at a subatmospheric pressure on the order of 38 millimeters of mercury.

Consistent with the desire to produce a dehydrated product having substantially the same characteristics and appearance as before drying, the amount of oxygen present within the drying chamber 11 should be maintained at no greater than 2% of the quantity of hydrogen gas therein. With the operation of the refrigeration system, the hydrogen gas is cooled to approximately –40°F to freeze any water vapor in the hydrogen stream and thereby prevent the hydrogen stream from carrying the vapor to the product undergoing drying. It should be appreciated that a stream of very dry, inert hydrogen having a temperature of −40°F leaves the ice accumulator **21** and enters the regenerator **32** of the heat exchange section H of the drying chamber **10**. The hydrogen stream upon entering the regenerator **32** is heated to approximately 15°F.

This heated hydrogen stream, then, enters the heater section of the heat exchange section wherein it is further heated by means of the propane burner **33** (and the exhaust from the diesel engine) to a temperature of 80°F. This heated dry gas is then conveyed through the storage section and passed over the meat patties stored on the trays.

The temperature of the hydrogen stream has been selected to cause the ice crystals formed therein to sublime. In addition, the velocity of this gas stream has been selected to place in balance the heat transfer rate due to the heated gas and the diffusion rate to allow the fluids of vaporization resulting from the sublimation to be simultaneously carried away from the surface of the meat patty.

The other aspect of this drying procedure is that the product is so controlled that it keeps itself frozen through its high rate of evaporation. This maintenance of the product in the frozen state is assured by the provision of the supporting trays and the structure of the storage sections to cause the product to receive heat solely from the drying gas. This includes the prevention of heat transfer to the product from radiation through the shell **11**, the storage structures, as well as conduction from the supporting structures.

It should also be noted that the use of hydrogen at the disclosed pressures, temperatures and velocities has been found to be at least twice as effective as the other inert gases including nitrogen and, therefore, is thought preferable. As a result of being conveyed through the storage section S the hydrogen stream is cooled to a temperature of about 30°F. This hydrogen stream passes back along the inner walls of the drying chamber **10** back to the refrigeration section R.

At the regenerator **32**, this hydrogen stream is further cooled to approximately −20°F due to the exposure from the very cold hydrogen stream leaving the refrigeration section. From the regenerator **32**, the gas flows through the refrigeration section where it is again cooled and recirculated through the axial passageway. The drying chamber **10** is operated in this drying cycle for approximately 1 hour.

An important aspect of the method is the control of the internal temperature of the product undergoing drying (the meat patty). It is important to control the internal product temperature to preserve the capillaries defined by the sublimed ice crystals. Accordingly, in terms of the meat patty under consideration, the internal temperature is governed by the lowest temperature at which the beef fat will melt.

This upper limit has been determined to be 80°F and accordingly the meat patty is never heated above the 80°F while the interior of the meat patty is not allowed to rise above 25°F until the moisture in the product has been reduced below a critical value of about 22% moisture. To maintain the desired temperatures, after the initial phase of the drying cycle, it is necessary to reduce and control the temperature of the drying gas. This is done by regulating the amount of heat supplied the heater proper from the diesel exhaust and the propane burner 33. From the graphical representation of the drying cycle, it was noted that after operating the system at the initial drying gas temperature (80°F) for approximately 1 hour, the gas temperature is reduced to 38°F. The reduction of the temperature of the gas entering the storage section S will also automatically maintain a balance of the temperature throughout the rest of the drying chamber 10, in view of the laws of heat balance and transfer.

The drying cycle continues at this reduced drying gas temperature for approximately 3½ hours. At the end of this interval, the meat patty should be sufficiently dried whereby the water or liquid remaining in the meat patty is below the level where heating of the product would not change the basic properties thereof.

Accordingly, to complete the dehydration as rapidly as possible the drying gas is increased in temperature to its initial drying temperature of 80°F. At the end of approximately 1½ hours the meat patty will have obtained its desired degree of dryness and the dehydration cycle will have been completed. It will then be seen that the complete dehydration of the meat patty will require approximately 7 hours.

Repeated Cycles of Gas Pressure Variation During Drying

J.D. Mellor; U.S. Patent 3,352,024; November 14, 1967; assigned to Commonwealth Scientific and Industrial Research Organization, Australia describes a method of freeze drying frozen food and similar heat-labile substances comprising the steps of subliming the ice by applying heat to the frozen food or substances while at the same time maintaining the food or substance in partial vacuum, and removing the water vapor formed by the sublimation.

The method is characterized in that the vacuum pressure is cyclically varied between predetermined higher and lower levels during drying. The optimum pressure and duration of the periods of low and high pressure, will depend upon the nature of the product, the gas composition, the dimensions of the chambers and the efficiency of the vacuum and refrigerating systems and can be determined by experiment, or assuming the nature of the gas and characteristics of the machinery are known, by calculation.

In general the duration of the near optimum pressure cycle can be determined by taking particular fractional values for (1) the transition from zero to steady state velocity of the water vapor to obtain the low pressure time interval, and of (2) the constant pressure thermal conductivity of the dry product to obtain the high pressure time interval.

For air as the gas the fractional values lie in the range of 0.85 to 0.93 (and usually will be 0.89) for both low and high pressure parts of the cycle. For gases having a higher thermal conductivity than air, such as helium, the duration of the near optimum pressure cycle for the same product will be shorter. The low pressure time interval will be determined by a fractional value of 0.85 to 0.93 as before since it depends very largely on the pumping speed. However, the high pressure time interval will be determined by values lower than the above range for air since this time interval depends upon the thermal conductivity of the gas. The corresponding range of time intervals for the aforementioned fractional values is generally a low pressure time interval of the cycle between 1 and 2½ minutes, and a high pressure time interval from 1 to 9 times as long as the low pressure intervals.

In the cyclic process it has been found necessary to reduce the heat input to the product in the final stages of drying to prevent overheating it, but early reduction in heat input is not so important as in the existing freeze drying process as constant pressure.

Example 1: Cooked Meat Product — A pilot freeze drying run was carried out under cyclic vacuum pressures in trays loaded with 3.3 pounds frozen cooked meat product per square foot of tray area. When the cyclic pressure was caused to vary from 0.4 to 20 torrs with the heaters at 252°F, the temperature of the refrigerated coil varied from -26° to -31°F, and that of the product from -4° to -11°F. The temperature of the heaters was caused to fall gradually to 60°F 2 hours before the end of the cyclic pressure run.

The high pressure interval of the cycle was set at about three times the low pressure interval which was longer than a minute. Another cyclic pressure run was also carried out under similar conditions but with the heater temperature at 280°F, this being the highest temperature that was possible without scorching the surface of the product.

For comparison, a further freeze drying run was carried out in trays loaded with 3.3 pounds frozen cooked meat product per square foot of tray area under a constant vacuum pressure of 0.4 torr. The temperature of the heaters was 255°F, the refrigerated coil -31°F and the product -4°F. All the temperatures were maintained fairly constant during the run. The temperature of the heaters was caused to fall gradually to 60°F 3 hours before the end of the constant pressure run.

Drying times of 3½ to 5 hours for the two heater temperatures of 252° and 282°F were obtained with cyclic pressures, as compared with 8 hours under constant pressure conditions. The cyclic freeze drying time represents a decrease of 37.5% against the comparative freeze drying time under constant pressure, and a decrease of 50% over some existing freeze drying times, using similar type heaters, and the same product.

Example 2: Whole Egg Pulp — Pilot freeze drying runs were carried out as in Example 1 with 2.9 pounds frozen whole egg pulp per square foot of tray area. At a constant pressure of 0.5 torr and a temperature of 201°F at the heaters, the refrigerated coil and product temperatures were -47° and -2°F respectively. When the cyclic pressure was varied from 0.35 to 20 torr with the heaters at 208°F, temperature variations in the coil and product were -44° to 54°F, and +1 to -6°F respectively. The heater temperatures were gradually reduced as before. The ratio of high to low pressure intervals in the cycle was about 2:1 with the low pressure interval slightly longer than before.

Drying times of 7¾ and 10 hours for cyclic and constant pressures respectively were obtained which represent a decrease of 25% in drying time of the cyclic process over the constant pressure process.

Example 2: Apple Rings — Pilot freeze drying runs were carried out as in Example 2 with 2 pounds frozen apple rings per square foot of tray area. At a constant pressure of 0.5 torr and a temperature of 201°F at the heaters, the refrigerated coil and product temperature were −47° and −7°F respectively. When the pressure was varied cyclically from 0.35 to 20 torr with the heaters at 214°F temperature variations in the coil and product were from −36° to −54°F and from 19° to 12°F respectively. The heater temperatures were gradually reduced to 60°F as before. The high and low pressure intervals in the cycle were each over 2 minutes and were below the optimum.

Drying times of 7¼ and 12½ hours for cyclic and constant pressures respectively were obtained. These represent a decrease of 42% in drying time by the cyclic process over the constant pressure process. This is a 15% decrease in drying time over an existing contact heating process at constant pressure for diced apples.

Freeze Drying at Atmospheric Pressure

H.T. Meryman; U.S. Patent 3,096,163; July 2, 1963; assigned to the U.S. Secretary of the Navy has developed a process for freeze drying at atmospheric pressure. This is applicable to large-scale freeze drying of a wide variety of foodstuffs, such as milk, juices, soups, coffee extract, and solids such as meats or vegetables.

In the process a stream of dry air or dry inert gas is flowed past the specimen or material being freeze dried so as to in effect sweep away water molecules which reach the surface of the specimen from its interior. The transfer of water vapor from the vicinity of the ice crystal at the drying boundary through the shell of the already dried material to the surface of the specimen is achieved by a very thorough drying of the air which flows across the specimen. In this manner, it is possible at atmospheric pressure to create a partial pressure of water vapor at the specimen surface which approaches zero and simultaneously to remove effectively water molecules as they reach the specimen surface.

In the process, the heat transfer may be introduced by conduction from the flowing drying gas. The drying gas can be flowed over all of the surfaces of the drying specimen so that the specimen is never shielded from the source of heat.

Temperature control of the air which flows through the drying chamber is easily achieved, and it is possible to therefore have absolute knowledge of the temperature of the specimen surface, which is not possible in a radiation system.

The material to be freeze dried should be granulated, and preferably relatively finely granulated. Granule size is a prime factor in determining the rate of freeze drying. The top limit for size will vary depending on the nature of the material being dried, but generally the granules should be no larger than half-inch cubes, and preferably appreciably smaller. With relatively large size granules, it is difficult to get the drying gas around the granules, in order to effect drying; moreover the time requirements needed to effect penetration are greatly increased.

The drying gas should of course, be nonreactive with the material being dried.

Freeze Drying of Other Foodstuffs

Where practical, air is the preferred drying gas. However, in situations where oxidation is a problem, an inert gas may be used as the drying gas. The drying gas may be dried by a variety of techniques, and then recirculated. Two suitable techniques for drying the drying gas are the use of a desiccant, and the use of a refrigerated condenser, although if desired, both the use of a desiccant and refrigerated condenser within a single system may be effected. If a desiccant is utilized, it should have a very low vapor pressure and high water capacity and be nonreactive with the drying gas. If a refrigerated condenser system is used to remove moisture from the drying gas, it should be associated with a subsequent heat exchanger for restoring the drying gas to the operating temperature.

Example: 57 grams of prefrozen granular beefsteak meat ground to hamburger proportions and disposed in a disk having a thickness of one and one-eighth inches was freeze dried in accordance with the process from a water content of 68.9 weight percent to a water content of 4.79 weight percent in 10 hours. The air temperature used to effect the drying was maintained between $2°$ and $4°C$, while the ground beefsteak was maintained during drying at a temperature of $-5°C$.

The air was desiccated by contact with a refrigerated condenser plate which lowered its temperature to approximately $-60°C$. From the desiccation stage, the drying air was warmed to the temperature of $2°$ to $4°C$ by heat exchange, and then contacted with the meat, after which the air was passed to the refrigerated condenser.

Controlled Humidity Freeze Drying Process

C.J. King; U.S. Patent 3,964,174; June 22, 1976; assigned to The Regents of the University of California found that if almost any food product is freeze dried in a limited manner so as to leave behind a moisture content in the range of 5 to 25% such a limited freeze dried product may be compressed to perhaps one-half to one-third or less of its original volume without crumbling or powdering, or appreciable harm to the tissue structures.

The compressed, limited freeze dried product is usually subsequently further dried to reduce the moisture content thereof to the usual 1 to 3%, after which the compressed freeze dried product is packaged and stored in the customary manner. Such products, upon reconstitution, regain most of their original volume and appearance and for all intents and purposes are comparable with uncompressed freeze dried products.

Freeze drying of foods is carried out under controlled humidity conditions whereby the dried products retain moisture in the 5 to 25% range and are thereby rendered suitable for compression. Humidity in the drying chamber is controlled by the presence of suitable hydrating inorganic salts that pass from a first low hydrated state to higher hydrated states during the drying process.

During the hydration transition, the salts regulate the relative humidity in the drying chamber and further may furnish thermal energy for transmission to the drying foods and sublimation of moisture therefrom at reduced atmospheric pressure.

Different types of food products require different amounts of residual moisture in order to undergo compression without destroying the desirable tissue structures and keeping qualities thereof. These desirable residual moisture levels must be determined empirically.

In order to dry the food products to the desired level of residual moisture, the drying is conducted, most usually, under reduced atmospheric pressures, with the relative humidity in the atmosphere being maintained by hydrated salts undergoing hydration reactions. Some useful salts are listed in Table 1.

TABLE 1

Hydrate	Transition (No. of H_2O)	Density (g/cc)	g H_2O/cc Hydrate	Relative Humidity
$MgSO_4$	(5-6)	~1.8	~0.15	41-43% at 25°C
$CaSO_4$	(½-2)	~2.4	~0.45	35% at 20°C
$Al_2(SO_4)_3$	(6-18)	~2	~1	30.3% at 20°C
$CaCl_2$	(4-6)	~1.5	~0.3	21% at 20°C
$CaCl_2$	(2-4)	0.835	0.20	14% at 20°C
$Ca(NO_3)_2$	(1-2)	~2	~0.2	10% at 20°C

The required relative humidity necessary to reduce the water content for a particular food product must also be empirically determined. Some work has been undertaken in this regard, and data are presented for several food products in Table 2 below, along with possible salts and their related hydration transitions that are capable of establishing the required relative humidity for limited freeze drying of the foods.

TABLE 2

Food	Moisture Level Desired, %	Relative Humidity Required, %	Suitable Salt
Cooked beef	8-14*	22	$CaCl_2 \cdot 4H_2O \rightarrow CaCl_2 \cdot 6H_2O$
Cooked peas	12	40	$MgSO_4 \cdot 5H_2O \rightarrow MgSO_4 \cdot 6H_2O$ or $CaSO_4 \cdot \frac{1}{2}H_2O \rightarrow CaSO_4 \cdot H_2O$
Cooked shrimp	9	8	$Ca(NO_3)_2 \cdot H_2O \rightarrow Ca(NO_3)_2 \cdot 2H_2O$ or $CaCl_2 \cdot H_2O \rightarrow CaCl_2 \cdot 2H_2O$
Canned tuna	12-17	32-52	$MgSO_4 \cdot 5H_2O \rightarrow MgSO_4 \cdot 6H_2O$ or $CaSO_4 \cdot \frac{1}{2}H_2O \rightarrow CaSO_4 \cdot 2H_2O$
Cooked chicken	10-15	20-45	$CaSO_4 \cdot \frac{1}{2}H_2O \rightarrow CaSO_4 \cdot 2H_2O$ or $Al_2(SO_4)_3 \cdot 6H_2O \rightarrow Al_2(SO_4)_3 \cdot 18H_2O$

*11 optimum

Operation of the Limited Freeze Drying Process: Limited freeze drying utilizing hydrating salts to maintain a desired environmental relative humidity can be carried out in suitable apparatus somewhat similar to that already used in conventional freeze drying suitably modified for the special requirements of limited freeze drying with hydrating salts.

Specifically, a vacuum chamber is provided with the necessary vacuum pumps capable of reducing the pressure within the chamber to a moderate vacuum in the range of from about 0.1 to 50 mm of mercury, absolute pressure. Means must be provided to convey moisture given up by the freeze drying food to the hydrating salts and at the same time convey thermal energy generated by the hydrating salt back to the food.

A particularly advantageous arrangement is a layered-bed vacuum chamber where a series of mesh or perforated trays are stacked, one over the other. The trays alternately hold the food product and hydrating salt. The trays are surrounded in the chamber by solid partitions so that circulating, reduced pressure gas in the pressure range 3 to 100 mm Hg (preferably 5 to 50 mm Hg) is forced vertically through each alternate food salt layer in turn.

Fan or blower means must be provided for circulation of the gas (most usually air, although other gases can be efficaciously used). Thus the evolving moisture is conveyed to the hydrating salt while the generated heat is conveyed to the freeze drying food.

The gas velocities are not critical, but must be sufficient to ensure as uniform a thermal distribution between salts and food product as possible and a similar uniformity of relative humidity in the chamber. Velocities of about 25 feet per second and higher have been found adequate for freeze drying beef and turkey meat loadings of 1 pound per square foot wherein the food products were in 1 centimeter cubes.

The quantity of hydrating salt necessary to dry a given quantity of food product can be calculated from the grams of water per cubic centimeter of hydrate taken up in a particular transition (see next to last column in Table 1). The number of grams of water to be removed from a given quantity of food product is calculated and sufficient hydrated salt is charged into the vacuum chamber to furnish the necessary water uptake capacity. In addition, a small excess is added to absolutely ensure that the desired relative humidity is maintained through the end of the freeze drying period.

After loading the food and desired hydrated salt into the vacuum chamber, a seal is effected and the pressure is reduced to the indicated 5 to 50 mm Hg level. The blowers are started and freeze drying commences and is continued until the ice is completely sublimed from the food. Drying times take on the order of perhaps 6 to 10 hours depending on the size of food pieces, their tissue nature, the hydrating salt, the atmospheric pressure, and the like.

Upon complete ice sublimation from the food product, the chamber atmosphere is increased to ambient pressure and the product is removed for further processing, i.e., compression and subsequent complete drying or packaging.

The salt, largely in the higher hydrated state, is processed to regenerate the lower hydrate. This operation can be quite conventional, generally comprising subjecting the higher hydrate to a controlled vacuum drying to drive off one or more molecules of water from the salt. The regenerated lower hydrate is then recovered for recycling in the freeze drying chamber.

Example — Samples of frozen beef and turkey (desired limited freeze dried moisture level, about 11%) were processed in a layered-bed, circulating-gas vacuum chamber utilizing calcium chloride hydrate as the hydrating drying salt. The conditions of the freeze drying run and the results thereof are given below.

Conditions of Freeze Drying Run

Circulating gas	Air
Pressure	30 mm Hg abs
Gas velocity	90 ft/sec
Food loading	1.0 lb/ft^2
Hydrating salt, start	$CaCl_2 \cdot H_2O$
finish	$CaCl_2 \cdot 5.7H_2O$ (average)
Drying time	8.5 hours
Piece sizes	1-cm cubes

Results of Freeze Drying Run

	Turkey	Beef
Final moisture content (dry basis)		
A. Piece-to-piece variation	11.2%	11.3%
	11.2%	9.1%
	11.8%	8.2%
	10.7%	9.4%
	10.1%	9.3%
B. Comparison of moisture content of outer layers with that of the center of the same piece		
Piece No. 1 — outer layers	9.8%	8.7%
center core	11.4%	10.6%
Piece No. 2 — outer layers	10.4%	10.2%
center core	11.6%	14.4%
Shrinkage upon freeze drying (loss of volume during freeze drying/frozen volume)		
Piece A	0.19	0.13
Piece B	0.14	0.23
Rehydration characteristics (weight after reconstituting in water at ambient temperature for 20 min and then blotting divided by weight before freeze drying)		
Piece A	1.06	0.90
Piece B	1.12	0.93
Compressibility of product	Good	Good

The average moisture content achieved from the example was 11% for turkey and 9.5% for beef, with average deviations for five pieces of 3% and 8% moisture for turkey and beef, respectively. The moisture content from piece to piece was quite uniform, while moisture distribution within a piece fell well within the desired range (8 to 15%). Shrinkage, rehydration, and compressibility characteristics were all good. These results, and taste tests, lead to the conclusion that product quality of the limited freeze dried turkey and beef was excellent.

Fluidized Bed Freeze Drying

The work of *I. Abelow and J. Wagman; U.S. Patent 3,436,837; April 8, 1969; assigned to the U.S. Secretary of the Army* relates to a method of preserving heat-sensitive materials by drying them from the frozen state in a fluidized bed created by a flow of dry air or other gases or gas mixtures. The process is carried out by freezing small pieces of the product to be dried including cubed meat, cooked peas, or other food material which can be diced or cut into small pieces. Products which are in the form of liquid suspensions or solutions of solids may be made into frozen droplets by pelleting or other means. The process is described with reference to Figure 10.6.

FIGURE 10.6: FLUIDIZED BED FREEZE DRYING

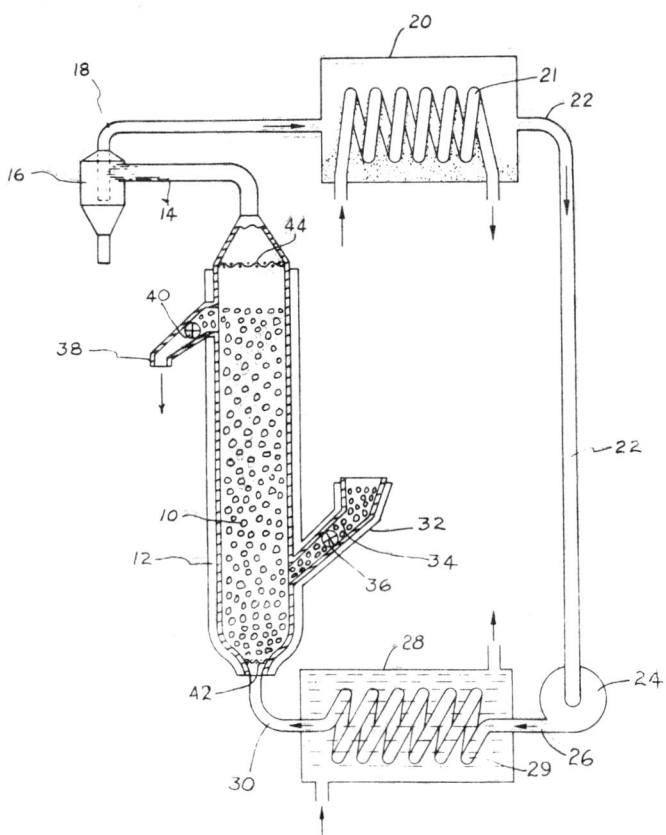

Source: U.S. Patent 3,436,837

In the drawing, **10** shows the drying column with jacket **12**. The top of the column is vented in conduit **14** to cyclone separator **16**. The take-off conduit **18** from the separator passes to the unit **20** which is a refrigerated cold trap, which is provided with a refrigerating unit **21** maintained at freezing temperatures for removing moisture from the gas. The dried gas then passes through conduit 22 to inlet of positive pressure blower 24, which compresses the dried gas and forces it successively through conduit 26, heat control unit 28, conduit 30, and through the bottom opening of column 10 with sufficient velocity and pressure to fluidize the bed of solid pieces in the column.

Supply hopper 32 receives the frozen material 34 to be dried, the rate of entry of which into the column is regulated by star valve 36. Dried pieces are removed from the column through dry product take-off 38, and the rate of removal of such dried material is regulated by star valve 40. A fine mesh screen 42 at the bottom of column 10 prevents material in the bed from dropping through the bottom opening of the column when the circulation of gas is interrupted, and fine mesh screen 44 prevents the solid pieces in the fluidized bed from being blown out through the top opening of the column but permits fines to pass through to the cyclone separator 16.

In operation, hopper 32 is charged with a frozen product to be dried and the column is supplied from hopper 32. Dried air or other gas or gas mixture is fed through this bed by means of blower 24 at such a rate and pressure that the bed is fluidized. Any fines that result from the operation are carried over through conduit 14 and are deposited in cyclone separator 16.

The gas which is laden with moisture continues on through refrigerated vapor trap 20 where the moisture is removed and the dew point reduced. The gas then continues through the blower 24 and on to heat control 28, shown as a coiled duct 29 submerged in a temperature bath. The temperature of the gas may be raised to further lower the relative humidity and to provide heat of sublimation, but never above the point required to maintain the material in the fluidized bed in the column in a frozen state.

In this column, the drier material because of lowered bulk density gradually migrates to the top of the fluidized bed. It is possible therefore to withdraw such dried material through product take-off 38 as it reaches the top of the fluidized column. The material to be dried is admitted into column 10 at the same rate as dried material is withdrawn through product take-off. By this means it is possible to make what is normally a batch operation into a continuous process.

Fluidized Bed of Solid Discrete Sodium Chloride Particles

W.H. Mink and H. Nack; U.S. Patent 3,239,942; March 15, 1966; assigned to The Battelle Development Corp. have developed a process for drying food by immersing it in a frozen condition in a fluidized bed of solid discrete particles at subatmospheric pressure. An apparatus for carrying out the process is described in Figure 10.7.

The fluidized bed comprises a bed of solid, discrete particles 11 subjected to an upward gaseous current 12, the size and weight of the particles and the velocity and nature of the current being so chosen that the force exerted by the current is sufficient to counterbalance the gravitational force on free particles and to expand the bed, thus allowing movement of the particles, but being insufficient to convert the bed into a stream of particles.

FIGURE 10.7: FLUIDIZED BED OF SOLID DISCRETE SODIUM CHLORIDE PARTICLES

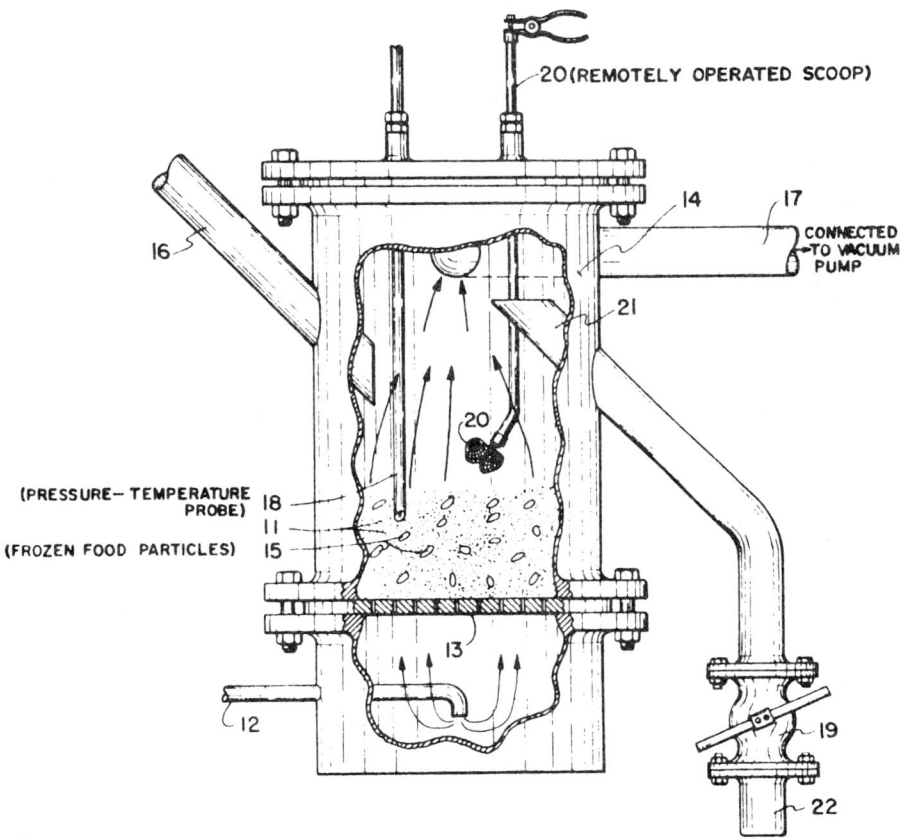

Source: U.S. Patent 3,239,942

Among the materials that are suitable as fluidized bed particles are sodium chloride, calcium chloride, tricalcium phosphate, limestone, monosodium glutamate, sugar, rice, beans, lentils, or any material that is a solid of appropriate particle size at the temperature of operation and is generally regarded as safe as a food additive or is, in fact, an edible food. Since sodium chloride is stable over most operating temperatures encountered, is available in various particle sizes, is generally added to most foods before eating, and is a readily available commodity at extremely low price, it is the most desirable of the potential bed materials in practicing the process.

The gas 12 used to fluidize the particles in the bed must not render the food to be treated inedible. Gases suitable for this purpose are air, nitrogen, carbon dioxide, and flue gas. Where it is desired to inhibit oxidation of the food being treated, a nonoxidizing gas should be used. Nitrogen and carbon dioxide are most suitable in this instance.

The food **15** to be freeze dried by this process is first frozen. The method works best when the food to be freeze dried has at least one dimension not greater than 2 inches. Liquids can be frozen by known techniques in appropriate size, or can even be reduced in size. The frozen food **15** is completely immersed in a fluidized bed of solid, discrete particles **11**. The frozen food may be fed either continuously or in batch fashion into the fluidized bed.

During the drying step of the process, the vapor pressure in the fluidized bed chamber cannot be greater than about 4 mm of mercury. This can be maintained by keeping the absolute pressure in the fluidized bed at not greater than about 4 mm of mercury, or by using a dry gas for fluidizing the bed of particles. The pressure can readily be reduced by means such as a tube **17** connected to a vacuum pump.

The only pressure limitation imposed on this process is that imposed by the phase diagram of water. The drying steps of this process are carried out at a pressure and temperature at which solid ice will sublime. However, practical considerations, such as the size and cost of vacuum equipment, dictate that the absolute pressure in the fluidized bed chamber not be reduced below about from 1 to 10 microns of mercury.

It has been observed that operating at a reduced pressure does not interfere with the fluidization of the bed particles and that fluidization of the bed particles does not prevent operating the system at a reduced pressure sufficient for freeze-drying. Of course, the depth of the fluidized bed must not be so great that the pressure drop from the top to the bottom of the bed results in a pressure at the bottom of the bed that exceeds about 4 mm of mercury. The critical bed depth for any given bed particle can readily be determined by means of a pressure-temperature probe **18** that can measure the pressure or temperature at any given bed depth.

While the frozen food is immersed in the fluidized bed of solid, discrete particles **11**, with vapor pressure in the fluidized bed chamber **14** of not more than about 4 mm of mercury, the heat of sublimation is supplied to the food to remove the ice as water vapor. The particles in the fluidized bed are heated by any known means.

The food to be dried is contacted on all sides by the bed particles **11**. It is this intimate contact of a solid with a solid that produces heat transfer rates that are vastly superior to those obtained by any other known method of freeze drying. Even when the only heat supplied is from the fluidizing gas at normal room temperature, about 20°C, drying times superior to those obtainable by known processes are achieved.

When additional heat is provided by elevating the temperature of the fluidized bed particles, drying times are reduced still further, without affecting taste or appearance of the food. For example, at a bed temperature of about 40°C, three-eighth-inch cubes of chicken can be dried to a residual moisture content of less than 2% in less than 2 hours. Times reported for prior processes range from 7 hours up to 4 hours.

When the residual moisture content of the foods being dried reaches the desired level, the food is removed from the fluidized bed chamber. While residual moisture contents not greater than about 7% are acceptable, the food is storable for a longer period of time without deterioration when the residual moisture content is not greater than about 2%.

Example: (a) Sodium chloride particles were fluidized by dried nitrogen gas. Chicken was segmented to a size having at least one dimension of the order of ¼ to ½ inch. These chicken pieces were frozen by conventional means. The frozen segments of chicken where then immersed in the fluidized bed of sodium chloride particles.

The system was closed and the absolute pressure in the fluidized bed chamber was reduced to approximately 2 mm of mercury. Under these conditions, the chicken pieces were reduced to a residual moisture content of not greater than 4% in approximately 6 hours. In this run, the fluidized bed chamber was maintained at approximately 20°C.

(b) The conditions of (a) were repeated with the one exception that heat was supplied to the sodium chloride particles in the fluidized bed. Sufficient heat was supplied to maintain the fluidized bed chamber at approximately 40°C. Under these conditions, the corresponding residual moisture content of the chicken segments was reached in less than 2 hours.

Bacon Which Does Not Spatter on Cooking

L.C. Hinnergardt; U.S. Patent 3,914,446; October 21, 1975; assigned to the U.S. Secretary of the Army describes a method of producing freeze-dried bacon slices characterized by freedom from spattering when being heated which comprises the steps of freezing raw bacon slices at about −10°F and then freeze vacuum dehydrating the frozen raw bacon slices to a moisture content of not more than 5% (preferably not more than 2%) while maintaining the platen temperature employed in the freeze vacuum dehydrating step at about 125°F and maintaining the pressure employed in the freeze vacuum dehydrating step at about 0.3 mm of mercury.

Example: Shingled slabs of raw sugar-cured bacon slices were frozen at a temperature of about −10°F, then freeze vacuum dehydrated in a conventional manner in a commercial freeze dryer at a pressure of about 0.3 mm of mercury using a platen heated to about 125°F until the moisture content of the bacon was reduced to about 0.73%.

The freeze dried raw bacon was vacuum packed at a pressure of 28 inches of mercury in metal cans which were hermetically sealed to exclude air. Cans of the freeze dried raw bacon were stored at various temperatures of 40°F, 70°F, and 100°F for 6 months. The freeze dried raw bacon was removed from the cans and grilled at about 350°F for about 1½ minutes on each side prior to testing of the bacon in terms of flavor, odor and appearance (overall). It did not spatter grease and the slices of bacon did not curl appreciably while being grilled as usually happens with fresh bacon.

The table gives the results of technological panel testing of the grilled freeze dried bacon initially, i.e., within 24 hours after freeze drying and after storage for 6 months at the three temperatures of 40°F, 70°F and 100°F. Eighteen trained food technologists were on the panel. Each value shown represents an average of the eighteen ratings of the panel on the Hedonic Scale, which involves the assignment of ratings from 1 to 9, a rating of 5 representing the borderline of acceptability while the higher the rating, the more acceptable, and the lower the rating, the less acceptable the product is with respect to the characteristic being rated, insofar as the technological panel members are concerned.

	Flavor	Odor	Appearance
Initial	7.3	6.9	6.5
6 months at 40°F	6.8	6.8	6.5
6 months at 70°F	6.8	7.0	6.6
6 months at 100°F	6.9	7.1	7.1

It is apparent from the above results that freeze dried raw bacon was quite storage stable for at least 6 months at storage temperatures as high as 100°F when packaged under vacuum in hermetically sealed gastight containers. The above results compare favorably with oven-baked fresh bacon cooked at about 350°F for about 30 minutes to attain about the same degree of crispness, the latter cooked bacon being rated an average of 7.6 for flavor, 7.6 for odor and 7.2 for appearance as rated by a comparable technological panel.

In addition to the above-described advantages of freeze dried raw bacon of being highly storage stable and of not spattering grease when being cooked, slices thereof separate more easily than slices of fresh raw bacon. Furthermore, it does not feel as greasy or oily as fresh raw bacon or prefried bacon and it does not require as high a concentration of sodium chloride as prefried bacon for preservation purposes.

If desired, the freeze dried raw bacon may be fried or oven-baked prior to packaging for storage purposes. If packaged under vacuum in hermetically sealed containers which are gastight, the freeze dried cooked bacon remains acceptable for considerable periods of storage time, particularly if it has been freeze dried to not more than 1.0% moisture and if it is packaged in gastight containers and stored at temperatures of 70°F or lower.

The freeze dried cooked bacon has the advantage that it is ready to be eaten as soon as the package is opened and it constitutes a good replacement for sliced prefried bacon prepared from fresh bacon. It appears less greasy than fried fresh bacon or sliced prefried bacon before or after packaging and storing thereof.

Mechanical and Chemical Tenderization of Steak Before Cooking

The process of *L.C. Hinnergardt; U.S. Patent 3,971,854; July 27, 1976; assigned to the U.S. Secretary of the Army* relates to a method of producing freeze dried cooked beef steak which is tender when rehydrated and which rehydrates rapidly to the extent of reabsorbing within 10 minutes at least about 95% of the moisture which it contained prior to being freeze dried and of reabsorbing within 2 minutes sufficient moisture to make the steak acceptable (at least about 90% of the moisture which it contained prior to being freeze dried).

Example: USDA Choice top rounds of beef, 7 to 10 days post-mortem, were individually wrapped in freezer paper and stored at −23°C for 7 days. The rounds were randomly divided into four lots and by random selection equilibrated to a temperature of 4.4°C. The semimembranous muscle was excised from each round. The muscles from each of the four lots were treated differently, as described below, in order to provide data for a statistical analysis of the effects of the four treatments on tenderness, rate of rehydration, and other characteristics of rehydrated freeze dried, cooked beef steaks obtained from the several lots of beef rounds.

Each semimembranous muscle from Lot No. 1 was mechanically tenderized by passing the muscle three times through a Tend-R-Rite Model TR-2 tenderizer at the slowest throughput speed at which the tenderizer operates.

This type of mechanical tenderizer comprises multiple parallel rows of cutting blades which are about one-eighth-inch wide and one-thirty-second-inch thick at the ends which completely penetrate the meat to be tenderized. The blades are staggered in successive rows so that the penetrations in the meat are about one-fourth inch apart on centers in the majority of cases, though at the outer edges of the penetration pattern they may be about one-half inch apart.

Then an aqueous solution containing 3.0% sodium tripolyphosphate (TPP) and 7.5% sodium chloride was pumped into the muscles to 10% of their weight, thus introducing into each muscle about 0.3% sodium tripolyphosphate and about 0.75% NaCl. This was accomplished using a Koch Tenderizer Injector (8127) equipped with four stainless steel needles spaced about 2.2 cm apart. A constant pump gauge pressure of 13.6 kg was maintained during pumping.

Each muscle from Lot No. 2 was only mechanically tenderized by passing the muscle through the Tend-R-Rite Model TR-2 tenderizer in the same manner as was done in the case of Lot No. 1 muscles. Each muscle from Lot No. 3 was only treated with the same aqueous solution of TPP and NaCl as in the case of Lot No. 1 muscles.

Each muscle from Lot No. 4 was given no mechanical tenderization treatment and no aqueous solution of TPP and NaCl treatment, thus being a control lot. The muscles were frozen at −23°C immediately following completion of the treatment or treatments described above, the control lot being maintained frozen until all of the muscles were ready to be cut into steaks. The muscles were then sliced, five slices 1.27 cm thick being obtained from the center portion of each muscle by cutting across the grain with a meat saw.

Individual steaks were cut from the slices using a 6.35 cm diameter die. Each steak was grilled at a temperature of 176°C for 3 minutes on each side. The grilled steaks were frozen to −23°C, then freeze vacuum dehydrated in a Stokes freeze dehydrator at a plate temperature of about 51.6°C and a chamber pressure of 0.3 to 0.5 mm Hg to a moisture content of less than 2.0%.

All of the freeze dried steaks were hermetically sealed under a vacuum of at least about 27 inches of mercury in No. 2½ metal cans after a nitrogen flushing of the cans containing the steaks, thus reducing the oxygen content to 2.0% or less, as determined in accordance with the method of Bishov, S.J. and Henick, A.S., *Journal of the American Oil Chemists' Society*, Vol. 43, page 477 (1966). The packaged freeze dried cooked beef steaks were stored for 1 month at room temperature prior to rehydration and testing thereof.

These steaks were evaluated with respect to their rates of rehydration in water at 60°C by submerging steak samples in water at about 60°C for time periods in increments of 2 minutes from 2 to 10 minutes. In the results below, each value represents the average of ten samples similarly treated and evaluated.

	Percent Moisture Uptake Rehydration Time (minutes)					Percent Rehydration Rehydration Time (minutes)				
	2	4	6	8	10	2	4	6	8	10
Lot No. 1 (mech + TPP/NaCl)	55	56	57	57	58	90	92	94	94	96
Lot No. 2 (mech only)	53	54	56	56	56	86	88	92	92	92
Lot No. 3 (TPP/NaCl only)	48	50	53	53	53	76	80	86	86	86
Lot No. 4 (neither treatment)	48	50	51	52	52	76	80	82	84	84

As a result of many studies of rehydration of freeze dried, cooked beef steaks, it has been found that to be acceptable the rehydrated steak must have a percent rehydration of at least 90%. This corresponds in general to a percent moisture uptake of at least 55%.

In addition, the combined use of mechanical tenderization and pumping into the beef muscle a mixture of TPP and NaCl resulted in a greater tenderization of the beef muscle. Cuttability also was greatly enhanced.

MILK PRODUCTS

Whole Milk Cakes

R.P. Ogden; U.S. Patent 3,297,455; January 10, 1967 describes a method of freeze drying liquid milk products. In the process fresh whole milk or homogenized fresh whole milk is first dehydrated by employing conventional procedures to approximately one-third of its volume, after which the resulting concentrate is quickly frozen into a thin sheet that is immediately fragmented into flakes or chips. The resulting frozen flakes or chips are then compacted into a cake or block which is then fully dried under vacuum conditions while maintained fully frozen to provide the finished product. The resulting cake or block is then applied to a suitable container which preferably is proportioned to receive several of the cakes or blocks. The actual milk drying procedures are performed when the milk itself is frozen.

Nonhygroscopic Honey-Milk Product

D. Torr; U.S. Patent 3,244,528; April 5, 1966 is concerned with the preparation of a dry, powdered, unadulterated honey-milk composition of nonhygroscopic character and which is capable of being used or stored while exposed to relatively high moisture-containing atmospheres without reversion of the honey constituent to the objectionably sticky form and without agglomeration or caking of the product.

Contrary to previous procedural efforts to obtain shelf-stable, noncaking, dry honey-milk products, the process involves the addition of water rather than the exclusion thereof prior to the final dehydration step. It has been found that the addition of water in quantities beyond that normally present in commercial honey will advantageously advance the extent of dehydration achievable for a honey-milk mixture, and further advantageously minimize any hygroscopic properties otherwise exhibited by such mixtures after drying.

Whether dried or fresh whole milk, skim milk, buttermilk, whey or mixtures of the foregoing were mixed with honey, the presence of or addition of water beyond that already present in the honey, appeared to contribute to the stability of the finally obtainable dried composition. The moisture content of commercially available honeys ranges between approximately 13 and 26%. Any good commercial honey can be used with the process.

Example 1: Honey-Powdered Milk — (a) A honey-milk solution at room temperature, having a 30% solids concentration was prepared, the solution containing 197 grams of honey (80%), 52.5 grams of powdered whole milk and 450.5 grams of water. The solution after being freeze dried yielded a fluffy, white, shelf-stable, noncaking, 75:25 honey-milk food product weighing 200 grams. The moisture content was determined to be less than 2%.

(b) A honey-milk solution at room temperature having a 30% solids concentration was prepared, the solution containing 28.1 grams of honey and 67.5 grams of whole milk powder. The solution after being freeze dried yielded a fluffy product as in (a) and weighing 100 grams. The product contained 25% honey and 75% whole milk solids. Moisture content: less than 2%.

Example 2: Honey-Fresh Milk — A mixture at room temperature, containing 100 grams of honey (approximately 18% moisture content) and 100 grams of fresh whole half-milk and half-cream was prepared. The solution was spray dried and yielded the desirable noncaking, shelf-stable product as obtained in the foregoing procedures.

In carrying out the above process using the freeze drying technique, the conditions were as follows: the prepared honey-milk solution was allowed to freeze for 2 hours at −45°C; the drying time, 12 to 18 hours; the pressure, 70 to 100 microns of mercury; shelf temperature, 125°F; ambient temperature, 85° to 90°F; condensate plate temperature, −40°F; and product temperature 125°F maximum.

Where spray drying was utilized, the conditions were as follows: the inlet temperature, 120° to 135°C; outlet temperature, 65° to 85°C; atomizer pressure, 5 kg per cm^2, drying rate of solution, 25 to 35 ml/min; and product temperature 55° to 75°C.

The initial honey-milk-water mixture prior to drying may be maintained at any temperature ranging between 32° to 110°F. The period of time between the mixing and drying step is of no consequence and drying may be accomplished promptly after the mixture is prepared.

Separate Drying of Serum Portion and Coagulated Residue

U. Hackenberg and K. Kautz; U.S. Patent 3,321,319; May 23, 1967; assigned to Leybold Hochvakuum Anlagen GmbH, Germany are concerned with the freeze drying of materials which are capable of separation into an aqueous serum portion and a coagulated residue, particularly milk and/or other dairy products.

A dairy product, such as milk, is separated into a serum or skim portion and a coagulated residue by mechanical means. Then the coagulated residue, preferably immediately following its separation, is subjected to cooling and then is subjected to a freeze drying step in accordance with conventional methods. The serum constituent is dried in its separated state, employing known drying methods. Thereafter the products of both drying steps are mixed together to form the final dried product.

When employing milk and/or dairy products as starting materials, the coagulation and the concurrent formation of their residue can be advantageously performed by means of acidification with acid bacteria in known manner. Alternatively, there can be employed the so-called "sweet" coagulation by bacteria or enzyme addition. In various applications, for example, the preparation of a yogurt or yogurt-curd powder, it is desirable to carry out the coagulation by acidification and enzyme addition in accordance with known methods.

Example 1: Yogurt — 5,000 liters of skim milk yogurt are separated by centrifuging into 1,000 kg of coagulated protein having about 20% dry weight, and 4,000 liters of milk serum having a dry weight content of 6%. The milk protein in the form of a coagulated residue, immediately following the separation, is cooled in a plate cooler to +6°C and then subjected to freeze drying.

The drying is carried out at 0.8 mm Hg to a residual moisture of 2%, and there is thus obtained 200 kg dry product, which can be mechanically pulverized if desired. The 4,000 liters of milk serum are evaporated to 500 liters in a vacuum evaporator and finally spray dried. The initial air temperature is 150°C and the exit temperature is 70°C. There is obtained a yield of 250 kg of dry substance in powder form.

Thereupon the products of both drying steps are admixed and there are obtained 450 kg of yogurt powder. When 100 grams of the product are made up to 1 liter with cold water, there is obtained 1 liter of full-strength yogurt, the biological and gustatory properties of which correspond to those of the starting product.

Example 2: Cream — 5,000 liters of cream having a fat content of 15% are treated with rennet in a ratio of 1:100,000 at 32°C and mixed with 250 liters of pure acid culture. The temperature is maintained constant for 5 hours and the coagulate is mechanically separated from the serum by sieving. There are obtained 2,000 kg of a fat-rich coagulate, which is immediately cooled to 4°C with a plate cooler. The coagulate is then freeze dried at 0.4 mm Hg, resulting in a yield of 860 kg of dry substance.

The 3,000 liters of serum are concentrated to about 50% solids in a vacuum evaporator and finally subjected to spray drying as described in Example 1. There is obtained a yield of 240 kg of dry substance. Both dry products are admixed, yielding 1,100 kg of dry product. When 220 grams of this dry product are made up to 1 liter with cold water there is obtained 1 liter of creamlike homogeneous liquid which can be processed and whipped like ordinary cream.

Example 3: Milk — 5,000 liters of milk with 25 to 500 liters, preferably 250 liters, of acid culture as generally obtainable in trade is warmed up to 26°C. 25 to 100 grams, preferably 50 grams, of rennet are then added. The mixture is left for 12 to 16 hours until an acid ratio of 26, according to Soxhlet-Henkel, is reached. There are thus obtained 830 kg protein residuals having about 20% dry weight and 4.170 kg of milk serum having a dry weight content of 5.5%.

The protein residuals are separated from the milk serum by means of centrifuging. The further treatment of the protein residuals is according to the normal freeze drying process, while a cheaper process such as spray drying is applied to the milk serum. Thereupon the products of both drying steps are admixed in a mechanical mixer in the proportions of the starting product.

JUICES

Alternately Raising and Lowering Temperatures of Hard Frozen Mass

N. Ganiaris; U.S. Patent 3,949,486; April 13, 1976; assigned to Struthers Scientific and International Corp. provides a process for the freeze concentration of a comestible product, comprising the steps of completely freezing an aqueous extract of a comestible product to form a hard frozen mass thereof, alternately raising and lowering the temperature to convert the hard frozen mass firstly into a soft frozen mass and then into another hard frozen mass, granulating the other hard frozen mass to obtain particles, and drying the particles obtained under vacuum.

With orange juice it is preferable to remove some water from the juice prior to formation of the first hard frozen mass. This may be done by partially freezing the juice and removing the ice crystals so formed. It is advisable to filter the juice to remove orange pulp therefrom before partially freezing, and then returning the same pulp to the concentrated juice prior to freezing.

Example: Freshly squeezed orange juice (8° to 14° Brix) is fed to a pulp separator where it is filtered to remove the orange pulp. It is then freeze concentrated to over 20° Brix and preferably in the range of 40° to 60° Brix. The concentrated orange juice (40° to 60° Brix) is mixed again with orange pulp from the pulp separator which pulp was removed prior to freeze concentration and then cooled slowly in a scraped surface heat exchanger or other device until it becomes a thick paste or slurry.

This thick paste of orange juice, pulp and discrete ice crystals is fed to the freezer where it is cooled further to a temperature below –35°F and preferably in the range of –60° to –80°F and kept at this temperature until it becomes a frozen solid mass.

The frozen material from the freezer is passed to a treatment zone where it is allowed to warm up to –35°F and become a soft frozen mass, which is then frozen again to become a hard frozen mass at -60° to -80°F. This technique of warming and refreezing is utilized several times in order to insure that all the water has been frozen. For orange juice, the warming and refreezing should be done at least twice but not more than eight times, preferably not more than six times.

After the final freezing, the hard frozen mass is fed to a granulator at -60° to -80°F and then screened. The particles from the screens having an average diameter of less than one-eighth inch are loaded into trays and the trays are transferred into a vacuum chamber. The coarse frozen particles from the screens are recycled to the gridding device or to the concentrated orange juice prior to freezing.

In the vacuum chamber, the pressure is lowered by means of a pump to less than 100 microns and preferably in the range of 60 to 80 microns. The vacuum chamber is fitted with a condensor the temperature of which is maintained at lower than -50°F and preferably -80° to -100°F. Heating plates which supply the required heat for the sublimation of ice from the particles are held at -20° to -10°F for a period of 3 to 6 hours; then the temperature of the heating plates is increased gradually to about 32°F over a period of 3 to 6 hours and finally the temperature is increased to 80°F over a period of 1 to 2 hours. At the end of the cycle, the product is heated to 120° to 140°F for a few minutes, preferably less than 30 minutes.

The dried product removed from the vacuum chamber has a moisture content of less than 3%, preferably less than 1%. If the product is held for 30 minutes at 140°F, the moisture content is reduced to less than 1%.

Use of Electromagnetic Induction Heating

A. van Gelder; U.S. Patent 3,253,344; May 31, 1966 provides a process which employs familiar techniques similar to those of freeze vacuum drying but where the dehydration is accomplished by applying an induction heated medium in close contact with the product, thus providing an increased evaporation rate and simplicity in the cleaning of the equipment over known methods.

The result is a dried powder-like end product which can be reconstituted with either cold or hot water into a form having substantially the same properties as the original product. For example, if the product to be dehydrated is orange juice, the dried powder would be reconstituted with cold water in order to provide a mixture having substantially the same taste, coloring, texture, and vitamin content as the original orange juice.

Initially it is desirable to prepare the product to bring it into a state of solution, suspension, colloidal suspension, and/or pulp consisting of finely divided solids and liquids. Thus, if the end product is dehydrated orange juice, the juice is prepared by conventional squeezing.

The second step of the process consists of dividing the product into small particles preferably by introducing the product into any type of commercially available spraying equipment by means of nozzles or rotating disks, bowls, or rotors of any size or shape capable of dividing the product into small particles.

The third step of the process is to freeze the small particles of the product to produce a powdered snow-like product. The third step may be combined with the second step by locating the spraying equipment inside of a chamber and subjecting the particles to a flow of cold gases having a temperature sufficient to freeze the particles substantially instantaneously upon contact therewith.

An alternate method of freezing is to bring the particles into intimate contact with cold surfaces which are maintained at a sufficient temperature to cause the particles to freeze while the particles are being constantly agitated to prevent their sticking to the surfaces.

The next step in the process is introducing the frozen particles into a low temperature mixing chamber containing small pieces of metal, preferably in the form of spheres or balls, coated with Teflon which is polyfluorovinyl resin, to prevent the frozen particles from sticking to the metal or the coating. The metal spheres constitute not more than 42% of the total volume.

The mixing chamber and metal spheres or balls are cooled to a temperature below the freezing point of the particles before the particles are introduced in order to prevent the particles from melting upon coming into contact with the spheres. The spheres and the frozen particles may be thoroughly agitated (though this is not necessary) to further prevent the frozen particles from sticking to the spheres during the period of residence.

It is also possible to combine this step with the previous step, in that the particles may be frozen while spheres are being cooled in the same mixer, during which time the mixer is agitated in order to keep the particles from sticking to the spheres until the particles are frozen, at which time the agitation of the mixer continues to thoroughly distribute the spheres throughout the frozen particles.

In the next step of the process, the frozen particles and spheres are introduced into a chamber in which a high vacuum is drawn. While the frozen particles and the spheres are being subjected to the high vacuum, the spheres are subjected to electromagnetic forces which will produce induction heat in the spheres, and this heat is transferred to the frozen particles by means of conduction and radiation to raise the temperature of the frozen particles to not more than 30°F.

This heating of the particles creates a differential vapor pressure between the particles and the vacuum which effects evaporation of the moisture in the particles and results in their dehydration. The intimate contact between the frozen particles and the spheres causes a uniform heat transfer from the spheres to the frozen particles, thereby resulting in rapid dehydration of the particles.

The final step in the process is releasing frozen particles and spheres from the vacuum chamber and screening the frozen particles from the spheres. The frozen particles are then packed or subjected to further treatment, as may be desired, and the spheres are preferably cleaned and returned to a suitable receptacle for reuse.

Increasing Surface Area via Use of Conductive Teflon-Coated Spheres

A. van Gelder; U.S. Patent 3,316,652; May 2, 1967; assigned to Sun-Freeze Inc.

provides a process inducing a greatly increased rate of low-temperature evaporation from liquids containing suspended solid particles by continuously maintaining these liquids and particles in intimate and close contact with an induction heated medium having a very large surface area and doing this in a low pressure environment thereby permitting a dehydrated product which upon reconstitution is not materially unlike the original substance from which the dehydration product was derived. The process is particularly suited to orange juice.

The continuous method of dehydration of liquids and liquids containing less than 25% suspended or colloidal solids comprises the steps of reducing the solids in the liquid to finely divided form, instantly freezing the material to produce a powder of frozen particles, mixing the frozen particles under continuous agitation with a predetermined volume of electrically conductive pellets which have previously been cooled to a temperature below the freezing point of the powder, moving the powder proportioned with the pellets in continuous tumbling engagement through a vacuum chamber in the presence of induction heating thereby raising the vapor pressure of the powder above the negative pressure causing dehydration of the powder, continuously separating the dehydrated powder from the pellets and continuously returning the pellets to the system.

Preferably, the pellets are Teflon-coated graphite, electrically conductive material for induction heating in geometric shapes.

The spheres are used for the purpose of obtaining a large surface area in direct contact with the frozen particles being processed. For example, when the pellets or spheres are one-half inch in diameter and arranged side by side in layers one cubic foot of these would represent a total surface area of 75.36 square feet.

The interstices available for the product would represent some 46% of each cubic foot occupied by the spheres. Should the ratio be changed so that 60% of each cubic foot would be occupied by the product then the possible surface area of the spheres for contact with the product would be 64.81 square feet. Thus, it is plain that by changing the proportion of the spheres per cubic foot, it is possible to mathematically change the surface area available for contact with the frozen particles. The maximum surface area of spheres for the maximum volume of frozen particles, i.e., not less than 60% is the operating requirement for this process.

The final steps in the process include the releasing of the dehydrated particles and spheres from the vacuum chamber. The dehydrated particles at room temperature and atmospheric pressure are separated from the spheres in a rotary screen or other suitable manner. The dehydrated product is packaged. The spheres are recovered from the rotary screen and are cleaned in any suitable manner and returned for continuous recycling through the system.

Fluidized Bed at Atmospheric Pressure

In accordance with a process developed by *G.J. Malecki; U.S. Patent 3,313,032; April 11, 1967; assigned to the U.S. Secretary of Agriculture* freeze drying is conducted not under vacuum but at ordinary (normal) atmospheric pressure. Significant advantages are faster rate of dehydration, more uniform dehydration of individual particles, and a substantial saving in equipment costs because the apparatus may be of lighter and simpler construction.

The method for freeze drying a material at essentially atmospheric pressure comprises: (a) freezing particles of the material; (b) forming a bed of the frozen particles, and (c) injecting a predried gas upwardly through the bed of frozen particles at essentially atmospheric pressure to maintain the bed in a fluidized condition and to dehydrate the particles, the temperature of the injected gas being high enough to be effective to cause sublimation of ice but not so high as to thaw the frozen particles. The process is described with reference to Figure 10.8.

FIGURE 10.8: FLUIDIZED BED AT ATMOSPHERIC PRESSURE

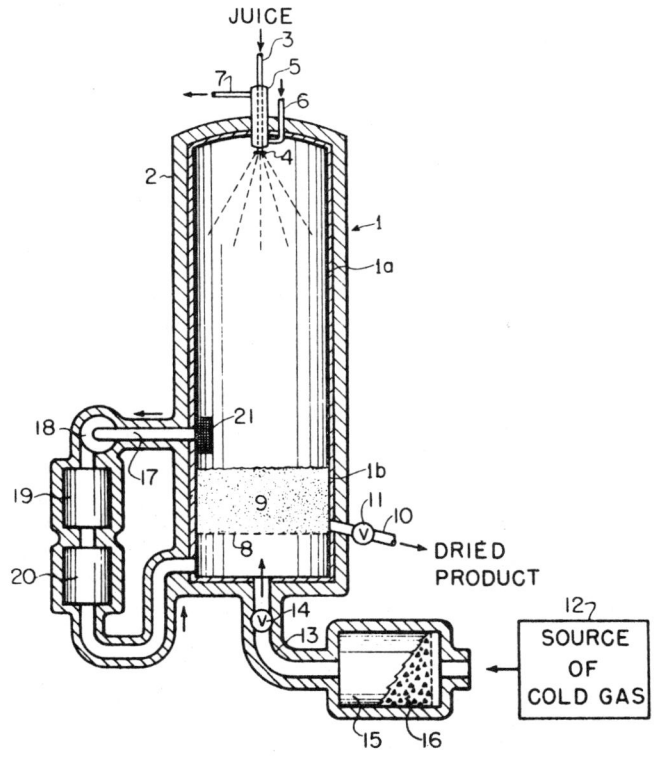

Source: U.S. Patent 3,313,032

The device includes a chamber, which provides an upper (freezing) zone **1a** and a lower (drying) zone **1b**. Numeral **2** designates thermal insulation which covers chamber **1** as well as the other elements of the device. For introducing the liquid to be dried, fruit juice, for example, there is provided a conduit **3** connecting with spray nozzle **4**. A heating jacket **5** is provided about the conduit to prevent the juice from freezing before it is atomized.

The liquid sprayed from nozzle 4 meets a column of cold gas in chamber 1a, whereby the droplets of liquid are converted into frozen particles. The size of the frozen particles can be regulated by the type of nozzle used, and is preferably maintained at 30 to 200 microns, although larger particles up to one-half inch in diameter can be used. As the frozen particles are formed they drop down from freezing chamber 1a to drying chamber 1b.

Positioned in the lower part of chamber 1 is a screen 8 for supporting a bed 9 of frozen particles and for permitting passage of an upward flow of gas. Dried product is removed from the system via conduit 10 and valve 11. Excess gas may be vented from the system together with dried product via the conduit. If necessary electrostatic precipitation, filtration, centrifugation, or a pneumatic cyclone system may be provided to separate the particles of product from the excess gas.

Cold nitrogen, or other gas, from source 12, is introduced into the system via conduit 13 and valve 14. Interposed in conduit 13 is a drier 15, provided with granules of drying agent 16, whereby the moisture is removed from entering gas.

For recirculating gas through bed 9, there is provided a recirculation conduit 17, pump 18, drier 19, and heat exchanger 20. In operation, part of the gas rising through bed 9 is propelled by pump 18 through drier 19 and heat exchanger 20 back to the base of chamber 1. Drier 19 removes from the gas stream the moisture which has been taken up from the particles being dehydrated. Drier 19 may be constructed similar to gas drier 15.

Heat exchanger 20 is not a critical item but may be provided to heat or cool the recirculating gas as necessary under specific conditions of operation. A screen or filter 21 is provided at the upper end of conduit 17 to prevent entrainment of particles in the gas stream. This filter may be provided with a conventional vibrator to remove deposits of solid material.

Example: Orange juice is sprayed from nozzle 4 into upper (freezing) chamber 1a. Cold nitrogen (or air, CO_2, N_2O, etc.) at about $-40°F$ is rising countercurrently from below fluidized bed 9 and in zone 1a, freezes the falling droplets of juice to solid particles. These solid particles drop down to zone 1b, forming bed 9.

The size of the droplets is regulated, preferably to about 0.05 to 0.1 mm in diameter, and the flow of gas is adjusted accordingly to maintain the fluidized state in bed 9, as well as to maintain proper freezing of the incoming liquid droplets. The rate of cooling gas flow should be large enough to freeze all the droplets to below the eutectic point of the material under treatment, but not so large as to carry away the majority of the smaller size droplets.

In order to be able to correlate the rate of flow of gas through the freezing chamber, there is provided the recirculation system of conduit 17 and associated components. Any excess or deficiency of gas going into the freezing chamber 1a can be removed or supplemented by this recirculation or by-pass system. During operation heat exchanger 20 may be activated to heat or cool the recirculating gas stream to the appropriate temperature level (to about $-40°F$, in this particular example).

The gas flowing through the bed effects a sublimation of ice from the frozen particles. The evolved vapor is carried away by the gas stream and eventually absorbed by drier **19**. The process may be continued until the orange juice particles are completely dried.

Fluidized Bed of Foodstuff in Particulate Form

This process of *C.E. Dryden, W.H. Mink and H. Nack; U.S. Patent 3,269,025; August 30, 1966; assigned to The Battelle Development Corp.* is an improvement over U.S. Patent 3,239,942 (reported elsewhere) which utilized a fluidized bed of auxiliary particles at subatmospheric pressure. The method eliminates the necessity of immersing the food in a fluidized bed of solid, discrete particles by eliminating auxiliary fluidized particles and by having the foodstuff, per se, in a particulate form making up the fluidized bed.

In these examples, which follow, there is employed a vertically positioned, generally cylindrical column enclosing a chamber for fluidization and vacuum freeze drying of frozen particulate materials. The column is provided, near its top, with a means for introduction of a frozen particulate material and for subsequent sealing of the column after introduction of the frozen particulate materials to be dried.

Also provided at the top of the column is a conduit to a means for evacuating and maintaining a suitable reduced pressure within the column. Near the bottom of the column, there is provided a porous, perforated distributor plate. Directly below the distributor plate is provided a means for introduction of a gas for fluidizing the frozen particulate material in that portion of the column directly above the porous distributor plate.

Example 1: Coffee — An aqueous solution containing approximately 20% by weight of soluble constituents of roasted coffee beans is frozen solid. Freezing is in a vacuum cabinet, under a reduced pressure, by contact with a surface chilled by a Freon evaporator. The frozen-solid solution, now a thin frozen-solid sheet, is crushed to a fine powder, while still frozen solid. The resulting frozen-solid crushed product is a fine, free-flowing, frozen-solid, coffee extract of a temperature lower than about –25°C.

This fine, free-flowing, frozen-solid coffee extract, while still at temperature lower than –25°C is charged to a precooled fluidization column of a temperature lower than about –25°C. The column is evacuated and a precooled (below –25°C) dry nitrogen gas is introduced to fluidize the charged, frozen-solid, crushed coffee extract.

During this vacuum freeze drying example, the introduced precooled nitrogen gas is of a temperature lower than about –25°C and the absolute pressure within the column during fluidization is maintained at less than 4 mm of mercury. Under these conditions, the fine, free-flowing frozen-solid coffee extract reaches a residual moisture content of less than about 2% by weight in less than about 48 hours. The resulting dried powder product is found to be storable under an inert atmosphere at about 25°C for extended periods of time up to at least several months, and, after such storage, readily dissolves in hot water to form a palatable, drinkable coffee.

Example 2: Orange Juice — A commercially available frozen orange juice concentrate is thawed sufficiently for removal from its container, then rapidly frozen at a temperature lower than -25°C, and crushed to a fine, free-flowing, frozen-solid, orange juice concentrate of a temperature of lower than about -25°C.

This crushed frozen-solid concentrate, while still frozen solid, is charged to a precooled fluidization column of a temperature lower than -25°C. The column is evacuated and the charged concentrate fluidized by introduction of a precooled substantially dry nitrogen gas of a temperature lower than about -25°C. Fluidization is carried forth for about 12 hr with the fluidized concentrate maintained at a temperature lower than -25°C and under a reduced absolute pressure of less than 4 mm of mercury. At this time the resulting product, a dehydrated orange juice concentrate, is removed from the column. The dehydrated orange juice concentrate is thawed and found to have a water content significantly lower than the commercially available orange juice concentrate. Rehydration of this dried concentrate is readily accomplished by mixing with water. The reconstituted orange juice is found by a taste panel to be substantially indistinguishable from a similarly reconstituted orange juice prepared from the original commercially available frozen orange juice concentrate.

Example 3: Carrots — About 150 parts of fresh carrots of a water content of 113 parts by weight of water are diced and chopped to rough cubical shaped bits of a size ranging from $\frac{1}{32}$ of an inch on a side to about $\frac{1}{8}$ of an inch on a side. These cubical shaped bits of carrots then are frozen solid by contact with liquid nitrogen. While at a temperature lower than -30°C, the frozen-solid carrot bits are charged to a precooled fluidization column of a temperature lower than -30°C.

The column is evacuated and the charged carrot bits fluidized by introducing a precooled dry nitrogen gas. During the initial several hours of fluidization, the carrots bits are fluidized under an absolute pressure of less than about 4 mm of mercury and during the entire fluidization the carrot bits are fluidized under a vapor pressure of less than about 4 mm of mercury. During the initial several hours of fluidization, the introduced precooled dry nitrogen gas is of a temperature of about -5°C or lower. After the initial fluidization the introduced nitrogen gas is gradually warmed and upon cessation of fluidization, the introduced nitrogen gas is of a temperature of 30°C.

Care is used as the introduced nitrogen gas is gradually increased in temperature by closely observing the fluidized bits to make certain that loss of the fluidized state does not result. After less than 4 hr of fluidization under these conditions, fluidization is discontinued and the carrot bits removed from the fluidization column. The dried carrot bits are found to have a residual moisture content of 5% by weight. The dried carrot bits are observed to have retained their original color and apparently not to have shriveled or shrunk in size. A portion of the dried carrot bits upon cooking from 5 to 7 min in water is found to taste substantially the same as similarly diced and chopped fresh carrot bits cooked in an equivalent manner.

If diced carrot bits of a lower residual moisture level are desired, fluidization can be continued in the column for a longer period before removing the diced bits or, the removed diced bits can be returned to the fluidization column and dried further by additional fluidizing under a vapor pressure of less than 4 mm of mercury.

Use of Inert Gas to Eliminate Oxidation of Cells of Citrus Fruits

R.C. Webster; U.S. Patent 3,365,310; January 23, 1968; assigned to Air Reduction Company is concerned with the rapid freezing and subsequent dehydration of citrus fruit cells so as to produce a dehydrated citrus fruit product which is in a substantially permanently preservable state.

The term cell is used to designate juice containing sacs, and the expression discrete juice cells is used to designate cells that have been detached from other cells and from the other parts of the fruit without breaking the walls of the cells.

The mature fruit is first peeled by any of the conventional peeling methods. This leaves some of the pithy lining of the peel on the outside surfaces of the section membranes (carpellary). The peeled fruit is then immersed in an extremely cold liquid, such as nitrous oxide of approximately –128°F, liquid nitrogen of approximately –320°F, or other liquids of comparable low temperatures. The temperature of the liquid used to freeze the fruit should be of cryogenic temperature, that is, a temperature lower than –100°F. The term immersed is used herein to designate a covering of the fruit with a liquid for a limited period of time.

Ordinarily, the immersion is accomplished by dipping the fruit in the liquid, a number of pieces of fruit being treated simultaneously by placing them in a wire mesh basket, but it can be accomplished by covering the fruit as a result of spraying liquid on it or by subjecting the fruit to a vapor of cryogenic temperature. The immersion referred to may also be a combination of the various freezing media just mentioned. The period of the immersion should be long enough to freeze the fruit to a solid condition; but it can be no longer. A period of nine seconds has been found sufficient with orange sections immersed in nitrous oxide at a temperature of –128°F.

When the fruit is withdrawn from contact with the immersion medium, it is found to have been shattered by thermal shock, that is, extensive separation of the fruit has taken place, along the segmented wall and between individual juice cells. Sharp, cracking sounds are heard as the fruit is withdrawn from the immersion media into an ambient atmosphere at room temperature.

It is theorized that this shattering phenomenon results from thermal shock. The sudden freezing followed by the sudden transferring from the extremely low temperature to a higher temperature contributes to the shattering. Agitation of the shattered sections, while still frozen, causes them to crumble further into individual or discrete cells, but the juice cells themselves are not broken and are distinguished from one another. Processing of the fruit while in this frozen condition and with the juice cells intact, permits them to be completely separated from each other and from the pithy material, carpellary membranes, seeds and vascular bundles.

The agitation or separation can be carried out by passing chunks of adhering cells on a belt under resilient rollers or between such rollers. This detaches the juice cells from one another and also from the membranes that cover the sections. The juice cells may then be separated from these carpellary membranes and from the pithy material, seeds and vascular bundles by means of shaking sieve screens that let the juice cells go through them and stop the other parts of the fruit.

The mass of discrete juice cells is thus made substantially free of other constituents. The frozen cells are then ready for dehydration.

According to the process, the citrus fruit to be dehydrated has been flash or rapidly frozen, which avoids cell destruction as the result of ice crystal growth, since the time of freezing and the time for growth of ice crystals are greatly reduced.

The flash frozen citrus fruit product is then dehydrated in known dehydration equipment. It is placed in a vacuum chamber, and, in a high vacuum, water in the product is transformed directly from ice form to a vapor, which is abstracted from the product and collected elsewhere, for example, on refrigerated plates located in the chamber. Controlled heating is applied to the citrus fruit product in the vacuum environment to create a temperature gradient allowing removal of the water from the product. The dried product can then be stored.

The frozen discrete citrus juice cells, in one preferred method of dehydration, are placed in a vacuum chamber environment below 4 mm Hg, usually in a batch type arrangement for sixteen or twenty-four hours. The moisture is abstracted from the individual discrete juice cells, from the boundary of ever receding high crystal zones of each cell. The possibility of chemical, enzymatic or microbiological action during drying is remote.

Heat is normally conducted to the product during the drying procedure by conduction and radiation. The dehydrated discrete citrus juice cells are dried to a point where approximately 98% or more of the water is removed.

Nitrogen or other inert gases may be used in the dehydration process during the break back process. As already stated, the chamber of the freeze dehydrator is under a high vacuum. There has been very little oxidation of the cells, due to the absence of oxygen. The cavities left between sublimating ice crystals are virtually void of air. If the vacuum is broken by breaking back with nitrogen or other inert gases, the cavities in the produce will be occupied with the gases and the oxygen included within the product practically negligible. The inclusion of the breaking back concept aids in the packaging of the dehydrated discrete citrus fruit cells since oxygen will be practically excluded from the cells. The nitrogen or other inert gas used may be the effluent of the nitrogen or other inert gas previously used during the flash freezing operation described above.

Use of a High Carbohydrate Additive in Processing Apple and Grape Juice

Certain fruit juices and syrups, such as apple juice, grape juice, honey, are difficult, if not impossible, to dry by usual methods. The difficulty seems to arise from their relatively high content of monosaccharides, notably levulose, which is highly soluble and highly hygroscopic. High temperatures cannot be employed without losing characteristic flavor. At low temperature the hygroscopicity of the sugars involved, together with their tendency to form viscous syrups, results in prolonging the time required for drying beyond practical limits.

R.M. Stern and A.B. Storrs; U.S. Patent 3,483,032; December 9, 1969; assigned to Great Lakes Biochemical Co., Inc. have developed a method of processing fruit juices such as apple and grape juice and syrups such as honey comprising the steps of adding a corn syrup solid having a dextrose equivalent of 45 or below, or lactose to the juice. The amount of material to be added comprises at least 75% of the levulose in the fruit juice, or at least 25% of the total monosaccharide content, whichever is higher. This is the minimum amount which can be expected to be beneficial. Best results will be obtained if the higher carbohydrates in the treated juice are adjusted to about 125% of the levulose content or about 75% of the total monosaccharide content, whichever is higher.

In making the tests outlined below, the drying technique employed was the so-called freeze drying process. The freeze drying process is well known and involves the sublimation of moisture in the frozen state in the material under conditions of high vacuum. The latent heat of sublimation is customarily applied by conduction from platens, which are heated during the course of the drying.

In the examples below, the material to be dried was frozen by reducing its temperature to about 10°F or below. The frozen material was placed on a platen in a vacuum chamber. The pressure was then reduced to at least 2 mm mercury or less and then heat was applied by the platen.

Example 1: Apple Juice — Untreated apple juice can be dried by freeze drying only with the greatest difficulty, if at all. It requires a long drying cycle and invariably dries only to the glassy state, which cannot be collected or harvested efficiently. While the end-product is soluble, it lacks the instant properties which are desired.

The following series of samples, using fresh, unconcentrated apple juice, demonstrate the favorable results which can be obtained by the use of high carbohydrate additives. In this series the drying equipment was operated at a platen temperature of 200°F for the first 3¼ hours after the initial pull-down to operating vacuum, followed by 2½ hours at a platen temperature of 125°F. Total cycle time was 6 hours.

> Control sample (no additive): did not dry completely; did not exhibit any puff or foam structure; was glassy with plastic, tacky body at end of cycle.
> 2% lactose: had very slight foam structure; was tacky and not completely dry.
> 4% lactose: slight foam structure; very slightly tacky; not quite completely dry.
> 8% lactose: dry; good foam structure; light, friable body.
> 16% lactose: dry; high foam structure; light, friable body.
> 2% corn syrup solids (24 DE): very slight foam; slightly tacky; not completely dry.
> 4% corn syrup solids (24 DE): slight foam; very slightly tacky; not quite completely dry.
> 8% corn syrup solids (24 DE): dry; good foam structure; light, friable body.
> 16% corn syrup solids (24 DE): dry; good foam structure; light, friable body.

All of the above samples were soluble when reconstituted with water, but those containing 8 and 16% added carbohydrates were more rapidly dissolved than those at lower levels.

Inasmuch as fresh apple juice contains approximately 6.6% levulose and about 1.6% dextrose, it will be seen that optimum results were achieved when the carbohydrate additive amounted to at least 120% of the levulose content, or at least 95% of the total monosaccharide content. The added carbohydrates shortened the time required for drying the juice, a matter of some economic importance.

Low DE corn syrup solids, in this case maltodextrin, gave good foam control at the 80% level and a good end-product. High DE corn syrup, 63 DE, containing about 40% dextrose, caused excessively wild foaming during drying at either 40 or 120% levels, although such of the end-product as could be recovered was dry and friable.

Example 2: Grape Juice — The sugars of grape juice are monosaccharides, fresh juice containing about 8% levulose and 14% dextrose. With this composition problems will be encountered in drying due to a lack of foam-forming or structural elements.

The addition of 20% lactose to grape juice gave very good results with respect to drying. As little as 4% lactose improved drying performance but not sufficiently to be of practical significance. Additions greater than 20% did not appear to have any increased advantages.

Maltodextrin, a low DE corn syrup product, had about the same effect as lactose except that about half again as much was required for the same degree of improvement.

POTATOES

Diced Potatoes—Freeze Drying Followed by Air Drying

H.W. Kruger; U.S. Patent 3,438,792; April 15, 1969; assigned to Lamb-Weston, Inc. describes a process for dehydrating food products wherein freeze-drying of the product is combined with air drying, thereby to result in a material reduction in the total time required for dehydration and to obtain a product that will reconstitute faster and be of high quality.

The product is frozen by either a quick freeze or a slow freeze method. Quick freezing is defined as that process where the temperature of the food passes through the zone of maximum ice crystal formation (32° to 25°F) in 30 minutes or less. A product can be quick frozen in an air stream of suitable temperature. It can also be dipped in or sprayed with liquid nitrogen, refrigerant or liquid carbon dioxide. Slow freezing is defined as freezing a product in an air stream above 0°F. The nature of the raw product, such as the sum of the total solids therein, will determine if slow freezing or quick freezing is to be used.

After the freezing step has been accomplished, the product is then subjected to freeze-drying for a period of time sufficient to remove water from the external regions to a depth which will prevent liquid solutions from the interior from depositing a film on the outer surfaces of the product. This depth will vary depending on the nature and size of the product, and in the case of 3/8 inch potato cubes may be about 1/10 inch.

Freeze Drying of Other Foodstuffs

The freeze drying time required may vary from about 30 minutes to one hour. Any of the usual forms of freeze drying may be used. For example, the product may be subjected to freeze drying in a vacuum at a pressure below 4.5 mm of Hg (4,500 microns), the vapor pressure of ice at 0°C, so as to keep the product from thawing while heat is applied to achieve sublimation.

Alternatively, the product may be subjected to freeze drying at atmospheric pressure in the air stream having a partial water vapor pressure below that which exists above ice in equilibrium at the temperature of the air stream, or the product may be subjected to freeze drying at intermediate pressures in an air stream having a partial pressure of water vapor below the vapor pressure of ice in equilibrium at the temperature of the air stream. The resulting evaporation forms a porous layer on the exterior of the product through which the moisture in the interior can readily pass. The interior of the product beneath this porous exterior layer, however, is not dehydrated.

A satisfactory way of accomplishing the freeze drying step is to place the pieces of the frozen product on product trays on which they are preferably spaced apart sufficiently to permit the moisture to escape. The heat necessary for sublimation is supplied by means of top and bottom heating plates. In some instances the product may be tumbled or otherwise agitated during freeze drying.

Finally the product is air-dried to a desired moisture content, such as by placing it in an air dryer and subjecting it to air of an elevated temperature. The time and temperature will depend on the product and may, for example, be between about 1 and 3 hours. For example, in the case of potatoes the temperature of the air may be between about 140° and 220°F. On the other hand, meat products should not be heated higher than 120°F.

During the air-drying step the moisture remaining in the interior migrates toward the freeze dried surface where the porous external layer formed on the product as a result of the freeze drying step readily permits it to escape. It is thus possible to achieve a final moisture content of 5% or less in a relatively short period of time.

The soluble solids that were present in the moisture in the center of the product are left in the porous external layer when the water vaporizes, but do not form an impenetrable case hardened layer.

Examination of the interior of product pieces dehydrated by the process discloses that the interior is not uniformly tightly packed and shriveled as in the case of fully air-dried products. Rather, the shriveling appears to be somewhat lattice-like. Voids are formed between the shriveled portions. This lattice-like structure is not deleterious to the texture of the reconstituted product.

Example: Potatoes having a specific gravity of 1.0825 and a total solids content of 21.3% were cut into 3/8 inch cubes (diced) and were placed in a solution of approximately 0.5% sodium bisulfite and citric acid for 45 minutes. They were then blanched in water at 162°F for 15 minutes.

Next they were placed in a freezing chamber and subjected to rapid freezing (in circulating air at –20°F) for one hour and five minutes. The potatoes were then subjected to freeze drying for 40 minutes in a vacuum chamber. They were placed on two product trays, each of which was placed between two heating plates. The current supplied to the apparatus was 4.9 amperes at 141 volts. Thermocouples were attached to both the heating plates and the product trays to measure the temperature as the freeze drying progressed.

Thermocouple 1 was on the upper heating plate; thermocouple 2 was on the upper product tray; thermocouple 3 was on the heating plate above the lower product tray; and thermocouple 4 was on the lower product tray. The vacuum pump was turned on at the start of the cycle. The data was as tabulated below:

Time, min	Vacuum, microns	Temperature, °F, at Thermocouples			
		1	2	3	4
0	--	116	42	116	34
10	330	151	59	145	48
20	350	196	101	190	81
30	400	220	131	214	105
40	400	241	162	237	132

The loss of water as a result of the rapid freezing and freeze drying amounted to 36.5%, based on the weight of potatoes that was initially subjected to rapid freezing. The potatoes were then removed from the vacuum chamber and air-dried for 120 minutes, the air temperature varying from 172° to 180°F. The final moisture content was found to be 1.7%. The bulk density was found to be 13.0 lb/ft^3. The thus dried pieces of potato were then reconstituted by placing them in boiling water for 1½ minutes and then allowing them to stand in the hot water for an additional one minute. All were found to be firm and to have good color, flavor and odor.

Diced Potatoes—Two Blanchings Prior to Freezing Step

This process of *J.L. Sloan; U.S. Patent 3,644,129; February 22, 1972; assigned to Lamb-Weston, Inc.* is particularly suited for use preparatory to the Kruger process (U.S. Patent 3,438,792 above) and results in a potato product that will demonstrate both quicker and better rehydration. Potato pieces so processed demonstrate improved physical integrity and texture, rehydrate readily and produce a reconstituted potato piece similar to a freshly boiled potato.

The process treats potato pieces by first blanching them, then cooling them, and then blanching them again. The potato pieces are individually quick frozen and then dehydrated according to the previous process. Specifically, the process comprises peeling the potatoes, cutting them into the desired physical shape, rinsing off the free starch, and then soaking them in a dilute solution of sodium bisulfite. The potato pieces are then blanched for from 5 to 30 minutes in water having a temperature of between 160° to 180°F. The pieces are then removed from the water and cooled for about 5 to 30 minutes in either air or water having a temperature of between 33° and 80°F and thereafter, blanched a second time for about 5 minutes in water having a temperature of about 160°F.

The thus preliminarily processed potato pieces are individually quick frozen at a temperature between −20° and 0°F. The pieces are freeze dried at subatmospheric pressure for from 10 to 60 minutes to experience a weight loss of between 10 and 50% of the weight of the frozen product. The product is then subjected to drying in air at atmospheric pressure at a temperature that may range between 140° to 170°F so as to reduce the moisture content to 7% or less of the dry weight of the product.

Example 1: Washington Russet Burbank potatoes were peeled, cut into 3/8 inch cubes and rinsed free of surface starch. Sulfite was applied to the potatoes by soaking them for 5 minutes in an 0.03% sodium bisulfite solution. The potatoes were then placed in 180°F water containing 0.5% sodium acid pyrophosphate, and the water was allowed to cool to 160°F, at which temperature the blanch water was maintained. The potato pieces were held in the 160°F blanch for a period of time sufficient to achieve a total time of 15 minutes in the water.

The potato pieces were then removed to a 40°F room on stainless steel wire mesh screens. Air at a temperature of between 40° and 45°F was then circulated over the potatoes for 10 minutes. After thus being cooled, the potatoes were reblanched in 160°F water for 5 minutes. The potatoes were then individually quick frozen by placing them in air at –10°F.

After preliminary preparation as above disclosed, the potato cubes were freeze dried for one hour to achieve a 32.4% weight loss and then air dried by placing them in air at a temperature of between 145° and 150°F for 2 hours and 10 min. When reconstituted, none of the potato pieces showed any sloughing, and only an insignificant number had hard centers. The pieces were similar to freshly boiled potatoes.

Example 2: Washington Russet Burbank potatoes were prepared again as in Example 1, except that they were cooled by immersing them in 60° to 65°F water between the two blanching steps. Although some leaching was experienced, a satisfactorily reconstituted product was achieved.

French Fried Potatoes—Improving Permeability of Crust After Frying

R. Menzi and C. Giddey; U.S. Patent 3,518,097; June 30, 1970; assigned to Georges Lesieur & Ses Fils, France are concerned with the preparation of dehydrated french fried potatoes which can be kept at ambient temperature in ordinary moistureproof wrapping, and which can then, because of their crispy state, be eaten as they are, or which can then be used for preparing hot and hydrated french fried potatoes by mere rehydration in cold or hot water and quick frying (the second frying stage of the conventional method).

This method comprises frying potato pieces of desired shape so as to cook the pieces and to form thereon either a crust which is permeable to water vapor and to water or a relatively impermeable crust which is subsequently rendered at least partially permeable, freezing the fried potato pieces to a temperature below –10°C, and lyophilizing the frozen potato pieces to reduce their water content to below 8%.

It is preferable for the frying operation to be carried out in a manner such as to produce a more resistant and hence a thicker crust which will impart greater rigidity to the fried potatoes. This will be the case when the oil temperature is greater than 110°C and the length of frying time is greater than 10 minutes.

An increase in crust thickness and hence in crust resistance can only be achieved at the expense of its permeability to water vapor and to water. The crust must therefore be made at least partially permeable after it is formed through frying. This permeabilization of the crust can be carried out either before or after freezing of the fried potatoes by making perforations therein with for instance a needle of suitable diameter. The crust could also be permeabilized by removing part of the crust, in particular by cutting off at least one of the ends in the case of slices.

Another method of permeabilizing the crust consists in subjecting the potatoes, when still hot, to the abrupt effect of a high vacuum. The vapor that is suddenly generated swells the potatoes and hence causes cracks to form in the crust. The cooling and freezing of the thus treated potatoes must take place immediately after swelling. Crust permeabilization through swelling has the advantage of increasing the porosity of the inner mass. The crust can also be permeabilized by first swelling the potatoes and then, once frozen, by making perforations in their crust or by removing a portion of the latter.

The freezing of the fried potatoes below $-10°C$, preferably between $-25°$ and $-35°C$, can be carried out either in an air stream or by the action of a high vacuum, corresponding to a pressure below 1 mm Hg, causing intensive evaporation and temperature lowering. In the second case, the potatoes must be subjected to the action of the vacuum while still hot, i.e., after frying and draining.

Example 1: Slices having a square cross section of 9 x 9 mm are prepared in known manner from raw potatoes. They can, if desired, be dipped into a bath of sulfurous, acetic, citric or ascorbic acid in order to stabilize the color. After being wiped dry, the slices are fried for 25 minutes in a bath of ground-nut oil having a temperature set at $105°C$.

Once removed from the bath and after draining off excess oil, the fried potatoes are frozen in an air stream to $-18°C$ and are then fed to the lyophilization chamber in which a vacuum is set up such as to produce ice evaporation at $-15°C$. Depending on the amount of heat imparted to the fried potatoes, the water content thereof will be reduced to 5 to 2% after about 8 to 12 hours.

Dehydrated fried potatoes are obtained which have a good appearance and which can be eaten as they are or which can be rehydrated and reheated by a quick fry in oil having a temperature lying between $170°$ and $200°C$. The complete rehydration, which can take place either in cold or hot water, is very quickly achieved due to the permeability to water of the crust and to the porosity of the inner mass. The firmness in the rehydrated state is rather slight so that potato slices have to be handled carefully during frying. Once fried they have a good texture.

Example 2: Slices prepared as in Example 1, are fried for 20 minutes in an oil bath at a temperature of $120°C$. After draining off the excess oil, the slices are placed, while still hot (about $50°C$), in a vacuum tunnel, having inlet and outlet air-locks, wherein the pressure is first progressively lowered, over 10 seconds, to 300 torrs, then abruptly, in 1 second, down to $6 \cdot 10^{-2}$ torr. The abrupt effect of the vacuum causes the slices to swell and the intensive evaporation, which then ensues, causes the temperature to drop to $-30°C$. The slices thus swollen and frozen are then placed in the lyophilization chamber and dehydrated as in Example 1. The dehydrated fried potatoes have similar properties to those produced according to Example 1 except that full rehydration thereof is achieved less quickly.

French Fried Potatoes—Density Fractionation Before Freezing

The process developed by *J.C. Smith, D.A. Butler, J.J. Opella and G.D. Porter; U.S. Patent 3,573,070; March 30, 1971; assigned to Uncle Ben's Inc.* relates to methods for dehydrating white potatoes whereby an entire potato may be processed without formation of hard, vitreous material thus promoting quick rehydration. Generally, the method includes initial preparation of the potato such as peeling, trimming and slicing with appropriate chemical rinsing followed by cooking to gelatinize the slices; cooling; freezing; freeze drying; and air drying.

In an alternative modification, prior to cooking the cut pieces may be subjected to density fractionation with isolation of the heavier pieces for further processing by cooking, cooling, freezing and air drying.

White potatoes are washed, etc., cut into pieces, rinsed and treated to prevent browning. The pieces are then cooked to gelatinize the starch by heating under substantially nondrying conditions at a temperature of about 165°F for approximately 25 minutes. The pieces are then cooled by a quench bath and placed on a porous stainless steel belt to be frozen under substantially nondrying conditions.

Cold, saturated air at a temperature preferably at least as low as 21°F is then passed through the bed of potato pieces for about 30 minutes so that moisture within the pieces is frozen. Relative humidity of the air is then reduced to preferably 50% or lower and temperature raised to about 31°F so that the moisture is removed by sublimation and the moisture concentration reduced to not greater than approximately 65 to 70%. Then the moisture concentration is reduced to about 10 to 12% by heating the moving air to the range of 115° to 130°F. The air preferably is always at about atmospheric pressure as it moves through the bed of potato pieces during each of the steps and it is usually desirable that the process be continuous such as by use of an endless belt passing through differing air zones to carry out each of the above steps.

Example 1: About 46.5 pounds of Russet Burbank potatoes from Washington State were washed, peeled, trimmed and cut into ⅜ inch cross-sectional french fry type slices. The slices were then dipped for 30 seconds into a 0.2% sodium bisulfite solution and placed into a cooker. Cooking by contact with saturated air at a temperature of about 165°F proceeded for 30 minutes whereupon the slices were removed and rinsed with tap water at 75°F followed by rinsing in cold water at 35°F to cool the slices.

The slices were then treated for browning by immersion in a solution containing 0.75% glucose, 0.75% disodium phosphate and 0.25% glycine. Then the slices were transferred for freezing to a freezing chamber where they were placed on a porous screen support with a 3½ inch bed depth. Saturated air at a temperature of 21°F and a velocity of about 400 to 450 feet per minute was passed generally perpendicularly and downwardly through the bed for 40 minutes. Then to accomplish sublimation, temperature of the air was raised to 31°F, the relative humidity adjusted to 50% and the velocity increased to about 500 feet per minute. After 13 hours at the adjusted flow rate, the weight of the potato slices was reduced to 50% that of the initial weight which is attributable to less of moisture. The slices were then dried by passing an air stream through the bed of potatoes for six hours, the air being at a temperature of 115°F and a velocity of about 400 feet per minute. Total moisture within the potato slices was thus reduced to about 10 to 12%. An outstanding dehydrated product resulted which had no observable vitreous matter.

The dehydrated potatoes when rehydrated in tap water had a rehydration ratio of about 4.4 which is the final weight of the pieces divided by the weight of the dehydrated pieces. When the rehydrated pieces were fried in vegetable oil, the slices had 19.1% fat and the fried pieces were quite edible and with good texture. By way of contrast, another sample of the same lot of potatoes was processed as above but partial drying by sublimation was omitted. The resulting product had extensive vitreous matter.

Since the core or relatively low specific gravity material of a potato is undesirable in that it has a high tendency to form vitreous material, such lower specific gravity material may be treated by a method wherein sublimation after freezing is carried out to remove a part of the moisture therein and establish structural stability of the cellular lattice work before being dried by heated air. The remainder of the potato which is of higher specific gravity may then be processed without necessity of the sublimation drying step.

Example 2: Raw potatoes may be peeled, trimmed, sliced, etc. Then the slices may be subjected to gravitational separation by placing the pieces in a brine bath having a specific gravity of approximately 0.01 less than the average specific gravity of the raw potato lot. Preferably about 20% of the lowest specific gravity material is removed from the brine bath and rinsed in sulfite solution followed by cooking so as to gelatinize the starch in the process. The slices are then cooled and frozen under substantially nondrying conditions whereupon the frozen pieces are dried partially by sublimation of the moisture and further dried by hot air.

The sinkers remaining in the brine bath consist of higher specific gravity material which may be drawn off and likewise rinsed in sulfite solution, cooked so as to gelatinize the starch, cooled and frozen under substantially nondrying conditions and then dried such as by contact with heated air to reduce moisture content thereof to about 10 to 12% resulting in a high quality, dehydrated potato product with practically no vitreous matter and with good texture and flavor. The combined embodiment permits use of the entire potato wherein only about a fifth of the bulk of the potato need be dried partially by sublimation of moisture as ice vapor to establish structural stability of the potato pieces.

Dehydrated Fried Potato Cake

K.R. Johnson, T.R. Schmidt and G.M. Cooper; U.S. Patent 3,489,575; Jan. 13, 1970; assigned to U.S. Secretary of the Army have developed a process of making dehydrated potato cakes comprising uniformly mixing precooked potato particles with a suspension of comminuted and gelatinized corn in water to form a mixture comprising 6 to 12% gelatinized corn and having a total water content of 75 to 80%, forming the mixture into cakes, frying the cakes to form a porous crust thereon, and freeze-vacuum-dehydrating the fried cakes to a water content below 3%. Instantly rehydratable storage-stable shape-sustaining dehydrated potato cakes are obtained, which upon rehydration have the appearance, texture and flavor of freshly prepared potato cakes.

Example: Raw U.S. No. 1 grade Idaho Russet potatoes are washed, peeled, cut into strips of cross-sectional dimensions of ¼" and cooked for 5 min in water at 200° to 210°F. The cooked potatoes are then drained and cooled on a screen (U.S. No. 8 mesh) for not less than 15 min at room temperature and a relative humidity of 30 to 40% to substantially eliminate surface water therefrom.

Seven parts by weight of commercial grade of degerminated corn meal having an initial water content of about 12% by weight are dispersed in 21 parts by weight of water and the following flavor ingredients are added: 1.25 parts of salt, 0.08 part of white pepper, 0.50 part of minced, dehydrated onion, 1.00 part of shortening, and 0.10 part of monosodium glutamate. The slurry of corn meal and flavoring ingredients is heated until the corn meal has completely gelatinized (approximately 185°F), forming a suspension of the corn meal. Then 69.07 parts by weight of drained, cooked potatoes are added to the corn meal suspension and the mixture is thoroughly mixed in a food mixer.

The resulting mixture is formed by hand or by machine into cakes approximately 3¾" x 2½" x ¾" in dimensions. These cakes are fried in deep fat at 375° to 400°F until the surfaces thereof develop a golden brown color. This usually requires approximately 3 minutes cooking time. The cakes are then drained free of excess fat, cooled, and then freeze-vacuum dehydrated and the frozen moisture (ice) is removed by sublimation.

Recommended freeze drying conditions for cakes on trays placed on heated platens of a conventional vacuum-freeze dryer are a vacuum of about 0.5 to 1.5 mm (0.75-1.00 mm) of mercury absolute, a platen temperature varied during the dehydration cycle beginning at about 130°F and being reduced to a temperature below 125°F by the end of the cycle, and a dehydration time of approximately 12 hours. The platen is heated by circulating a heating fluid through tubes therein.

In an alternate drying procedure, the potato cakes are placed on suspended trays between radiant heating plates at a plate temperature of 160°F under the same vacuum conditions as above, and for a dehydration time of about 5 to 6 hours. The water content of the dehydrated product is below 3%, and may be as low as a trace; about 2 to 3% water content is presently considered most desirable for good storage stability and about 2.5% water content is preferred.

If it is intended to store the dehydrated potato cakes over an extended period of time, they should be packed in the absence of atmospheric oxygen, e.g., canned with a high vacuum (say of the order of about 26 to 27 inches of mercury) or in the presence of an inert gas, such as carbon dioxide or (preferably) nitrogen. In lieu of a can, flexible plastic waterproof hermetically heat-sealable pouches, e.g., of polyethylene terephthalate film may be employed, with the exclusion of atmospheric oxygen. Samples of potato cakes produced in accordance with this example and stored in the absence of oxygen at semitropical temperatures of the order of 100°F for 6 months have been found completely acceptable in appearance.

Rehydration is virtually instantaneous upon addition of hot water (e.g., about 170° to 212°F), which can be done in a sauce pan or canteen cup, if desired.

For optimal restoration of the characteristics of a freshly prepared fried potato cake, a cake of the thickness of that of the example should be rehydrated, for about 30 to 60 seconds and then the excess, unabsorbed water should be poured off so that the potato cake will not become soggy.

VEGETABLES

Mushrooms—Reducing Enzymatic Degradation and Discoloration

J.W. Tarvin and J.R. Fisher; U.S. Patent 3,033,690; May 8, 1962; assigned to Armour & Company have developed a process for freeze drying mushrooms which reduces enzymatic degradation and accompanying discoloration during the freeze drying procedure and also substantially prevents product shrinkage. The process comprises rapidly chilling the mushrooms to a temperature below about -20°C, placing the chilled mushrooms under vacuum below about 2 mm Hg, and, while maintaining the vacuum zone at a pressure below 2 mm Hg, increasing the temperature of the vacuum zone until a product temperature of 70°C is obtained.

Preferably, the mushrooms are cooled rapidly to -30°C at a rate such that there is apparently little breakdown of the cell walls within the mushrooms by the freezing of the water within the cells. The frozen mushrooms are then placed under a vacuum below about 1 mm Hg and heated to about 85°C; the vacuum is broken and the mushrooms are removed from the oven.

To accomplish the drying of the mushrooms, the oven is heated from room temperature to about 85°C over a period of two hours after the desired vacuum is obtained, care being taken to insure that the vacuum remains at the desired level. This temperature is maintained until the product heats to between about 70° and 85°C at which temperature the mushroom contains less than about 2% residual moisture by weight.

The freezing step can be carried out in any equipment designed to cool materials to the desired temperatures. The drying step is carried out in any suitable vacuum oven, preferably an oven having hollow shelves and a jacket for the circulation of heat transfer agents for the heating and cooling of the vacuum zone.

Example: A 2.1 kg portion of a light strain of *Agaricus campestris* mushrooms having a 3.7 to 7.5 cm diameter, closed veils and light colored interior gills was diced to about 0.5 mm dimensions, washed with cold water, and placed in a stainless steel tray. The tray of mushrooms was placed in a walk-in freezer maintained at a temperature below -30°C and rapidly chilled to about -20°C over a period of three hours.

After chilling, the mushrooms were removed from the freezer and placed in a cooled, jacketed, drying oven. The oven was sealed and a vacuum of 1 mm drawn with a four stage steam ejector system. After about twenty minutes the desired vacuum was obtained and water having an initial temperature of about 20°C was circulated through the platens and jacket of the oven. Over a two hour period the temperature of the water was increased to 85°C. The product was removed from the oven when dry (after about 22 hours) and found to have an improved appearance as to shrinkage and color.

Carrots—Color Stabilization with Ascorbic and Erythorbic Acids

Freeze dried carrots exhibit a whitening or fading in color and quality. *D.M. Gottlieb and J.R. Linaberry; U.S. Patent 3,894,157; July 8, 1975; assigned to General Foods Corporation* found that treating frozen carrot dice prior to freezing and freeze drying with ascorbic or erythorbic acid, the color values are fixed.

A dilute solution of ascorbic acid or erythorbic acid is infused into a decorticated (peeled) subdivided carrot which has been blanched and partially cooked, whereupon the coated and subdivided carrot piece is frozen preferably at a slow rate and then freeze dried to a moisture content of 2 to 3%.

It is believed that by reason of the blanching and cooking, physical changes are induced in the carrot piece which permit infusion of the antioxidant to the inter- and intracellular structure of a carrot so that oxidation of carotenoids is retarded, which infusion is further enhanced by the change in turgor caused by freezing slow enough to grow large nonamorphous crystals after application of the acid solution.

Example: Carrots were trimmed, peeled, cleaned and then diced to produce pieces measuring 1/8 x 1/4 x 1/4 inch. The dice were then fed in continuous, uniform manner to a rotary immersion water blancher adjusted to blanch at a water temperature of 210°F for 4 minutes. The dice were then discharged from the blancher onto an inspection belt to remove defects. The defect-free dices were then hydrocooled in refrigerated water and dewatered in a dewatering shaker immediately ahead of the system for applying ascorbic acid solution.

The dice were sprayed with a 3% ascorbic acid solution at a ratio of about one part of solution to 10 parts of product and under conditions sufficient to effect substantially uniform coating and absorption of the solution onto the dice, the rate of application being such as to also permit any unabsorbed solution to run off prior to freezing. The dice were turned over (rotated) by a double oscillating device on a moving belt during spray impingement. The dice were then evenly fed through a freezer to achieve a product temperature of -10° to -20°F. The product was frozen over a period of approximately five minutes.

The frozen product was loaded onto freeze drying trays in a cold storage transfer room maintained at 10°F at an approximate loading of 2 lb/ft^2. The pieces were then freeze dried in a batch freeze dryer over a cycle time of approximately 12 hours to a final moisture content of 2.5% (broadly 2.0 to 3.0%) and under a vacuum of 300 to 500 microns; the freeze dryer had a maximum platen temperature of 170°F generated by the heat transfer medium used to supply the latent heat of sublimation, and the product temperature never exceeded 135°F. The product thus produced could be stored for long periods (1 year) at a temperature less than 40°F and for shorter intervals (6 months) at room temperature (70°F).

Avocado—Heating Single Particle Layer of Frozen Material

F.K. Baerwald; U.S. Patent 3,445,247; May 20, 1969; assigned to Basic Vegetable Products, Inc. found that if a mass of frozen, finely divided material, is spread in a single-particle layer on a surface in a vacuum chamber, with the particles in the layer being separated in discrete units, and if, in addition, these units of frozen material are heated by heating of the contacting surface, that heat transfer and transfer of water vapor from these particles is rapid or instantaneous. An arrangement suitable for use in the process is shown in Figure 10.9.

FIGURE 10.9: HEATING SINGLE PARTICLE LAYERS OF FROZEN MATERIAL

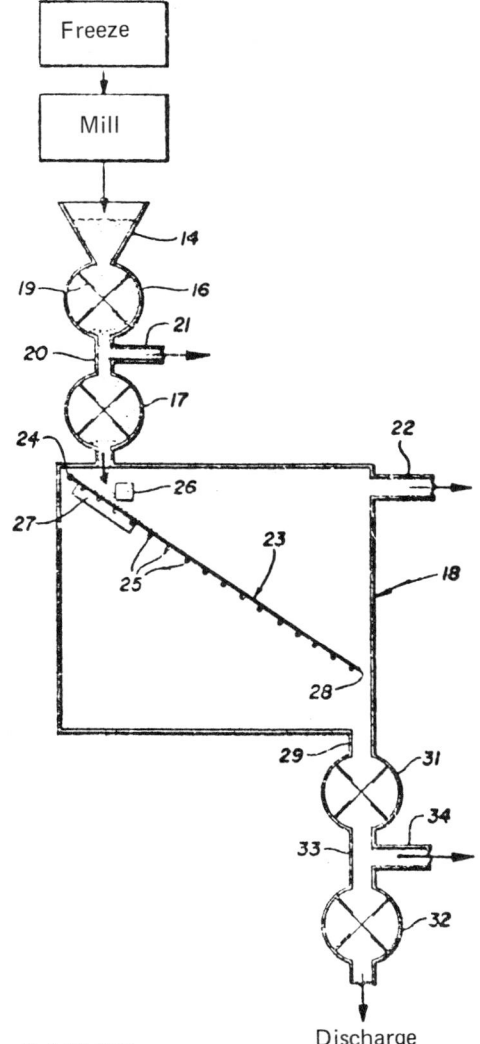

Source: U.S. Patent 3,445,247

Finely divided frozen material is introduced into a feed hopper **14** from which it passes through air lock feeders **16** and **17** into a vacuum tank **18**. The air lock feeders **16** and **17** are connected by a passageway **20** which is in communication with a source of vacuum and a condenser through the piping **21**.

Vacuum tank **18** is likewise in communication with a source of vacuum and a condenser, not shown, through the piping **22**. A cascade plate **23** is pivotally mounted at **24** within the tank **18** so that its degree of slope, or angularity, may be readily adjusted to any point between horizontal and 45°, for example.

Operation of the unit is as follows. The position of the cascade plate **23** if first adjusted so that the desired degree of angularity is achieved. The position of the spreading device **26** with respect to the upper surface of the member **23** may likewise be adjusted depending upon the particle size of the material to be dried.

The vacuum pump and condenser will be placed in operation so that the desired degree of vacuum is drawn. The heating element **25** will also be operated so that the surface of the plate **23** will be brought to the desired temperature, preferably a temperature well below that which might scorch or otherwise damage the product.

When the operator has determined that the desired degree of vacuum exists within the tank **18** and that the surface of the heating element is at the desired temperature, finely divided frozen material is introduced into the feed hopper **14** and is passed through the air lock feeders **16** and **17** into the vacuum chamber **18**.

As the finely divided frozen material falls upon the upper surface of the cascade plate **23**, it is spread by the spreading device **26** into a uniformly thin layer over the upper surface of the plate **23**. The material then tumbles or rolls down the plate **23** with the result that substantially the entire surface of each particle will engage the heated cascade plate **23**. The ice within the finely divided particles will be vaporized immediately and the vapor will be withdrawn. The finely divided dry material will then pass through the air lock outlets **31** and **32** to suitable packaging or other equipment.

The vibration unit **27** may or may not be utilized. In the event it is utilized, the plate **23** may be adjusted to a less steep angle with the result that the material may remain on the plate for a slightly longer period of time.

Example 1: Ripe avocados were peeled and halved and the pits removed. The halves were frozen, milled and freeze dried and the resulting product was used in the preparation of spreads, salad dressings and aspics. This product is unique inasmuch as it is not possible to preserve avocado by any other method of food preservation.

Example 2: Green vegetables, such as spinach, green beans, asparagus or peas, after washing and trimming were first slightly cooked, then quickly frozen, milled and freeze dried. The resulting powder is reconstitutable instantaneously to a puree by the addition of hot water and the resulting puree has the natural flavor and color, and is indistinguishable from, of freshly prepared and cooked puree of vegetables.

Peas—Alternate Vacuum Freezing Dehydration and Air Dehydration Steps

F.G. Lamb; U.S. Patent 3,218,725; November 23, 1965; assigned to Lamb-Weston, Inc. provides a cycle or sequence of process steps wherein vacuum dehydration and freezing are alternately employed with air drying in such manner that the dehydration can be carried through to any predetermined and desired limits within the range of 30 to 95% removal of the original moisture content of the product.

There is produced an end product which reconstitutes, upon hydration, far more quickly and has substantial preservation of the natural cellular formation of the product.

The process may be summarized as involving these essential phases. Firstly, subjection of the product to a vacuum of that magnitude which will partially dehydrate the product to at least 3 to 10% of its original moisture content, which magnitude of vacuum is preferably of that degree that will lower the temperature of the product to the freeze point but not reduce the temperature thereof to below the freeze point. It is preferred that the vacuum employed in this vacuum dehydration step be of the order of about 4.6 mm of mercury. At this reduced pressure the product temperature will be lowered to about 32°F, or just above, and the product will remain unfrozen.

Secondly, the product is permitted to remain at this predetermined amount of reduced pressure for a period of time to accomplish the desired amount of dehydration. As set forth in more detail in U.S. Patent 3,219,463 the purpose of holding the product at this temperature and pressure for a designated period of time is to remove at least, as the required minimum, that amount of moisture which will permit subsequent freezing without cellular damage.

Removal of at least 10% of the original moisture content will assure expansion space for the ice formed when the temperature is reduced to below freezing by further reduction of pressure. Where the primary objective is as now, dehydration rather than mere freezing, in contrast to the U.S. Patent 3,219,463, then of course the vacuum, in this predehydration phase, is applied for a maximum or optimum period to withdraw a maximum amount of moisture from the product.

Thirdly, after the so-called wait period for at least that period of time to permit subsequent freezing without damage and preferably, in this instance, to remove the maximum amount of moisture, the vacuum is broken and the product removed from the vacuum zone. It is then subjected to air drying or drying by other appropriate methods.

Fourthly, since it is obvious that air drying warms the product to a significant degree dependent upon the temperature of the air utilized and the length of time of application thereof, it may be preferable to subject the product to one or more additional vacuum-freezing stages as set forth above. This would be particularly true if the product is to be ultimately distributed, not simply as a dehydrated, dry material, but as a dehydro-frozen product.

In such event the same procedure is followed as in the first step, the product is readmitted to the vacuum zone in its warm state (due to the air drying phase), vacuum is applied to lower the temperature to the freeze point without freezing for a period of time to further dehydrate to the desired degree, then lowered sufficiently to ultimately freeze the material. This is accomplished by pressure decrease to from about 0.2 to 5 mm of mercury pressure. The product is now dehydro-frozen and can be placed in permanent cold storage without further application of other freezing methods.

The process is applicable without restriction to almost any type of food product, particularly those which have a porous exterior, it being appreciated that some foods, such as grapes and peas, preferably have their skins punctured for ready access to the interior thereof in order for the vacuum procedure to have its desired effect.

Example: A quantity of fresh peas are slit and placed within a vessel able to withstand internal and external pressures of about 20 psi. The vessel is constructed with a 100 pound capacity. The vessel is pressure sealed and steam introduced until the peas reach a temperature of between 200° to 212°F, and that temperature maintained for a period of about 5 minutes. Such steam treatment blanches the peas and also introduces an amount of heat beneficial in the next step, consisting of vacuum application.

The steam is turned off and a vacuum applied to the interior of the vessel in an amount of 4.6 mm mercury pressure. The charge is submitted to such vacuum for a period of about 8 minutes, the temperature being reduced during this period of time to about 33°F.

The vessel is kept at this pressure and the charge maintained at the stated temperature for a period of about 10 minutes. At the end of this time, the peas are dehydrated or moisture removed therefrom, in an amount of about 15% of the original moisture content. The purpose of the instant step in the procedure is not only to obtain sufficient dehydration but, as a result of such dehydration to thereby provide sufficient space for the expansion of the residual water when it is frozen by the next quick freeze phase.

The pressure is then reduced to 1 mm of mercury, such reduced pressure lowering the temperature of the product to 0°F. Pressure is then increased to normal and the product removed from the vacuum zone. It is then subjected to air drying at a temperature of about 180°F, which air is passed through the product by known media. When the product reaches approximately the air temperature it is then again placed in the vacuum zone, and vacuum applied of the same order, 4.6 mm of mercury.

The product is again subjected to such lowered pressure for a period of 8 min, at the end of which period pressure is lowered to 1 mm of mercury and the product again frozen. It is then removed from the vacuum zone and placed, as a dehydro-frozen product, into permanent cold storage.

Green Beans—Simultaneous Compaction and Freeze Drying

A.R. Rahman; U.S. Patent 3,984,577; October 5, 1976; assigned to U.S. Secretary of the Army describes a method of simultaneously compacting and freeze-vacuum-dehydrating particulate foods, such as vegetables and meats, so that the dehydrated foods will have substantially greater bulk densities (from 0.5-1.2 g/cc) than they do when freeze-vacuum-dehydrated without compaction, and so that upon rehydration the foods will be restored to substantially the same distinct particulate states in which they existed prior to compaction and freeze-vacuum-dehydration.

It is preferable to employ a platen temperature of about 100° to 150°F and a total gas and vapor pressure in the vacuum chamber of about 0.1 to 2.0 mm of mercury. When these conditions prevail, the frozen food should be subjected to a mechanical pressure of at least about 20 psi in order to realize a satisfactory degree of compaction of the food by the time its moisture content is reduced to less than about 4.0% by weight.

The compaction ratios attained in accordance with the process will be from about 2:1 to 16:1. Compaction ratio is the ratio of the volume occupied by the frozen food particles prior to freeze-vacuum-dehydration and compaction to the volume occupied by the compacted freeze-vacuum-dehydrated food.

Vegetables and meats (including fish as well as other edible animal products) that occur naturally in particulate form or are susceptible to subdivision into particulate forms having at least one dimension not greater than about ¾ inch are suitable for simultaneous compaction and freeze-vacuum-dehydration. The process is particularly applicable to carrots, peas, green beans and diced beef.

Example 1: Simultaneous Compaction and Freeze-Vacuum-Dehydration of Diced Carrots — Fresh whole carrots were washed, scraped substantially free of skin, rinsed to remove the remnants of the scraped skin, and cut to produce dice of about 3/8 inch length on each side. The diced carrots were blanched by contacting them with water, heated to about 205°F, for 5 minutes.

The blanched, diced carrots were drained free of surface water. The diced carrots were individually quick frozen in a blast freezer. The individually frozen, diced carrots were placed in a cylindrically shaped, perforated vessel having perforations uniformly spaced apart throughout its side walls and bottom, the perforations being approximately 1/8 inch in diameter.

Enough of the frozen diced carrots to cover an area of 1.3 square inches were placed in the vessel and a perforated cover was placed on top of the sample of diced carrots within the vessel. A 26 pound weight was placed on top of the cover, providing a pressure of about 20 psi on the sample of frozen, diced carrots. The perforated vessel containing the frozen, diced carrots under the 20 pounds per square inch pressure was placed on a platen inside of a freeze dryer and freeze-vacuum-dehydration of the diced carrots was carried out at a platen temperature of about 120°F and a total gas and vapor pressure within the freeze dryer of about 0.2 mm of mercury.

After approximately 12 hours, the freeze dryer was opened and the freeze-vacuum-dehydrated carrots were found to be in a compacted form. The carrots in this compacted form represented a compaction ratio of about 12.5:1, based on the original volume of the frozen, diced carrots. The compacted, freeze dried carrots were found to have a bulk density of 0.5 g/cc compared with a bulk density of freeze dried diced carrots which were not compacted while being freeze dried of 0.04 g/cc. The moisture content of the compacted, freeze-vacuum-dehydrated carrots was found to be about 2.0% by weight.

Freeze Drying of Other Foodstuffs

The compacted, freeze dried carrots were rehydrated by placing one or more compacted, freeze dried portions thereof in an excess of boiling water over that required for rehydration of the carrots, continuing to boil the water for 15 min, then allowing the carrots to stand and simmer for 15 minutes more in water at a temperature of about 190°F. The rehydrated diced carrots were strained to remove excess water and found to be quite acceptable and to be closely comparable in shape and size to the original diced carrots.

Example 2: Simultaneous Compaction and Freeze-Vacuum-Dehydration of Diced Beef — Longissimus dorsi muscles of beef were frozen, then sawed to form cubes ½ inch in length on each side and maintained frozen. The diced beef was placed in a cylindrically shaped, perforated vessel and a 26 pound weight was placed on top of the perforated cover, providing a pressure of about 25 psi on the sample of frozen, diced beef.

The perforated vessel containing the frozen, diced beef under the 25 psi pressure was placed on a platen inside of a freeze dryer and freeze-vacuum-dehydration of the diced beef was carried out at a platen temperature of about 120°F and a total gas and vapor pressure within the freeze dryer of about 0.2 mm of mercury.

After approximately 16 hours, the freeze dryer was opened and the freeze-vacuum-dehydrated beef was found to be in a compacted form. This compacted form of the beef represented a compaction ratio of about 3.0:1, based on the original volume of the frozen, diced beef. The compacted, freeze dried beef was found to have a bulk density of 1.2 g/cc compared with a bulk density of freeze dried beef which was not compacted while being freeze dried of 0.4 g/cc. The moisture content of the compacted, freeze-vacuum-dehydrated beef was found to be about 2.0% by weight. The compacted, freeze dried beef was rehydrated in a similar manner to that described above.

Salads—Use of Critical Proportion of Emulsion Type Dressing

The process of *J.M. Tuomy and R.G. Young; U.S. Patent 3,264,121; August 2, 1966; assigned to U.S. Secretary of the Army* relates to vacuum-freeze-dehydrated salads containing emulsion type food dressings which, when properly protected from oxygen and moisture in storage, will maintain their high consumer acceptability under adverse storage conditions. These vacuum-freeze-dehydrated salad-type rations are edible, compact, high in caloric value and very nutritious under survival condition in any climate when eaten with or without prior rehydration.

One of the problems concerning the preparation of frozen salads or any other type of frozen food product containing mayonnaise or other emulsion type food dressings has been the tendency of the emulsion to break down under freezing conditions with separation of the oil resulting. This is avoided by utilizing critical proportions of emulsion type food dressing. The term emulsion type food dressing includes any known commercial mayonnaise product, as well as food dressings of the emulsion type, having an oil content of from about 35 to about 80%.

The critical proportions of emulsion type food dressing in the salads range from approximately 4 parts per 100 total parts of salad in the case of salads containing high proportions of vegetable ingredients to approximately 20 parts per 100 total parts of salad in the case of salads containing relatively low proportions of vegetable, but high proportions of meat. If meat as well as vegetables are included in the salad, the emulsion type food dressing may be present in proportions between 4 and 20 parts per 100 total parts of salad, the proportion of emulsion type food dressing being determined by the relative proportions of vegetable and meat.

Example 1: Potato Salad, German Style — White potatoes were cooked by boiling until tender, but still firm. They were then diced and 15 lb, 5 oz of the diced potatoes were mixed with 14 oz of cooked diced onions, ½ oz onion powder, 1/8 oz pepper, 1 lb, 2 oz dehydrated vinegar, 4 oz salt, 3 oz sugar, 1 lb, 2 oz of prefried and diced bacon, 3 oz bacon fat, and 14 oz of mayonnaise having an oil content of about 80% by weight. The order of addition of the ingredients is relatively immaterial.

After being thoroughly mixed, the potato salad was cast into individual molds having internal dimensions of about 4 x 2½ x ½ inch. The molds were provided with false bottoms so that by inverting the mold containing the salad, the salad could be removed from the mold onto a tray by pushing the false bottom to the opening in the mold.

The molded blocks of salad on trays were subjected to deep freeze conditions in a blast freezer at −10°F until frozen. They were then vacuum-freeze-dehydrated at 750 to 1,000 microns of mercury pressure with a plate temperature of 110°F until their moisture content had been reduced to about 1.4%. The vacuum-freeze-dehydrated blocks of potato salad were packed in cans in a nitrogen atmosphere and hermetically sealed, and then stored for one year at 70°F.

At the end of this storage period, the dehydrated potato salad was rehydrated by adding enough tap water to rehydrate the salad and stirring well for 15 sec. The rehydrated potato salad was eaten with a fork in the normal manner and found to be highly taste-acceptable. Some of the blocks of dehydrated potato salad were eaten without prior rehydration and were also found to be highly taste-acceptable.

Example 2: Chicken Salad — Chicken was raw boned, then cooked in a form, then chilled and diced to a convenient size, e.g., 3/8 x 3/8 x 1/4 inch. Celery was blanched and then diced to a convenient size, e.g., ¼ x ¼ x ¼ inch. Bacon was cut into small pieces and fried to a crisp state. Then three pounds of cooked diced chicken, 9 oz of blanched diced celery, 4.5 oz of prefried bacon in small pieces, and 9 oz of mayonnaise having an oil content of about 80% by weight were thoroughly mixed, then frozen at about −10°F in a block. The frozen block was then sawed into smaller blocks having dimensions of about 4 x 2½ x ½ inches. The small blocks were then placed on a tray while still frozen and were vacuum-freeze-dehydrated in the same manner as the potato salad of Example 1 to a moisture content of about 1.0%.

The vacuum-freeze-dehydrated chicken salad was packed in cans in a nitrogen atmosphere and the cans were hermetically sealed and stored for a year at 70°F, then tested in the same manner as the potato salad of Example 1. The vacuum-freeze-dehydrated chicken salad was found to be highly acceptable when eaten dry or after rehydration.

Low Calorie Dietary Fruit or Vegetable Foodstuff

C.H. Barr, Sr., C.H. Barr, Jr., and J.W. Barr; U.S. Patent 3,360,374; December 26, 1967; assigned to Courtland Laboratories describe a process for producing a dietary food product comprising preparing a mixture of a foodstuff and methylcellulose, the foodstuff being substantially free from amino acid, adding an amino acid to the mixture and freeze drying the mixture. The amino acid is added in sufficient amount to substantially prevent the mixture from sticking to the mouth.

The amount of methylcellulose used in the product may be varied within any limit depending upon the amount of calorie intake desired. It has been found that the use of methylcellulose greatly facilitates the freeze drying procedure. The consumable amino acid may be added in dried form to the other ingredients after they have been freeze dried. The amino acid may also be added before freeze drying in either dried or solution form.

Any type of freeze drying production may be used. It is preferred to use a pump-condenser system because pumps without condensers require the use of extremely large pumps and the use of steam jets introduces the hazard of backfiring. When a condenser system is used, the condenser surfaces should have extremely cold surfaces and a large area. The temperature of the condenser surfaces should be $-10°C$ or lower at 100 microns McLeod Gauge pressure. This temperature functions satisfactorily to condense the sublimed gases to the frozen state without the condensation of liquid.

In essence, any temperature which is sufficiently low to condense the sublimed gases to the frozen state without the condensation of liquid is operable. These temperatures will, of course, vary in accordance with the composition of the gases and the pressure. The radiated heat used to sublime the gases from the frozen dietary food mixtures preferably has a temperature of about $110°F$, but it has been found that a temperature as high as about $375°F$ inside shelves where the food-filler mixture is suspended or elevated slightly over the shelves is operable.

The amount of heat which is applied is a function of the condenser capacity to condense the sublimed gases to the frozen state. It has been found preferable to commence the freeze drying procedure with the material which is to be freeze dried at a temperature of about $-40°C$. The temperature of this material will rise during the freeze drying procedure and care must be taken to prevent the frozen liquids in the material from melting, i.e., the temperature of the material must, in most cases, be maintained below $0°C$.

The temperature at which the foodstuff-methylcellulose mixture will sublimate properly will vary somewhat with the composition thereof and the pressures employed. In general, it has been found that the pressure should not exceed 250 microns McLeod Gauge.

The process provides a convenient method for the control of the water content present in the diet. The dietary food produced normally contains on the order of 1% or less by weight of water. Accordingly, the desired control of water intake may be simply accomplished by using the prescribed measured amount of water to reconstitute the dietary product or by taking the prescribed amount of water in conjunction with the dietary product when it is consumed in the dried state.

The dietary product is of further benefit in controlling the cholesterol content of the diet. This may be accomplished by adjusting the fat content of the dietary product to any desired predetermined level.

Example: A mixture of 33 1/3% by volume of cooked mashed carrots, 33 1/3% by volume of methylcellulose and 33 1/3% by volume of glycine was prepared. The mixture was then formed into a disk approximately 4 inches in diameter. The disk was first frozen and then freeze dried by subjecting it to a vacuum of about 100 microns (McLeod Gauge). The vacuum was developed by means of a mechanical pump. Radiated heat at a temperature of about 100°F was applied to the disk to cause the frozen liquids contained therein to sublime. This product was then removed from the vacuum. It was found that the product, if hermetically sealed, can be stored indefinitely without regard to the ambient temperature.

Before eating, the disk can be soaked in water whereby the methylcellulose is caused to absorb water. This imparts weight to the disk and enhances the transfer of flavor to the methylcellulose. Alternatively, the product can be eaten in dried form in the same manner as a cracker. The product takes on water in the stomach and swelling occurs as a result of the absorption of water, discouraging hunger.

FRUIT

Blueberries—Frozen Berry Punctured Before Freeze Drying

The process of *R.K. Scharschmidt and R.E. Kenyon; U.S. Patent 3,467,530; September 16, 1969; assigned to General Foods Corporation* is founded upon the discovery that when blueberries are perforated by one or more needles or like penetrating protuberances, hereinafter referred to for convenience as pins, in a manner wherein the pins puncture not only the outer integument or skin portions but also a substantial amount of the interior core portion, the thus perforated blueberry can be freeze dried to a form which readily rehydrates.

Preparatory to puncturing the blueberry it is frozen, ideally at a very slow rate, preferably in the neighborhood of 10 to 20 hours, to a temperature of 0°F and below. It has been found that when a frozen blueberry is so punctured the freeze dried product does not shrivel or otherwise undergo change in the physical appearance of the outer integument portion and instead dries in a substantially unwrinkled bulbous spherical shape.

The thus punctured and freeze dried fruit will rehydrate in an aqueous preparation such as milk or cream in a period of 30 seconds to 3 minutes, depending upon such factors as size of the berry, degree of maturity, particular freezing rate employed prior to puncturing, climatic variations, etc. This rehydration rate is to be contrasted with that of blueberries which have been simply freeze dried under comparable freezing and freeze drying conditions but which have not been so punctured, the rate being 30 minutes and more.

Puncturing of the blueberry in the frozen form facilitates the transfer of frozen water from the core portion in the vapor form. As a result the total elapsed time of freeze drying the thus punctured blueberry down to the desired stable moisture level of less than 3% will be in the neighborhood of 18 hours or less, which will be in the order of 2/3 or less of the time normally required to freeze dry an unpunctured blueberry.

The pin referred to herein above should be so located with respect to the means whereby the berry is punctured that it not only ruptures the integument or skin portion but also penetrates a substantial part of the core portion; typically each pin puncture will cause the core portion to be ruptured along a depth of penetration of approximately 1/3 to 2/3 of the mean diameter of the blueberry.

Since fruit will vary in diameter it is preferred to grade the fruit according to size so that there will be a minimum variation in the average particle size of the blueberry. Thereafter the graded berry will be introduced to the space between a pair of oppositely rotating rolls designed to afford in a predictable and controllable manner the degree of penetration.

The rolls will have spaced annular grooves therein along the longitudinal length of the rolls, the grooves serving to orient the berries as they are deposited coaxially from above onto the pinch between two adjacent rolls; this roll will also preferably contain longitudinal grooves. The annular rings will thus run parallel to one another and serve to orient the blueberries so that they are introduced in a controllable manner to the locus of the free ends of pins suitably mounted on the adjacent roll and adapted to intersect the blueberries in the annular grooves of the first described roll. The longitudinal grooves are located around the cylindrical surface of the roll so as to assure the provision of a positive feed of blueberries to the point at which the berries are punctured by the pins.

The adjacent roll will be adapted to puncture the blueberries located on the first described roll, comprising a cylinder having a plurality of longitudinal rows of parallel pins, adjacent pins being spaced from one another in each row on feeders less than the anticipated mean diameter of the blueberry, and the pins having a height from the surface of the cylinder sufficient to assure penetration of the blueberry by the free end of the pin.

The loci of the free ends of the pins form annuli which are spaced from the rubber roll and annular grooves therein but are sufficiently proximate to the surface of the rubber coated roll to assure that the free ends of the pins will puncture the blueberries. Preferably the blueberry perforating rolls will be duplicated two or more times by locating a second and third series of like rolls beneath the first set.

The blueberries in practice are frozen preferably over a total elapsed time of 17 to 19 hours, during which time the blueberries are reduced gradually from ambient room temperature to the freezing point plateau of the water in the blueberry and eventually further reduced in temperature to 0° to 10°F.

The solidified frozen blueberries will then be fed to the described apparatus, passing between the first, second and third sets of rolls. Preferably the prefrozen blueberries will be tempered, that is, allowed to rise in temperature to above 0°F and typically to a temperature 15° to 25°F, whereafter the blueberries will feed to the perforating apparatus. As a result of this tempering the puncturing of the berries by the pin means will be facilitated and the blueberries themselves will be less prone to shatter incident to penetration. Thereafter the frozen blueberries will be introduced to any conventional freeze drying apparatus such as a shelf type vacuum freeze dryer wherein blueberries will be loaded to a bed depth of ½" to 1". The blueberries are freeze dried by following any conventional freeze drying profile.

Fruit for Dry Cereal Mix—Use of Slow Freezing Technique

W.L. Vollink, R.E. Kenyon, S. Barnett and H. Bowden; U.S. Patent 3,395,022; July 30, 1968; assigned to General Foods Corporation found that improved freeze dried fruit capable of rehydration in milk within 30 to 90 seconds can be produced by a process which comprises cooling the fruit to just above the freezing point of the water present in the fruit, then slowly freezing this water to develop a growth of ice crystals which expand the cellular structure of the fruit, and freeze drying the slowly frozen fruit to a terminal moisture content of less than 3%.

Slow-freezing involves transforming the moisture in the fruit into a growth of large ice crystals over a period of at least two hours, preferably 5 to 8 hours, in order to expand and partially rupture the cellular walls or tissues of the fruit.

The frozen product may be sliced prior to freeze drying in order to accelerate the drying time. Prior to slicing, the fruit is warmed to a semifrozen state or a product temperature of between 10° to 25°F. The specific temperature will depend on the particular fruit. This is necessary to facilitate the cutting operation and prevent fragmentation of the fruit due to its brittle character at a product temperature of about 0°F or below. The fruit is preferably sliced into 3/16" to ¼" thick sections, the fines are screened out and the sliced fruit is then freeze dried immediately or again cooled to below 0°F for storage purposes.

Example 1: Strawberries — A charge of freshly picked strawberries was washed and then graded for uniformity. Strawberries having a particle size of about ¾" to 1" in diameter were then arranged in monolayers on freezing trays and the freezing trays were placed in a large freezing room having an ambient temperature of 0°F.

The strawberries took about ½ hour to cool from room temperature to about 28°F and about 6 hours for the water to be completely frozen at 28°F. The strawberries were then allowed to cool to 0°F. A cooling curve in which temperature was plotted against time showed a temperature profile wherein an initially relatively steep slope down to the freezing point of the moisture in the fruit was followed by a flat line or plateau during the actual change from a liquid moisture state to a frozen moisture state, the flat line again sloping rapidly when substantially all of the water was frozen and the product temperature lowered to 0°F.

The strawberries remained in the plateau for 6 hours and in this time developed a growth of large ice crystals to partially rupture the cellular walls of the strawberries. The frozen strawberries were then stored at 0°F to protect the strawberries against enzymatic or bacterial degradation arranged in monolayer fashion on solid aluminum trays and freeze dried in 10 lb charges in a freeze dryer under a vacuum of 100 microns Hg, a platen temperature of 80°F and a condenser temperature of –60°F for about 20 hours until a terminal moisture content of 1.5% was attained.

Freeze Drying of Other Foodstuffs

The freeze dried strawberries (moisture content 1.5%) were then combined with corn flakes (moisture content 2%) at a level of about 8 to 10% by weight strawberries and 90 to 92% corn flakes. The blending operation was conducted in a packing room having a relative humidity of 30% in a period of less than 10 to 15 minutes thereby limiting moisture pick-up of the strawberries to less than 1%.

The strawberry corn flakes cereal was then packaged in a water resistant, wax-laminated foil liner which was placed inside a ship-board shell and enclosed with a wax-laminated overwrap. The packaged product having a terminal moisture content of less than 2.5% for the strawberries and about 2% for the corn flakes was stored in this form at 70°F and 50% relative humidity for 3 to 6 months without any degradation in product quality.

At the end of this period the strawberries were found to reconstitute in milk or cream in about 60 seconds to a flavor, texture, and appearance close to that of fresh berries. The breakfast cereal could then be eaten within 1 to 5 minutes with the berries being fully reconstituted but not mushy and the cereal still in a crisp form.

Example 2: Peaches — Peaches were lye peeled, cut into halves, pitted, and were then frozen according to the Example 1 procedure (being kept in the plateau for about 6 hours). The frozen peaches were stored at 0°F. The peaches were warmed to 25°F in preparation for the cutting operation and then cut into equal sized wedges having a thickness of about 3/16" slices. The peaches were refrozen to 0°F, stored and then freeze dried to a terminal moisture of 1 to 2%.

The peaches were then blended with corn flakes similar to Example 1, packaged, stored for 3 months and were found to reconstitute with milk in about 80 seconds to a flavor, texture and quality similar to fresh peaches. The peaches did not become soggy within the normal 5 to 10 minute period for consumption of a breakfast cereal.

Example 3: Bananas — Bananas were peeled, arranged in monolayer fashion on freezing trays, and frozen in a freezing room having a temperature of 10°F. The bananas remained in the plateau for about 8 hours and were then cooled to 0°F. The bananas were warmed to 24°F, and sliced into ¼" cross section. The bananas were then freeze dried to a terminal moisture of 1.5%, blended with wheat flakes, packaged and stored according to the procedure of Example 1. After storage the bananas were found to reconstitute in milk or cream within 80 to 90 seconds while giving a texture, taste and quality similar to fresh bananas.

Preliminary Intense Heat Treatment of Frozen Surface

R.L. Macy, Jr.; U.S. Patent 3,462,281; August 19, 1969; assigned to Kellogg Company is concerned with the production of dry, partially puffed food items with acceptable natural flavor, appearance, texture and rehydration and eating characteristics substantially similar to the fresh foods from which they originated. These are storage stable under room temperature conditions for reasonable lengths of time. The dry products can be used readily as additives to other food products such as cereals, soups, etc., or they can be eaten directly as snacks. This is attained by subjecting pieces of the foodstuff to be freeze dried to a preliminary treatment, namely a brief intense heat treatment of the surfaces of the frozen pieces to slightly melt them, following which the pieces are promptly placed in the vacuum drying chamber of the freeze drying apparatus and therein subjected to vacuum, before raising the temperature therein.

The process is generally carried out by slicing food items of relatively large volume to slices of about one-eighth to one-half inch in thickness. Other food items which are naturally in this thickness range do not require slicing. The pieces of the food item are then spread in a single layer on trays and frozen to a temperature of about $-20°$ to $-40°F$. These frozen items are then briefly subjected to relatively intense, uniform heat which causes the outermost layers or surfaces of the food items to melt without causing the interiors of individual food pieces to thaw appreciably.

The frozen centers of the individual food pieces maintain the natural shape and integrity of the individual food pieces. These heat treated food pieces are then transferred, before further melting occurs, to the vacuum chamber of a freeze drying apparatus and the pressure therein reduced to less than 100 microns mercury, as is conventional in freeze drying. This causes the melted surfaces of the food items to foam slightly. The reduced pressure is maintained until evaporation of moisture from the melted surfaces of the food pieces causes the food to refreeze at the surface.

The trays of food pieces are then permitted to remain in the vacuum chamber of the freeze drying apparatus and conventional freeze drying technique is then continued to finish drying the food items accompanied by conventional heating of the vacuum chamber to raise the temperature therein to a degree sufficient to accomplish sublimation while retaining the pieces in frozen condition so that the frozen water content of the food pieces as ice is directly sublimed off and condensed in an adjacent chamber, and the dried pieces are recovered at a moisture of approximately 2%.

Example 1: Bananas — Fresh ripe bananas are sliced to pieces of $3/16$ inch in thickness and frozen at $-40°F$ while spread out on a tray. The frozen banana slices are then passed through intense heating means such as an apparatus having closely spaced rows of infrared light arranged 4 inches above and 4 inches below the frozen slices, at a rate of about 6 seconds per inch or for a total period of about 60 seconds.

These heat treated slices are then transferred to a vacuum chamber of a conventional freeze drying apparatus and the pressure lowered therein to less than 100 microns mercury before further melting occurs. Reduced pressure is then maintained for about 10 minutes before the platen temperature of the radiant heating means in the chamber is raised to $190°F$ for conventional freeze drying finishing. Thereafter the freeze dried product is recovered at a moisture content of about 2%. The product was readily reconstituted on addition of milk.

Example 2: Oysters — Fresh oysters are spread on trays and quickly frozen at $-20°F$. The frozen product was then heated as in Example 1 with a rate of passage of the frozen product through the infrared heating apparatus of about 9 seconds per inch for a total of about 90 seconds. The surface-melted frozen pieces were then placed on trays in the vacuum chamber of a freeze drying apparatus, dried therein first under vacuum without heat until refrozen as above, and then finish dried in the conventional freeze drying technique. The dried

Freeze Drying of Other Foodstuffs

product was readily reconstituted on cooking in water. In lieu of the infrared heating means described in the foregoing, any intense uniform zone of heating may be employed to cause surface melting. The rate of passage of the product through the heated zone may be varied to accommodate any specific product. The amount of foaming on the surface of the individual food pieces may be varied using measures in the range of 0 to 100 millimeters mercury absolute, until refrozen, and before freeze drying.

ARTIFICIALLY SWEETENED FRUIT

Frozen Fruit—Sweetened, Refrozen and Freeze Dried

D.J. Ewalt and R.E. Kenyon; U.S. Patent 3,501,319; March 17, 1970 described a process for enhancing the flavor of freeze dried fruit by freezing the fruit, adjusting the temperature of the frozen surface of the fruit to just below the melting point and contacting the frozen fruit with a water solution containing a flavor enhancer. This effects the melting of the liquefiable constituents at the fruit surface with the resultant intermingling of the flavoring solutes with the fruit constituents present in the melted fruit surface. The temperature of the fruit is reduced to a point well below the crystallization temperature of the water solution thereby refreezing the solution, after which the frozen fruit is freeze dried.

Among the useful flavor-enhancing agents that can be surface applied are the artificial sweeteners which include sodium and calcium cyclamates and saccharin. The flavor-enhancing solution is preferably formulated to contain material such as an alcohol like propylene glycol, or ethyl alcohol, or a low molecular weight saccharide such as invert sugar.

The preferred freezing point depressant is invert sugar. Generally, the flavor-enhancing solution may contain anywhere from 7.5 to 35% of such an oligosaccharide. In the case of propylene glycol specifically, another preferred additive, a concentration of 5 to 15% by weight of the flavor-enhancing solution can be employed to advantage.

The flavor-enhancing solution may be applied at a ratio of 1.5 to 4.5 pounds per 100 pounds of frozen fruit and like plant material. Typically, a flavor-enhancing solution containing an artificial sweetener at such a concentration can be applied to any number of fruits which have been tempered to a product temperature ranging from 18° to 22°F.

Example 1: A five gallon batch of sweetening solution was prepared by mixing together the following ingredients:

   (a) 2.7 pounds of a 7 to 1 blend of sodium cyclamate and sodium saccharin;
   (b) 2.0 pounds of propylene glycol;
   (c) 36.85 pounds of hot water (130°F).

Whole frozen fruit (e.g., strawberries) at 0° to –5°F was tempered in a tunnel freezer for 12.5 to 15 minutes on a belt using an air temperature of 22° to 27°F, the fruit being metered onto the belt continuously at a bed depth such as assured that the fruit emerged from the belt after the elapsed period of 12.5 to 15 min with an internal pulp temperature of 20° to 22°F. The thus tempered fruit, typically strawberries, but commonly any one of a variety of popular fruits such as peaches, pears and the like at the aforesaid internal pulp temperature have the surface portions thereof substantially solid. The sweetening or flavor-enhancing solution was sprayed onto the product, preferably after it had been directed into rotating slicers or halved.

In the case of strawberries the sliced fruit was passed into a 3 foot diameter reel revolving at 12 rpm, nozzles being positioned within the reel so that at least the final 25% of the reel length was available to tumble the sweetened fruit and adequately coat all of the product with the sweetening solution sprayed thereon. The sweetening solution was sprayed on at the rate of 37.5 pounds per 2,500 pounds of frozen fruit. By virtue of the reduced freezing point of the sweetening solution, the solution penetrates the fruit and intermixes with fruit pigments prior to refreezing, thereby assuring that a sufficient length of time is provided for the sweetener solution to harden on the surface and further assuring that increased sweetener penetration of the fruit surface is obtained. In this way, a high level of addition of sweetening solution may be practiced without encountering a crystallizable residue of sweetener on the surface of the fruit at any such high level of addition.

Immediately after addition of the sweetening solution, the fruit was recooled to a temperature from 0° to −10°F by passing the fruit through a refreeze tunnel using an air temperature of −20° to 35°F and a residence time of 5 minutes. The refrozen fruit was then freeze dried in a conventional manner. The freeze dried fruit did not have a white residue of sweetener on the surface thereof and had an appearance which was substantially unchanged as compared to a freeze dried fruit having no sweetener added.

By virtue of the generally solid condition of the interior of the fruit, despite the addition of the sweetening solution to the surface thereof and the minimal surface softening that occurs incident to surface liquefaction, the fruit is in a substantially discrete condition and the pieces are not deformed during their handling in the refreezing tunnel. As a consequence the fruit has an appearance, both in the dried and the reconstituted form, that is acceptable in that it is free of surface deformation and is not clustered.

Freeze Dried Fruit Impregnated to High Level of Sweetness

N.A. Lemaire and R.D. Peterson; U.S. Patent 3,356,512; December 5, 1967; assigned to Kellogg Company found that proper application of artificial sweeteners to freeze dried fruits will result in negligible moisture increase and still give proper impregnation of the sweetener and retain the storage stability of the product.

Moreover, artificial sweetening agents applied to freeze dried fruits do not result in glazing, as does sucrose, since the artificial sweetener is able to penetrate the fruit due to the inherent porosity of freeze dried fruit. Further, sweetener when applied to freeze dried fruit where penetration has occurred, does not dissipate but is carried into the fruit by the rehydration medium, further resulting in a uniformly sweetened fruit product.

The limit of permissible rehydration of freeze dried fruit with aqueous sweetener has been found to be approximately 6%, thus generally prohibiting the use of aqueous solutions of sucrose when it is desired to sweeten the freeze dried fruit to a level of 40% by weight of sucrose equivalent of sweetness.

However, artificial sweetening agents such as sodium cyclamate, calcium cyclamate, saccharin sodium, cyclohexylsulfamic acid and mixtures thereof, which have sweetening power many times greater than that of sucrose, can be applied in an amount adequate to bring the fruit up to the desired level of sweetening without incorporation of more than 6% by weight of moisture. This small amount of moisture can be substantially removed by heat to the original moisture content of the freeze dried fruit, i.e., below about 2.0% by weight, without detriment to the rehydration properties of the freeze dried fruit and without necessity for refreeze-drying.

Example 1: Fresh sliced peaches having a moisture content of about 88% by weight were spread out and disposed on shelves in a freeze drying chamber having a shelf temperature of about 250°F for about 2 to 3 hours, followed by gradual reduction to a temperature on the order of about 120°F for a total drying cycle of about 8 to 10 hours.

The surface temperature of the product was not permitted to rise above 110°F and the pressure in the vacuum chamber was approximately 200 to 600 microns mercury. The ice that sublimed off was condensed in an adjacent chamber wherein the condenser temperature was –30° to –40°F.

The porous freeze dried fruit is then sweetened by spraying thereover a hot concentrated aqueous solution of artificial sweetener in an amount to bring the fruit up to the desired sweetness level and which will not increase the moisture content beyond approximately 6% by weight, followed by heat drying to remove excess moisture and to bring it back to substantially the original freeze dried moisture content.

Preferably the freeze dried fruit pieces are first subjected to heating as by forcing warm air at approximately 10% relative humidity through a bed of the freeze dried pieces until the fruit temperature has reached approximately 130° to 150°F, at which time there is a marked and unexpected decrease in friability of the fruit product without increasing the moisture content.

In another specific variation of the process the aqueous solution of sweetener is sprayed onto the freeze dried fruit at a temperature above about 212°F such as for example 220°F, to result in a flashing on and into the fruit of the sweetener with substantial evaporation of the moisture so that no further heating or drying may be necessary, although such further drying may be practiced if found necessary to reduce the moisture content to below about 2% by weight.

Example 2: The following amounts of freeze dried fruit and sweetening solution provide a product having a 40% relative sucrose equivalent of sweetness.

Freeze dried peach slices	100 g
Calcium cyclamate	1.114 g
Saccharin sodium	0.0832 g
Water	4.45 ml

DESSERTS AND CANDY CENTERS

Sour Cream Fruit Product

This process of *M. Laskin; U.S. Patent 3,472,663; October 14, 1969; assigned to W.R. Grace & Co.* relates to freeze dried sour cream-fruit or berry mixtures (especially sour cream-strawberry mixtures); to methods for preparing them; and to candy and/or cookie items incorporating such products as centerpieces, fillers and the like.

The initial cooling or freezing step should ordinarily be carried out at or about atmospheric pressure to avoid foaming of the frozen product. The product should be cooled, while at or about atmospheric pressure, to the point where it is solidly frozen, particularly in having a hard frozen surface which, in handling and forming, is not predisposed to melt, thaw or soften. In this connection a temperature of –10°F or preferably –20°F or below is used.

Once the product has been cooled to the desired point, i.e., where the entire monolith including all surfaces is frozen hard, evaporative cooling may be, and usually is, used to maintain the temperature of the undehydrated frozen portions in an unthawed state during the dehydration step.

The dehydration (freeze drying) process is carried out under low pressures, i.e., below the vapor pressure of ice at the particular temperature of the product undergoing dehydration. Preferably the pressure should be below about 1.5 millimeters of mercury, and preferably below about 1.0 millimeter of mercury (absolute). The product is heated during the dehydration step of this process by means of radiant energy which is emanated from the surface of the heat exchanger platen or surface.

A broad range of heat exchanger or platen temperatures that can be employed is from about 250° to 90°F, and preferably between 160° to 90°F. The product temperature, at least with respect to the frozen undehydrated portion, should be maintained at a point where no substantial thawing takes place, i.e., about –10°F or less during the drying step.

The freeze drying can be carried out over a period of from about 1 or 2 to about 20 hours. In the freeze drying dehydration method the product should be cut, molded or otherwise formed into monolithic slabs, balls, pieces, or sheets, less than one inch thick, and preferably about ½" or less in thickness. This is typically done before placement in the freeze drying apparatus, and usually before it is frozen into a hard solid monolith to preclude foaming.

Ordinarily, the thinner slabs permit the more rapid diffusion of vapor from the core of frozen material when it is being dehydrated, as well as presenting a broader sublimation interface. The products are dried to a low moisture content, preferably less than 4% moisture (weight basis) and most preferably less than 1½% moisture.

Example: A sweetened strawberry puree was prepared by intensive mixing of 300 parts by weight of cleaned fresh strawberries with 100 parts by weight of cane sugar, in a conventional blender. This puree was then blended in various proportions with a commercially available natural sour cream.

The blends were 50/50 and 25/75 parts by weight of sour cream/sweetened strawberry puree. Blending is readily accomplished in conventional manner using, e.g., hand stirring or a home-type blender. The strawberry/sour cream blends were initially frozen solid to temperatures of about -10°F at atmospheric pressure to form monolithic slabs about 4 inches in diameter and ½ inch or less thick.

These slabs were then placed in a freeze dryer chamber (Freeze Dry Pilot, Model UPFD-X) and the chamber pressure drawn down to a vacuum to about 0.3 mm of mercury (absolute) over a time period of about 8 minutes (pull down time). At the end of the pull down the product temperature was about -20°F. The heat in the dryer platens was applied according to the following schedule of platen temperature and time while the pressure was maintained at or below about 0.5 millimeter of mercury absolute.

Platen Temperature, °F	Time
200	3 hr, 15 min
175	45 min
150	2 hr
125	1 hr, 25 min

The characteristics of the dried monoliths indicated better drying and better body with increasing concentration of strawberry puree. The freeze dried monolithic products were eminently suited for (among other things) use in chocolate coated candy bars and the like, either alone or together with such items as vanilla cookie wafers or the like.

Puddings, Jellies, Pie Fillings

Here, *M. Laskin; U.S. Patent 3,419,402; December 31, 1968; assigned to W.R. Grace & Co.* is concerned with freeze dried cake and pie fillings, puddings, and jellies (such as pectin fruit jellies); methods for preparing them; and candy and/or cookie items incorporating such products as centerpieces, fillers and the like. The conditions for freezing and dehydration (freeze drying) have been described above in U.S. Patent 3,472,663.

Example: Commercially available flavored (butterscotch, vanilla, chocolate, lemon) pudding and pie filling powders (3 to 4 ounces each) containing sugar, dextrose, cornstarch, and flavorants as needed (e.g., cocoa for the chocolate sample) were purchased. These were made into puddings in accordance with manufacturers' instructions, e.g., gradually dissolving in 2 cups of milk and cooking over medium heat with constant stirring until pudding thickens and begins to boil, followed by chilling in suitable containers.

The samples were initially frozen solid to temperatures of -10° to -20°F at atmospheric pressure and then cut into monolithic slabs and pieces about 5/16" thick. These were placed in the trays of a freeze dryer chamber and the vacuum was turned on. The chamber pressure was drawn down to a vacuum of about 0.25 millimeter of mercury (absolute) over a time period of about 10 minutes (the pull down time). At the end of the pull down the sample temperatures were all about -20°F. The heat in the dryer platens was then applied according to the following schedule of platen temperature and time: 250°F, 1 hour; 200°F, 3 hours; 125°F, 30 minutes. After this the vacuum was released and the products removed. Total treatment time (including pull down) was 4 hours and 40 minutes. All samples dried very well. Moisture contents for each were below about 1%. The body and texture were good. The freeze dried monoliths were very well suited for coating with chocolate, caramel or other candy coating to make a candy bar or like candy piece.

The candy coating (e.g., chocolate) will act as an effective moisture barrier surrounding the freeze dried monolithic core or centerpiece, thus precluding rehydration by atmospheric moisture. The products could also be stored, as such, or after grinding to a fine powder in hermetically sealed moistureproof containers without refrigeration for extended periods of time without significant degradation.

Ice Milk Confections

This process of *M. Laskin; U.S. Patent 3,464,834; September 2, 1969; assigned to W.R. Grace & Co.* relates to a freeze dried, storage stable frozen ice milk and cream confection product that may be used as a confection per se or as the centerpiece of a candy-coated confection product, or may be rehydrated to form the original frozen milk and cream confection by the mere addition of water and freezing. Frozen ice milk confections include ice cream mixes, ice cream products, ice milks, soft ices, high fat ices, sherbets, frozen custards and the like (including synthetic ice cream). The conditions for freezing and dehydration (freeze drying) have been described previously in U.S. Patent 3,472,663.

Example: (a) Vanilla ice cream of 10% butterfat content was cut into monolithic slabs of one-half inch in thickness. The slabs were preliminarily cooled at atmospheric pressure to -5°F. The product was then placed in a freeze dryer chamber and subjected to a vacuum [pressure 0.41 mm Hg (absolute)] for a period of one hour and fifteen minutes. During this time the product temperature dropped to about -22° to -25°F due to sublimation cooling. The heat in the dryer platens was applied according to the following schedule of chamber pressure, platen temperature and time for each sequence.

Platen Temperature (°F)	Pressure [mm Hg (abs)]	Time
150	0.68	50 min
125	0.49	30 min
100	0.45-0.5	15 hr, 40 min

The product was removed from the freeze dryer after a total treatment time of 18 hours 15 minutes as a monolithic slab of about 1% moisture content. The product is readily stored in hermetically sealed, moistureproof containers without refrigeration.

(b) Pieces of the slab were coated with chocolate to make a candy bar or piece type of confection. Other sections of the dried slab were cut into small pieces, coated with caramel and then with a layer of chocolate. It should be understood that the candy coating of chocolate and/or caramel acts as a moisture vapor barrier surrounding the candy piece core.

(c) A portion of the slab was ground to a fine powder and stored in containers for a period of twelve weeks. At the end of this time the product in the comminuted state was mixed with water in the ratio of two parts by weight freeze dried product to three parts by weight of water, and the mixture placed in trays and frozen. When solidified the material had essentially the smooth texture, taste, body and fine quality of the original ice cream.

Gelatin Based Desserts

Here, *M. Laskin; U.S. Patent 3,483,000; December 9, 1969; assigned to W.R. Grace & Co.* provides a method of dehydrating gelatin compositions to provide a stable food product which does not require refrigeration and which generally retains the original gel structure in an essentially dehydrated state. This permits the ready adaptation of gelatin products into purees, jellies and similar confection and pastry fillings or toppings, and confection centerpieces.

It is necessary to freeze or chill the gelatin product prior to commencing the drying or lyophilization step. Freezing may be accomplished by conventional techniques or may be accomplished by evaporative cooling. The evaporative cooling involves applying a high vacuum to the chilled gelatin product and the resultant sublimation of moisture reduces the product temperature.

The pressure at which the freeze drying or lyophilization process is carried out is below the vapor pressure of ice at the temperature employed. For most purposes the pressure is below about 1.5 mm of mercury preferably below about 1.0 mm of mercury (absolute).

Preferably, higher temperatures are used in the inital stages of the drying process and they are reduced in a stepwise fashion as the dehydration approaches completion. Thus platen temperature may range from 250°F down to about 90°F and preferably from 160° to 90°F. The product temperature however should be maintained at a point where it remains in the frozen or unthawed state. The nature of gelatin is such as to cause melting or foaming if thawing is permitted. The gelatin dessert in a dehydrated state has a density that is quite low, broadly stated in the order of from 0.03 to 0.2 gram per cc. The freeze drying process is usually accomplished over a period of time, ranging from about 1 to 24 hours.

Example: (a) Strawberry gelatin dessert of the sugar free type was prepared in the conventional manner by mixing one-half ounce of dry flavored gelatin with one pint of boiling water. The gelatin was dissolved and the mixture poured into flat pans to a depth of about one-half inch, permitted to cool, and placed in a refrigerated chamber. When setting or gelation was complete the trays were further chilled to a temperature of about –10°F.

(b) An orange gelatin dessert was prepared using 3 ounces of orange gelatin (sugar sweetened), 8 ounces freeze dried peaches and one pint water. The frozen gelatin products in thin sheet form were placed in a lyophilization apparatus for freeze drying.

An initial temperature of –16°F at atmospheric pressure was gradually increased in 1½ hours to 18°F at 0.32 atm and overnight to 105°F at 0.30 atm for Example (a). Temperatures for Example (b) ranged from –24°F at atmospheric pressure to 105°F at 0.30 atm. There was no heat in platens during startup of runs.

Gelatin of Improved Solubility in Cold Water

T.M.N. Hobday and G.W. Jordan; U.S. Patent 3,892,876; July 1, 1975; assigned to P. Leiner & Sons (Wales) Limited and Ranks Hovis McDougall Limited, England found that the solubility of gelatin in water at temperatures of the order of 15° to 20°C is substantially improved by effecting a rapid freezing of an aqueous gelatin sol containing at least 10% w/w of gelatin, and then subjecting the frozen gelatin sol to an accelerated freeze drying process before any melting of the frozen aqueous gelatin sol is permitted.

A warm aqueous gelatin sol having a gelatin concentration of at least 10% w/w, and preferably of the order of 20% w/w or more, is taken direct from a gelatin manufacturing process, and fed to an appropriate nozzle from which the gelatin sol is sprayed on to the surface of a continuously rotating refrigerated drum from a fixed spray gun.

The temperature of the gelatin sol in the nozzle from which it is sprayed is in the range of 40° to 60°C, and is preferably of the order of 45° to 50°C. In any particular process the temperature of the gelatin sol emitted from the nozzle should be maintained substantially constant.

The gelatin sol is sprayed on to the surface of the refrigerated drum in a spray pattern which is elliptical, with the minor axis of the ellipse small in comparison with the major axis, and preferably line contact between the spray and the surface of the drum is approached by using a spray having a fan shape.

Advantageously, however, the spray gun has a central nozzle from which the sol is ejected and an external air nozzle consisting of two jets disposed on opposite sides of the central nozzle to assist in atomization of the gelatin sol and to provide a suitably shaped spray pattern. The spray should be a fine spray, but not as fine as a mist.

The surface of the refrigerated drum is maintained at a temperature in a range from −10° to −60°C, preferably in a range from −20° to −50°C, by circulating refrigerating fluid inside the drum. Sufficiently rapid freezing of the aqueous gelatin sol as it is sprayed onto the surface of the refrigerated drum is assured when the temperature of the surface of the drum is maintained at about −20°C.

The time within which the aqueous gelatin sol must be frozen is dependent on the concentration of the sol, quicker freezing being essential for higher concentrations. Generally the freezing of the aqueous gelatin sol should take place in a time not longer than five seconds, and preferably within two or three seconds. For an aqueous gelatin sol containing 20 percent by weight of gelatin freezing is achieved in a time of the order of one second from the time at which the spray is formed.

The gelatin sol freezes on the surface of the refrigerated drum as the individual particles impinge on the surface of the drum, so that the frozen gelatin sol is in the form of a discontinuous film or lace-like network of frozen material. The thickness of the frozen gelatin sol on the surface of the drum is preferably not greater than 0.050 inch, because it has been found that if the gelatin sol is applied to the surface of the drum in a manner which produces too great a thickness of the frozen material, a proportion of insoluble gelatin, probably in the form of a gel, is present on the outer part of the frozen gelatin sol.

A doctor blade, which is continuously in contact with the surface of the rotating drum, removes the frozen gelatin sol which falls directly into a chamber maintained at a temperature not higher than -20°C, and preferably between -30° and -50°C, so that the frozen gelatin sol is not permitted to melt.

The frozen gelatin sol is then subjected to an accelerated freeze drying process to a level of about 5% moisture after which the particle size of the dried gelatin product is reduced using an impact mill, so that the final gelatin product has a particle size less than 30 mesh, and preferably less than 100 mesh. The final product is found to have a solubility of at least 50% when two parts of the product are introduced into 100 parts of water at 20°C.

The aqueous gelatin sol which is sprayed on to the refrigerated surface may include other edible ingriedients for a dessert product. For example, for a table jelly or a mousse, or as aspic, there would be included in the aqueous gelatin sol sugars and artificial sweeteners, natural and artificial flavoring agents, coloring agents, fruit acids, buffering agents, emulsifying agents and thickening agents.

When two parts by weight (in grams) of gelatin, are added gradually to 100 parts by volume (in milliliters) of cold water at 20°C, while stirring, the gelatin is soluble in the cold water to give a solution of gelatin which will set to give a firm jelly. The process of forming this jelly from the treated gelatin is clearly a two stage process.

The gelatin is added gradually to the water with stirring, and the stirring is continued for about three minutes to dissolve the material. After standing for about 30 minutes in a domestic refrigerator, the solution sets to give the firm jelly. The gelatin is found to have 90% or more of the gelling property which is present in a sample of similar gelatin which has not been submitted to the method and which is used with hot water to form a jelly.

TEA

Extract Chilled and Filtered Before Freeze Concentration to Remove Tars

R.G. Reimus and A. Saporito; U.S. Patent 3,432,308; March 11, 1969; assigned to Struthers Scientific and International Corporation describe a process for the concentration of tea extract in which a mixture of ice and tea extract is separated. The separation process comprises chilling a tea extract having a solids concentration of approximately 7% by weight to a temperature above the freezing point of the tea extract and storing the extract in a temperature range above the freezing point until a substantial amount of precipitation occurs, separating the insoluble precipitate from the aqueous extract, forming ice in the separated liquid extract and separating ice from the liquid extract to leave a more concentrated liquid extract.

Example: A tea extract containing 6.8% tea solids, at above 80°F, after extraction, is cooled in a heat exchanger to 36°F. The cooled material is then filtered through a filter press and the resulting tar-free extract is freeze concentrated in a tubular crystallizer. The resulting ice-concentrated tea slurry is easily separated on a rotating basket screen-type centrifuge.

Isolation of Cream of Tea for Later Use

During the freeze concentration of tea solutions, large quantities of solids, known as cream of tea, precipitate as the tea solution is brought down to the freezing point. This precipitate is a valuable constituent of the tea which loses flavor and taste without it.

N. Ganiaris; U.S. Patent 3,598,608; August 10, 1971; assigned to Struthers Scientific and International Corporation provides a process for the freeze concentration of tea solution in which there is no loss of flavor in the concentrated tea solution.

This process cools a tea solution, separates the low temperature precipitates from the tea solution, mixes these precipitates with hot water to dissolve them, freeze concentrates the tea solution, and mixes the dissolved precipitates and hot water with the concentrated tea solution. The resulting concentrated tea solution, which has lost none of its solids content, may then be freeze dried. This process is described with reference to Figure 10.10.

FIGURE 10.10: ISOLATION OF CREAM OF TEA FOR LATER USE

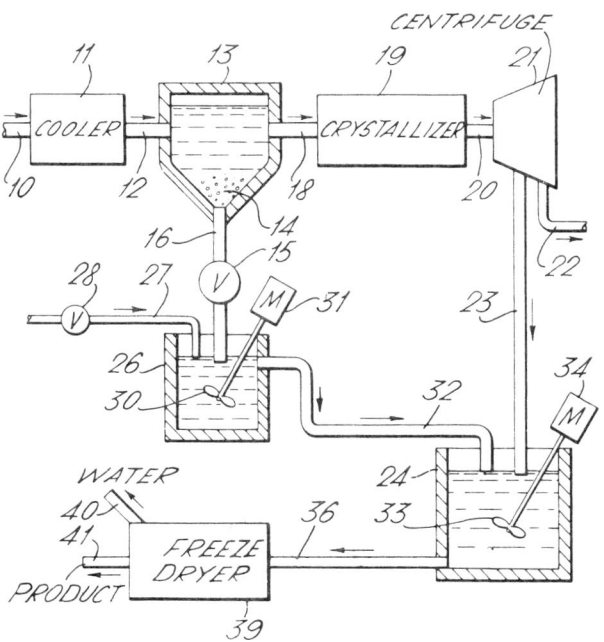

Source: U.S. Patent 3,598,608

Referring to the drawing in detail, an aqueous 5 to 10% solution of tea solids at a temperature of about 75°F enters the system through feed pipe **10** to pass into cooler **11**. In the cooler, the tea solution is cooled to between 32° and 35°F and then is passed through pipe **12** into the precipitate separator **13**. During residence of the cold tea solution in the separator, cream of tea **14** precipitates out of the cold solution to be withdrawn through pipe **16**. Valve **15**, which is set according to the rate of feed through pipe **10**, ensures that pipe **16** drains mostly precipitate.

Cold tea cleared of precipitate passes through pipe **18** into crystallizer **19** in which ice crystals are grown in the tea solution to concentrate it. A slurry of ice crystals and tea solution passes through pipe **20** into a centrifuge **21** or an equivalent piece of apparatus where ice crystals are separated to pass from the system through pipe **22**. Tea solution at a concentration of 36 to 40% at a temperature of about 26°F passes from centrifuge **21** through pipe **23** into mixing tank **24**.

Precipitates from pipe **16** flow into tank **26** into which a small flow of water at 120°F is introduced through pipe **27**. This small flow of water may be metered by a valve **28**. An agitator **30** driven by motor **31** dissolves the precipitate in the hot water which flows as a solution through pipe **32** into tank **24**. Agitator **33** driven by motor **34** mixes the dissolved solution of precipitates with the 36 to 40% solution of tea solids at about 26°F. This mixed and reconstituted solution leaves tank **24** through pipe **36** as a 30 to 35% solution of tea solids at a temperature of from 35° to 40°F. This solution may then be freeze dried in freeze dryer **39**. Water is removed from freeze dryer **39** through pipe **40** and the product, which may be instant tea, through pipe **41**.

Sublimation at Reduced Pressure in Gaseous Medium

L. Rey, J.-P. Bouldoires and D. Rovero; U.S. Patent 3,873,745; March 25, 1975; assigned to Societe d'Assistance Technique pour Produits Nestle SA, Switzerland are concerned with a lyophilization process which allows sublimation of the solvent to be effected at low pressures while avoiding contamination. The process for separating a solvent from a substance in solution or in suspension in this solvent or containing this solvent, comprises solidifying the solvent and then separating it from the substance by sublimation at reduced pressure in a lyophilization chamber, in which an atmosphere is created within the chamber consisting essentially of a gaseous medium capable of existing in condensed state at a temperature above −196°C and of having at that temperature a saturated vapor pressure not exceeding 2 torrs.

The medium is then condensed on a cooled surface the temperature of which is such that the saturated vapor pressure of the medium does not exceed 2 torrs. The solvent is then separated from the substance by sublimation and is condensed on the cooled surface. By the expression "atmosphere consisting essentially of a gaseous medium" is meant, an atmosphere obtained by replacing air initially present in the chamber by the gaseous medium, by a series of purges. The gaseous medium may be carbon dioxide, nitrogen pentoxide, nitrogen dioxide, sulfur dioxide, ammonia, nitrous oxide, hydrogen sulfide or acetylene.

The process may be applied with particular advantage to the preparation of aromatic extracts. Thus a solution of aromatic constituents may be prepared from tea or coffee, by impregnating this substance with a solvent, freezing the solvent and separating the solvent from this substance by the process. The solvent may be water. The resulting condensate, which is uncontaminated, may then be recovered, preferably in liquid state after having been warmed, and provides a solution of the aromatic constituents of the material treated. This solution may then be used as an aromatizing agent. It may be sprayed on a powdered extract of the substance treated, for example at a level of 2% by weight, or it may be added to a liquid or pasty extract of this substance.

Example: An aqueous tea extract having a solids concentration of 3% is frozen, particulated, and loaded on the trays of a lyophilization chamber. An atmosphere consisting essentially of carbon dioxide is established in this chamber, as well as in the enclosure containing a tubular condenser, by effecting five successive purges, each purge consisting of evacuating the atmosphere of these enclosures with a liquid-seal pump down to a pressure of 20 torrs, followed by establishing an atmosphere of CO_2 by introducing this gas at a pressure of 1 atmosphere.

After the last evacuation of the atmosphere from the two enclosures, these are isolated from the pump and CO_2 source, and liquid nitrogen is fed to the tubular condenser. The pressure in the enclosure containing the condenser decreases to a value of the order of 10^{-4} torr. The lyophilization chamber is then connected to the condenser and the pressure in this chamber decreases, also reaching around 10^{-4} torr. The frozen product is then heated to induce sublimation, and this sublimation is continued for 6 hours to obtain a powdered tea extract.

OTHER FOOD PRODUCTS

Rice—Use of Two Cycles of Consecutive Thawing and Freezing

The process of *C.S. Huber; U.S. Patent 3,692,533; September 19, 1972; assigned to the National Aeronautics and Space Administration* is directed to a manner of preparing rice which provides a product which is quickly prepared, relatively lightweight, easily stored, and conveniently handled. The final product is a dehydrated rice granule which is substantially reduced in weight and which can be quickly prepared and served, requiring only perhaps a minute of exposure to hot water.

The rice is cooked in water over low heat for a sufficient interval of time to become tender and edible. The tender and edible rice is frozen and subsequently exposed to two cycles of consecutive thawing and freezing. The freezing is preferably done at quite low temperatures, $-10°$ to $-20°C$.

After the two complete cycles of thawing and freezing are finished, the ice crystals (water) in the kernels of rice are removed by the freeze drying process. The granular rice is placed on trays within a freeze drying chamber. Heated platens are located above and below the trays which contain the granular rice. The chamber pressure is reduced to less than 250 microns. At this pressure, heat is applied to the heating platens above and below the trays. The platen temperature is maintained at approximately $50°C$ during the drying cycle. The ice crystals within the grains of rice are converted into water vapor without passing through the liquid state and, of course, the vapor condenses on refrigerated coils in the chamber maintained at a very low temperature, such as in the range of −60°C. The moisture content (by weight) of the rice granules is less than 3.0% at the conclusion of the drying cycle. The product which remains is the granular rice, minus the water, and is a product which is easy to package, requires no refrigeration, and if properly packaged, can be stored indefinitely without undesirable effects. The dehydrated granules can be readily reconstituted by placing them in water having temperatures in the range of 50° to 100°C. At 100°C, water will reconstitute the rice within one minute. At lesser temperatures, the process requires somewhat more time, but is still accomplished within 1 or 2 minutes. The rehydrability of freeze dried rice has been modified by the consecutive freezing and thawing cycles.

Peanut Butter

M. Laskin; U.S. Patent 3,396,041; August 6, 1968; assigned to W.R. Grace & Co. found that cellular freeze dried, rehydratable peanut butter-containing products can be prepared by thoroughly blending one part by weight of peanut butter with about 0.2 to 1.5 parts (preferably 0.3 to 0.75 part) by weight of water and, where desired or indicated, a sufficient amount (typically this will be from 3 to 10%, by weight, of the total weight) of a suitable emulsifier to provide a substantially uniform dispersion of water through out the blend; quickly freezing the blended materials at or about atmospheric pressure to a hard solid; and then freeze drying the solid frozen blend by dehydration at low pressures and at temperatures at which the frozen portions of the product remain in the frozen state until dehydration is substantially complete.

If desired the freeze dried peanut butter-containing products may include sweetening agents such as sugar, glucose or the like to enhance such confectionary utility. The sweetening agents, when used, may comprise from 5 to 40% by weight based on the total weight of the materials which are to be blended, frozen and dried.

In one modification, a very light, highly cellular product (one in which the volume of the dehydrated product is at least 1.5 times the volume of the peanut butter incorporated therein) is made by uniformly admixing peanut butter in suitable desired proportions with a previously prepared highly aerated blend of water and emulsifier; rapidly freezing the admixture to temperatures sufficiently low to retain its volume and shape, and then freeze drying as above described. The amount of peanut butter in the admixtures can range from 40 to 60% of the combined weight of peanut butter, water and emulsifier.

Prior to introduction into the freeze drying chamber, the peanut butter-containing product must be solidly frozen so as to have a hard frozen surface which, in handling and forming, is not predisposed to melt, thaw or soften. In this connection, a temperature of −10°F or preferably −20°F or below is typically used.

It has been found, especially with respect to the highly cellular products, that a rapid freeze to a very low temperature is particularly preferred. For this purpose, freezing with liquid nitrogen or liquid nitrogen vapors to temperatures as low as −50° to −200°F or lower has been found to be very satisfactory. The dehydration step is carried out at a pressure below 1.5 mm of mercury (absolute), and preferably below about 1.0 mm of mercury (absolute).

A broad range of heat exchanger or platen temperatures that can be employed is from 250° to 90°F, and preferably no higher than 160°F. The product temperature, at least with respect to the frozen undehydrated portion, should be maintained at a point where no substantial thawing takes place, i.e., -10°F or less during the drying step. The time of the drying step is ordinarily over a period of from 1 to 20 hours.

The peanut butter-containing product should be molded or otherwise formed into monolithic slabs, balls, pieces, or sheets, preferably less than 25 mm thick, and in the most preferred instance 12 mm or less in thickness, before placement in the freeze drying apparatus. The products are dried to a low moisture content of less than 4% moisture (weight basis), preferably less than 1½% moisture.

Example: One part of a commercially available smooth-style peanut butter spread was blended with 0.75 part by weight water. About 6.7% by weight (based on total weight) of a commercially available emulsifier-containing edible vegetable base was added to aid in thoroughly dispersing the water throughout the blend. About 20% by weight of sugar (based on the total weight of all materials) was also added. The resultant blend, a thick, whipped peanut butter, was subdivided into bite-size chunks and preliminarily cooled at atmospheric pressure to a temperature between about -10° and -15°F.

The product was placed in a freeze dryer chamber and the chamber pressure was drawn down to 0.5 mm of mercury (absolute) over a time period of 5 minutes (hereinafter referred to as the pull-down time). At the end of the pull-down, the product temperature was about -20°F. The heat in the dryer platens was applied for 1 hour at 150°F and 5 hours, 17 minutes at 125°F and the pressure was maintained at or below about 0.5 mm of mercury (absolute).

The product was removed from the freeze dryer unit after a total treatment time (including pull-down) of 6 hours and 22 minutes. The freeze dried product had a cellular or foam-like structure, a moisture content of less than 1%, very good body and texture, and excellent flavor. This product can be stored, as such, in sealed containers resistant to moisture vapor transmissions without refrigeration for extended periods of time without significant degradation. Some of the freeze dried chunks were coated with an enrobing layer of chocolate to make a candy bar or candy drop confection. The chocolate candy coating acts as an effective moisture barrier surrounding the freeze dried monolithic core or centerpiece, thus precluding rehydration by atmospheric moisture.

Oyster and Shrimp Soup Material

The process developed by *S. Ohtaki; U.S. Patent 3,256,098; June 14, 1966; assigned to Nisshin Kako KK, Japan* relates to a method for producing powdered oyster and shrimp soup materials. These products have the inherent and characteristic taste of each particular raw material and are very easily preserved.

The method comprises the following steps:

(1) Raw oyster or raw or dried shrimp is sprinkled with 2-3 wt % of NaCl and stirred. The impurities on the surface are washed out with the waste.
(2) The resultant washed oyster or shrimp is boiled for a short time (10-15 min) in 2-3% NaCl which is adjusted to a pH 5.5-6.0 by addition of a phosphate buffer. The oyster or shrimp is removed immediately after boiling, and the solution filtered.

(3) The boiled oyster or shrimp is then ground and a composition consisting of decomposing enzymes, mainly protease, together with a 2-3% NaCl solution is added to the ground oyster or shrimp. The mixture thus obtained is decomposed by the enzymes at pH 5.5-6.0 and a temperature of 45°-60°C, for 1-2 hours. The enzyme action is then stopped by heating at 90°C for 10 minutes.

(4) The two extracts are combined, and auxiliary ingredients are added. The mixture is homogenized and dried to less than 8% water content.

Foodstuff Extracts—Addition of High Molecular Weight Organic Substances

The process of *K. Neumann; U.S. Patent 3,013,341; December 19, 1961; assigned to National Research Corporation* finds application in freeze drying aqueous solutions of foodstuff extracts where the solution of foodstuff is maintained during drying at a temperature sufficiently low to accomplish freezing of water therein and, at a temperature sufficiently high to provide for sublimation of water substantially directly from the solid state to the vapor state with substantially no conversion of water to the liquid state during freeze drying.

The process comprises adding to the extract solution prior to freezing a sufficient amount of a water-soluble high molecular weight organic substance selected from the group consisting of gelatin, agar agar, pectin and polyvinylpyrrolidine to form a 0.2 to 2% concentration of the substance which substantially increases the eutectic point of the resulting solution. The solution thus formed is frozen and then subjected to freeze drying. An economical drying of substances with low eutectic point is possible; the melting point of this substance is raised by addition of water-soluble, high molecular weight materials.

Example: Meat extract must be kept at −45°C in an already known process if a condensation of the product would be avoided. According to the present process, there resulted a considerable increase of the temperature limit by the addition of gelatin in a concentration of 0.2%. As a result condensation and foaming first occurred at −32°C. Increasing the gelatin addition to 0.5% permits reaching a drying temperature of −21°C. With addition of 2% gelatin a drying temperature even up to −13.5°C can be reached. If one considers the relatively high expenditure necessary to maintain the low temperature of the icy material, for example, −40°C, then the economies obtained by the process are evident.

Liquid Foodstuffs—Thin Film of Liquid in Revolving Apparatus

G. Vigano; U.S. Patent 3,257,731; June 28, 1966 describes a process for the freeze drying of liquid substances which consists in spreading on the inner surface of a cylindrical container or sublimator under vacuum a thin layer of material in the form of a fluid film in a thickness about $3/10$ of a millimeter. This film is preheated from the outside, to a temperature of +15°C, while at the inside the temperature of the frozen film is under −20°C. The film is then heated briefly also from the inside, to ensure the sublimation of the frozen product without melting. Finally the resultant dried layer is scraped off and falls down in the form of a powder.

COMPANY INDEX

The company names listed below are given exactly as they appear in the patents, despite name changes, mergers and acquisitions which have, at times, resulted in the revision of a company name.

Abbott Laboratories - 132
Afico SA - 167
Air Reduction Co. - 353
American Sterilizer Co. - 127
Armour & Co. - 364
A/S Atlas - 178
Basic Vegetable Products, Inc. - 366
Battelle Development Corp. - 336, 351
Bell Telephone Laboratories, Inc. - 13
Beverly Refrigeration, Inc. - 199
Centre National de la Recherche Scientifique - 115
Commonwealth Scientific and Industrial Research Organization - 328
Courtland Laboratories - 373
Cryo-Maid, Inc. - 170
Cryodry Corp. - 320
Curtiss-Wright Corp. - 300
J.P. Devine Manufacturing Co. - 120
Edwards High Vacuum International Ltd. - 24
FMC Corp. - 26, 39, 73, 75, 105, 183, 187, 281, 293, 307
Cie Francaise Thomson-Houston - 143
General Foods Corp. - 100, 103, 245, 250, 257, 258, 259, 269, 271, 272, 273, 277, 285, 287, 289, 290, 291, 292, 295, 296, 299, 308, 310, 311, 365, 374, 376
General Foods, Ltd. - 247, 248, 294
Georges Lesieur & Ses Fils - 359
Gerber Products Co. - 107
W.R. Grace & Co. - 382, 383, 384, 385, 391
Great Lakes Biochemical Co., Inc. - 354
H.J. Heinz Company Ltd. - 204
Hills Bros. Coffee, Inc. - 263
Hupp Corp. - 31, 34
Kellogg Company - 377, 380
Kraus-Maffei AG - 49, 279
Lamb-Weston, Inc. - 356, 358, 368
P. Leiner & Sons (Wales) Ltd. - 386
Leybold-Anlagen Holding AG - 140, 146, 154, 160
Leybold-Heraeus-Verwaltung GmbH - 19, 137, 139, 162, 177
Leybold-Hochvakuum-Anlagen GmbH - 86, 344
Arthur D. Little, Inc. - 152
Matsushita Electric Works, Ltd. - 159
Mitchell Engineering Ltd. - 59
National Aeronautics and Space Administration - 390
National Research Corp. - 393
New Brunswick Scientific Corp. - 241
A/S Niro Atomizer - 174
Nisshin Kako KK - 392
Pennsalt Chemicals Corp. - 80, 125, 135, 175
Pennwalt Corp. - 136
Pillsbury Company - 10, 109
Procter & Gamble Co. - 274, 302
Ranks Hovis McDougall Ltd. - 386
Raytheon Co. - 57

Regents of the University of California - 331
SEC NV Seffinga Engineering Co. - 150, 157
Sintef - 322
Societe d'Assistance Technique Pour Produits, Nestle SA - 88, 118, 261, 262, 306, 389
Societe d'Utilisation Scientifique et Industrielle du Froid Usifroid - 83
Standard Brands, Inc. - 304
Struthers Patent Corp. - 251, 253
Struthers Scientific and International Corp. - 249, 253, 256, 267, 345, 387, 388
Sun-Freeze, Inc. - 20, 347
Uncle Ben's Inc. - 361
Unilever NV - 50
United Fruit Co. - 64, 66, 69
U.S. Secretary of Agriculture - 78, 195, 197, 348
U.S. Secretary of the Army - 314, 334, 339, 340, 362, 369, 371
U.S. Secretary of the Navy - 330
University of Illinois Foundation - 313
Vickers-Armstrongs (Engineers) Ltd. - 129
Vickers Limited - 91
Virtis Company, Inc. - 112, 172

INVENTOR INDEX

Abbott, J.A. - 183
Abelow, I. - 334
Achucarro, J.L. - 5
Anderson, J.L. - 261
Andre, J.R. - 302
Baer, A. - 143
Baerwald, F.K. - 366
Bally, J. - 88
Bardsley, R.F. - 103
Barnett, S. - 289, 294, 376
Barr, C.H., Jr. - 373
Barr, C.H., Sr. - 373
Barr, J.W. - 373
Bender, C.E. - 172
Bilenker, E.N. - 300
Blake, J.H. - 75, 187
Bonteil, R.C. - 159
Bouldoires, J.-P. - 88, 118, 389
Bowden, H. - 376
Bower, H.S. - 103
Brewster, M.L. - 17, 181
Brouwer, J. - 50
Buchzik, C.M. - 307
Butler, D.A. - 361
Carbonell, R.J. - 304
Casten, J.W. - 263
Catelli, C. - 204
Chaplow, R.A. - 247, 248
Chauffard, F. - 262
Christison, W.E. - 199
Clark, J.P., III - 78
Clinton, W.P. - 283, 287, 296, 299
Cole, R.D. - 127

Cooper, G.M. - 362
Copson, D.A. - 57
Cox, G.R. - 120
D'Alessandro, D.E. - 135
Dantoni, J.L. - 54
de Buhr, J.G. - 66
De George, R. - 269
Derby, R.R. - 281
Devik, O.G. - 322
Dousset, M. - 262
Dryden, C.E. - 351
Dwyer, D.E. - 277, 295
Dwyer, D.E., Jr. - 250, 290
Easton, H.T. - 295
Ehlers, H. - 36
Eilenberg, H. - 19, 137, 139, 140, 160, 162, 177
Elerath, B.E. - 100, 258, 259, 172, 292
Eolkin, D. - 107
Ewald, J.F., Jr. - 100
Ewalt, D.J. - 379
Fisher, J.R. - 364
Folsom, T.R. - 26
Fraser, D.S. - 112, 172
Freedman, D. - 240
Fuentevilla, M.E. - 125, 175
Gaedtke, J.A. - 261
Ganiaris, N. - 251, 253, 267, 345, 388
Giddey, C. - 359
Gidlow, R.G. - 10, 109
Gidner, R.R. - 127

Inventor Index

Ginnette, L.F. - 281
Gottesman, M. - 245, 254
Gottfried, H. - 210
Gottlieb, D.M. - 365
Grover, K.M. - 267
Guerard, A.S. - 3
Guggenheim, H. - 271
Hackenberg, U. - 36, 97, 146, 154, 201, 344
Hair, E.R. - 274
Hamilton, W.H. - 80, 135
Harper, J.C. - 71
Haugh, R.R. - 314
Hedrick, R.H. - 256
Hernandez, L.A., Jr. - 42, 94
Hertzendorf, M.S. - 311
Hinnergardt, L.C. - 339, 340
Hobday, T.M.N. - 386
Hodgman, R.A. - 247, 248
Huber, C.S. - 390
Illich, G.M., Jr. - 132
Jacobs, G.E. - 296
Janovtchik, V.J. - 204
Jeppson, M.R. - 320
Johnson, J.W. - 283, 296, 308
Johnson, K.R. - 362
Jordan, G.W. - 386
Kaleda, W.W. - 308, 310
Kamps, H. - 154
Kan, B. - 64, 69
Katz, S. - 103, 257, 290
Kaufman, V.F. - 195, 197
Kautz, K. - 344
Kenyon, R.E. - 373, 376, 379
Kessler, H.G. - 49
King, C.J. - 331
King, C.J., III - 78
Kraut, T. - 299
Kruger, H.W. - 356
Kwiat, E.V. - 313
Lamb, F.G. - 368
Laskin, M. - 382, 283, 384, 385, 391
Lemaire, N.A. - 380
Linaberry, J.R. - 365
Lind, V.W. - 190
Liobis, V.A. - 240
Lorentzen, J. - 178
Lowe, E. - 195
Ludwig, W.B. - 100
Lutz, G.J. - 285
Mace, R.C. - 234
Macy, R.L., Jr. - 377

Mahlmann, J. - 287, 291
Mak, T.T. - 294
Malecki, G.J. - 348
Manaresi, F. - 277
Mason, P.B. - 59
Mehrlich, F.P. - 314
Meier, R.W. - 245
Mellor, J.D. - 328
Menzi, R. - 359
Mercer, J.L. - 164, 170
Meryman, H.T. - 330
Meyer, F.W. - 296
Mink, W.H. - 336, 351
Mishkin, A.R. - 261, 265
Moore, F.J. - 234
Morand, A. - 306
Muller, J.G. - 256
Nack, H. - 336, 351
Nelson, A.I. - 313
Nerge, W. - 36
Neumann, K. - 86, 90, 393
Oetjen, G.-W. - 19, 177
Ogden, R.P. - 342
Ohtaki, S. - 392
Oldenkamp, H.A. - 39, 293
Opella, J.J. - 361
Oppenheimer, F. - 317
Pascal, F.D. - 254
Passey, A.D. - 214
Pelmulder, J.P. - 75, 187
Peterson, R.D. - 380
Pfluger, R.A. - 296, 311
Pictet, G. - 306
Pitchon, E. - 200, 245, 259
Ponzoni, G.B. - 283, 287
Porta, P.D. - 24
Porter, G.D. - 361
Rader, E.L. - 228, 237
Rahman, A.R. - 369
Reichert, A.W. - 127
Reimus, R.G. - 249, 387
Rey, L.P. - 115, 389
Rey, L.R. - 262, 306
Ridge, R.A.J. - 129
Rieutord, L.M.A. - 83
Rink, H. - 154
Rockwell, W.C. - 195, 197
Rothmayr, W. - 167
Rovero, D. - 389
Rowell, L.A. - 164
Sahara, Y. - 222
Saporito, A. - 249, 387
Sauer, H.A. - 13

Scharschmidt, R.K. - 374
Schimpfle, J. - 279
Schmidt, T.R. - 362
Schmitz, F.-J. - 19, 140, 160
Schulman, M. - 311
Schwartzberg, H. - 273
Seffinga, G. - 150, 157
Seligman, M. - 148
Shimabuku, S.H. - 263
Simon, H.R. - 289
Sloan, J.L. - 358
Small, R.F. - 39
Smith, H.L., Jr. - 31, 34
Smith, J.C. - 361
Stern, R.M. - 354
Stinchfield, R.M. - 152
Storrs, A.B. - 354
Strang, D.A. - 274, 302
Symbolik, W.S. - 265
Tangsrud, N. - 322
Tarvin, J.W. - 364
Thale, A. - 174

Thompson, H.P. - 91
Thompson, T.N. - 172
Thuse, E. - 73, 75, 105, 183, 187, 281
Togashi, H.J. - 170
Tooby, G. - 323
Torr, D. - 342
Tribout, M. - 143
Tuomy, J.M. - 371
Ullrich, W.P. - 199
van Gelder, A. - 20, 346, 347
van Olphen, G.C.W. - 29
Veldstra, J. - 50
Vigano, G. - 393
Vollink, W.L. - 376
Wagman, J. - 334
Waltrich, P.F. - 136
Watson, C.D. - 293
Webster, R.C. - 353
Wehrmann, G.A.H. - 52
Young, R.G. - 371

U.S. PATENT NUMBER INDEX

2,949,364 - 300
2,994,132 - 90
3,013,341 - 393
3,020,645 - 57
3,033,690 - 364
3,077,036 - 86
3,078,586 - 115
3,088,222 - 234
3,096,163 - 330
3,132,929 - 105
3,132,930 - 183
3,146,077 - 175
3,169,070 - 314
3,176,408 - 125
3,178,829 - 120
3,192,643 - 83
3,199,217 - 39
3,199,221 - 314
3,210,861 - 107
3,218,725 - 368
3,218,727 - 190
3,218,731 - 152
3,222,796 - 320
3,230,633 - 80
3,233,333 - 317
3,234,658 - 97
3,239,942 - 336
3,242,575 - 277
3,243,892 - 199
3,244,528 - 342
3,244,529 - 283
3,244,533 - 299
3,247,602 - 135

3,253,344 - 346
3,253,420 - 269
3,255,534 - 69
3,256,098 - 392
3,257,731 - 393
3,259,991 - 132
3,262,212 - 66
3,263,335 - 64
3,264,121 - 371
3,264,745 - 157
3,266,169 - 31
3,269,025 - 351
3,270,428 - 29
3,270,433 - 36
3,270,434 - 154
3,271,873 - 71
3,271,874 - 317
3,273,259 - 201
3,276,139 - 170
3,280,471 - 127
3,281,954 - 143
3,281,956 - 59
3,283,522 - 253
3,286,365 - 146
3,286,366 - 148
3,289,314 - 24
3,297,455 - 342
3,298,108 - 150
3,299,525 - 73
3,303,578 - 195
3,304,617 - 320
3,308,552 - 197
3,311,991 - 109

3,313,032 - 348
3,316,652 - 347
3,318,012 - 91
3,319,343 - 174
3,321,319 - 344
3,323,225 - 5
3,324,565 - 34
3,335,575 - 256
3,343,273 - 129
3,352,024 - 328
3,356,512 - 380
3,360,374 - 373
3,362,835 - 281
3,364,591 - 160
3,365,310 - 353
3,373,042 - 259
3,376,652 - 42
3,381,302 - 249
3,382,584 - 75
3,382,585 - 187
3,382,586 - 178
3,391,466 - 50
3,395,022 - 376
3,396,041 - 391
3,399,061 - 285
3,401,466 - 181
3,401,468 - 140
3,404,007 - 256
3,408,919 - 271
3,419,402 - 383
3,431,655 - 267
3,432,308 - 387
3,436,837 - 334

3,438,784 - 287
3,438,792 - 356
3,443,324 - 100
3,443,961 - 308
3,443,962 - 272
3,443,963 - 289
3,445,247 - 366
3,446,635 - 313
3,448,527 - 94
3,453,741 - 78
3,458,941 - 10
3,460,269 - 49
3,462,281 - 377
3,464,834 - 384
3,465,452 - 167
3,467,530 - 374
3,468,672 - 273
3,469,327 - 204
3,472,663 - 382
3,477,137 - 20
3,482,326 - 17
3,482,988 - 310
3,482,990 - 311
3,483,000 - 385
3,483,032 - 354
3,484,946 - 13
3,486,907 - 274
3,487,554 - 323
3,488,860 - 172
3,489,575 - 362
3,501,319 - 379

3,513,559 - 162
3,516,170 - 240
3,518,097 - 359
3,529,362 - 137
3,531,295 - 253
3,531,871 - 222
3,532,506 - 306
3,543,411 - 103
3,545,097 - 136
3,554,761 - 304
3,556,818 - 293
3,564,727 - 112
3,565,635 - 291
3,573,060 - 263
3,573,070 - 361
3,573,929 - 261
3,574,950 - 54
3,574,951 - 177
3,579,360 - 262
3,583,075 - 26
3,590,496 - 139
3,598,608 - 388
3,601,901 - 237
3,612,411 - 19
3,616,542 - 228
3,619,204 - 257
3,620,034 - 251
3,620,776 - 265
3,625,704 - 302
3,637,398 - 258
3,644,129 - 358 Appl. (Pub.)

3,648,379 - 164
3,653,929 - 277
3,655,398 - 245
3,672,917 - 307
3,673,698 - 3
3,682,650 - 295
3,684,532 - 254
3,692,533 - 390
3,728,798 - 52
3,731,392 - 210
3,733,716 - 159
3,765,910 - 247
3,804,960 - 294
3,811,199 - 188
3,843,823 - 248
3,845,230 - 250
3,873,745 - 389
3,882,610 - 88
3,892,876 - 386
3,894,157 - 365
3,903,312 - 296
3,909,957 - 214
3,914,446 - 339
3,936,952 - 279
3,949,486 - 345
3,961,424 - 292
3,964,174 - 331
3,966,979 - 290
3,971,854 - 340
3,984,577 - 369
B 363,337 - 322

NOTICE

Nothing contained in this Review shall be construed to constitute a permission or recommendation to practice any invention covered by any patent without a license from the patent owners. Further, neither the author nor the publisher assumes any liability with respect to the use of, or for damages resulting from the use of, any information, apparatus, method or process described in this Review.

TEA AND SOLUBLE TEA PRODUCTS MANUFACTURE 1977

by Nicholas D. Pintauro

Food Technology Review No. 38

In the United States most tea (loose and in bags) is fermented black tea from Eastern Asia. Imports of green or unfermented tea from China and Japan are on the increase to meet the demand for the manufacture of instant teas.

The subject areas in this publication are organized in logical sequence describing all steps in the manufacture of all types of tea products. The opening chapters deal with post-harvest handling and conditioning of tea leaves. Later chapters cover extraction, aroma recovery, and removal of troublesome tea creams. Finally, there is a chapter on tea bag packaging. Nonleaf tea products include:

Instant Tea: Spraydried tea brew, may contain carbohydrates.
Tea Mix: Instant tea with sugar or other sweeteners and flavors.
Canned or Bottled Tea: Ready-to-drink product made from reconstituted soluble tea or liquid concentrate. No refrigeration needed before opening.
Cold-Packed Tea: Ready-to-drink beverage in plastic or paper container. Refrigeration required.
Liquid Concentrate: Concentrated tea product to which water must be added. Shelf-stable in cans or bottles.
Frozen Concentrate: Similar to liquid concentrate. Packaged, stored and reconstituted like frozen fruit juice.

114 processes. A much shortened, partial table of contents follows here. Numbers in parentheses indicate the number of processes per topic.

1. WITHERING & ROLLING (6)
Separating Foreign Matter
Controlled Withering & Drying
Curling, Tearing, Cutting
High Speed Cutter
Rolling to Control Moisture
Rolling Additive

2. SORTING, FERMENTING, "FIRING" (13)
Fiber Separating
Stalk Removal
AFICO Maturing Process
Enzyme Conversion
Suppressed Thearubigens
Oxidative Conversion
Heat and Oxygen
Ozone Process
Hydrogen Peroxide
Potassium Permanganate
Tea Gum Preservative

3. EXTRACTION PROCESSES (17)
Preliminary Leaching
Preliminary Cold Extraction
Countercurrent Extraction
SELTZER and SAPORITO Processes
Extraction under CO_2
Ammonia Extraction
Acetone Extraction
Green Tea Extraction & Conversion
Hard Water Tolerance
Buffers vs. Ion Exchange
Addition of Ascorbic Acid
Irradiation Treatment

4. TANNIN-CAFFEINE PRECIPITATE—TEA CREAM REMOVAL (20)
Stepwise Addition of Gelatin
Suspending Agents
Cream Separation Methods
Oxidation and Bleaching
Tannase Treatment
Calcium Treatments
Use of Polyvinylpyrrolidone
Selective Tannin Removal
Low Temperature Method

5. FILTRATION & CONCENTRATION (9)
Centrifuge Designs
Use of Hydrophilic Membranes
Freeze Concentration Processes

6. DEHYDRATION (14)
Vacuum Drying
Freezedrying Processes
Green Tea Blends

7. AROMA RECOVERY (14)
Recovery During "Firing"
Steam Stripping
Inert Gas Stripping
Aroma Fixation

8. AGGLOMERATION & AROMATIZATION (12)
Various Agglomeration Processes
Black Tea Flavor
Aromatic Extracts
Fruit Extracts
Flavor Protection

9. TEA BAGS (9)
Bag Construction
Single Chamber Bags
Construction from Tubes
Tag Interlock
Slit Opening
Perforation Line
Tea Bag Receptacles
Vending Infusion Apparatus

ISBN 0-8155-0645-7

266 pages

DOUGHS AND BAKED GOODS
Chemical, Air, and Non-Leavened
1975

by D. J. De Renzo

Food Technology Review No. 26

This book emphasizes commercial baking techniques not employing yeast. Chemical leavening agents are preferred for sweet bakery products, and the necessary carbon dioxide is usually generated by the chemical action of some acid salt upon sodium bicarbonate.

Refrigerated biscuit doughs and similar products represent a line of semifinished baked goods which are discussed in this book perhaps for the first time in such detail. Here the leavening systems increase in chemical complexity, as premature release of carbon dioxide and interaction with dough conditioners and preservatives must be prevented.

It is perfectly possible to prepare simulated yeast-baked goods with the explicit directions given under flavoring processes. The thorough discussions of shortening compositions and emulsifiers in two large chapters are essential to the modern baker, as are preservation and antistaling methods.

The concluding chapter is on fillings, toppings and coatings which so often play a decisive role in the consumer acceptance of novelty baked goods.

This is a very informative volume for progressive bakery managers and food technologists specializing in baked goods. Almost 300 processes are described. A partial and very condensed table of contents follows here. Numbers in parentheses indicate the number of processes per topic or per chapter. Chapter headings are given, followed by examples of important subtitles.

1. CHEMICALLY LEAVENED PRODUCTS (75)
Continuous Process for
   Hard Sweet Dough
Spraydried Flour-Sugar-Shortening
Cake Baking with Microwaves
Lactalbumin Phosphate Instead
   of Nonfat Dry Milk & Egg White
Shelf Stable Doughnuts
High Protein Cookies
Partially Baked Goods for Toasting
Low Calorie Products

2. AIR-LEAVENED PRODUCTS (22)
Angel Food Cake Manufacture
One-Step Premix with Leavening
Instant Cream Puffs
Pregelatinized Starch
   Plus Powdered Fat
Starch + Fatty Antifoaming Agent
Fish Proteins for Egg Whites
No-Bake Cake Products
Raising Dough by Air Injection

3. NON-LEAVENED PRODUCTS (43)
Pie Doughs
Forming and Packaging Pie Crusts
Dihydroxyacetone for Browning
Improving Sheet Strength with Glycerol
Ready-to-Bake Pie Crust Dough
Tortilla Manufacture
Batters
Manufacture of Ice-Cream Cones

4. REFRIGERATED PRODUCTS (38)
Eliminating Phosphate Crystallization
Filled Dough Composites
Leavening System Comprising
   Ions and Polyphosphates
Frozen Cake Mixes & Batters

5. LEAVENING AGENTS (12)
Baking Acids & Carbon Dioxide Sources
α-Glucoheptono-γ-Lactone
Polyol Carbonates
Agglomerated Sodium
   Aluminum Phosphate

6. SHORTENING COMPOSITIONS (29)
Fluid Shortenings
Plastic Shortening

7. EMULSIFIERS (18)
Lactylated Mixed Esters
Ethoxylated Glycerides
Acidic Lipid Alkyl Carbonates

8. PRESERVATION METHODS (17)
Retardation of Staling
Use of Fatigued Gluten
Mold Inhibitors
Sterilization vs. Protective Coatings

9. FLAVORING (24)
Yeast Flavors
Reaction Product of Piperidine,
   Proline & Dextrose
Yeast Autolysates
Sourdough Flavor

10. FILLINGS, TOPPINGS, COATINGS (21)
Stable Freeze-Thaw Fillings
Cold & Heat Resistant Gels
Reduced Density Icings
Powdered Coatings
Glazes
Glaze Dry Mix

ISBN 0-8155-0590-6   435 pages